AF745235

RADIO ASTRONOMY AT THE FRINGE

COVER ILLUSTRATION:

Image of the radio source 3C 84 (at the center of NGC 1275), taken with the Very Long Baseline Array at 23 GHz. It shows the bright core region and the structures north and south of the core thought to be related to the far and near side jets respectively. The contour levels start with -1, 1, 2, 2.83, 4, 5.66 mJy beam^{-1} and increase from there by factors of $\sqrt{2}$. The peak flux density is 1.9 Jy beam^{-1}. The resolution is 0.49 x 0.35 mas elongated in position angle -4.3 deg.

From R.C. Walker, these proceedings, page 113.

A SERIES OF BOOKS ON RECENT DEVELOPMENTS IN ASTRONOMY AND ASTROPHYSICS

Publisher

THE ASTRONOMICAL SOCIETY OF THE PACIFIC

390 Ashton Avenue, San Francisco, California, USA 94112-1722
Phone: (415) 337-1100 E-Mail: orders@astrosociety.org
Fax: (415) 337-5205 Web Site: www.astrosociety.org

ASP CONFERENCE SERIES - EDITORIAL STAFF
Managing Editor: D. H. McNamara
Production Manager: Enid L. Livingston
Production Assistant: Andrea Weaver

PO Box 24463, Room 211 - KMB, Brigham Young University, Provo, Utah, 84602-4463
Phone: (801) 422-2111 Fax: (801) 422-0624 E-Mail: pasp@byu.edu

LaTeX-Computer Consultant: T. J. Mahoney (Spain) – tjm@ll.iac.es

ASP CONFERENCE SERIES PUBLICATION COMMITTEE:
Joss Bland-Hawthorn
George Jacoby
James B. Kaler
J. Davy Kirkpatrick

A listing of all of the ASP Conference Series Volumes and IAU Volumes published by The ASP may be found at the back of this volume

ASTRONOMICAL SOCIETY OF THE PACIFIC
CONFERENCE SERIES VOLUMES

Volume 300

RADIO ASTRONOMY AT THE FRINGE

Proceedings of a Conference
in honor of
Kenneth I. Kellermann
on the occasion of his 65th Birthday
held in Green Bank, West Virginia, USA
10-12 October 2002

Edited by

J. A. Zensus
Max-Planck-Institut für Radioastronomie
and
National Radio Astronomy Observatory

M. H. Cohen
California Institute for Technology

E. Ros
Max-Planck-Institut für Radioastronomie

DUNELM UNIV
1 4 JUN 2005

520.27
RAD

2163366

© 2003 by Astronomical Society of the Pacific. All Rights Reserved

No part of the material protected by this copyright notice may be reproduced or utilized in any form or by any means – graphic, electronic, or mechanical including photocopying, taping, recording or by any information storage and retrieval system, without written permission from The Astronomical Society of the Pacific .

Library of Congress Cataloging in Publication Data
Main entry under title

Card Number: 2003107126
ISBN: 1-58381-147-8

ASP Conference Series - First Edition

Printed in United States of America by Sheridan Books, Ann Arbor, Michigan

Radio Astronomy at the Fringe
ASP Conference Series, Vol. 300, 2003
J.A. Zensus, M.H. Cohen, & E. Ros, eds.

Contents

Part 3. Jets

Part 4. Variability and Brightness Temperature

Preface

A symposium entitled "Radio Astronomy at the Fringe" was held in honor of Kenneth I. Kellermann's 65th birthday on October 10–12, 2002, at the National Radio Astronomy Observatory in Green Bank, West Virginia. Some seventy astronomers, radio engineers, and astrophysicists attended the meeting.

For a few years several of Ken's friends had been promoting a symposium in his honor, but he resisted. Then suddenly, as his 65th birthday loomed, he relented. Barry Clark, Marshall Cohen, Ron Ekers, Ed Fomalont, Dave Jauncey, Leonid Matveyenko, Jim Moran, Dave Shaffer, Peter Shaver, and Anton Zensus formed the Scientific Organizing Committee and put together the program for the conference. We are grateful to these colleagues and to many others who helped make this meeting a success, especially to Rick Fisher who led the local organizing team.

This book presents the proceedings from the conference. As such, however, it can only give a glimpse of the informal and stimulating atmosphere that persisted throughout the meeting. The presentations and discussion focused on compact radio sources, especially as studied with Very Long Baseline Interferometry, and on cosmological studies using radio source surveys. About two-thirds of the papers in this volume are on the former subject, and one-third on the latter. These have been the two main themes of Ken's research, although he has done a substantial amount in other areas as well. A few of the papers in this volume are historical in nature, but the rest taken together provide an account of the current state of research in these areas.

The symposium also provided an opportunity for some of Ken's collaborators to come to Green Bank not only to attend the meeting but in addition to have their own private mini-workshop. This was successful and added a few days of intense scientific discussion to the more genial air of the main event.

Ken Kellermann's career is closely related to Green Bank, where he lived and worked for many years, although several participants commented that he often traveled around the globe in pursuit of excellent observing opportunities. Many other locations for the meeting were considered, some prestigious and befitting this highly-regarded scientist and some in far-away places, including the honoree's first choice, Reykjavik. In the end, however, the organizers could not think of any more suitable venue than Green Bank, which is famous among radio astronomers for its mixture of a stimulating observatory environment and West Virginia hospitality. Furthermore, as in Ken's earlier years at the NRAO, the observatory is once again going through an exciting period of bringing on-line a major new radio astronomy instrument, the Robert C. Byrd Telescope. This new telescope will be an important tool for much of the research discussed at the symposium and as a welcome aside, the participants could learn first-hand about its current capabilities.

Ken has long been an internationalist, and over the years he has collaborated with many scientists from many countries, including especially the United States, Australia, Germany, and Russia. So many of Ken's long-time colleagues did come that the Green Bank cafeteria simply could not hold them all with their spouses, resulting in two conference dinners, unheard of in this community. It was especially enjoyable to have Ken's mother Rae, his wife Michele, and daughter Sarah and her husband Tim present at these events, as they, too, have known some of the participants for many years. We take all this as an indication of the respect and affection that Ken enjoys among this scientific community, many of whom are his personal friends. One of these, Ivan I. K. Pauliny-Toth, could not come owing to ill health, and three colleagues from Russia, Leonid Matveyenko, Nikolay Kardashev, and Yuri Kovalev were held back by the vagaries of U.S. visa requirements. We are, however, pleased that the written contributions of these Russian colleagues are included here.

Bonn and Pasadena, April 2003

J. Anton Zensus
Marshall H. Cohen
Eduardo Ros

Acknowledgments

The organizers and editors of the symposium are grateful to the following for their contribution: Dr. Rick Fisher in Green Bank for his excellent leadership in the local organization of the meeting, and his colleagues, especially Ms. Becky Warner; Dr. Paul vanden Bout, Dr. Miller Goss, and Dr. Fred Lo for making the conference and proceedings possible; Dr. Phillip Jewell and his staff in Green Bank for their hospitality; and Mr. Matthias Kadler for assistance in editing these proceedings. These proceedings were also supported by the Max-Planck-Gesellschaft and Alexander v. Humboldt Foundation through a Max-Planck-Forschungspreis to J. A. Zensus.

The editing of this volume has made use of NASA's Astrophysics Data System and the SIMBAD database, operated at CDS, Strasbourg, France.

The USA National Radio Astronomy Observatory is operated by Associated Universities Inc., under Cooperative Agreement with the USA National Science Foundation.

We thank the authors, editors, and publishers of the following works for permission to reproduce copyright material:
Astronomy & Astrophysics: pages 120 & 289; copyright © 1999 & 2003 by the European Southern Observatory.
The Astrophysical Journal: pages 76, 100, 109, 129, 188, 203, 204, 125, 216, & 217; copyright © 1969, 1999, 2002, & 2003 by the American Astronomical Society.
Monthly Notices of the Royal Astronomical Society: pages 59 & 62; copyright © 2001 & 2003 by the Royal Astronomical Society, London.
New Astronomy Reviews: page 59; copyright © 1999 by Elsevier Science B. V.
Publications of the Astronomical Society of Japan: page 195; copyright © 2000 by the Astronomical Society of Japan.

Workshop Participants

Margo Aller, Department of Physics, University of Michigan, 500 East University Avenue, Ann Arbor, MI 48109-1120, USA ⟨ margo@astro.lsa.umich.edu ⟩

Hugh Aller, Department of Physics, University of Michigan, 500 East University Avenue, Ann Arbor, MI 48109-1121, USA ⟨ hugh@astro.lsa.umich.edu ⟩

Glenn Berge, 421 Mariposa Ave., Sierra Madre, CA 91204-2307, USA ⟨ gib@venus1.gps.caltech.edu ⟩

Roger Blandford, California Institute of Technology, Mail Code 130-33, 1200 East California Boulevard, Pasadena, California 91125, USA ⟨ rdb@caltech.edu ⟩

Roy Booth, Onsala Space Observatory, 439 92 Onsala, Sweden ⟨ roy@oso.chalmers.se ⟩

Alan Bridle, National Radio Astronomy Observatory, 520 Edgemont Road, Charlottesville, VA 22903-2475, USA ⟨ abridle@nrao.edu ⟩

John Broderick, Department of Physics, Virginia Polythecnic Institute and State University, Blacksburg, VA 24061, USA ⟨ john.broderick@vt.edu ⟩

Geoff Burbidge, Center for Astrophysics and Space Sciences, University of California-San Diego, Box 0424, University of California, San Diego, CA 92093-0424, USA ⟨ gburbidge@ucsd.edu ⟩

Barry Clark, National Radio Astronomy Observatory, Array Operations Center, P.O. Box O, 1003 Lopezville Road, Socorro, NM 87801, USA ⟨ bclark@aoc.nrao.edu ⟩

Marshall Cohen, California Institute of Technology, Department of Astronomy, MS 105-24, 1200 East California Boulevard, Pasadena, CA 91125, USA ⟨ mhc@astro.caltech.edu ⟩

Jim Condon, National Radio Astronomy Observatory, 520 Edgemont Road, Charlottesville, VA 22903-2475, USA ⟨ jcondon@nrao.edu ⟩

Bill Cotton, National Radio Astronomy Observatory, 520 Edgemont Road, Charlottesville, VA 22903-2475, USA ⟨ bcotton@nrao.edu ⟩

Vivek Dhawan, National Radio Astronomy Observatory, Array Operations Center, P.O. Box O, 1003 Lopezville Road, Socorro, NM 87801, USA ⟨ vdhawan@aoc.nrao.edu ⟩

Ron Ekers, Australia Telescope National Facility, CSIRO, Post Office Box 76, Epping, NSW 2121, Australia ⟨ rekers@atnf.csiro.au ⟩

CARLA FANTI, Istituto di Radioastronomia, Via P. Gobetti 101, I-40129 Bologna, Italy ⟨cfanti@ira.cnr.it⟩

ROBERTO FANTI, Istituto di Radioastronomia, Via P. Gobetti 101, I-40129 Bologna, Italy ⟨rfanti@ira.cnr.it⟩

RICK FISHER, National Radio Astronomy Observatory, PO Box 2, Green Bank, WV 24944, USA ⟨rfisher@gb.nrao.edu⟩

ED FOMALONT, National Radio Astronomy Observatory, 520 Edgemont Road, Charlottesville, VA 22903-2475, USA ⟨efomalon@nrao.edu⟩

DENISE GABUZDA, Physics Department, University College Cork, Cork, Ireland ⟨gabuzda@phys.ucc.ie⟩

BARRY GELDZAHLER, NASA Headquarters, Space Sciences SE, 300 E St. SW, Washington DC 20546-0001, USA ⟨bgeldzah@hq.nasa.gov⟩

MILLER GOSS, National Radio Astronomy Observatory, Array Operations Center, P.O. Box O, 1003 Lopezville Road, Socorro, NM 87801, USA ⟨mgoss@aoc.nrao.edu⟩

LEONID GURVITS, Joint Institute for VLBI in Europe, Postbus 2, NL-7990 AA Dwingeloo, The Netherlands ⟨lgurvits@jive.nl⟩

DAN HARRIS, HEA–Smithsonian Astrophysical Observatory, MS-3, 60 Garden Street, Cambridge, MA 02138, USA ⟨harris@cfa.harvard.edu⟩

DAVE HEESCHEN, 702 Copa d'Oro, Marathon, FL 33050, USA ⟨dheeschen@earthlink.net⟩

HISASHI HIRABAYASHI, Institute of Space and Astronautical Science, Sagamihara, Kanagawa 229, Japan ⟨hirax@vsop.isas.ac.jp⟩

JAN HÖGBOM, Stockholm Observatory, SCFAB, SE-10691 Stockholm, Sweden ⟨janh@ebox.tninet.se⟩

DAN HOMAN, National Radio Astronomy Observatory, 520 Edgemont Road, Charlottesville, VA 22903-2475, USA ⟨dhoman@nrao.edu⟩

BILL HOWARD, 1653 Quail Hollow Court, Mclean, VA 22101-3234, USA ⟨wehoward@sigmaxi.org⟩

MAKOTO INOUE, Nobeyama Radio Observatory, Nobeyama, Minamisaku, Nagano 384-13, Japan ⟨inoue@nro.nao.ac.jp⟩

CAROLE JACKSON, Research School of Astronomy & Astrophysics, The Australian National University, Mount Stromlo Observatory, Canberra, Australia ⟨cjackson@mso.anu.edu.au⟩

DAVID JAUNCEY, Australia Telescope National Facility, CSIRO, P.O. Box 76, Epping NSW 2121, Australia ⟨David.Jauncey@csiro.au⟩

DAYTON L. JONES, Jet Propulsion Laboratory, California Institute of Technology, Mail Code 238-332, 4800 Oak Grove Drive, Pasadena, CA 91109, USA ⟨dj@sgra.jpl.nasa.gov⟩

SVETLANA JORSTAD, Institute for Astrophysical Research, Boston University, 725 Commonwealth Avenue, Boston, MA 02215-1401, USA ⟨jorstad@bu.edu⟩

KEN KELLERMANN, National Radio Astronomy Observatory, 520 Edgemont Road, Charlottesville, VA 22903-2475, USA ⟨kkellerm@nrao.edu⟩

LUCYNA KEDZIORA-CHUDCZER, Australia Telescope National Facility, CSIRO, P.O. Box 76, Epping NSW 2121, Australia ⟨Lucyna.Kedziora-Chudczer@atnf.csiro.au⟩

LEONIA KOGAN, National Radio Astronomy Observatory, Array Operations Center, P.O. Box O, 1003 Lopezville Road, Socorro, NM 87801, USA ⟨lkogan@aoc.nrao.edu⟩

YURI KOVALEV, Astro Space Center of Lebedev Physical Institute, Profsoyuznaya 84/32, 117997 Moscow, Russia ⟨yyk@asc.rssi.ru⟩

GOPAL-KRISHNA, National Centre for Radio Astrophysics/TIFR, Pune University Campus, Post Bag No. 3, Ganeshkhind, Pune 411007, India ⟨krishna@ncra.tifr.res.in⟩

ROBERT LAING, Space Science and Technology Department, CLRC, Rutherford Appleton Laboratory, Chilton, Didcot, Oxon OX11 0QX, United Kingdom ⟨rlaing@astro.ox.ac.uk⟩

MATT L. LISTER, National Radio Astronomy Observatory, 520 Edgemont Road, Charlottesville, VA 22903-2474, USA ⟨mlister@nrao.edu⟩

FRED LO, National Radio Astronomy Observatory, 520 Edgemont Road, Charlottesville, VA 22903-2475, USA ⟨flo@nrao.edu⟩

ALAN MARSCHER, Institute for Astrophysical Research, Boston University, 725 Commonwealth Avenue, Boston, MA 02215-1401, USA ⟨marscher@bu.edu⟩

GEORGE MILEY, Sterrenwacht, Leiden University, P.O. Box 9513, NL-2300 RA Leiden, The Netherlands ⟨miley@strw.leidenuniv.nl⟩

JAMES M. MORAN, Smithsonian Astrophysical Observatory, 60 Garden Street, Cambridge, MA 02138, USA ⟨moran@cfa.harvard.edu⟩

MARCUS PRICE, Institute for Astrophysics, Department of Physics and Astronomy, University of New Mexico, 800 Yale Blvd. NE, Albuquerque, NM, USA ⟨rmprice@as.unm.edu⟩

V. RADHAKRISHNAN, Raman Research Institute, C. V. Raman Avenue, Bangalore - 560 080, India ⟨rad@rri.res.in⟩

JON ROMNEY, National Radio Astronomy Observatory, Array Operations Center, P.O. Box O, 1003 Lopezville Road, Socorro, NM 87801, USA ⟨jromney@aoc.nrao.edu⟩

EDUARDO ROS, Max-Planck-Institut für Radioastronomie, Auf dem Hügel 69, 53121 Bonn, Germany ⟨ros@mpifr-bonn.mpg.de⟩

MICHAEL RUSSO, California Institute of Technology, Department of Astronomy, MS 105-24, 1200 East California Boulevard, Pasadena, CA 91125, USA ⟨mrusso@ugcs.caltech.edu⟩

RICHARD T. SCHILIZZI, Joint Institute for VLBI in Europe, Postbus 2, NL-7990 AA Dwingeloo, The Netherlands ⟨schilizzi@jive.nl⟩

MAARTEN SCHMIDT, Palomar Observatory, California Institute of Technology, MS 320-47, Pasadena, CA 91125, USA ⟨mxs@deimos.caltech.edu⟩

GEORGE SEIELSTAD, Aerospace, University of North Dakota, Jon D. Odegard School of Aerospace Sciences, University & Tulane, Grand Forks, ND 58202-9007, USA ⟨gseielst@aero.und.edu⟩

DAVID SHAFFER, Radio Metrics Inc., 1742 Saddleback Court, Henderson, NV 89014, USA ⟨shaffer@nevada.edu⟩

PETER SHAVER, European Southern Observatory, Karl Schwarzschild Straße 2, D-85748 Garching bei München, Germany ⟨pshaver@eso.org⟩

SLAVA SLYSH, Arecibo Observatory, HC3 Box 53995, Arecibo PR 00612, Puerto Rico ⟨vslysh@naic.edu⟩

GEORGE W. SWENSON, University of Illinois at Urbana-Champaign, Department of Electrical & Computer Engineering, 1406 West Green St., Urbana, IL 61801-2918, USA ⟨gswenson@uiuc.edu⟩

JIM ULVESTAD, National Radio Astronomy Observatory, Array Operations Center, P.O. Box O, 1003 Lopezville Road, Socorro, NM 87801, USA ⟨julvesta@nrao.edu⟩

ESKO VALTAOJA, Tuorla Observatory, University of Turku, FIN-21500 Pikkiö, Finland ⟨esko.valtaoja@utu.fi⟩

PAUL VANDEN BOUT, National Radio Astronomy Observatory, 520 Edgemont Road, Charlottesville, VA 22903-2475, USA ⟨pvandenb@nrao.edu⟩

HARRY VAN DER LAAN, Astronomical Institute, Utrecht University, PO Box 80 000, NL-3508 TA Utrecht, The Netherlands ⟨vdlaan@astro.uu.nl⟩

RENÉ VERMEULEN, Netherlands Foundation for Research in Astronomy, Postbus 2, NL-7990 AA Dwingeloo, The Netherlands ⟨rvermeulen@astron.nl⟩

R. CRAIG WALKER, National Radio Astronomy Observatory, Array Operations Center, P.O. Box O, 1003 Lopezville Road, Socorro, NM 87801, USA ⟨cwalker@nrao.edu⟩

JASPER WALL, Department of Astrophysics, University of Oxford, Nuclear and Astrophysics Laboratory, Keble Road, Oxford OX1 3H, United Kingdom ⟨jvw@astro.ox.ac.uk⟩

JACK WELCH, Radio Astronomy Laboratory and Department of Astronomy, University of California at Berkeley, Berkeley, CA 94720, USA ⟨wwelch@astro.berkeley.edu⟩

SANDER WEINREB, Jet Propulsion Laboratory, California Institute of Technology, Mail Code 238-332, 4800 Oak Grove Drive, Pasadena, CA 91109, USA ⟨sweinreb@caltech.edu⟩

JOAN WROBEL, National Radio Astronomy Observatory, Array Operations Center, P.O. Box O, 1003 Lopezville Road, Socorro, NM 87801, USA ⟨jwrobel@nrao.edu⟩

J. ANTON ZENSUS, Max-Planck-Institut für Radioastronomie, Auf dem Hügel 69, 53121 Bonn, Germany ⟨azensus@mpifr-bonn.mpg.de⟩

Radio Astronomy at the Fringe
ASP Conference Series, Vol. 300, 2003
J.A. Zensus, M.H. Cohen, & E. Ros, eds.

A Review of the History of VLBI

B. G. Clark
National Radio Astronomy Observatory, P.O. Box O, Socorro, NM 87801, USA

Abstract. I briefly review the history of VLBI. I start with a review of the lines of evidence that lead to the belief that very long baseline interferometry was a desirable and necessary development of instrumentation. I then briefly mention the first efforts to construct VLBI instruments. I finally look back over the past thirty years and call out several themes that have developed over the years.

Much of the early history of VLBI has been reviewed by Kellermann & Moran (2001) and by Kellermann & Cohen (1988). These accounts have left me with little substantive to add, but only a viewpoint. And even in viewpoint, I restrict myself to things that particularly interest me. First, I discuss the reasons why, in the mid 1960's, three different radio astronomy groups, in the Soviet Union, Canada, and the US, felt that the development of a long baseline interferometer was a natural and pressing thing to do. I make a very few remarks about the development of the VLBI equipment, adding little to what is discussed in the references above. And I finally mention the areas in which it seems to me that VLBI interferometry has made substantive contributions to knowledge, and a few highlights in each area.

1. Why we needed VLBI

In the mid 1960s there were several indications that there were features of radio sources of angular size too small to resolve using conventional, connected element, radio interferometers of convenient size. First, large, connected element interferometers continued to find unresolved sources at ever increasing baselines. Second, the variability of radio quasars was discovered, and if one believed the cosmological distances to the quasars, then the variation time scale times the speed of light divided by the distance gave a very small indicated angular size. At this time also, the first of the Gigahertz peaked spectrum objects were being studied, and if the spectral turnover was interpreted as being due to synchrotron self absorption, they again indicated a very small angular size. And finally, the phenomenon of interplanetary scintillation of radio sources had been discovered, and while this was sensitive to angular scales rather larger than those probed by the other phenomena mentioned, it did indicate that sub-arcsecond structure was common in radio sources.

Once the sizes and structures of a few radio source's had been determined by various groups, the energetic group at the University of Manchester developed the technique of radio linked interferometers, and pushed them to ever increasing baselines. The natural limit of the radio link technology of the time was just over 100 km. They pushed their 158 MHz interferometer to this distance in the early 1960s. Allen et al. (1962) report measurements of many sources on this 70,000 wavelength baseline. Having reached the natural baseline length supported by their technology, they then proceeded to raise the operating frequency, reporting measurements at 408 MHz, which gave a baseline of 180,000 wavelengths (Anderson et al. 1965). Operation at still higher frequencies required a large antenna with a sensitive receiver, and the radio link interferometer was next taken to the Royal Radar Establishment's antenna at Malvern, giving a baseline of 500,000 wavelengths at 20 cm (Adgie et al. 1965). This baseline was taken to yet higher frequencies, giving baselines of one million and two million wavelengths at, respectively, 11 cm and 6 cm wavelengths (Palmer et al. 1967). This last development was not yet known by the time that tape recorder based interferometry was undertaken in the US and Canada, but the prevalence of fringes in the earlier experiments already indicated that to properly explore and characterize the small scale structure already seen would involve the use of interferometers of such baseline that the tape recorder technique would be the more practical approach.

Variability has been claimed since the earliest days of radio astronomy. The earliest claims were later proven to the the result of ionospheric scintillation. The first modern claim of variation was that of Sholomitskii (1965) of the source CTA 102. The claim was greeted with some skepticism. However, a definitive paper demonstrating the variability of 3C 273 was soon published by Dent et al. (1965). This paper calculated the angular size required by the measured timescale of the variation, and rather apologetically noted that the varying region must be as small as two milliarcseconds.

Yet another clue to small angular size was given by source spectra. It was early realized that a power law distribution of relativistic electrons would result in a power law emitted spectrum of synchrotron radiation, but with a low frequency turnover due to the source becoming optically thick to its own radiation. If the spectral turnover known in a few sources such as PKS 1934−63 (Kellermann 1965) were identified as this synchrotron self absorption effect, very small, milliarcsecond, angular sizes were indicated.

Objects of small angular size have an interesting observational property. If they are viewed through an irregular ionized screen, they will vary due to the relative motion of source and screen. This scintillation reaches a maximum at the radio frequency at which the phase variation through the scattering screen approaches a wavelength. It occurs for objects whose angular size is less than that of the first Fresnel zone. It was the low frequency ionospheric scintillation of Cygnus A that first indicated that it was an object of relatively small angular size, and thus began the study of discrete radio sources. The equivalent phenomenon of scintillation caused by the ionized interplanetary medium was first noted by Hewish et al. (1964). These authors noted that one can, in theory, derive actual brightness distributions from the scintillation properties of the radio source. The technique at convenient frequencies is sensitive to source sizes of about one second of arc. Hewish et al. suggested that it could be extended to look for smaller structure by observing at lower frequencies. (In retrospect, this would not have worked because synchrotron self absorption causes regions of small angular size to have very low flux densities at low frequencies.)

Cohen (1965), following up the work of Hewish et al., reported that scintillation in the interplanetary medium is a common phenomenon, and that therefore sub-arcsecond structure is common. There would be no shortage of sources for VLBI investigation.

Scintillation in the interstellar medium has become the most recent emergence of the phenomenon, and, as in the case of the interplanetary scintillation, is used for an indication of the angular size of objects smaller than the resolution of contemporary interferometers. Interstellar scintillation occurs for objects smaller than a few microarcseconds. Imaging with scintillations is more difficult, although Gwinn et al. (2000) have deduced the structure of the emitting region of a pulsar through analysis of scintillation.

2. Building the first tape recorder interferometers

The tape recorder interferometer appears to have first been discussed in the Lebedev Institute in Moscow. Despite attracting great interest in the radio astronomy group there, the idea was not taken up by the astronomical establishment in the Soviet Union, and the eventual publications were in journals which were not being translated, and the ideas did not readily pass to the West. The subject was broached with Sir Alfred Charles Bernard Lovell on a visit in 1964 (see discussion in Kellermann & Moran, 2001), and an agreement that was reached that the Lebedev group would work with the Manchester group to develop such an instrument. Work proceeded very slowly, however, because modern recording technology was not readily available to the Lebedev group, and because the Manchester group was working hard on the series of connected interferometers mentioned above. As resources became available, the Manchester group did complete a tape recorder interferometer back end, but not before the NRAO and Canadian efforts came to fruition, and the system was never deployed in practice.

Interest in long baseline interferometry arose in another quarter entirely at this time. The planet Jupiter emits short bursts at decameter wavelengths, and the question arose as to whether the bursts were a planet-wide event, or came from a small region on the planet. To settle the question required resolution of about one arcsecond at ten meter wavelength, a baseline of a few hundred kilometers. The Jupiter bursts are very intense, though of short duration and low duty cycle. The post-detection correlation technique (Brown & Twiss 1954) was well known at that time, and is well adapted for intense bursts. The bursts are so intense, and the required bandwidths so narrow, that a post-detection correlation interferometer could be realized using ordinary audio tape recorders (Carr et al. 1965). This group quickly realized that at this low frequency, a local oscillator could easily be made that would keep in phase between the two elements, and they went on to make observations with Michelson interferometry, though they did not publish the results until after the NRAO and Canadian interferometers were working (Brown et al. 1968).

The science drivers discussed above attracted the attention of groups at the Hertzberg Institute in Canada and at the NRAO in the US. The state of the art in stable oscillator construction was such that coherence could be attained over most of the radio astronomy bands. Readily affordable oscillators based on the 6 GHz transition in rubidium vapor were seen to have sufficient stability to preserve coherence up to 1 GHz or more, and the expensive and finicky hydrogen maser could preserve coherence for many seconds at frequencies of 30 GHz. The time was ripe for the development of the tape recording interferometer.

The Canadian group began development of a system for use at 408 MHz, based on studio-type television tape recorders, implementing the delay tracking and fringe rotation functions in mostly analog form. After a couple of false starts, they achieved trans-Canada fringes at 408 MHz in 1967, reported by Broten et al. (1967).

I had similar ideas about how to implement the tape-recorder interferometer, and spent a day wandering around an electronics show in 1965, learning as much as I could about the state of the art in TV and instrumentation recorders. (The latter interested me more than the former, because of the TV recorder's built in dependence on the TV vertical and horizontal sync signals, which the Canadians actually very nicely exploited as an integral part of their system.) However, Sander Weinreb, newly the head of NRAO electronics, quickly informed us that digital was the way to do things, and that as much as possible we should exploit mass produced off-the-shelf digital technology for the recording system. It was quickly decided that the system would be based on a standard computer tape transport, with a correlator implemented in software. It was my job to provide the latter. Initial interferometer operation was to be demonstrated at 610 MHz, and then we were to see how high in frequency we could go. After a couple of false starts, the system produced fringes in 1967, first on the short baseline between the 140 foot and 85 foot radio telescopes in Green Bank. The next step in the quickly extending series of baselines was to observe with the interferometer formed by the Green Bank 140 foot telescope and Naval Research Laboratory 85 foot at Maryland Point. Then, at the urging of the Haystack group, who were interested in the interstellar maser OH line, we observed between the Green Bank 140 foot and the Haystack Observatory 120 foot telescope in Massachusetts (Clark et al. 1968a). We finally finished the sequence of rapidly increasing baselines with an observation between Green Bank and the 85 foot telescope at Hat Creek in California (Clark et al. 1968b), achieving transcontinental fringes at 18 cm wavelength.

The MIT/Haystack group, being interested in interferometric spectroscopy, independently developed a correlator for spectroscopy, using the NRAO hardware.

Soon after the first demonstration of the NRAO recording system it was clear that there was enough high resolution science to justify a more capable and easier to use system. The narrow bandwidth (360 kHz, limited by the 720 kbit s^{-1} that the computer tape transport could write) and the short running time of the tape (about 3 minutes) were clearly severe disadvantages. At the time, the first affordable helical scan TV tape recorders we available, and we decided to use them as a basis for a new generation of tape recorder interferometer. They could record a 2 MHz bandwidth (recorded digitally as 4 Mbit s^{-1}) on reels of tape that lasted as long as three hours. These early helical scan recorders were difficult to maintain and align. Making sure that the recordings at the telescope could be played back at the correlator was always a problem. These early recorders were replaced by later generation devices, eventually with home video cassette recorders (VCRs), which were much more reliable and much less touchy. But the basic design, dubbed the Mark II interferometer, remained the workhorse of radio astronomy VLBI for many years.

The initial Mark II correlator at NRAO correlated a single baseline. We realized, of course, that multi-station observations could be conducted to advantage, but reasoned that the number of baselines would tend to be about the reciprocal of the percentage of time that large radio telescopes would devote to long baseline interferometry, so that a single baseline correlator running full time could keep up with the observations. The single baseline quickly proved inadequate in practice, especially a few years later, when advances in image reconstruction techniques demonstrated the importance of a large number of simultaneous baselines. The NRAO Mark II correlator was expanded to three baselines, and a large, 16 station Mark II correlator was built at Caltech.

The possibilities of using VLBI for astrometry and geophysics were realized quite early on (Gold 1967). After early experience with the NRAO Mark I system, the geodesists realized that phase connection was a very difficult thing to do indeed, and that a much easier route to precision measurement was through measuring delay. For that, one needs a wide spanned bandwidth, which is difficult to attain with the Mark I system. The group at Haystack observatory began development of the Mark III system specifically for geodesy, with fourteen frequency channels digitized and written to the fourteen independent heads of a high speed instrumentation-type tape transport. The fourteen channels could be positioned to cover a very wide spanned bandwidth. This system quickly became the standard for geodesy. For astronomical purposes, however, the figure of merit tends to be the number of bits recorded on a dollar's worth of medium. This number was much the same for Mark II and Mark III, so the easier-to-implement Mark II remained in use until the design and retrofit of Mark IIIa, which uses narrow heads to lay down parallel written tracks, so that up to fourteen times as much data could be written to a single tape as Mark III. The Mark IIIa and the very similar VLBA and Mark IV formats quickly became the standards.

Mark II was extended by the Canadians to have comparable bandwidths to the Mark III/Mark IV/VLBA recorders, by operating several VCRs in parallel, each recording at 8 Mbit s^{-1}. This system is called the S2, with a further development, using studio quality VCRs, called S3.

The Haystack group meanwhile has noted that the cost to store a bit on computer disk storage is dropping much more rapidly than the cost to store a bit on tape. Within a couple of years, off-the-shelf computer disks will become the most cost effective media, even with the cost of motor, spindle, actuator included in the disk unit. There are also substantial operational advantages to disk units: easier maintenance of the mechanical parts, effectively instantaneous start and stop, essentially error-free operation. Haystack is now designing a system to use disks, to be called Mark V. A preliminary version of the system has been built and has been used for astronomical observations.

3. Science Themes of VLBI

As discussed in Section 1, VLBI as we know it was developed for studies of the quasars and bright radio galaxies — what are they, how do they work? But many uses not conceived at the birth of the technique have found the technique to be of importance. With the increase in sensitivity that came with the broader recorded bandwidths of the Mark III and successor systems, not only the bright quasars can be studied, but also the low luminosity active galactic nuclei. The early work on astronomical masers has been extended to many interesting results, in studies of OH, water, and SiO masers. And the early work on astrometry and geodesy continues. The accuracy of VLBI baseline determinations remains somewhat better than any other long-distance technique, and only VLBI gives the absolute orientation of the Earth in an inertial coordinate system to high accuracy.

Interestingly, although study of quasars and other AGN was the primary motivation for development of the modern VLBI systems, it has perhaps not provided the number of interesting results that some other themes have done. The early years did yield exciting results - that the quasars really were milliarcsecond objects, that they are elongated, not round, and that the elongation is due to the core-jet nature of the emission. Of course, perhaps the most exciting early result was the discovery of superluminal motion by Whitney et al. (1971), quickly confirmed by other groups (for instance, Cohen et al. 1971). Since that time there has been a great deal of phenomenology, but little increase in understanding. Decades of effort have gone into the question of whether jet motion is ballistic, or follows an established path (answer: sometimes), but this has not yet resulted in a firm understanding of the processes at work. The most interesting modern work on AGN has been outside the main thrust of work on the brightest objects. This includes the discovery of the compact symmetric objects (Readhead et al. 1996), detection of the inner collimation region in M 87 (Biretta et al. 2002), and the quotidian use of VLBI as a means of separating active galaxies into starburst, AGN, and starburst+AGN systems. And our own AGN, Sgr A*, offers interesting possibilities, as we struggle to measure its intrinsic size, which will quite likely contribute some interesting physical insight if we manage to resolve it well (Falcke et al. 2000).

The astronomical masers have been interesting for both the intrinsic small sizes of the maser spots (Moran et al. 1968; Burke et al. 1971; and other authors), and for the applications of masers for inferring distances and motions. Maser proper motions have been used to give dynamic parallaxes for objects in our galaxy (a technique pioneered by Genzel et al. 1981). Shock excited masers have been used to outline the shock fronts in outflows from proto-stellar objects, and to measure their motions (for instance, Wootten et al. 1999). Perhaps one of the most interesting astronomical results of the whole decade of the 1990's was the detection and characterization of the Keplerian molecular disk surrounding the massive black hole at the nucleus of NGC 4258, outlined in a series of papers starting with that by Miyoshi et al. (1995). The SiO masers in the extended atmospheres of the AGB stars provide a picture of motion of the atmospheric material which already demonstrates that the outer atmospheres of these stars have more complex dynamics than was conceived prior to the measurement (Diamond et al. 1998).

Despite their low flux densities, active stars have proved a fertile area of research. The flaring regions in their outer atmospheres can be readily resolved, and comparisons to solar activity (which is thousands of times weaker) are leading to understanding of how these active stars radiate in the radio regime. And the loop back to quasars is closed by the discovery of the micro-quasars, stellar mass objects that demonstrate the same phenomena as quasars, on a vastly smaller spatial scale and vastly shorter time scale.

The precision measurements for astrometry and geodesy have been vigorously pursued by the Haystack, NASA-Goddard, and Naval Observatory groups. An early successful use of VLBI astrometry was the accurate determination of the relativistic displacement of radio source positions due to the gravitational influence of the Sun (Counselman et al. 1973). VLBI quickly grew to be one of the most accurate techniques in both astrometry and geophysics (for instance, Ma et al. 1986; 1998). The accurate determination of the Earth's orientation, and the length of day, have lead to truly remarkable results in geophysics (for instance Marcus et al. 1998). The progress of radio measurements of position has been one of the great success stories of precision measurement. In 1960 there were few radio positions determined with sufficient accuracy to permit identification with even bright or unusual objects known optically. Forty years later, VLBI measurements of positions of extragalactic objects establish the celestial reference frame, in inertial coordinates, leaving optical techniques struggling to provide sufficient accuracy to register the images made at radio and optical wavelengths. In that forty years, the accuracy of common radio positions has improved almost a million fold.

References

Adgie, R.L., et al., 1965, Nature, 208, 275
Allen, et al., 1962, MNRAS, 124, 477
Anderson, B. et al., 1965, Nature, 205, 375
Biretta, J. A., Junor, W., & Livio, M., 2002, New Astron. Reviews, 46, 239
Broten, N. W., et al. 1967, Nature, 215, 38
Brown, G. W., Carr, T. D., & Block, W. F 1968, Astrophysical Letters, 1, 89
Brown, R. H., & Twiss, R. Q. 1954, Philosophical Magazine, 45, 663
Burke, B. F., et al. 1970, ApJ, 160, L63
Carr, T. D., et al., 1965, IEEE NEREM Record, 7, 222
Clark, B. G., et al. 1968a, ApJ, 153, 705
Clark, B. G., et al. 1968b, ApJ, 153, L67
Cohen, M.H., 1965, Nature, 208, 277
Cohen, M. H., et al., 1971, ApJ, 170, 207
Counselman, C. C., et al. 1974, Phys. Rev. Lett, 33, 1621
Dent, W. A. 1965, Science 148, 1458
Diamond, P. J., & Kemball, A. J., 1998, BAAS, 30, 1349
Falcke, H., Melia, F., & Agol, E., 2000, ApJ, 528, L13
Genzel, R., et al., 1981, ApJ, 244, 884
Gold, T., 1967, Science, 157, 302
Gwinn, C. R., et al. 2000, ApJ, 531, 902
Hewish, A., Scott, P. F., & Wills, D., 1964, Nature, 203, 1214

Kellermann, K. I., 1965, AJ, 70, 681
Kellermann, K. I., & Cohen, M. H. 1988, JRASC, 82, 248
Kellerman, K. I., & Moran, J. M., 2001, ARA&A, 39, 457
Ma, C., et al. 1986, AJ, 92, 1020
Ma, C., et al. 1998, AJ, 116, 516
Marcus, S. L., et al. 1998, Science, 281, 1656
Miyoshi, M., et al. 1995, Nature, 373, 127
Moran, J. M., et al. 1967, ApJ, 148, L69
Moran, J. M., et al. 1968, ApJ, 152, L97
Palmer, H. P., et al. 1967, Nature, 213, 789
Readhead, A. C. S., et al. 1996, ApJ, 460, 612
Sholomitskii, G. B., 1965, AZh, 42, 673 (translation in Soviet Ast., 9, 516)
Whitney, A. R., et al. 1971, Science, 173, 225
Wootten, A., et al. 1999, BAAS, 31, 898

Radio Astronomy at the Fringe
ASP Conference Series, Vol. 300, 2003
J.A. Zensus, M.H. Cohen, & E. Ros, eds.

Very Long Baseline Interferometry in the USSR

L. I. MATVEYENKO
Space Research Institute, Profsojuznaja 84/32, 117977, Moscow, Russia

Abstract. A method of interferometry using independent local oscillators, registration of the radio and pilot signals on tape recorders, and correlation of these data was proposed by me in Spring 1962, to the Soviet DSN. It was reported at FIAN' and GAISh' seminars (September 1962), and presented to the Patent Bureau. A paper on the long-baseline radio interferometer was sent to Radiofizika (December 1963). In the Summer 1963, during a visit to the USSR, B. Lovell became acquainted with "independent" interferometry and DSN equipment, including atomic standards and tape recorders. A memorandum about a DSN-Jodrell Bank experiment at 32 cm was prepared. On 23 February 1968 K. Kellermann and M. Cohen sent a letter to V. Vitkevich proposing a USA–USSR VLBI experiment. We visited NRAO in January 1969 to adjust the parameters of the experiment, and then discussed with W. Welch (Berkeley) possible H_2O sources as candidates for VLBI, and we discussed with B. Burke (MIT) the equipment for a Simeiz-Haystack experiment at 1.35 cm. I review the Global VLBI Network, studies of AGN objects, maser sources, and Venus balloons—*VEGA*.

1. Introduction

The history of very long baseline interferometry (VLBI) is connected with space explorations. At the end of the 1950s the Lunniks were launched. A contact of one of them with the Moon's surface had been measured by radio astronomers of the Lebedev Physical Institute (FIAN), with a radio interferometer at the Crimea Radio Astronomy Station. At this time, near Evpatorija, in the Crimea, the Deep Space Network was being built. A radio interferometer was the main instrument of the measuring system. Figure 1 shows one of the antennas. Matveyenko (FIAN) actively participated in building a radio interferometer at 8 and 32 cm. The measurement accuracy was to be $0\rlap{.}''1$. We did not know positions of radio sources with this accuracy, and invited G. S. Khromov and G. B. Sholomitsky (aspirants of I. Shklovsky, Sternberg Astronomical Institute —GAISh) to solve this problem. (Matveyenko 1981a). The main interest was connected with objects with a peculiar spectrum – those with extra high frequency emission (Sholomitsky et al. 1964). The peculiarity should be the result of ejection of relativistic particles (Razin 1960, Kardashev 1962, Slysh 1963, Kellermann & Pauliny-Toth 1968). Shklovsky proposed that in the central part of the Crab Nebula is located a relativistic particle ejector which should be a compact radio source. He recommended that I study this problem by the occultation method. We observed the Crab Nebula occultation at centimeter - meter radio wavelengths in 1964. The diffraction pattern from the compact radio source (future pulsar) was discovered at 32 cm. We detected variable radio source in the SE part of the Nebula, (Matveenko 1981a; Matveyenko & Kostenko 1979). This was the first variable cosmic radio source.

Observations of occultations of the 3C 273 source, at 8 and 32 cm at the DSN antennas, showed that the compact core, the ejector, has high frequency emission, and that the jet is the low frequency component (Sholomitsky et al. 1966). The ejected stream of relativistic particles should be unstable and the radio emission should be variable. Sholomitsky (1965) showed that source CTA 102 can be variable. The variability of radio sources was confirmed by Dent, 1965. For studies of the cores we needed super angular resolution.

Figure 1. The DSN array of eight 16 m parabolic mirrors—left; 70 m antenna—right

1.1. "Independent" radio interferometry

Occultations of radio sources are unique events. But not all sources are occulted, and angular resolution of the occultation method with available interferometers was not enough for studies of the cores. The solution was found in Spring 1962 - a radio interferometer with "independent" elements: separate coherent local oscillators, registration of the radio and the pilot signals on the tape recorders and correlation of these data. In this case the baseline is not limited, an element of the radio interferometer can be carried away from the Earth. The DSN had ammonia type atomic frequency standards, and tape recorders with 100 kHz bandwidth. Gus'kov G. Ya. (Head of DSN) suggested to begin experiment between DSN antennas. That Autumn I presented a method for VLBI at the seminar of the Radio Astronomy Laboratory of FIAN, Pushino. The seminar (Chairman V. V. Vitkevich) did not support publication. I had only one possibility—to connect with GAISh' astronomers and made report at GAISh' seminar. Chairman D. Ya. Martinov decided to patent the invention and we sent the proposal to the Patent Bureau.

In 1963 B. Lovell was in the Summer Astronomy School at the Crimea Astrophysical Observatory (CrAO) and visited the DSN. The DSN director charged me (as head of the Radio Astronomy section) to meet B. Lovell. We invited I. Shklovsky and G. Khromov at meeting for translation. Shklovsky proposed that I explain to B. Lovell the new method of interferometry. Being a radio physicist he highly appreciated the method, but expressed certain doubts in the efficiency of the high angular resolution as the known point sources, stars, had low-level radio emission (low temperatures). The strong sources of synchrotron emission had big size with brightness temperatures around $T_b \approx 10^7$ K. The compact sources, quasars and masers, were just unknown. Shklovsky perceptively noted that the brightest compact radio sources can be unknown because there was no tool to measure them. The near future showed that he was right. We prepared a Memorandum on the Evpatoria-Jodrell Bank interferometer at wavelength 32 cm. B. Lovell said that he would discuss details of the experiment with experts and would send a letter. I am proposing that he discussed this experiment with Jodrell Bank and Canadian colleagues. But DSN or FIAN did not receive any answer. At the end of 1993 Dr. H. Palmer was in CrAO and discussed a VLBI experiment Jodrell Bank – Simeiz at 75 cm wavelength (Lovell 1973).

In December 1963 the Patent Bureau allowed publication and the article, "On the long-baseline radio interferometer" by L. I. Matveyenko, N. S. Kardashev and G. B. Sholomitsky, was sent to Radiofizika.

2. Kellermann & Cohen's proposal

VLBI was implemented in 1967 by USA scientists and independently by Canadian teams (Bare et al. 1967; Broten et al. 1967). On 23 February 1968 M. H. Cohen (Caltech) and K. I. Kellermann (NRAO) offered to V. V. Vitkevitch (FIAN) to conduct an experiment with an intercontinental baseline, between the 22 m antenna (Pushino) and 43 m antenna (Green Bank) at a 3 cm wavelength, for studying compact radio sources with super high angular resolution. V. V. Vitkevitch gave me the letter and added: "This your idea, you are to do it". He had serious doubts for realization of the experiment. These were the years of the cold war and the implementation of this sophisticated experiment was limited due to not only technical reasons. The FIAN authority and support of M. V. Keldysh enabled us to overcome these difficulties. On 23 July 1968 we informed our USA colleagues about our agreement. For many reasons we had to change the FIAN radio telescope to the 22-meter dish (RT-22) at the CrAO. A. B. Severny, CrAO director, supported us, and we began to prepare the antenna at 3 cm. After solving the organization problems, NRAO invited Matveyenko and Moiseev to Green Bank to discuss details of the VLBI experiment. We arrived there on January 13, 1969. The main question for selection of sources was sensitivity; this involved the angular resolution and brightness temperature of the objects. We solved all technical questions and, for guarantee, decided to use two wavelengths $\lambda = 2.8$ cm and $\lambda = 6$ cm.

3. The Crimea–Green Bank observations

In September 1969 B. Clark, K. I. Kellermann, and J. Payne came to the USSR with equipment at 3 cm and 6 cm, the MK-1 terminal, and the hot HP Rb frequency standard. A hydrogen frequency standard was used at GB. The HP Rb standard (clock) served to synchronize time at the two stations and it was therefore kept warm while being taken from the USA to Crimea and back. Synchronization had an accuracy of a few μsec. Results of testing observations were written on magnetic tapes and sent to NRAO for correlation. The fringes were seen, and we began observations on the main program. Data processing was carried out on computers IBM 360/60 (NRAO, Charlottesville) and 360/75 (Caltech and GSFC). The CR–GB interferometer attained its maximum possible angular resolution. The baseline was measured with an accuracy of about 50 m. The 2.8 cm observations detected compact structures $\sim$ 0.4 mas in 3C 273 and 4C 39.25. This work demonstrated the feasibility of global radio interferometry at a short wavelength. It showed that compact radio objects have complex structure, and that we needed more stations which were more sensitive, to continue these studies (Broderick et al. 1970).

The Institute for Space Research (IKI) was organized to bring a highly efficient structure to space studies. It had a department for astrophysics, headed by I. Shklovsky, who invited Matveyenko to head the VLBI Laboratory. In July 1969 the VLBI field was transfered to IKI; L. I. Kogan was head of the processing part and V. I. Kostenko was head of astrophysics.

In June 1971 an experiment was conducted at $\lambda = 3.55$ cm using the 64 m NASA/DSN antenna at Goldstone with the telescopes at Crimea and Green Bank. The antennas formed three interferometers, with fringe sizes 2, 0.9, and 0.7 mas. Sensitivity also was enhanced. A low-noise maser amplifier was built for the Crimea radio telescope, and the antenna was equipped with a Cassegrain illumination system. The NRAO designed a new wide-band recording system – the MK-2 based on a TV recorder and a special correlator (Clark et al. 1972). At Goldstone there was a maser preamplifier and a hydrogen-maser frequency standard. M. Cohen and B. Clark brought to Simeiz a special highly stable (on a short time scale) crystal oscillator stabilized by a rubidium frequency standard.

In this experiment we discovered very bright compact sources with $T_b \sim 10^{12}$ K. This confirms the theory of Kellermann & Pauliny-Toth (1969). The angular size of an outburst region in the source OJ 287 was found to be ≤ 0.2 mas and brightness temperature $T_b \geq 4 \times 10^{12}$ K. (Clark et al. 1973).

Additional antennas became involved in observations and formed a global VLBI Network. The wavelength decreased to 1.35 cm, which is optimum for studies of AGN objects and H_2O maser sources. We developed low noise maser preamplifier $T_n \sim 25$ K, increased the antenna effective area $S = 210$ m^2, equiped it with the hydrogen maser frequency standard (Batchelor 1976; Pauliny-Toth et al. 1978).

AGN objects were observed systematically over a wide band of radio wavelengths (Kellermann et al. 1971; Matveyenko et al. 1973, 1980, 1981a; Pauliny-Toth et al. 1978; Romney et al 1984). It was determined that quasars were the galaxies' nuclei at different stages of activity. A compact object is a nozzle which forms a self-focusing outflow - a jet of relativistic plasma. The jet helix structure of the typical QSS 3C 345 and 1803+784 is a result of precession. The core is surrounded by thermal plasma which is observed in emission lines. The changing opacity determines the low frequency variability of the synchrotron emission of the core. Changing of absorption with distance from the core was measured in typical QSS 3C 345 and 1803+784 (Matveyenko 1993; Britzen et al. 2001). The map of the last object was obtained with a global VLBI network (20 antennas) in a broad dynamic range (shown at Fig. 2).

4. The H_2O maser sources

On February 7–10, 1969 we were at Ch. Townes' seminar (Berkeley). W. Welch reported the discovery of water vapor sources which radiated in very strong narrow lines. The sizes of these sources were not known. It was proposed that these sources have thermal emission. But very strong emission in narrow lines (50 kHz) showed that these sources could be objects for experiments with the USSR-USA baseline. A few days later we discussed with B. Burke (MIT) and Cosmo Papa the technical possibility of observations of H_2O sources (λ=1.35 cm) with the Simeiz–Haystack baseline. A mixer receiver with a Rb standard could be enough. Test observations began inside the USA and then we did observations on the CR-HSK baseline.

In May 1971 B. Burke, J. M. Moran, S. Knowles, and Cosmo Papa came to USSR. The objects Orion, W 3, and W 49 were measured with the interferometer CR–HST, with fringe size 0.3 mas. These studies showed that the brightness temperatures of H_2O sources are $T_b \geq 10^{12}$ K and correspond to a maser mechanism. The H_2O outburst was observed in the gas-dust complex W 49. The source was compact and very bright $T_b \geq 10^{16}$K (Burke et al. 1972).

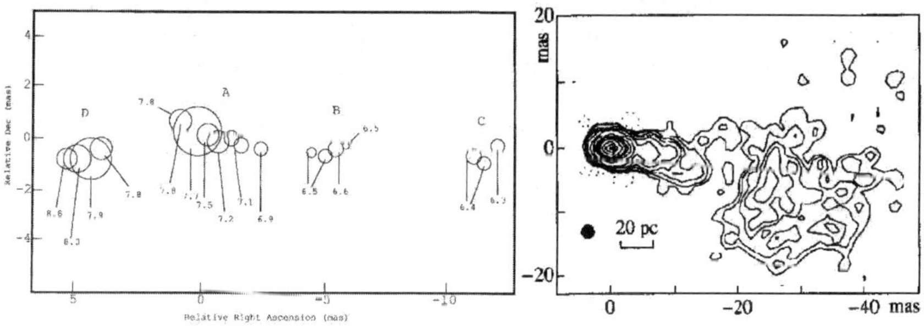

Figure 2. The structure of the H_2O super maser region in Orion KL, the size of circle indicates the logarithm of the component brightness, maximum is 10^{17} K - left; The map of 1803+784 object at 18 cm - right

The studies of maser sources continued with a global VLBI Network (Ablyazov et al. 1974; Batchelor 1976; Genzel 1978; Matveenko et al. 1988; Dhavan et al. 1992). Studies of the Orion gas-dust complex showed that star formation is a collective process, with eight active zones (Genzel 1978). During 1979–1987 high activity was accompanied by H_2O super maser emission, $F \leq 8 \times 10^6$ Jy (Matveenko 1981b). The region has highly organized structure - a chain of four groups of compact components. The group velocities increase along the chain from 6.45 $\mathrm{km\,s^{-1}}$ to 8.75 $\mathrm{km\,s^{-1}}$, as shown in Figure 2. The source sizes are ~ 0.1 AU. The strongest components, with $v = 7.7\,\mathrm{km\,s^{-1}}$, have $T_b = 5 \times 10^{16}$ K. This structure corresponds to a thin accretion disk divided into rings, observed on edge. Rotation velocity of the strongest ring is 5 $\mathrm{km\,s^{-1}}$ and it is expanding at 3.8 $\mathrm{km\,s^{-1}}$. The maser ring is a halo of expanding H_2O molecules. The active zone is surrounded by an envelope, which amplifies the 7.65 $\mathrm{km\,s^{-1}}$ emission by more than two orders of magnitude, (Matveyenko et al. 1988, 1992). The nozzle and bipolar outflow were detected.

5. Space–Ground VLBI

The first results of studying the H_2O maser sources showed that a further increase in angular resolution would be necessary. In February 1972 a decision was made to create a space-based interferometer at 1.35 cm. V. P. Mishin was responsible for the project as a whole, L. I. Matveyenko was the scientific supervisor and Valery Kostenko was the technical supervisor. The prototype of the space radio telescope had a parabolic mirror with diameter of 3.1 m (to fit the fairing). The accuracy of the mirror surface was ≤ 0.05 mm. A space telescope had to be put in orbit with apogee 12,000 km, (Kostenko & Matveyenko 1982). The mirror's frame was coated with thermal insulation and its reflecting surface was fixed on thin invar needles. The secondary 30 cm mirror had three degrees of freedom for the focal point, temperature compensation, and correction for pointing errors of the fly bus. The low-noise receiver was cooled by releasing heat into space. Data processing was based on Soviet computers (Kogan & Matveenko 1976). The $\lambda = 1.35$ cm is optimal for studies of AGN cores, small scattering (Matveyenko 1984), and absorption in H II regions. The maser sources also are important for checking of the base line and synchronizing time.

USA scientists were interested in this project. M. V. Keldysh received a proposal from NRAO, MIT, and JPL. to jointly create a space-ground interferometer. NASA considered the possibility of KRAS-3 delivery to orbit by the shuttle. After the Soyuz-Apollo project was completed there was a need for a new impressive mission. From the scientific point of view our project outstripped all the other proposals. By that time the global VLBI network operated successfully. New more versatile antennas were added to the network, including a 70 m antenna at Evpatoria, 100 m at Effelsberg, 32 m at Medicina and Noto, 10 m at Bangalore. The Complex Long Term Program Sciences-Technical Cooperation between India and USSR considered building a 70 m antenna in the equatorial region of India. A 10 m Space antenna—the *RadioAstron* project, was studied (Sagdeev 1984). But the country moved to Perestroika, and that was changed our plans. However, Russian decisions were approved, and the VSOP project showed that our solution was right. We can only regret that these unique studies could not be supported.

6. *VEGA* project

In 1974 a decision on studying Venus' atmosphere with freely floating balloons was made at a Soviet-French meeting—the *VEGA* mission. The balloons' trajectories were to be measured using the global VLBI network. The balloons transmitted at a wavelength of 18 cm; the hydroxyl line is a natural imitator of the transmitter. There was a task to upgrade the Soviet VLBI network; i.e., to create the feeds, low noise receivers, coherent local oscillators based on hydrogen frequency standards, and a system for data registration and processing of MK-2 tapes (Matveyenko et al. 1986).

The *VEGA* project was implemented by wide international cooperation. A network of 20 antennas was used : 70 m antennas at Ussuriisk and Evpatoria; 25 m at Ulan Ude; 22 m at Simeiz and Pushino; 64 m at Bear Lake; 100 m at Effelsberg; 76 m at Jodrell Bank; 64 m at Tidbinbilla, Madrid, and Goldstone;

300 m at Arecibo, 40 m at Owens Valley; 18 m at North Liberty; 26 m at Penticton, Fort Davis, and Hartebeesthoek; 37 m at Haystack; and 14 m at Atibaya.

Balloons *VEGA–1/2* were dropped into the Venus atmosphere on June 11 and 15, 1985 and were tracked for 46 hours. The transmitters' power was 1 W. The balloons moved parallel to the equator with velocities 69 $\mathrm{km\,s^{-1}}$ and 67 $\mathrm{km\,s^{-1}}$. The measurement accuracy was so high that we registered the transmitter swinging on a rope 12 m long. The balloon *VEGA-2*, when close to the Aphrodite mountains, changed its trajectory and drifted to the north at 3.5 $\mathrm{m\,s^{-1}}$ (Sagdeev et al. 1992).

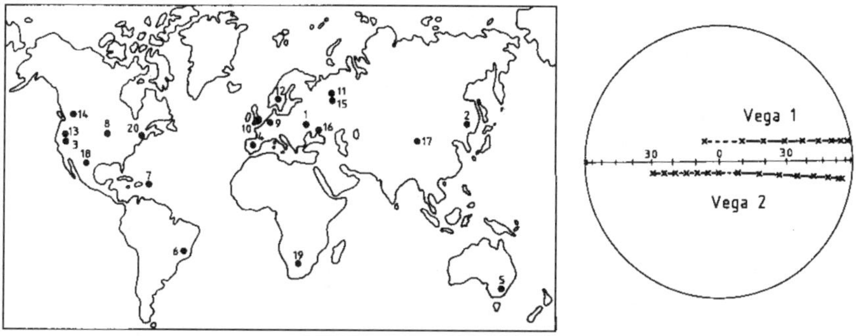

Figure 3. The global VLBI network of 20 stations - left; Trajectories of the balloons across Venus, points are marked at 4h intervals - right

7. Conclusion

VLBI, like Galileo's telescope, opened a new page of astronomy. Star formation regions and AGN objects became visible. Ken remarked that VLBI has joined not only radio telescopes, but people of different countries and continents. We have better understood not only the Universe, but ourselves. We thank Ken, one of pioneers of VLBI, and his colleagues for their activity and contribution in establishing and developing VLBI. We are looking forward for further VLBI progress, Space–Ground observations, and new scientific results.

References

Ablyazov, V. S., et al. 1974, Radiofizika, 17, 10, 1431

Bare, C. C., et al. 1967, Science, 157, 14, 189

Batchelor, R., et al. 1976, Sov.Astron.Lett., 2, 10, 467

Britzen, S., et al. 2001, Astronomy Letters, 27, 1, 1

Broderick, J. J., et al. 1970, AZh, 47, 4, 784

Broten, N. W., et al. 1967, Science, 156, 1592

Burke, B. F., et al. 1972, AZh, 49, 3, 465

Clark, B. G., et al. 1972, AZh, 49, 4, 700

Clark, B. G., et al. 1973, ApJ, 182, L57

Dent, W. A., 1965, Science , 148, 1458

Dhavan, V., et al. 1992, Sov. Astron. Lett., 18, 4, 391
Genzel, R., et al. 1978, A&A, 66, 13
Kardashev, N. S., 1962, AZh, 39. 393
Kellermann, K. I., et al. 1971, ApJ, 169, 1
Kellermann, K. I., & Pauliny-Toth, I. I. K. 1968, ARA&A, 6, 417
Kellermann, K. I., & Pauliny-Toth, I. I. K. 1969, ApJ, 155, L71
Kogan, L. R., et al. 1974, in the Soviet–French Symposium on "EOS–Venus", 7
Kogan, L. R., & Matveenko L. I. 1976, Radiofizika, 9, 1265
Kostenko, Val. I., & Matveenko, L. I. 1982, Kosmicheskie issledovanija, 20, 149
Lovell, A. C. B. 1973, in Out of the Zenith: Jodrell Bank, (Oxford: Oxford Univ. Press; Harper&Row)
Matveenko, L. I. 1993, Soviet Ast. Lett.., 19, 4, 291
Matveenko, L. I. 1977, in Itogi Nauki i Tekhniki, 1
Matveenko, L. I. 1981a, in Space Phys. rev., ed. R. A. Syunyaev, Vol. 1, 83
Matveenko, L. I. 1981b, Soviet Ast. Lett., 7, 54
Matveenko, L. I., et al. 1973, AZh, 50, 6, 1157
Matveenko, L. I., et al. 1980, Soviet Ast. Lett., 6, 2, 77
Matveenko, L. I., et al. 1992, in Proc. of Conference Held in Arlington, Virginia, 176
Matveenko, L. I., Graham, D. A., & Diamond, P. J. 1988, Sov. Astron. Lett., 6, 468
Matveenko, L. I., & Kostenko, V. I. 1979, Austral. J. Phys., 32, 105
Matveenko, L. I., Kardashev, N. S., & Sholomitsky, G. B. 1965, Radiofizika, 8, 4, 651
Matveyenko, L. I. 1984, in Proc. of Workshop on *Quasat*, 119
Matveyenko, L. I., et al. 1986, Sov. Astron. Lett., 12, 1, 59
Pauliny-Toth, I. I. K., et al. 1978, Sov. Astron. Lett., 4, 2, 64
Razin V. A. 1960, Radiofizika, 3, 584
Romney, J., et al. 1984, ApJ, 99, 289
Sagdeev, R. Z. 1984, in Proc. of Workshop on Quasat, 19
Sagdeev, R. Z., et al. 1992, A&A, 254, 387
Sholomitsky, G. B. 1965, AZh, 42, 673
Sholomitsky, G. B., et al. 1964, AZh, 41, 823
Sholomitsky, G. B., et al. 1966, AZh, 9, 1135
Slysh, V. 1963, Nature, 199, 682

Radio Astronomy at the Fringe
ASP Conference Series, Vol. 300, 2003
J.A. Zensus, M.H. Cohen, & E. Ros, eds.

Early Work in Imaging

J. A. Högbom
Stockholm Observatory, AlbaNova University Center, SCFAB, SE-106 91 Stockholm, Sweden

Abstract. In this paper I will focus on two stories in which I have been personally involved: Earth rotation synthesis and the CLEAN algorithm.

1. Introduction

It's an excellent idea to arrange a workshop at this occasion; for many of us an opportunity to meet with friends from that ancient generation to which I myself belong. Marshall H. Cohen first suggested that I speak about "CLEAN and Aperture Synthesis". Such a title gives the impression of a thoughtful and systematic presentation of the development of these ideas throughout history. Fortunately he changed his mind. Many people, including Ken Kellermann (Kellermann & Moran 2001), have done the proper research and traced the development of radio image formation—from the famous cliff interferometer paper by McCready, Pawsey & Payne-Scott (1947) over the Mill's Cross, the Christiansen solar arrays, Ryle's aperture synthesis, the VLA and so on, all the way to the sophisticated image making procedures used today. I cannot improve on these fine papers. But here, in the history session, and with a more flexible title, I may be excused to do some micro history and focus on two stories in which I have been personally involved: Earth rotation synthesis and the CLEAN algorithm.

2. Earth Rotation Synthesis

I came to Cambridge in 1955 to do thesis work and stayed there until 1959. It was in a time when one began to realize that Radio Astronomy had the potential of becoming much more than just a poor cousin of traditional optical astronomy. Angular resolution was perhaps not doomed to remain vastly inferior to that of optical astronomy just because nobody could build a steerable paraboloid several miles in diameter.

It was a turbulent period at Cambridge. The 2C survey had just been published and come under fire. I understood that there were some very unpleasant people down in Sydney, but not much more. Tony Hewish was my supervisor and I used simple two-antenna interferometers to observe the scattering of radio waves by the solar corona and to position radio bursts on the solar disc. Much of the heated arguments about source statistics passed me by.

However, I attended Ratcliffe's now famous lecture series on Fourier analysis and like many others I became excited by the beauty of it all and, in particular, by its application to aperture synthesis.

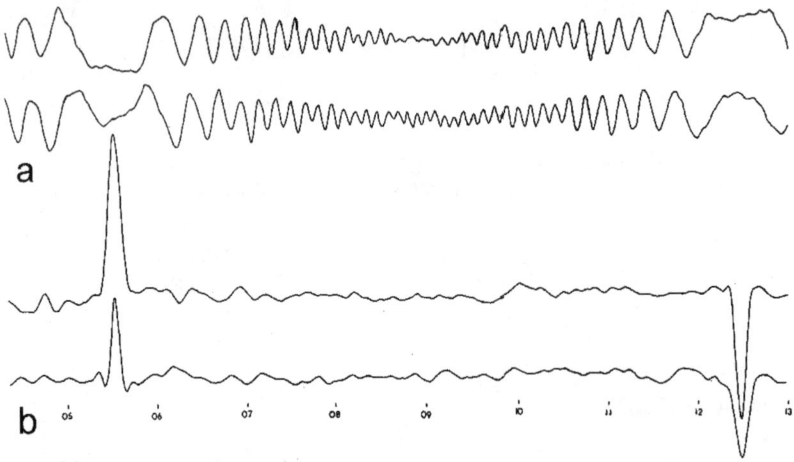

Figure 1. a) The simple small antenna N-S interferometer: Cos- and Sin- recordings showing the Crab Nebula and M 87. b) Synthesised N-S interferometer: Cos- and Sin- recordings.

The NS and EW simple interferometers that I used for the positioning of solar bursts were kept running during the night to get amplitude and phase calibrations from the Crab Nebula and M 87. Virtually all telescopes at Cambridge were meridian passage instruments. All interferometer records coming off the pen recorders in the receiver hut looked much the same - a regular sine wave under some envelope determined by the antenna dimensions. My NS interferometer on the other hand produced something very different, a non regular periodicity with an obvious central fringe (Fig. 1a). I understood of course how that could be, but I was puzzled by some properties of the pattern. First, in contrast to my EW interferometer pattern, there was an obvious central fringe, and the variable fringe period seemed to make it possible to separate the two sources even if they had been close enough to overlap. Convolving the record with the theoretical source response - which is what you should do when looking for a particular pattern - would clearly give a strong dominant peak just at the meridian passage of the source and little else. The same operation on the EW interferometer record would just produce a similar uninteresting sine wave with no obvious central peak. Why this apparently superior amount of information available from the NS interferometer?

I did the celestial sphere geometry and soon realized that convolving the pattern with the theoretical response was much the same as synthesising the output from an EW (curved) line antenna traced out by the rotation of the Earth. I gave a preliminary write-up of my ideas to Ryle in July 1958 and somewhat sheepishly asked him to "read it in his bath". Apparently he did, because the next time we met he said that he had had an "interesting bath". As far as I can remember, he made no further comment. I also showed it to Tony Hewish who helped me by running the convolution—now synthesis—calculations on the Cambridge EDSAC II computer (Fig. 1b).

I thought at the time that Earth rotation synthesis was a fun idea but of limited use. The trick, I imagined, required very small antennas with wide beams: the source must stay in the beam long enough for a significant synthesis. Why couldn't I see that fully steerable antennas would make all the difference between a fun idea and something really useful? Well, I just didn't. Fully steerable antennas did not exist at Cambridge! That shows the difference between us ordinary guys and people like Martin Ryle.

My thesis work on the Sun produced 78 typewritten pages and looked rather thin. Previous radio astronomy theses in the library were quite solid and none contained less than 100 pages. I searched for something more to say about the Sun—but failed. In the end I crossed my fingers and added two chapters on a completely different subject: aperture synthesis theory and the Earth rotation experiment. That widened my thesis to a more comforting 104 pages (Högbom 1959). That's why the Earth rotation story ever got written down!

I did of course not 'invent' Earth rotation synthesis. I 'rediscovered the wheel' (or at least some part of the wheel!). Martin Ryle had it all thought out long before—including the one mile telescope design. But he kept it to himself, worrying that others with plenty of money—read the Americans—would exploit his ideas before he had a chance to do much himself. I was in the Netherlands working on the Benelux Cross project when in 1962 he published a short Nature article containing a simple drawing of the future One Mile Telescope (Ryle 1962). I saw the article and felt a bit stupid not to have thought about such a beautiful application. From there on I began propagating for a super-synthesis array, possibly as a complement to a shortened cross for extended objects. As time went by, the cross faded out of the project altogether to make place for the Westerbork Synthesis Telescope completed in 1970 (Högbom & Brouw 1974).

3. The CLEAN Algorithm

The second story I'd like to tell is the story about CLEAN and, in fact, that story started right here in Green Bank in 1968. I had time on the 3-antenna interferometer for a program of mapping the structure and polarisation of some 60 sources selected mainly from the 3CR catalogue. The plan was to make short "snapshot mode" observations at six hour angles and with a few different antenna spacing arrangements. That should give me 50-60 complex visibilities distributed over the (u, v)-plane for each source. At the time I had no firm idea how to reduce the data, but naïvely imagined that it shouldn't be too much of a problem to determine source structures from such measurements.

Back in Leiden I realized that there was indeed a problem! Straight interpolation between the measurements is just mathematically wrong as everybody knows. Model fitting was messy and unreliable. I tried an iteration scheme, flipping between the visibility and map domains. Starting with the direct transform of the individual measurements—i.e., what is now known as the 'dirty map', I shaved off negative values in the map domain as being impossible, returned to the visibility domain, restored the visibilities to their measured values and so on. It half- worked in some simple cases but never very well.

During these trials I found myself looking at 'dirty maps' of many sources including some calibration point sources. It was then a small step to ask: if I subtract a full theoretical point source pattern, a suitably scaled and positioned 'dirty beam', from the map then there should be nothing left - unless of course there is something else out there! Often there was and I went on subtracting. Returning to the map only the nice part - the central lobe - of each subtracted pattern was a temptation I couldn't resist and it actually seemed to work! From there it rolled on. When things got messy it often helped to do things cautiously, a little at a time, i.e., using a reduced 'loop gain'. So CLEAN had a very simple-minded beginning but in the end turned out to be more useful than I had ever expected. Dave Rogstad was in Groningen at the time and we had some good discussions. He used it with his Caltech interferometer measurements of M 101 and published the first CLEANed map (Rogstad & Shostack 1971).

It interesting to note that the paper describing the CLEAN algorithm was later presented just where the story started: in Charlottesville, at the 1972 NRAO symposium on 'The Collection and Analysis of Astrophysical Data' (Högbom 1974). Now 30 years later, it is good to be back here again and to look upon this whole story as part of history!

References

Högbom, J. A. 1959, Ph.D. Thesis, Cambridge University, UK

Högbom, J. A. 1974, A&AS, 15, 417

Högbom, J. A., & Brouw, W. N. 1974, A&A, 33, 289

Kellermann, K. I., & Moran, J. M. 2001, ARA&A, 39, 457

McCready, L. L., Pawsey, J. L., & Payne-Scott, R. 1947, Proc. Roy. Soc. London, Series A, 190, 357

Rogstad, D. H., Shostack, G. S. 1971, A&A, 13, 99

Ryle, M. 1962. Nature, 194, 517

Radio Astronomy at the Fringe
ASP Conference Series, Vol. 300, 2003
J.A. Zensus, M.H. Cohen, & E. Ros, eds.

CTD 93 – Spanning the Years

D. B. SHAFFER
Community College of Southern Nevada, 700 College Drive – H3C, Henderson, Nevada 89015, USA

Abstract. Observations of the radio galaxy CTD 93 by Ken Kellermann and collaborators span almost 40 years, from the beginning of his radio astronomy career to the present. This source was first cataloged by Ken and has lately been getting his attention again. Initially thought to be a compact double source, it now appears that CTD 93 is a core-jet object with a very long, highly collimated jet.

The Owens Valley Radio Observatory (OVRO), operated by Caltech, performed several source surveys in its early days with the then-new 90-foot antennas: the CTA, CTB, and CTD surveys (Caltech "A", Caltech "B," etc.) (I'm told by Ken that there was no "C" survey. Was there no desire to compete with Cambridge nomenclature?) The "D" survey was carried out by Ken and Dick Read at a frequency of 1421 MHz (Kellermann & Read 1965). The only publication of the results is in the OVRO series. Apparently at this early date, it was not appreciated that sources might be variable - the soon-to-be-famous variable source observer and Dick give no dates for the OVRO observations! The best we can assume is that the observations took place sometime before the 1965 publication of the survey. Fortunately, it appears that CTD 93 has very little, if any, flux density variability.

The CTD survey contains 143 sources. CTD 93, the ninety-third source listed, is the strongest "new" source in the survey, and the 10th strongest source overall. All of the CTD sources stronger than CTD 93 (and many of the weaker ones, too) are 3C objects. Given its strength, it is perhaps a bit surprising that CTD 93 did not receive more attention, but I guess lots of other sources were relatively new and exciting at this time, too.

As far as I can tell, other than some flux density measurements, and the occasional "rediscovery" of the source in various surveys of the late 1960s (DA 403, OS 211, VRO 26.16.01), there were few, if any, targeted observations of CTD 93 for any reason. The spectrum of the source peaks at about 1 GHz, and is very similar to that of 1934–63 which was observed and analyzed by Ken during his stay in Australia (Kellermann 1966).

My association with CTD 93 began when I was selecting objects for the first observing run of the US VLBI Network, in a joint proposal with Ken. By 1975, it was appreciated that compact sources could have complicated structure. I was hoping to find "new" complex sources and especially wanted to see what they looked like at lower frequencies. (The early excitement about superluminal sources came from observations at 8 GHz and above.) It seemed to me that for transcontinental USA baselines, the most interesting sources would be those that had a peak in their spectrum at about the frequency of observation, and I was planning 18 cm observations. Hence, CTD 93, which was known to produce fringes at higher frequencies (Kellermann et al. 1971) and had been shown by George Purcell in his thesis (Purcell 1973) to have interesting structure at 50 cm seemed like an obvious candidate for more thorough observations.

We did find CTD 93 to have very extended structure. The visibility data (this was just before map making!) showed it to be a wide-spaced double, with a separation between components of about 50 mas. At that point, CTD 93 was among the biggest of the compact sources whose structure had been well-established by VLBI. After this brief fling of 18 cm observing, I returned to higher frequency observations. I did suggest to Bob Phillips, however, that CTD 93 and similar sources might be interesting objects to study - which he did for his thesis.

It was on the basis of Bob's observations of CTD 93 and several other sources with similar peaked spectra that the notion of compact double sources was born. (I think it is fair to say that Bob's results contributed significantly to the idea that "GHz Peaked Sources" might be interesting objects, although neither he nor I had the foresight to come up with the catchy name!) Bob and collaborators (including me—I was still interested in this source: Phillips & Shaffer 1983) observed CTD 93 and the others of its ilk at higher frequencies, and the observations always seemed consistent with these sources being compact, or "baby," double sources - objects that would grow up some day to be real double sources like those that radio astronomers had been studying for years. There was little evidence of complex structure within either component.

By the early 80's, this remained the status of the compact doubles. Bob and I became part of the NASA Crustal Dynamics Project (CDP) VLBI group, and had little time for further studies of individual sources. The CDP wanted to observe the most compact objects, not extended sources like CTD 93!

With the advent of the VLBA and its much better sensitivity at higher frequencies and improved (u, v)-coverage, it seemed to me that the time was ripe to revisit CTD 93. (Besides, I finally had some free time for such a project.) In particular, based on the results from the Caltech VLBI group that nearly every compact symmetric object (CSO) that they observed had an obvious nucleus (Taylor, Readhead, & Pearson 1996), I figured it was time to find the nucleus of CTD 93, thereby putting to rest a question which had nagged me for years: was the nucleus in the middle of the source, as would be expected if it were indeed a "compact double," or would it be somewhere else - one of the components being the plausible alternative. (Part of the usually unspoken logic here was that compact doubles were a subset of the CSOs and that the nucleus *had* to be in the middle. I was not convinced.)

Since I had not had much experience observing with the VLBA—or more relevant—*reducing* VLBA data, I needed a knowledgeable collaborator for my search for the nucleus. Ken seemed like an obvious choice: he had been a Co-I on the early 18 cm observations, was a good friend, and most importantly, knew how to analyze VLBA data! (Ken did, in fact, initiate me into VLBA processing during a visit to the AOC after our CTD 93 observations. As Ken pointed out, I was one of the few folks he knew that he could invite to Socorro to cool off!)

We proposed a comprehensive attack on CTD 93. We planned to observe it in detail at 18 cm to look for any changes since our data of twenty years previously, and observe it in detail at 15 GHz to look for the core (the cores of the Taylor et al. (1996) CSOs had all been obvious in 8 or 15 GHz maps, with one notable exception: 1031+567, which looks *remarkably* like CTD 93). We also included a few scans at 50, 13, and 3.8 cm. Initially, we had planned a few scans at 6 cm as well, but learned that Tim Cornwell had made a high dynamic range map of CTD 93 at 6 cm as part of VLBA testing and imaging technique investigations. So, we skipped 6 cm and subsequently subsumed Tim and his 6 cm map into our analysis.

The results of these observations, in our opinion, ruined the compact double theory for CTD 93 (Shaffer, Kellermann, & Cornwell 1999). Our 15 GHz results (Fig. 1) show CTD 93 to be a core-jet source. The core appears to reside in the northern component, not somewhere in between the components.

At 18 cm, we found no evidence of structure change from our earlier observations. We could not make a very sensitive test of this, though, because there were practically no baselines in common between the two experiments. Our 1976 S2 observations, with the US VLBI Network, used seven antennas: Haystack, NRL, NRAO (GB), VRO, HRAS, OVRO, and Hat Creek. Of these, two are gone (HRAS—taken down deliberately and Hat Creek—blown over!) and three are no longer used for (low frequency) VLBI: Haystack, NRL, and VRO. Only two sites, HRAS and OVRO, produced new data for baselines essentially the same as observed previously. To within half a milliarcsecond or so, there did not seem to be any change in the 18 cm structure of CTD 93 over 20 years. At a redshift of 0.473, this corresponds to $0.5c$ or less for component motion.

In order to take an even deeper look at the structure at 15 GHz and to investigate our proposed core in the northern component at a higher frequency, we observed CTD 93 again with the VLBA, at 15 and 22 GHz late in 1996. At this time, Matt Lister joined our project and analyzed the new observations. The two 15 GHz maps (Figs. 1 and 2) overlay each other remarkably well, with no evidence of structure change, although this time with only a year and a half between experiments. The 22 GHz structure is very similar to that at 15 GHz (Fig. 3). In particular, the northern component contains the most compact region of the source. There is no evidence of any other similarly compact (core) component. Once again, the 22 GHz results are consistent with our core-jet interpretation of CTD 93. There does not seem to be a centrally-located core as we would expect for a compact double/CSO.

What else is there to do? Clearly, higher frequency observations should be made, to refine our understanding of the nucleus of CTD 93. It would also be most interesting to compare optical and radio structure, to see if the optical core aligns with the northern radio component or falls somewhere else. Unfortunately, milliarcsecond optical/radio astrometry for this object, a 19th magnitude galaxy, will be pretty difficult. However, I think it will be possible some day, and Ken and I can share the results while we rock away at the old radio astronomers home!

I met Ken in the summer of 1966, when I came to Green Bank as a summer student to work for him and Ivan Pauliny-Toth. That was the first time I had ever heard of radio astronomy. I want to thank Ken, Ivan, and all the NRAO staff of that epoch for introducing me to my future vocation.
Thanks to Matt Lister and Ken for the latest versions of the CTD 93 maps shown here.

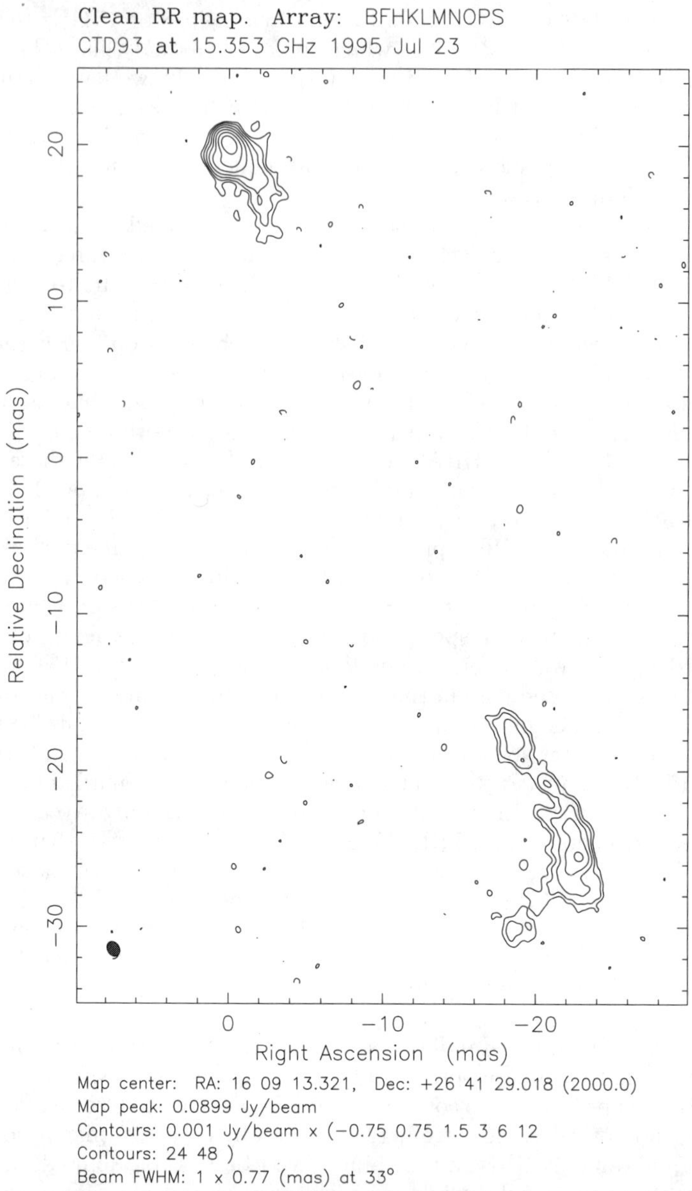

Figure 1. Map of CTD 93 at 15 GHz in July 1995

References

Kellermann, K. I. 1966, Australian J. Phys., 19, 195

Kellermann, K. I., et al. 1971, ApJ, 169, 1

Kellermann, K. I. & Read, R. B. 1965, Publ. Owens Valley Radio Obs., 1, 1

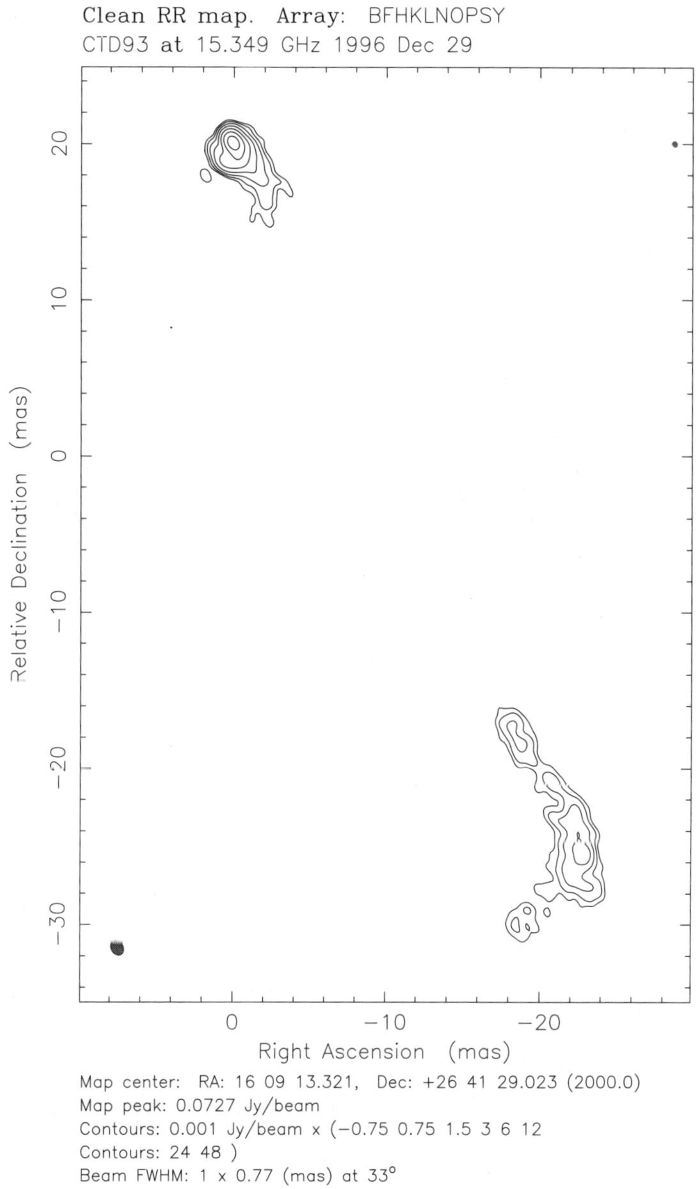

Figure 2. Map of CTD 93 at 15 GHz in December 1996

Phillips, R. B. & Shaffer, D. B. 1983, ApJ, 271, 32
Purcell, G. H. 1973, Ph.D. thesis, (Pasadena: California Institute of Technology)
Shaffer, D. B., Kellermann, K. I., & Cornwell, T. J. 1999, ApJ, 515, 558
Taylor, G. B., Readhead, A. C. S., & Pearson, T. J. 1996, ApJ, 463, 95

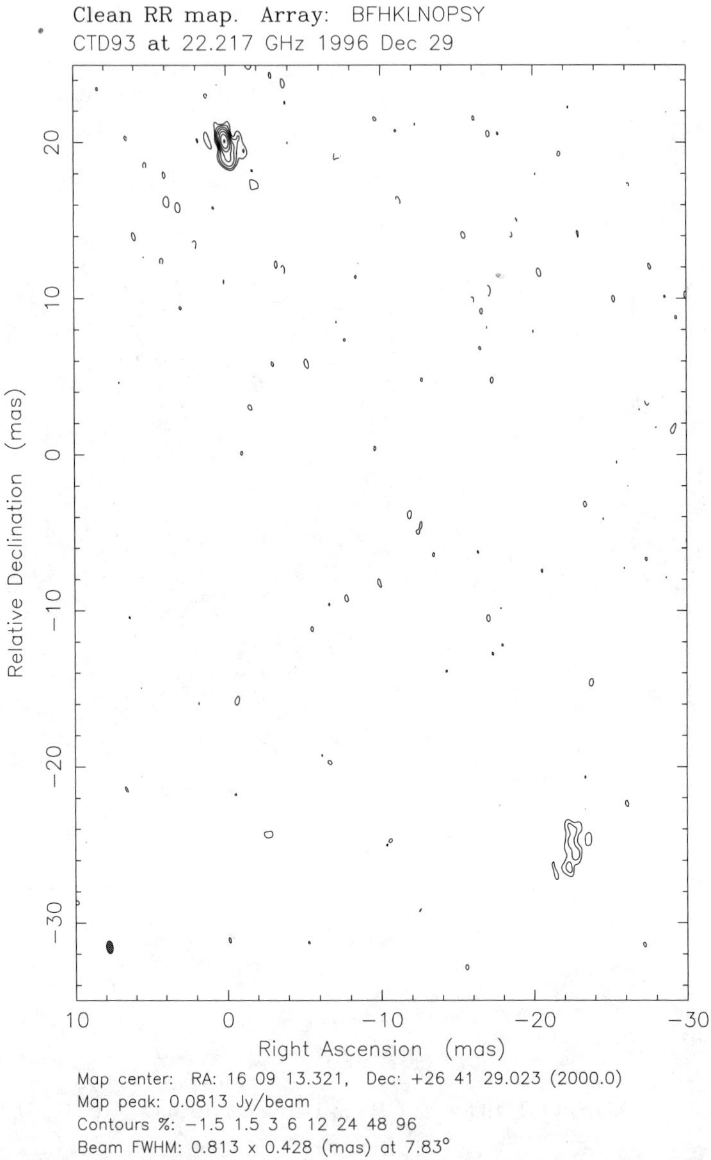

Figure 3. Map of CTD 93 at 22 GHz in December 1996

Radio Astronomy at the Fringe
ASP Conference Series, Vol. 300, 2003
J.A. Zensus, M.H. Cohen, & E. Ros, eds.

The 2 cm VLBA Survey

J. A. Zensus, E. Ros, M. Kadler
Max-Planck-Institut für Radioastronomie, Auf dem Hügel 69, D-53121 Bonn, Germany

K. I. Kellermann, M. L. Lister, D. C. Homan
National Radio Astronomy Observatory, 520 Edgemont Road, Charlottesville, VA 22903, USA

M. H. Cohen
Department of Astronomy, Mail Stop 105-24, California Institute of Technology, Pasadena, CA 91125, USA

R. C. Vermeulen
Netherlands Foundation for Research in Astronomy, Postbus 2, NL-7990 AA Dwingeloo, The Netherlands

Abstract. We report on the status of the 2 cm VLBA Survey, a program to study the kinematics of parsec-scale jets in active galactic nuclei (AGN). Over 200 sources have been observed at multiple epochs since 1994 with the Very Long Baseline Array (VLBA) at 15 GHz. The survey is restricted to flat-spectrum, bright sources ($S_{15\,\mathrm{GHz}} \geq 1.5$ Jy for $\delta > 0°$ and $S_{15\,\mathrm{GHz}} \geq 2$ Jy for $-20° < \delta < 0°$) with strong VLBI cores. More than 1000 images have been made so far and can be accessed from on-line archives at `http://www.nrao.edu/2cmsurvey` and `http://www.cv.nrao.edu/~mlister/MOJAVE`.
We present the distribution of apparent velocities measured for the 86 sources for which we have sufficient multi-epoch data, and we discuss the differences between the various classes of AGN (QSOs, BL Lacs, and galaxies).
We have initiated a new long-term project, MOJAVE (**Mo**nitoring of **J**ets in **A**GN with **V**LBA **E**xperiments), based on a statistically complete flux-limited sample which also includes studies of linear and circular polarization.

1. Introduction

The term Active Galactic Nucleus (AGN) refers to the energetic phenomena found in the central regions (nuclei) of galaxies which cannot be attributed directly to stars. Emission is associated with the existence of a super-massive black hole with an accretion disc from which bi-directional relativistic outflows ('jets') emerge. AGN radiate throughout the complete electromagnetic spectrum and their radio morphologies can usually be classified as being either 'extended' (e.g., spatially resolved) or 'compact' (e.g., unresolved at 1″ resolution). The extended sources can be as large as Megaparsecs and are transparent to their own radio synchrotron emission, whereas, due to self-absorption, the compact sources can have flat or even inverted spectra. Very Long Baseline Interferometry

(VLBI) probes the compact regions typically with resolutions of the order of 1 milliarcsecond (mas), which correspond to parsec scales (see Zensus 1997). VLBI is thus the key tool to study the relativistic outflows. Parsec-scale jets often show apparent superluminal motions, when those motions are oriented nearly along the line of sight (see sect. 3.1).

There have been various surveys to study the nature of AGN physics (see Table 1 and Green, Khachikian, & Sanders 2001). In spite of the extensive body of observational data and the sophisticated theoretical models which have been discussed in the last decades, major questions remain unanswered related to the nature of AGN jets. For example, it is not clear whether the flows are ballistic, that is, if individual features have straight trajectories, or if they follow the curvature of the jet. It also remains unclear if the changes in speed or direction of features as they propagate down the jet are typical or if they occur only under special circumstances. The different classes of AGN may present different characteristic Lorentz factors.

Table 1. Selected AGN Surveys

	Name	Refs.
Radio		
VLBI ($\lambda\lambda$ 18/6 cm)	Pearson-Readhead & Caltech-Jodrell Bank	Pearson & Readhead (1988); Taylor et al. (1994, 1996); Polatidis et al. (1995); Thakkar et al. (1995); Xu et al. (1995); Henstock et al. (1995); Vermeulen et al., these proceedings, page 43
VLBI (13/3.6/2)	Radio Optical Reference Frame	Fey, Clegg, & Fomalont (1996); Fey & Charlot (1997, 2000)
VLBI (6)	VSOP Pre-Launch	Fomalont et al. (2000)
VLBI (0.7)		Jorstad et al. (2001)
VLBI (0.3)	MPIfR	Lobanov et al. (2000)
Variability (6/3.6/2)	UMRAO	Aller et al. these proceedings, page 159
Variability	Effelsberg	Kraus et al. (2003)
Variability (1.3)	Northern 2 Jy	Valtaoja, Lähteenmäki, & Teräsranta (1992)
X-Ray		
ROSAT		Brinkmann et al. (1997); Lamer, Brunner, & Staubert (1996); Ciliegi & Maccacaro (1997); Miyaji, Hasinger, & Schmidt (2000)
*Beppo*SAX		Padovani et al. (2002)
γ-Ray		
EGRET	3rd Catalog	Hartman et al. (1999)

Individual, selected sources have been observed intensively in order to address these questions (e.g., 3C 345: Ros, Zensus, & Lobanov 2000; 3C 273: Lobanov & Zensus 2001; 3C 120: Gómez, Marscher, & Alberdi 1999; 3C 279: Homan et al. 2003, and these proceedings, page 35; 4C +12.50: Lister et al. 2003a). In particular, variability studies, multi-frequency monitoring, and observations of polarized emission have been important in understanding the physical conditions within individual sources. Our intent here is to study a representative sample of objects, carrying out a long-term, systematic monitoring of relativistic motion in AGN jets on parsec scales. We seek a significant improvement over previous surveys in terms of image resolution and fidelity, sample size, and statistical completeness. The goals of our program are to test and characterize the kinematics of these AGN jets and to determine how these are related to other source properties. We try to provide the basis for a comprehensive physical theory to relate the distribution of Lorentz factors and bends and other morphological characteristics, the internal temperatures, and the spectral behavior.

2. The Survey

Since 1994, we have been using the NRAO Very Long Baseline Array (VLBA) at 15 GHz (2 cm) to study the relativistic outflows in a sample of AGN (see description in Kellermann et al. 1998a and Zensus et al. 2002). Our sample is based on the 1 Jy catalog (Kühr 1981), and considers sources with flat spectra ($\alpha > -0.5$ for $S_\nu \propto \nu^{+\alpha}$ above 500 MHz) and flux densities $S_{15\,\mathrm{GHz}} > 1.5$ Jy for $\delta > 0°$, and $S_{15\,\mathrm{GHz}} > 2$ Jy for $-20° < \delta < 0°$. Some additional sources have been added, particularly those classified as compact symmetric objects or lobe-dominated sources with a strong core. The current 2 cm survey source list (complete up to February 2003) is shown in Table 2. The sources observed so far are tabulated under the column labeled with 'O'. The redshift distribution for the different classes of objects is presented in Figure 1 (a).

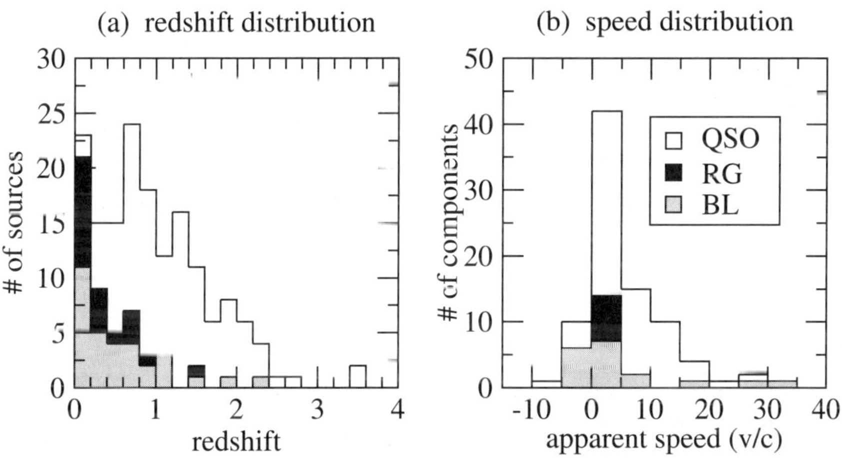

Figure 1. **(a)** Redshift distribution for the different classes of objects from the 2 cm survey. **(b)** Apparent speed distribution for the brightest components for which 3 or more data points were measured and a well-determined speed (18 BL Lac objects, 7 radio galaxies, and 61 QSOs).

We have completed more than 40 observing sessions, usually obtaining 24 images per run. We maintain a data bank of images from these observations, accessible at `http://www.nrao.edu/2cmsurvey`. Each source has been observed at least once per year. The observations typically consisted of multiple snapshots spread over 8 hours with a total integration time of 50 minutes. The root-mean-square noise level in each image is typically less than 1 mJy/beam. In other words, the dynamic ranges achieved are better than 1000:1, sufficient to

Table 2. 2 cm Survey Source List (Status February 2003)

Name	O[a]	K[b]	M[c]	Name	O	K	M	Name	O	K	M	Name	O	K	M
0003−066	√	√	√	0615+820	√			1148−001	√			1730−130	√	√	√
0007+106	√			0642+449	√	√	√	1150+812			√	1739+522	√		√
0016+731	√	√		0648−165			√	1155+251	√			1741−038	√		√
0026+346	√	√		0707+476	√	√		1156+295	√	√	√	1749+096	√		√
0035+413	√	√		0710+439	√	√		1213−172			♣	1749+701	√	√	
0048−097	√			0716+714	√	√		1219+044			♣	1751+288			√
0055+300	√	√		0723−008	√			1219+285	√	√		1758+388	√	√	√
0059+581			√	0727−115	√	√	√	1222+216			√	1800+440	√	√	√
0106+013	√	√	√	0730+504			♣	1226+023	√	√	√	1803+784	√	√	√
0109+224			♣	0735+178	√	√	√	1228+126	√	√	√	1807+698	√	√	
0112−017	√	√		0736+017	√	√	√	1253−055	√	√	√	1823+568	√	√	√
0119+041	√	√		0738+313	√	√	√	1302−102	√	√		1828+487	√	√	√
0119+115	√		♣	0742+103	√	√		1308+326	√	√	√	1845+797	√	√	
0133+476	√	√	√	0745+241	√	√		1323+321	√			1901+319	√	√	
0149+218	√	√		0748+126	√	√	√	1324+224			♣	1921−293	√		
0153+744	√	√		0754+100	√	√	√	1328+254	√			1928+738	√	√	√
0201+113	√			0804+499	√	√		1328+307	√			1936−155			♣
0202+149	√	√	√	0805−077			♣	1334−127	√	√	√	1954+513	√		
0202+319	√	√	√	0808+019	√			1345+125	√	√		1957+405	√	√	√
0212+735	√	√	√	0814+425	√	√		1354+196	√	√		1958−179			♣
0215+015			♣	0821+394			♣	1354−152	√	√		2005+403	√	√	√
0218+357	√			0823+033	√	√		1404+286	√	√		2007+776	√	√	
0221+067	√	√		0827+243			√	1413+135	√	√	√	2021+317	√	√	√
0224+671			♣	0829+046	√	√		1417+385			♣	2021+614	√	√	√
0234+285	√	√	√	0834−201	√	√		1418+546			♣	2037+511			√
0235+164	√		√	0836+710	√		√	1424+366	√			2113+293	√		
0238−084	√	√	√	0850+581	√	√		1458+718	√	√		2121+053	√		√
0300+470			♣	0851+202	√	√	√	1502+106	√	√	√	2128+048	√		
0310+013	√			0859+470	√			1504+377	√			2128−123	√	√	√
0316+161	√			0859−140	√	√		1504−167	√	√	√	2131−021	√	√	√
0316+413			√	0906+015	√		√	1508−055	√	√		2134+004	√	√	√
0333+321	√	√	√	0917+449	√	√		1510−089	√	√	√	2136+141	√	√	√
0336−019	√	√	√	0917+624			♣	1511−100	√			2144+092	√	√	
0355+508	√	√		0923+392	√	√	√	1514−241	√			2145+067	√		√
0403−132			♣	0945+408	√	√	√	1519−273	√			2155−152	√		√
0405−385	√			0953+254	√	√		1532+016	√	√		2200+420	√	√	√
0415+379	√	√	√	0954+658			♣	1538+149	√			2201+315	√	√	√
0420+022	√			0955+476			√	1546+027	√	√	√	2209+236	√	√	√
0420−014	√	√	√	1012+232	√	√		1548+056	√	√	√	2216−038			√
0422+004			√	1015+359	√	√		1555+001	√	√		2223−052	√	√	√
0430+052	√	√	√	1032−199	√			1606+106	√	√	√	2227−088	√		√
0440−003	√	√		1036+054			√	1611+343	√	√	√	2230+114	√	√	√
0446+112			√	1038+064			√	1622−253	√			2234+282	√	√	
0454+844	√	√		1045−188			♣	1633+382	√	√	√	2243−123	√	√	√
0458−020	√	√	√	1049+215	√	√		1637+574			√	2251+158	√	√	√
0521−365	√			1055+018	√	√	√	1638+398	√		√	2255−282	√		
0528+134	√	√	√	1055+201	√	√		1641+399	√	√	√	2318+049	√	√	
0529+075			♣	1101+384	√	√		1642+690	√	√		2331+073			♣
0529+483			♣	1124−186	√		√	1652+398	√	√		2345−167	√	√	√
0552+398	√		√	1127−145	√	√	√	1655+077	√	√	√	2351+456			√
0602+673	√			1128+385	√	√		1656+053	√	√					
0605−085	√	√	√	1144+402			♣	1656+477	√	√					
0607−157	√	√	√	1145−071	√			1726+455			√				

[a]: Included originally in the 2 cm survey (selection criteria, sect. 2).
[b]: Kinematic analysis performed. (Excludes sources with only one observing epoch or unresolved structure.)
[c]: Included in the MOJAVE sample (see sect. 4). √: confirmed, ♣: candidate.

determine the kinematics of the features in the sources by comparing images made at different epochs. The typical synthesized beam is 1 mas ×0.5 mas in position angle P.A. = 0°. Automatic imaging procedures proved to be sufficient in many of the cases (applying loops of CLEAN and phase self-calibration) using the DIFMAP software package.

The results of the survey have been published in several papers and conference proceedings. A description of the survey including contour plots for each of the sources is given in Kellermann et al. (1998a) and Zensus et al. (2002). General results from the Survey were presented in Kellermann et al. (1997; 1998b; 1999a; 2000; 2003), Kellermann (2002; and these proceedings, page 185), Cohen et al. (these proceedings, page 177), Ros et al. (2002), Ros (2003), and Lister et al. (2003b). Studies on individual sources have been presented: Kellermann et al. (1999b), Vermeulen et al. (2003a; 2003b; and these proceedings, page 43): NGC 1052, Homan (2002), Homan et al. (2003; and these proceedings, page 35): 3C 279; Lister, Kellermann, & Pauliny-Toth (2002), Lister et al. (2003a; 2003b): 4C +12.50; and Lister, Kellermann, & Pauliny-Toth (2002): 2134+004 and OQ 208.

3. Superluminal motions and kinematics

To determine the kinematics, we cross-identified the enhanced emission regions (or 'components') between different epochs. We parameterized their positions using the task JMFIT in AIPS, and with that we established the component motions for more than 100 sources in our sample (see Table 2, column labeled with 'K'). The apparent transverse velocity $\beta_{\rm app}$ of a component is given by the proper motion of the component along the line between it and the "core" (the brightest component), together with the metric distance for our assumed cosmology[1]). Our observations have also revealed non-radial motions, which are present in a ~30% (16 of 56) of the components (see Homan 2002 and Homan et al., these proceedings, page 35, for the case of 3C 279).

Given the relatively large number of sources in our sample, we are able to make a statistical analysis of the source kinematics. The distribution of $\beta_{\rm app}$ for the brightest component for the different classes of sources is shown in Figure 1 (b), for those with 3 epochs or more measured, and a well-determined speed. The $\beta_{\rm app}$ median values for each class of sources are 1.72 for BL Lac objects, 0.23 for radio galaxies, and 4.15 for QSOs.

In addition to $\beta_{\rm app}$, we also determine the (isotropic) luminosity, L, from the observed flux density S, and a lower limit to the brightness temperature $T_{\rm b}$ from the peak brightness in the image and the synthesized beam. However, a better limit to $T_{\rm b}$ can be obtained, for the compact sources, by model-fitting the visibilities in (u, v)-space (see the example of 1622−253 in Zensus et al. 2002, reaching values higher than the 10^{12} K limit from Kellermann & Pauliny-Toth 1969). From all these parameters describing the source, we can estimate the relativistic parameters: γ, the Lorentz factor, and δ, the Doppler factor. In addition we can estimate θ, the angle to the line of sight, which is important in

[1]We used a value of $H_\circ = 65\,{\rm km\,s^{-1}\,Mpc^{-1}}$ and $\Omega_{\rm m} = 0.3$, $\Omega_\Lambda = 0.7$.

understanding the geometry of these objects. These values can be compared with those measured at other frequencies, or with other techniques (e.g., variability of the flux density) to study the source physics, including tests of the general paradigm of relativistic beams.

4. Monitoring of Jets in AGN with VLBA Experiments: MOJAVE

Since we established our initial observing list in 1994, we have become aware of more sources which meet the selection criteria described in sect. 2. Also, we have found that other sources which we have been observing do not strictly meet those criteria. Therefore, we have begun a new observing program, MOJAVE, based on a complete VLBI flux density-limited sample of $\sim$ 100 sources covering the same sky region as the original 2 cm survey. The MOJAVE observations, which began in August 2002, (see `http://www.cv.nrao.edu/~mlister/MOJAVE`) will include observations of the changes in linear and circular polarization with time as well as the total power changes.

Our initial selection strategy for MOJAVE involved compiling a list of all northern AGN that have had a measured 2 cm *single dish* flux density greater than 1.5/2 Jy. We began by consulting the Kühr 1 Jy catalog (Kühr 1981; Stickel, Meisenheimer, & Kühr 1994), the VLA calibrator manual, and the JVAS survey. The latter is complete to within $2^\circ\!.5$ of the galactic plane. Since these combined catalogs are only complete at a wavelength of 6 cm, we included sources from the 22 GHz survey of Moellenbrock et al. (1996) and the high-frequency peaked spectrum samples of Teräsranta et al. (2001) and Dallacasa et al. (2000) that met our flux density criteria. We also included variable AGN from the U. Michigan and RATAN monitoring programs that have displayed sufficient 2 cm flux density at any epoch.

Since we are interested in a VLBI core-selected sample, we eliminated from our candidate list any source whose measured 2 cm VLBA flux had never exceeded 1.5/2 Jy at four epochs over a four year period beginning in 1995. The majority of sources in the previous 2 cm survey have been observed at least on four occasions in order to obtain reliable jet component speeds. For those sources with fewer than four VLBA epochs available, we checked if there were four RATAN or UMRAO flux measurements since 1995 with a similar time spacing to our VLBA epochs, and again rejected those that had never exceeded 1.5/2 Jy. Currently the MOJAVE sample list (column "M" in Table 2) consists of 99 confirmed AGN that meet these strict criteria. We maintain a list of 23 MOJAVE candidates for which we are obtaining VLBA or single-dish observing data that will span the requisite four year time range. The analysis of our VLBA data is currently underway, and new maps are being made available on our website at regular intervals as each epoch is reduced.

J.A.Z. was supported for this research through a Max-Planck-Research Award. The NRAO and the VLBA are operated by Associated Universities, Inc., under cooperative agreement with the U.S. National Science Foundation.

References

Brinkmann, W., et al. 1997, A&A, 323, 739

Ciliegi, P., & Maccacaro, T. 1997, MNRAS, 292, 338

Dallacasa, D., Stanghellini, C., Centonza, M., & Fanti, R. 2000, A&A, 363, 887

Fey, A. L., Clegg, A. W., & Fomalont, E. B. 1996, ApJS, 105, 299

Fey, A. L., & Charlot, P. 1997, ApJS, 111, 95

Fey, A. L., & Charlot, P. 2000, ApJS, 128, 17

Fomalont, E. B., et al. 2000, ApJS, 131, 95

Gómez, J. L., Marscher, A. P., & Alberdi, A. 1999, ApJ, 521, L29

Green, R. F., Khachikian, E. Ye., & Sanders, D. B. (ed.) 2002, ASP Conf. Proc. Vol. 284, AGN Surveys, (San Francisco: ASP)

Hartman, R. C., et al. 1999, ApJ, 123, 79

Henstock, D. R., et al. 1995, ApJS, 100, 1

Homan, D. C. 2002, BAAS, 34, 1245

Homan, D. C., et al. 2003, ApJ, in press

Jorstad, S. G., et al. 2001, ApJS, 134, 181

Kellermann, K. I. 2002, PASA, 19, 77

Kellermann, K. I., & Pauliny-Toth, I. I. K. 1969, ApJ, 155, L71

Kellermann, K. I., et al. 1998a, ApJ, 115, 1295

Kellermann, K. I., et al. 2003, in ASP Conf. Ser. Vol. 299, High Energy Blazar Astronomy, ed. L. O. Takalo & E. Valtaoja, (San Francisco: ASP) in press, (astro-ph/0211398)

Kellermann, K. I., Vermeulen, R. C., Cohen, M. H., & Zensus, J. A. 1997, BAAS, 29, 1257

Kellermann, K. I., Vermeulen, R. C., Zensus, J. A., & Cohen, M. H. 1998b, in ASP Conf. Ser. Vol. 144, Radio Emission from Galactic and Extragalactic Compact Sources, ed. J. A. Zensus, G. B. Taylor, & J. M. Wrobel, (San Francisco: ASP) 153

Kellermann, K. I., et al. 1999a, New Astron. Reviews, 43, 757

Kellermann, K. I., Vermeulen, R. C., Cohen, M. H., & Zensus, J. A. 1999b, BAAS, 31, 856

Kellermann, K. I., Vermeulen, R. C., Cohen, M. H., & Zensus, J. A. 2000, in Astrophysical Phenomena Revealed by Space VLBI, ed. H. Hirabayashi, P. G. Edwards, & D. W. Murphy, (Sagamihara, Japan: ISAS), 159

Kovalev, Y. Y., Kovalev, Yu. A., Nizhelsky, N. A., & Bogdantsov, A. B. 2002, PASA, 19, 83

Kraus, A., et al. 2003, A&A, 401, 161

Kühr, H., Witzel, A., Pauliny-Toth, I. I. K., & Nauber, U. 1981, A&AS, 45, 367

Lamer, G., Brunner, H., & Staubert, R. 1996, A&A, 311, 384

Lister, M. L., et al. 2003a, ApJ, 584, 135

Lister, M. L., et al. 2003b, in ASP Conf. Ser., ed. S. Collin, F. Combes, & I. Shlosman, (San Francisco: ASP), in press

Lister, M. L., Kellermann, K. I., & Pauliny-Toth, I. I. K. 2002, in Proc. of the 6th European VLBI Network Symp., ed. E. Ros, R. W. Porcas, A. P. Lobanov, & J. A. Zensus, (Bonn: MPIfR), 135

Lobanov, A. P., et al. 2000, A&A, 364, 391

Lobanov, A. P., & Zensus, J. A. 2001, Science, 294, 128

Miyaji, T., Hasinger, G., & Schmidt, M. 2000, A&A, 353, 25

Moellenbrock, G. A., et al. 1996, AJ, 111, 2174

Padovani, P., Costamante, L., Ghisellini, G., Giommi, P., & Perlman, E. 2002, ApJ, 581, 895

Pearson, T. J., & Readhead, A. C. S. 1988, ApJ, 328, 114

Polatidis, A. G., et al. 1995, ApJS, 98, 1

Ros, E., Zensus, J. A., & Lobanov, A. P. 2000, A&A, 354, 55

Ros, E. 2003, in Highlights in Spanish Astrophysics (III), ed. J. Gallego, J. Zamorano, & N. Cardiel, (Dordrecht: Kluwer), in press

Ros, E., et al. 2002, in Proc. of the 6th European VLBI Network Symp., ed. E. Ros, R. W. Porcas, A. P. Lobanov, & J. A. Zensus, (Bonn: MPIfR), 105

Stickel, M., Meisenheimer, K., & Kühr, H. 1994, A&AS, 105, 211

Taylor, G. B., et al. 1994, ApJS, 95, 345

Taylor, G. B., et al. 1996, ApJS, 107, 37

Teräsranta, H., Urpo, S., Wiren, S., & Valtonen, M. 2001, A&A, 368, 431

Thakkar, D. D., et al. 1995, ApJS, 98, 33

Valtaoja, E., Lähteenmäki, A., & Teräsranta, H. 1992, A&AS, 95, 73

Vermeulen, R. C., et al. 2003a, A&A, 401, 113

Vermeulen, R. C., et al. 2003b, PASA, 20, 65

Xu, W., et al. 1995, ApJS, 99, 297

Zensus, J. A. 1997, ARA&A, 35, 607

Zensus, J. A., et al. 2002, AJ, 124, 662

Radio Astronomy at the Fringe
ASP Conference Series, Vol. 300, 2003
J.A. Zensus, M.H. Cohen, & E. Ros, eds.

Jet Collimation: 3C 279 Caught in the Act

D. C. HOMAN
National Radio Astronomy Observatory, 520 Edgemont Road, Charlottesville, VA 22903-2475, USA

Abstract. We analyze a change in trajectory in a jet component of the quasar 3C 279. The component changes in speed and direction, and we find it to be moving with a Lorentz factor $\gamma \gtrsim 15$ at an initial angle of $\lesssim 1.2°$ to the line of sight. The new trajectory of the component parallels that of the original superluminal feature from the early 1970s. We argue that the trajectory change is a collimation event occurring at $\gtrsim 1$ kiloparsec (deprojected). Comparison to independent constraints on the Doppler factor suggest this component is near equipartition values for magnetic field and particle energy.

1. Introduction

Faster-than-light or "superluminal" motion was originally predicted by Rees (1966) and first observed by Whitney et al. (1971) and Cohen et al. (1971) in the powerful radio quasar 3C 279 ($z = 0.536$). The illusion of superluminal speed results from highly relativistic motion toward the observer, with the moving material chasing the radiation it emits, creating a compression in the time sequence of events as seen by the observer. In radio loud quasars, like 3C 279, a pair of highly collimated, relativistic plasma jets stream outward from the nucleus, and if one of these jets points within a small angle, θ, of our line of sight, then pattern motion at some intrinsic speed, $\beta = v/c < 1$, will appear to move across the sky with an apparent speed given by:

$$\beta_{\text{app}} = \frac{\beta \sin\theta}{1 - \beta\cos\theta}, \tag{1}$$

which can greatly exceed unity.

The radio jet in 3C 279 is a particularly interesting example with components ejected at multiple position angles and apparent speeds. The motion of the original superluminal feature, as summarized by Cotton et al. (1979) who analyzed the six available epochs spanning the period 1970–1972, was along a position angle of $-142° \pm 2°$ at $0.5 \pm 0.1\,\text{mas}\,\text{yr}^{-1}$. For a modern cosmology of $H_\circ = 65\,\text{km}\,\text{s}^{-1}\,\text{Mpc}^{-1}$, $\Omega_{\text{M}} = 0.3$, and $\Omega_\Lambda = 0.7$, this corresponds to $\beta_{\text{app}} = 17 \pm 3$. During the 1980s, Unwin et al. (1989) tracked another major component from 3C 279 which they called "C3". Component C3 was found to move along a position angle of $-134°$ with a much smaller speed of only $0.12 \pm 0.02\,\text{mas}\,\text{yr}^{-1}$ corresponding to $\beta_{\text{app}} = 4 \pm 1$.

At the very end of 1984, Unwin et al. (1989) mark the emergence of a new component they call C4. Component C4 is an extremely strong and compact feature which has been tracked up to the present day by a number of programs; however, the most detailed and complete results to date were presented by Wehrle et al. (2001). They followed C4 from 1991 to the end of 1997 with 22 GHz observations and from 1995 to 1997 at 43 GHz. They find the component to be steadily moving along a position angle of $-114°$ with a speed of $0.26 \pm 0.01\,\mathrm{mas\,yr^{-1}}$ which corresponds to 8.9 ± 0.3 times the speed of light in our choice of cosmology. They extrapolate an origin epoch for the component of 1984.7 ± 0.3, consistent with the first observation of C4 by Unwin et al. (1989). Wehrle et al. (2001) do analyze a dip in speed of C4 around 1994 which coincides with small changes (a few degrees) of the trajectory on the sky; however, the component resumes it previous motion and trajectory by the end of 1995.

Here we present results from the VLBA 2 cm survey (see Zensus et al., these proceedings, page 27) showing that after 1998, component C4 distinctly changes its trajectory on the sky and its apparent speed. This change in speed and direction puts C4 on an essentially parallel track to the original superluminal component with essentially the same speed as observed for that original component.

2. Observations and Results

Our observations of 3C 279 from the VLBA 2 cm survey consist of 11 epochs spanning the interval from 1995.5 to 2002.3. We have also incorporated 2 cm VLBA results from the Brandeis University parsec-scale jet monitoring project (Homan et al. 2001, 2002; Ojha et al. to be submitted) which adds six closely spaced epochs during 1996 and a seventh epoch at the end of 1997. We measured component positions both in the image plane, by fitting point sources at the peak locations of the core and C4, and in the (u, v)-plane by fitting 2-3 point components to the core region and a single elliptical Gaussian to C4. Image plane positions are used in Figures 1 and 2, and positions from both approaches are used in computing proper motion.

Figure 1 shows an image of 3C 279 from our March 1997 epoch. Superimposed on this image is the position of component C4 from each of the 18 epochs we compiled. Approximate positions for the original superluminal component are also included on the figure for comparison. Figure 2 is a more detailed plot of the component position versus epoch and also depicts the deconvolved FWHM Gaussian sizes of C4 in each epoch.

Vector proper motions were computed from the (x, y) component positions. The motions were computed over two different intervals, one prior to the change in trajectory and one after. We were unable to decide whether epoch 1998.2 should be classified with the motion prior to the change or after the change, and we found that it made little difference which choice we made. Table 1 summarizes the vector motions from a variety of approaches. From these results we take the motion prior to the "bend" to be along a position angle of $-114° \pm 1°$ at $0.25 \pm 0.02\,\mathrm{mas\,yr^{-1}}$ corresponding to 8.5 ± 0.7 times the speed of light. After the bend we take the motion to be along a position angle of $-139° \pm 1°$ at $0.40 \pm 0.02\,\mathrm{mas\,yr^{-1}}$ which corresponds to 13.7 ± 0.7 times the speed of light.

Figure 3 depicts the change in component flux density, size along the $-114°$ position angle, and brightness temperature as a function of epoch. Prior to the

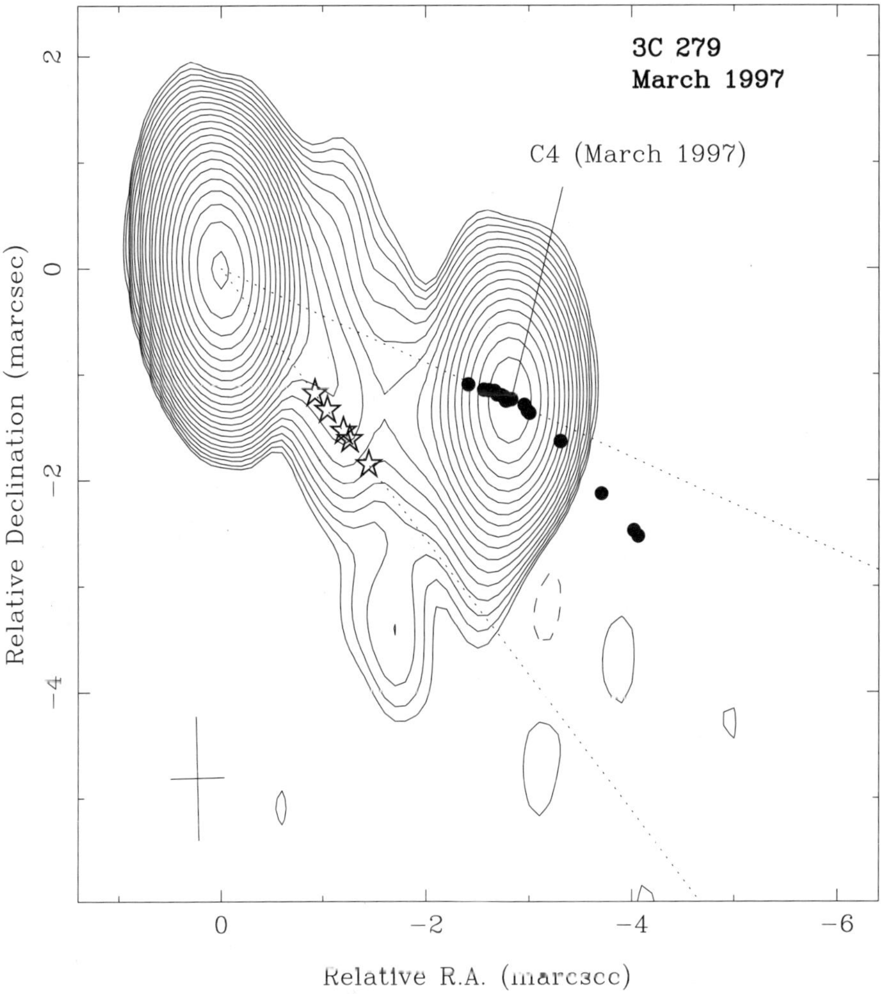

Figure 1. A λ2 cm, VLBA image of 3C 279 in March 1997. Overlayed on this image are component positions for C4 (solid dots) measured from 2 cm VLBA maps taken from the period 1995.5 to 2002.3. Open stars indicate the approximate locations measured for the original superluminal component (distances and position angle taken from Cotton et al. 1979). Dashed lines mark the −142° and −114° position angles respectively. *See the 2 cm Survey web page:* `http://www.cv.nrao.edu/2cmsurvey/` *for a movie of 3C 279.*

sharp change in motion sometime in 1998, the component increases in flux density by 50% and decreases in size along the direction of motion by a factor of two. These events seem to come before the change in motion, suggesting the component is already interacting with its environment while it is still propagating along the original direction of motion.

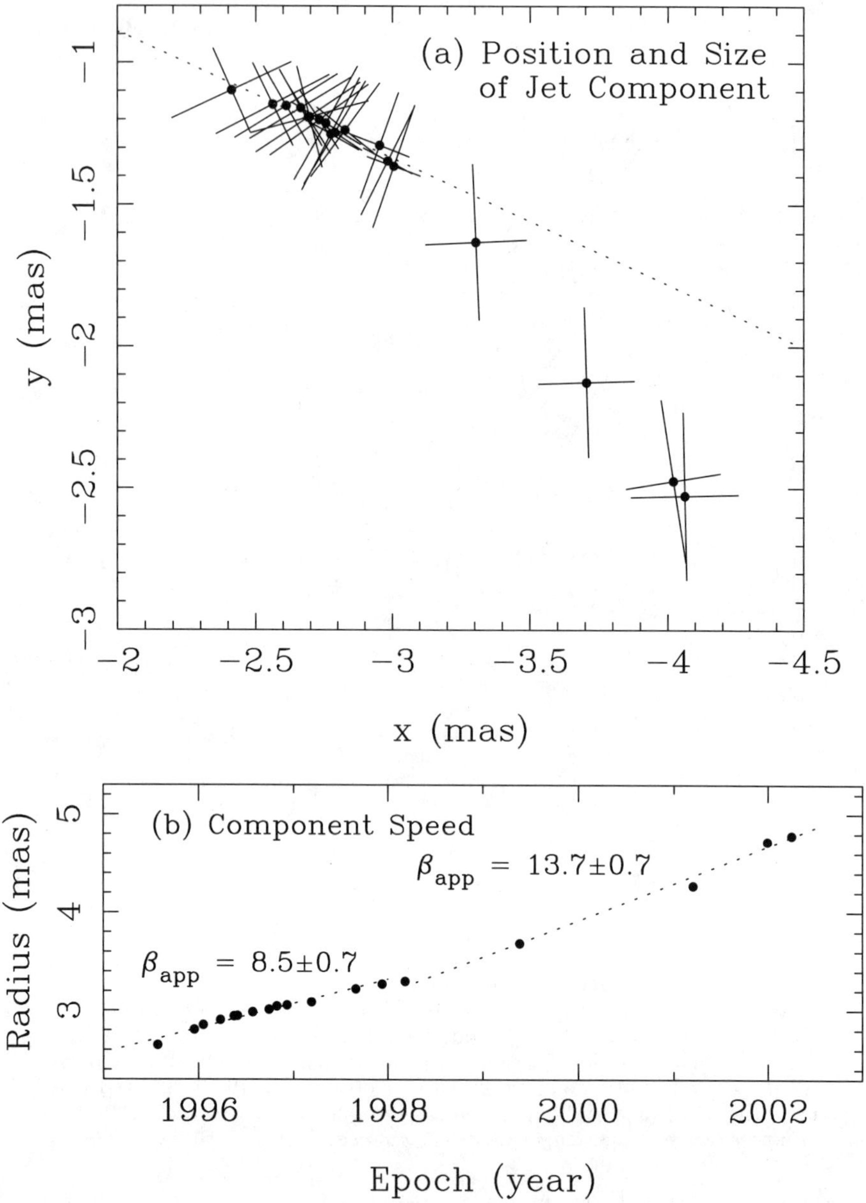

Figure 2. Panel (a): Plot of the (x, y) component position and deconvolved FWHM sizes of the fitted elliptical Gaussian. Panel (b): Plot of radial component position versus time with lines representing the projection of the fitted vector motion. Speeds derived from the vector motion fits are indicated.

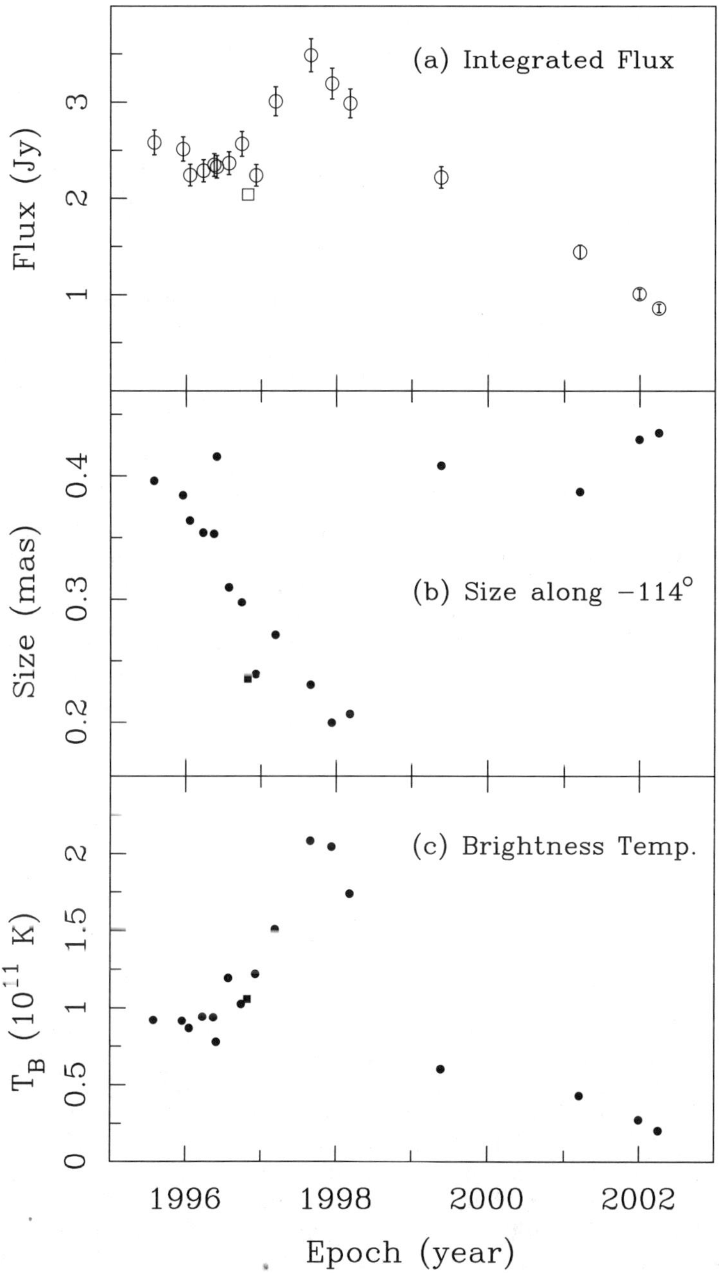

Figure 3. Plots of component flux density (panel (a)), FWHM size along the −114° position angle (panel (b)), and brightness temperature (panel (c)) versus epoch. The square symbols represent a poorly calibrated epoch with flux density ∼ 10% low.

Table 1. Vector motion of C4 computed for various ranges of epochs (N = number) for positions fit either the image or (u,v)-plane.

Epoch Range	N	Positions	μ (mas yr^{-1})	PA (deg)
1995.57–1997.94	13	Image	0.24 ± 0.01	-113.3 ± 1.8
1995.57–1998.18	14	Image	0.23 ± 0.01	-114.9 ± 1.6
1995.57–1997.94	13	(u,v)	0.27 ± 0.01	-113.5 ± 1.4
1995.57–1998.18	14	(u,v)	0.26 ± 0.01	-115.6 ± 1.4
1998.18–2002.25	5	Image	0.39 ± 0.02	-138.2 ± 2.3
1999.39–2002.25	4	Image	0.42 ± 0.02	-139.7 ± 3.4
1998.18–2002.25	5	(u,v)	0.39 ± 0.01	-138.1 ± 0.9
1999.39–2002.25	4	(u,v)	0.40 ± 0.01	-139.2 ± 0.8

3. Analysis

The apparent speed of this component has increased by more than 50% and the trajectory on the sky has changed by 25°. According to equation 1, the apparent speed can increase either due to a increase in the intrinsic speed, β, or by a change in angle to the line of sight. We consider an increase in β to be unlikely, as something must impart additional momentum to the component; however, a change in angle to the line of sight is quite consistent with the observed change in angle on the plane of the sky, so in this analysis we take β to be fixed and allow only the angle, θ, to change.

Figure 4 plots the Doppler factor, $\delta = 1/\gamma(1-\beta\cos\theta)$ where $\gamma = 1/\sqrt{1-\beta^2}$, versus the apparent motion, $\beta_{\rm app}$. Lines of constant speed (Lorentz factor, γ) and lines of constant angle are shown on the figure. $\gamma = 10$ is clearly too small and cannot accommodate speeds greater than $10c$. The lines for $\gamma = 15$ or $\gamma = 20$ are more reasonable, and the change in speed from $8.5c$ to $14c$ can be accommodated by either an increase or decrease in the angle of the motion with the line of sight.

To decide between the two possibilities of increasing or decreasing θ, we must look at the corresponding change in Doppler factor which will scale the flux density of the jet component like $S \propto \delta^{3.5}$, appropriate for a discrete component with a spectral index of 0.5. For a decreasing θ (i.e., a trajectory bent towards the line of sight), the Doppler factor more than doubles between $8.5c$ and $14c$ along the $\gamma = 15$ curve. This would predict an increase in flux density of the component of more than an order of magnitude; however, from Figure 2b, it is clear that the flux density of the component changes by no more than a factor of two, and after 1998 the flux density only decreases. For increasing θ (i.e., a trajectory bent away from the line of sight) the situation is much better, and the Doppler factor decreases by approximately 20% along the $\gamma = 15$ curve. This corresponds to a decrease in component flux density of a factor of two.

From this analysis assuming fixed component speed, we estimate $\gamma \gtrsim 15$ with an initial angle to the line of sight of $\theta \lesssim 1.2°$ increasing in the bend to become $\theta \lesssim 2.5°$. The bend in the plane of the sky appears to be 25°; however,

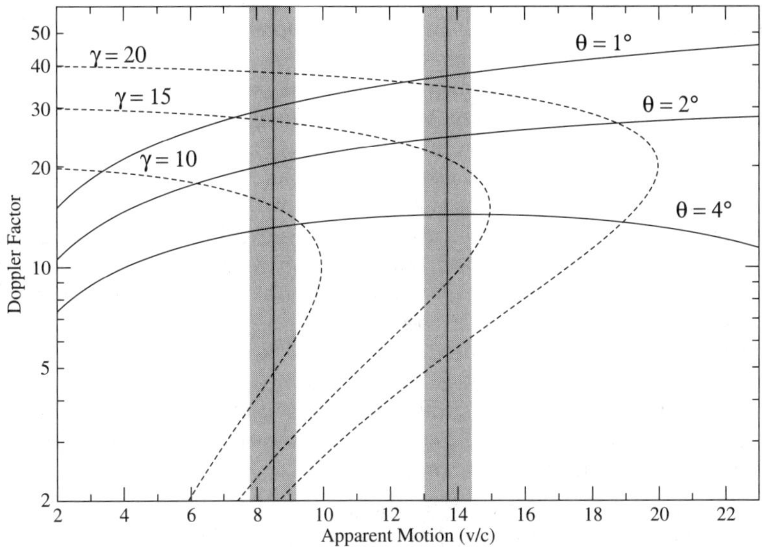

Figure 4. Plot of Doppler factor versus apparent speed. Dashed contours indicate lines of constant bulk Lorentz factor, and solid contours indicate lines of constant angle to the line of sight. Vertical lines with gray envelopes indicate the component speed before and after the bend.

deprojected, this bend is only $\sim 0.5^\circ - 1^\circ$, consistent in size with the bend along the line of sight. It is interesting to note that our value of $\gamma \gtrsim 15$ agrees well with that of Cohen et al. (these proceedings, page 177) who estimate $\gamma = 19$ for 3C 279 from an largely independent argument using variability Doppler factor.

4. Discussion

4.1. Doppler Factor and Equipartition

From the angles and Lorentz factors estimated above, we can derive approximate limits on the Doppler beaming factor of this jet feature before and after the bend. We find $\delta \gtrsim 27$ prior to the change in trajectory and $\delta \gtrsim 21$ after the change.

It is interesting to compare these values to the completely independent estimate of the Doppler factor made by Homan & Wardle (2000). They fit a spectral turnover for this individual VLBI component in epoch 1997.94, just before the change in trajectory described here. They found $\delta = 18.2^{+5.6}_{-2.8}\eta^{1/7}$ where η is the ratio of energy in the magnetic field to energy in the radiating particles: $\eta = U_B/U_p$.

The reasonable agreement between these two independent determinations of the Doppler factor of component C4, would seem to argue that the radiating particles and magnetic field must be near the equipartition value of $\eta \sim 1$, with departures from equipartition in favor of excess magnetic field energy.

4.2. What Caused the Bend?

We argue above that C4 interacts with its environment prior to the change in trajectory, resulting in decreased angular size along the $-114°$ position angle and increased flux density. It seems clear, therefore that C4 has been deflected onto its new trajectory. A random collision would be unlikely to deflect C4 in just the right fashion to have essentially the same speed *and* direction on the sky as the original superluminal component detected in 1971. It is important to stress that the trajectory of C4 is *parallel* to that of the original component; C4 is not following the actual path of that component (see Fig. 1).

We conclude that the change in trajectory is a collimation event resulting from the interaction of C4 with the boundary between the jet outflow and the interstellar medium. The exact nature of this boundary and interaction, we don't know; however, it interesting to note that this collimation event occurred at a deprojected distance from the nucleus of $\gtrsim 1$ kpc.

We also speculate that the large range of component speeds and angles observed in 3C 279 result only from small differences in the initial angle of ejection. Intrinsically, these components all may have very similar Lorentz factors, as do C4 and the original superluminal component.

This work is based on the VLBA 2 cm Survey by K. I. Kellermann, J. A. Zensus, M. H. Cohen, R. C. Vermeulen, E. Ros, M. L. Lister, M. Kadler, M. Russo and others. We thank the Brandeis group for sharing their monitoring data. This work has been supported by the National Radio Astronomy Observatory which is a facility of the National Science Foundation operated under cooperative agreement by Associated Universities, Inc.

References

Cohen, M. H., et al. 1971, ApJ, 170, 207
Cotton, W. D., et al. 1979, ApJ, 229, L115
Homan, D. C., et al. 2001, ApJ, 549, 840
Homan, D. C., et al. 2002, ApJ, 568, 99
Homan, D. C., & Wardle, J. F. C. 2000, ApJ, 535, 575
Rees, M. J., 1966, Nature, 211, 468
Unwin, S. C., et al. 1989, ApJ, 340, 117
Wehrle, A. E., et al. 2001, ApJS, 133, 297
Whitney, A. R., et al. 1971, Science, 173, 225

Radio Astronomy at the Fringe
ASP Conference Series, Vol. 300, 2003
J.A. Zensus, M.H. Cohen, & E. Ros, eds.

Motion Statistics in the CJ Survey – The Status in October 2002

R. C. VERMEULEN
Astron, P.O. Box 2, 7990 AA, Dwingeloo, The Netherlands

S. BRITZEN
Astron, P.O. Box 2, 7990 AA, Dwingeloo, The Netherlands
Landessternwarte, Königstuhl, 69117 Heidelberg, Germany

G. B. TAYLOR
National Radio Astronomy Observatory, PO Box O, Socorro, NM 87801, USA

T. J. PEARSON, A. C. S. READHEAD
California Institute of Technology, Pasadena, CA 91125, USA

P. N. WILKINSON, I. W. A. BROWNE
Jodrell Bank Observatory, University of Manchester, Macclesfield, Cheshire SK11 9DL, UK

Abstract. In the Caltech-Jodrell Survey of bright flat-spectrum radio sources, 3–5 epochs have now been observed for nearly all 293 sources; a uniquely large sample. The derivation of component motions has not yet been completed; it is complicated by, for example, variability and jet curvature. Nevertheless, some basic results are clear. The average apparent velocity in CJF quasars is near 3*c*, while for radio galaxies and BL Lacs it is around 1*c*. The distribution of velocities is broad, indicating a broad distribution of jet Lorentz factors, with many low values, and/or a decoupling of the Lorentz factors between the cores and the moving jet components, due to bending, speed changes, or due to pattern motions (shocks). A hint has emerged at this workshop that the average apparent velocities might be lower at lower radio frequencies; this will need careful verification.

1. Introduction

Superluminal motion is one of the many astronomical research topics in which Ken Kellermann has been deeply involved ever since it arose. Ken was of course already playing a central role in the exploration of variability in extragalactic radio sources (e.g., Pauliny-Toth & Kellermann 1966; Kellermann 1966). Then, with the development of VLBI (one of the earlier papers on the subject being Kellermann et al. 1968), came the gradual establishment of the fact that, as predicted by Rees (1967), evolving, relativistically moving components not only cause rapid flux density variability, but indeed also display superluminal velocities, due to geometrical effects when the motion is directed close to our line of sight. It is historically interesting to see the development of this interpretation in papers (co)authored by Ken through the late 1960s, the 1970s, and even the early 1980s, for example Cohen et al. (1971), Shaffer et al. (1972), Kellermann et al. (1973), Kellermann et al. (1974), and Cohen et al. (1977); perhaps this can be said to have culminated in Ken's 1985 review (Kellermann 1985).

In the early days, there was a focus on some very interesting individual cases of rapidly varying and very bright sources, including quasars such as 3C 273 and 3C 279, as well as galaxies such as 3C 84 and 3C 120. But, clearly, the statistics of a significantly sized sample are needed in order to obtain a more systematic overview of jet properties, such as the range of Lorentz gamma factors, and, for example, to probe the occurrence of pattern motions distinct from the bulk flow (e.g., Vermeulen & Cohen 1994). A more statistical approach is also needed to study how the properties of relativistic jets are related to the other physical parameters of the radio sources and their hosts, and also to test unification models, which explain apparent differences between object classes as being due to viewing them at different angles to the line of sight. Ken himself long ago recognised the need for large surveys, and, when the VLBA, of which he has of course been a long-term advocate, started operations in the 1990s, he initiated the 2 cm VLBA survey (Kellermann et al. 1998), as described elsewhere in these proceedings.

Meanwhile, the 6 cm Pearson-Readhead (PR) survey (Pearson & Readhead 1981), which had already been started on the Global VLBI Network in the late 1970s, had been extended into the first and second Caltech-Jodrell Bank surveys (CJ1: Polatidis et al. 1995, CJ2: Taylor et al. 1994). In order to allow unified use of these samples, we defined the Caltech-Jodrell Flat-spectrum (CJF) sample of 293 sources (Taylor et al. 1996). The present paper is a brief status update on the effort to measure the motions in the jets of all of the CJF sources.

2. CJF

CJF is a complete flux-limited sample of 293 flat-spectrum radio sources, drawn from the 6 cm and 20 cm Green Bank Surveys. The CJF sample is now 97% optically identified, and spectroscopic redshifts are available for 92% of the sources. Most of them are quasars, at redshifts ranging from $z = 0.263$ to $z = 3.886$. There are also a significant number of low redshift objects, but in addition there are some 25 galaxies and 20 BL Lac objects at $z > 0.6$; enough to allow a meaningful comparison of their motion statistics with that of quasars of the same redshift and luminosity. Between March 1990 and November 2000 there were numerous multi-antenna global VLBI and VLBA snapshot runs at 6 cm. For every source, including those in the PR subsample, this has resulted in 3–5 "modern" VLBI datasets, spanning 4–8 years.

3. Measuring jet component motions

All observing epochs for all sources are being analysed in the same standardised way (e.g., Britzen et al. 1999, 2001). We show in Figure 1 all of the currently available motions for all of the source components, using $H_\circ = 65\,\mathrm{km\,s^{-1}\,Mpc^{-1}}$

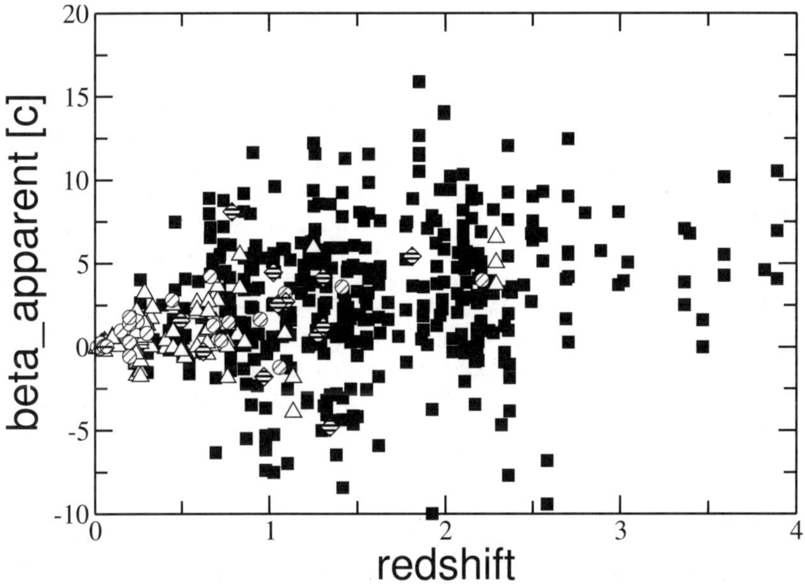

Figure 1. Overview of all of the motions as currently measured, for 597 components in 262 CJF sources, calculated with $H_\circ = 65$ km s^{-1} Mpc^{-1} and $q_\circ = 0.5$. Squares are quasars, circles are BL Lac Objects, triangles are galaxies, and diamonds are unclassified optical hosts.

and $q_\circ = 0.5$. The source structures can usually be described fairly well by using a limited number of Gaussian image components, fitted to the (u, v)-data using DIFMAP. However, it is not trivial to derive meaningful, consistent motions from the positions of these Gaussians.

The flux density and indeed the morphology of a feature clearly recognisable at one epoch may have changed considerably in the next epoch. This can lead to difficulties in "connecting" features between epochs. Expansion or brightness profile changes in resolved knots can also affect the centroid positions obtained. Furthermore, while in numerous cases the "core" is an unmistakable and presumably fixed reference point for the positions of the other features, there are also sources in which new jet features were emerging from the core (often accompanied by radio flaring), and these events can lead to a temporary shift in the "core" centroid. In some sources there is no component which, by its compactness and/or spectral index, is clearly the non-moving core; evidently, sometimes there is either little radio emission produced in the immediate vicinity of the central engine, or, more likely, the core emission is faint at 5 GHz due to either synchrotron self-absorption or free-free absorption.

We think that most and quite possibly all of the negative motions (i.e., inward towards the core) in Figure 1 are due to component misidentifications or to brightness changes which do not necessarily correspond to motion of a well-defined component. It would obviously be of great interest to find and then study a case of significant inward motion, but so far none of the candidates has survived close scrutiny.

Some of the components also clearly accelerate, decelerate, or follow a curved trajectory; these are not well characterised by a single separation velocity from the core. We plan to make a proper statistical description of the non-linear jet motions encountered, which we hope will prove to be very valuable in confronting the various models for jet acceleration, collimation, and propagation, and (shock) models for the generation of radio emitting particles inside those jets.

We have found that many of the ambiguities are substantially ameliorated for those sources where 4 or 5 rather than 3 epochs are available. However, there are no indications of any systematic velocity differences relating to the number of epochs, the time intervals between the epochs, or the total time spanned between the first and the last epoch. The measurements shown in Figure 1 also do not show selection effects, due to either resolution limitations (against low velocities at high redshifts), or to time sampling limitations (against high velocities at low redshifts).

4. Results

We are still in the process of defining an adequately homogeneous sub-sample of the highest quality motion measurements in which the impact of the effects described above is minimised. But a number of interesting conclusions can already be drawn from the current motion dataset. This includes 262 sources (the other 41 CJF sources have either no redshift or no jet components that could be monitored), with a total of 597 components.

The quasars have a mean apparent velocity of $2.9c$. This is also close to the peak in the distribution, with lower apparent velocities being somewhat less common, as are higher values; most of the quasar apparent velocities are below $8c$, but the tail of the distribution extends up to $16c$. These statistics confirm the finding of Vermeulen & Cohen (1994) that many of the superluminal velocities reported in the first studies of the 1970s and 1980s have turned out to be amongst the fastest ones in the larger population now available. Apart from the selection effect that faster motions may have been more amenable to study early on, many of the very brightest sources are also amongst the most luminous ones, and we also confirm the strong correlation between apparent velocity and 5 GHz luminosity (see below) already noted in Vermeulen (1995).

There is a pronounced difference between the apparent velocity distributions of the quasars on the one hand, and the galaxies and BL Lac objects on the other hand. Whereas the mean apparent velocity in the quasar jets is $2.9c$, the galaxies have a mean apparent velocity of $0.9c$, and in fact here the distribution peaks at the lowest values, $\leq 1c$, with most of the apparent velocities below $2c$, although there is a tail in the distribution which goes all the way up to $10c$. The BL Lacs may be slightly faster than the galaxies on average, with a mean apparent velocity of $1.4c$, but again the distribution peaks at the lowest values, $\leq 1c$. Most of the BL Lacs have apparent velocities below $4c$, and the distribution then cuts off steeply, with no values larger than $6c$. In general, such results are in line with previous work (e.g., Vermeulen & Cohen 1994), and with the predictions of beaming and unification models. But we have a much more extensive sample than hitherto available, with different source types at comparable redshifts and luminosities. Before coming to final conclusions, we are carefully re-examining the optical identifications. For example, using very deep optical images some objects may have been labelled as galaxies on the basis of faint extended emission which would not have been visible in less sensitive images. We also have the impression that it is important for the final statistics to differentiate not only between broad-lined (quasar-like) and narrow-lined (galaxy-like) objects, but also between between strong-lined and weak-lined (FR-I BL Lac ?) objects.

Doppler favouritism in simple beaming models combines with the solid angle available to predict a sample in which the apparent velocity distribution is sharply peaked at the maximum possible value ($\beta\gamma$); such models would predict that the measurements shown in Figure 1 would crowd against an upper velocity envelope (e.g., Vermeulen & Cohen (1994). This is obviously not the case, and the actual apparent velocity distribution may instead be dominated by a distribution of Lorentz γ factors which spans a very broad range over the sample of radio sources (see also Vermeulen & Cohen 1994). There is evidence for this in the very strong correlation with luminosity of the mean apparent velocity, which was already found in a subset of the CJF dataset (Vermeulen 1995), in the sense that no high velocities exist at low luminosity. The correlation can be attributed to a form of Malmquist bias in a sample selected on beamed radio emission (e.g., Lister & Marscher 1997).

Perhaps it will prove to be possible to turn this relationship around, and to deduce, from the observed distributions of apparent velocity, luminosity, and redshift, the luminosity distribution of the parent population from which the CJF sample arose through beaming. But, we first have to study carefully whether the Lorentz factors responsible for the selection of the sources, through relativistic boosting of the radio cores, are indeed tightly related to the Lorentz factors corresponding to the motions of the components. The latter could reflect patterns or shocks moving through the fluid of the jets. And even if the components move at the bulk flow velocity, bending of the jets will give different apparent speeds.

By compiling the statistics of the apparent velocities as a function of distance from the core, we also plan to investigate whether there is evidence for jet acceleration or deceleration between the core and the moving components. Perhaps related to this, very interesting hints have emerged at this Green Bank workshop that the average apparent velocity observed might decrease when going to lower observing frequency. The apparent velocities for CJF at 5 GHz, reported above, seem to be about a factor of two slower than those reported in these proceedings for the 15 GHz VLBA Survey (Zensus et al., these proceedings, page 27), and those in turn might be another factor of two lower than the 43 GHz velocities reported at this workshop by Marscher (these proceedings, page 133). Most of these measurements are as yet preliminary, and the differences between the selections of the various samples will also have to be carefully understood, before it will be warranted to incorporate correlations between observing frequency and apparent velocity firmly into the models. But several options do spring to mind immediately. Jet bending, or deceleration along the jets might be involved. Alternatively, each jet might incorporate a range of flow velocities, possibly in a "spine-sheath" or "onion-skin" geometry (perhaps akin to some models for FR-I sources, for instance in Laing et al. 1999). A decrease in average apparent velocity with decreasing frequency might then result from an increase of the average size and/or distance from the core of the components observed when going to lower frequency. In turn this could plausibly be connected to changing turnover frequencies for features with synchrotron self-absorbed spectra.

And so we have again found at this workshop that Ken Kellermann plays a central role in superluminal motion studies, in which a novel feature now is that we will be able to compare two large surveys: the 6 cm CJF and the 2 cm VLBA Survey, which Ken initiated, and in which he provides important momentum and inspiration.

References

Britzen, S., et al. 1999, in ASP Conf. Ser. Vol. 159, BL Lac Phenomenon, ed. L. O. Takalo, A. Sillanpää, (San Francisco: ASP), 431

Britzen, S., et al. 2001, in IAU Symp. 205, Galaxies and Their Constituents at the Highest Angular Resolutions, ed. R. T. Schilizzi, S. Vogel, F. Paresce, & M. Elvis (San Francisco: ASP), 106

Cohen, M. H., et al. 1971, ApJ, 170, 207

Cohen, M. H., et al. 1977, Nature, 268, 405

Kellermann, K. I. 1966, ApJ, 146, 621

Kellermann, K. I. et al. 1968, ApJ, 153, L209

Kellermann, K. I., et al. 1973, ApJ, 183, L51

Kellermann, K. I., et al. 1974, ApJ, 189, L19

Kellermann, K. I. 1985, Com. Ap., 11, 69

Kellermann, K. I., Vermeulen, R. C., Zensus, J. A., & Cohen, M. H. 1998, AJ, 115, 1295

Laing, R. A., Parma, P., de Ruiter, H. R., & Fanti, R. 1999, MNRAS, 306, 513–530.

Lister, M. L., & Marscher, A. P. 1997, ApJ, 476, 572

Pauliny-Toth, I. I. K., & Kellermann, K. I. 1966, ApJ, 146, 634

Polatidis, A. G., et al. 1995, ApJS, 98, 1

Rees, M. J. 1967, MNRAS, 135, 345

Shaffer, D. B., Cohen, M. H., Jauncey, D. L., & Kellermann, K. I. 1972, ApJ, 173, L147

Taylor, G. B., et al. 1994, ApJS, 95, 345

Vermeulen, R. C. 1995, Proc. Natl. Acad. Sci. USA, 92, 11385 (correction in Proc. Natl. Acad. Sci. USA, 93, 6846)

Vermeulen, R. C., & Cohen, M. H. 1994, ApJ, 430, 467

Radio Astronomy at the Fringe
ASP Conference Series, Vol. 300, 2003
J.A. Zensus, M.H. Cohen, & E. Ros, eds.

Studying Circular Polarization with the VLBA

D. C. HOMAN
National Radio Astronomy Observatory, 520 Edgemont Road, Charlottesville, VA 22903, USA

J. F. C. WARDLE
Physics Department, MS057, Brandeis University, Waltham, MA 02454, USA

Abstract. Very Long Baseline Array (VLBA) studies of circular polarization provide a new probe of the physics of extragalactic radio jets. In addition to a brief review of past results from VLBA circular polarization studies, we present preliminary new results on the famous radio galaxy 3C 84. These images reveal very strong circular polarization in a complex distribution with both positive and negative signs. We interpret a strong spot of positive circular polarization as originating from an unusual region of highly ordered magnetic field in the jet. This region cannot be too much smaller than our beam, ~ 0.5 light years.

1. Introduction

Numerical simulations of synchrotron emitting radio jets by Jones (1988) showed that *local* levels of circular polarization (CP) could be quite strong, exceeding 0.5% in some cases. This was in contrast to the low resolution measurements available at the time, summarized by Weiler & de Pater (1983) who considered 0.1% integrated CP to be strongly polarized. Jones (1988) outlined both the complexity and promise of circular polarization as a probe of jet physics and emphasized the importance of high resolution (VLBA) studies of circular polarization.

By seeing directly where the circular polarization arises in a jet, we can study its local fractional level and spectrum, and compare these results to local measures of linear polarization and Faraday rotation. In principle we can use these measures to probe both the three dimensional magnetic field structure as well as the lower end of the particle energy distribution in the jet. These interpretations hinge upon our ability to determine the dominant emission mechanism responsible for producing the circular polarization. Faraday conversion of linear to circular polarization is generally expected to be the dominant process (Jones 1988; Wardle & Homan 2001; Ruszkowski & Begelman 2002; Beckert & Falcke 2002); however, it is only in two sources (3C 279 and PKS 0607−157) where there is supporting observational evidence for Faraday conversion (Wardle et al. 1998; Homan & Wardle 2002). A viable alternative in most sources is intrinsic circular polarization as part of the direct synchrotron emission.

Faraday conversion, particularly if driven by Faraday rotation, requires a significant population of low energy, relativistic particles in the radio jet. In the case of 3C 279, we were able to measure a number of the observational constraints listed above and concluded that the CP resulted from the Faraday conversion process, implying a low cutoff in the particle energy distribution ($\gamma_{min} < 20$) and therefore suggesting a predominantly e^+e^- plasma for the jet on kinetic luminosity grounds (Wardle et al. 1998). Recent radiative transfer simulations by Ruszkowski & Begelman (2002) and Beckert & Falcke (2002), suggest that our derived cutoff in the particle energy distribution may not be a strict upper limit, relaxing the need for an e^+e^- jet. The alternative to Faraday conversion, intrinsic circular polarization, requires a charge imbalance in the radiating particles, and significant amounts of intrinsic CP almost certainly require a jet plasma dominated by "normal" matter, e^-p^+.

In this article we discuss the study of circular polarization in jets using the VLBA. In sect. 2 we briefly review some key results and issues from our earlier works. In sect. 3 we present new, multi-frequency VLBA images of the radio galaxy 3C 84. These images reveal very strong circular polarization in a complex distribution. We interpret a strong spot of positive circular polarization as originating from a region of highly ordered magnetic field which is not too much smaller than our beam, ~ 0.5 light years.

2. What have we learned?

Our circular polarization observations of 3C 279 (discussed above) were part of a multi-epoch VLBA monitoring campaign to study rapid changes (epochs spaced at 2 month intervals) in total intensity and polarization of the most highly active blazars. Five (of six) epochs from this campaign were suitable for circular polarization detection at 15 GHz, and these were re-analyzed to search for circular polarization, yielding our results on 3C 279 as well as multi-epoch detections on 3C 84, PKS 0528+134, and 3C 273 (Homan & Wardle 1999).

We detected strong *local* fractional CP in these jets, ranging from $0.3 - 1\%$ and appearing near (but not always coincident with) the VLBI core. In the case of 3C 273, the appearance of CP in the jet core coincided with the appearance of small amounts of linear polarization and a change in opacity in the core from very opaque to more optically thin as a new jet component began to emerge. In our final epoch, the circular polarization spanned the core region of 3C 273 but appeared most strongly associated with the newly emerging component. Interestingly, the sign of the circular polarization remained consistent in each source across all epochs; three of these sources were undergoing strong core outbursts which varied their total intensity and linear polarization. This suggested a magnetic field structure, responsible for setting the sign of the circular polarization, which is persistent on a timescales at least as long as an individual outburst event.

In Homan, Attridge, & Wardle (2001) we presented results of 5 GHz VLBA circular polarization observations of 40 sources. We detected circular polarization in 11 objects at levels of $0.14 - 0.46\%$ local fractional circular polarization, all coincident with the VLBI cores. Our results were very similar to the integrated measurements of Rayner, Norris, & Sault (2000) who studied a different sample of objects, also at 5 GHz, with the Australian Compact Telescope Array (ATCA). Other than a strong tendency for levels of linear polarization to exceed circular, both studies found apparently no relation between observed linear and circular polarization in either the integrated (Rayner, Norris, & Sault 2000) or parsec-scale core (Homan, Attridge, & Wardle 2001) measurements. Both intrinsic circular polarization and Faraday conversion require high degrees of field order to produce net circular polarization, and therefore both should have a correlation with linear polarization. However, we note that it is possible to destroy any correlation between linear and circular polarization through either some kind of Faraday depolarization or by simply having the circular polarization produced in a region much smaller than even our VLBI beam.

The apparent sign persistence of circular polarization in a given source was first recognized by Komesaroff et al. (1984) who found that changes in sign were rare over their ~ 5 year monitoring window. Our shorter timescale results (discussed above) were consistent with this, and we investigated the issue further in Homan, Attridge, & Wardle (2001) by comparing all recently reported CP measurements to those made ~ 20 years ago. We found that 5 out of 6 sources that were detected both today and in the past had the same sign of circular polarization. While the statistics were only suggestive, we postulated the existence of a long timescale, persistent magnetic field structure, perhaps tied to the black hole/accretion disk system, which sets the sign of circular polarization in a particular source.

Some doubt has been cast upon long timescale sign persistence as a general property of circularly polarized sources (Aller, Aller, & Plotkin 2002; McCormick, Spencer, & Fender 2002; Homan & Wardle 2002), and given that stochastic effects may play a significant role in CP production in at least some sources, long term monitoring campaigns may be the best way to establish the extent to which individual sources have a preferred sign of CP and on what timescale it persists. Bower et al. (2002) have already begun this process by using archived VLA observations to show that Sgr A* has maintained the same sign of circular polarization over 18 years.

3. New CP Images of 3C 84

We observed the radio galaxy 3C 84 in December of 1997 (epoch 1997.94) with the VLBA at four frequencies: 5, 8, 15, and 22 GHz. Our nearly full track observations were divided amongst the observing bands, giving excellent (u, v)-plane coverage and approximately 2 hours on-source per frequency. Here we discuss the CP results from our 15 and 22 GHz bands. Total intensity (Stokes I) images are presented in Figure 1.

3.1. Brief Calibration Discussion

To calibrate the antenna gains to detect circular polarization, we used a modification of the *zero-V* self-calibration technique described by Homan & Wardle (1999). *Zero-V* self-calibration assumes there is no real circular polarization in the data and calibrates the $RR = I + V$ and $LL = I - V$ correlations versus a pure Stokes I model. The derived antenna gains will act to suppress any real Stokes V in the data; however, except in the case of a point source, the gain "corrections" cannot remove the CP signal completely because the antenna gains are multiplicative while V is additive in the baseline correlations.

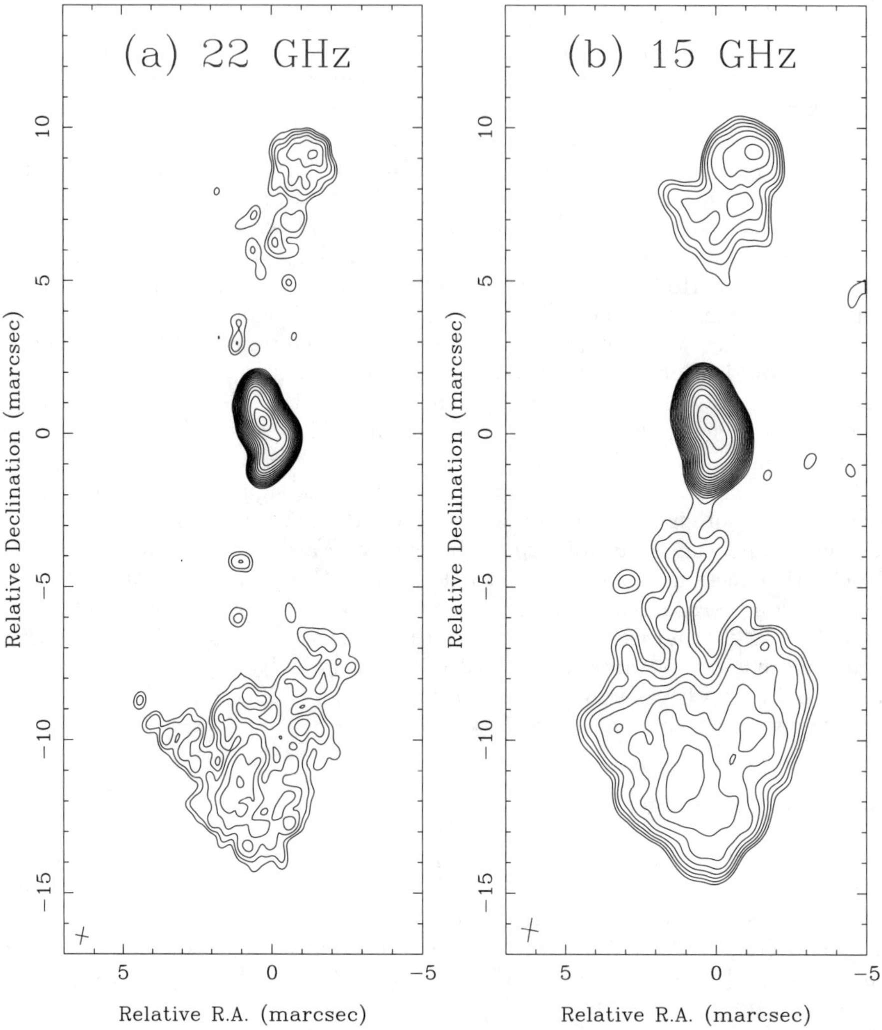

Figure 1. VLBA Stokes I images of 3C 84 at 22 GHz (a) and 15 GHz (b). The base contour is 10 mJy/beam, and the contours increase in steps of $\sqrt{2}$. The full-width-half-maximum (FWHM) dimensions of the naturally weighted beams are depicted by a cross figure in the lower left-hand corner of each panel.

For an extended source, like 3C 84, the effect of *zero-V* self-calibration on real Stokes V will be to distort it by reducing its level and inducing V of the opposite sign on other strong structure in the source. In principle, one can reconstruct the original circular polarization distribution through careful modeling procedures, and that is what we have done. For the 15 and 22 GHz results presented here, this reconstruction process is simple and robust.

We tried a number of algorithms for reconstructing the true circularly polarized signal from the distorted version produced from *zero-V* self-calibration, and all gave very similar results. Figure 2 shows the core region of 3C 84, the distorted CP-signal from *zero-V* self-calibration, and the reconstruction of the true CP distribution. The reconstruction was done by assuming the peak CP in the distorted map was real, subtracting this CP peak from the original data (prior to self-calibration assuming *zero-V*), calibrating the new subtracted data-set assuming *zero-V*, and examining the resulting residual map. After one iteration, the residual V map still showed signs of real CP which had been distorted by assuming *zero-V*[1]. So this process was repeated through a number of iterations until the final residual map was flat, we then restored all the subtracted components to the calibrated data. We ran a large number of tests on this algorithm using simulated data, and we found the reconstruction to be robust, particularly for the bright positive components in Figure 2. The negative components are real; however, our tests showed that the reconstruction algorithm may reduce their amplitude by $\sim 10 - 20\%$. See Homan & Wardle (in preparation) for a more detailed discussion of this technique.

3.2. Results

Our circular polarization images of 3C 84 are displayed in Figure 2, and we find very strong CP at both 15 and 22 GHz. Both frequencies show both positive and negative CP at the same locations in the jet. All the circularly polarized emission is located south of the map peak, with the change in sign from positive to negative occurring where the jet bends to the east. As discussed above, the positive CP signal is the strongest and most robust in each image, and we focus our analysis on that emission; however, we note that the negative CP is also real and quite strong, although we may have underestimated its amplitude. This negative circular polarization is $\simeq -0.5$ to -1.0% of the local Stokes I at each frequency.

The bright knot of positive circular polarization is unresolved at both frequencies and is located in a part of the jet that has a smoothly declining brightness profile with no distinct features in total intensity. This part of the jet is optically thin with a spectral index of $\alpha = -1$ to -1.5 ($S \propto \nu^{+\alpha}$) for the total intensity between 15 and 22 GHz. The fractional local circular polarization for this spot is $m_c = +3.2 \pm 0.1\%$ at 15 GHz and $m_c = +2.3 \pm 0.2\%$ at 22 GHz. The values are taken at the location of the Stokes V peak, and we can find the spectral index of the fractional circular polarization: $m_c \propto \nu^{-0.9\pm0.3}$. Note that

[1]We expect this because the original subtraction could not have been large enough. The peak that was subtracted had already been suppressed by some factor in the initial calibration assuming *zero-V*.

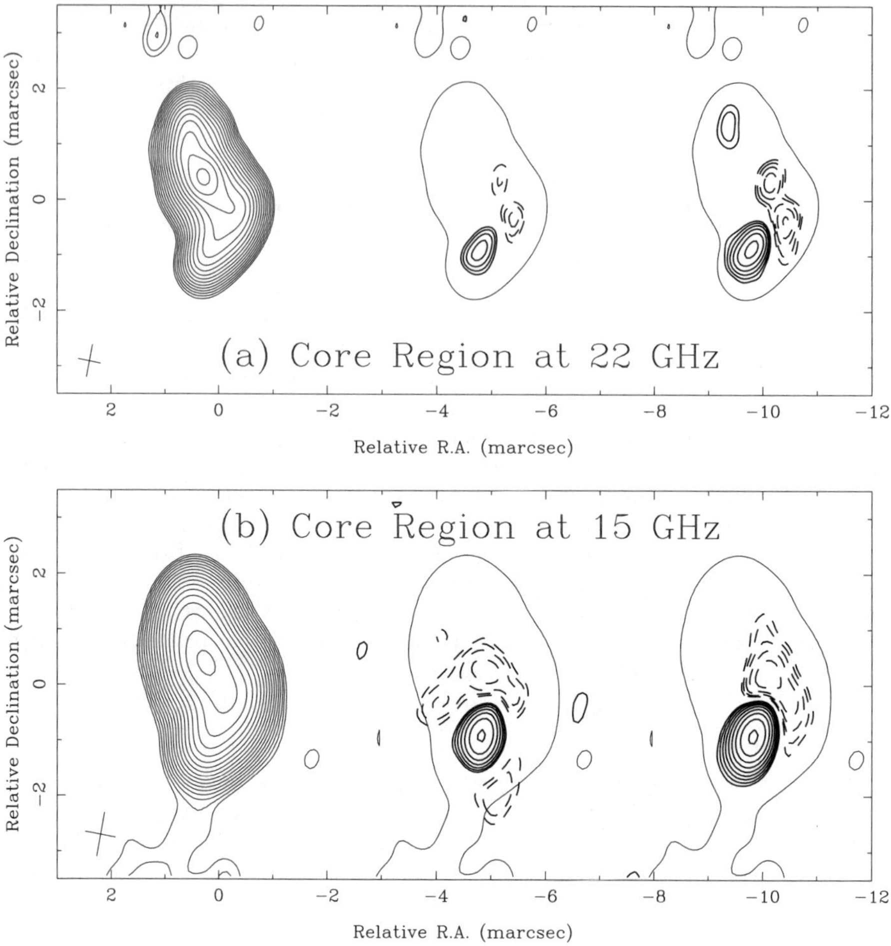

Figure 2. Core region of 3C 84 at 22 GHz (a) and 15 GHz (b). *Left:* Stokes I with contours beginning at 10 mJy/beam and increasing in steps of $\sqrt{2}$. *Center:* Stokes V, calibrated assuming *zero-*V, creating some distortion to the true V distribution. *Right:* Stokes V, reconstruction of the true distribution using the algorithm described in §3.1. For all Stokes-V images the contours begin at 2 mJy/beam and increase in steps of $\sqrt{2}$. A single Stokes I contour is placed around the V images to show registration. The FWHM dimensions of the naturally weighted beams are depicted by a cross figure in the lower left-hand corner of each panel.

the data were weighted and tapered at the two frequencies to give an effectively matched resolution for these measurements.

3.3. Discussion

The spot of positive circular polarization is extremely strong at +3% of the *local* Stokes I at 15 GHz. Considering the continuous nature of the Stokes I

distribution and the unresolved Stokes V, there has likely been some beam dilution of the circular polarization which may be distinctly higher than the observed 3%.

There are several points to make about this strong, positive spot of circular polarization. (1) *The positive sign differs from that in other parts of the jet.* It differs from the negative circular polarization seen upstream at 15 and 22 GHz, and our 5 GHz results (in prep.) suggest a predominately negative sign of CP further out in the jet. 3C 84 also appears to have negative CP at 18 cm (W. D. Cotton, private communication), and negative CP was repeatedly detected in the integrated measurements cataloged by Weiler and de Pater (1983). (2) *The spot appears unresolved.* This, combined with the unusual sign suggests that whatever characteristics are responsible for generating large quantities of positive circular polarization are not common to the jet as a whole. (3) *The spot is long lived and changes slowly.* We first observed this same spot of circular polarization at 15 GHz in January of 1996 (nearly 2 years before the observations presented here) where it had essentially the same flux (Homan & Wardle 1999).

Circular polarization intrinsic to the synchrotron radiation itself has a spectral index not too different from what we observe, $m_c \propto \nu^{-0.5}$, but would require a very strong magnetic field to explain the fractional circular polarization. For intrinsic CP, $m_c \sim 1/\gamma$, so electrons radiating with $\gamma < 100$ are required. But at 15 GHz, this corresponds to magnetic fields of > 0.5 G. For these parameters the half-life for electron energy loss is less than six months, demanding either in-situ particle acceleration or a continuous source of fresh particles. Doppler boosting doesn't help here, as the jet is only mildly relativistic (see Walker, these proceedings, page 113).

In conventional radiative transfer models, very high levels of circular polarization are most easily generated via the Faraday conversion process which converts linearly polarization to circular polarization. This conversion process can be either (1) pure conversion from different field orientations at the back and front of the jet, or (2) conversion driven by internal Faraday rotation. Pure conversion, case (1), is discarded because it would predict an optically thin spectrum of $m_c \propto \nu^{-3}$ (e.g., Jones & O'Dell 1977), which is not observed. Rotation driven conversion, case (2), has a more complicated spectrum and could be consistent with our observations. If this mechanism is dominant, it could significantly reduce the requirement of high field strengths in this region, although a highly ordered field is still required.

We detect no significant linear polarization in this part of the jet of 3C 84, making it difficult to distinguish between the two conventional mechanisms: intrinsic CP and rotation driven conversion. However, we note that these mechanisms should operate anywhere in the jet where the conditions are favorable, and the unusual strength, sign, and confined nature of the circular polarization imply local conditions that are not typical of the jet as a whole. We believe the most likely scenario is that the positive spot of circular polarization is generated by an unusual region of highly ordered field. The scale size of this region cannot be too much smaller than our beam, ~ 0.5 ly, or the circular polarization would be canceled and diluted by other parts of the source.

3C 84, at a redshift of $z = 0.017$, shows the strongest fractional circular polarization yet detected by the VLBA. Much more powerful quasars rarely show local circular polarization stronger than 0.3%; however, these sources are much further away and at such distances 3C 84 would have 0.1–0.2% local circular polarization due to blending with the core. In this light, it is interesting that the intraday variable source, PKS 1519−273, has been shown to have circular polarization at the 2–4% level in a scintillating component 15–35 μas in size (Macquart et al. 2000). This corresponds to a region of 0.2–1.0 ly in size ($z \geq 0.2$), comparable in scale to the region we suggest is responsible for the strong circular polarization in 3C 84.

This work has been supported by the National Radio Astronomy Observatory and by NSF grant AST 99-00723. The Very Long Baseline Array is operated by the National Radio Astronomy Observatory which is a facility of the National Science Foundation operated under cooperative agreement by Associated Universities, Inc.

References

Aller, H. D., Aller, M. F., & Plotkin, R. 2003, in: Circular Polarization of Relativistic Jet Sources, ed. R. P. Fender & J.-P. Macquart, Ap&SS, in press

Beckert, T., & Falcke, H. 2002, A&A, 388, 1106

Bower, G. C., et al. 2002, ApJ, 571, 843

Homan, D. C., Attridge, J. M., & Wardle, J. F. C. 2001, ApJ, 556, 113

Homan, D. C., & Wardle, J. F. C. 1999, AJ, 118, 1942

Homan, D. C., & Wardle, J. F. C. 2003, in: Circular Polarization of Relativistic Jet Sources, ed. R. P. Fender & J.-P. Macquart, Ap&SS, in press (astro-ph/0211183)

Jones, T. W. 1988, ApJ, 332, 678

Komesaroff, M. M., et al. 1984, MNRAS, 208, 409

Macquart, J.-P., et al. 2000, ApJ, 538, 623

McCormick, D. G., Spencer, R. E., & Fender, R. P. 2003, in: Circular Polarization of Relativistic Jet Sources, ed. R. P. Fender & J.-P. Macquart, Ap&SS, in press

Rayner, D. P., Norris, R. P., & Sault, R. J. 2000, MNRAS, 319, 484

Ruszkowski, M. & Begelman, M. C. 2002, ApJ, 573, 485

Wardle, J. F. C., & Homan, D. C. 2002 in ASP Conf. Ser. Vol. 250, Particles and Fields in Radio Galaxies, ed. R. Laing & K. Blundell, (San Francisco: ASP), 152

Wardle, J. F. C., et al. 1998 Nature, 395, 457

Weiler, K. W.,& de Pater, I. 1983, ApJS, 52, 293

Radio Astronomy at the Fringe
ASP Conference Series, Vol. 300, 2003
J.A. Zensus, M.H. Cohen, & E. Ros, eds.

What We've Learned from VSOP Polarization Observations

D. C. Gabuzda
Physics Department, University College Cork, Cork, Ireland

Abstract. Many would have said that it was a bit of a gamble hoping to obtain much useful information from VSOP polarization observations, due to the small size of the orbiting antenna and the ability to record only one circular polarization for baselines to *HALCA*. In fact, these observations have been among the most interesting and informative of all those obtained by the VSOP project. They have confirmed that the 6 cm core polarization of compact BL Lac objects is dominated by the contribution of emerging jet components, and have provided compelling evidence that the dominant magnetic fields of these sources may be toroidal. In addition, they have enabled us for the first time to directly observe the 90-degree flip in the polarization position angle in the transition from the optically thick to optically thin regimes. Space polarization observations clearly have a bright future in VLBI.

1. Introduction

There is a distinct tendency for the VLBI jets of BL Lac objects to display predominantly transverse magnetic (**B**) field structures. This was initially widely interpreted as evidence for transverse shocks in which the **B** field component perpendicular to the direction of propagation of the shock was enhanced by compression (Laing 1980; Gabuzda et al. 1992). It was hoped that VSOP polarization observations might be able to shed light on conditions in the region of formation of these compact shock components.

It has also been known for a long time that the degrees of polarization m of the centimeter-wavelength VLBI cores of BL Lac objects are modest, but appreciable, typically ranging from $m \simeq 2-6\%$ and sometimes reaching as high as $m \simeq 9\%$ (Gabuzda, Pushkarev, & Cawthorne 2000 and references therein). The polarization angles χ observed for the 6 cm VLBI cores of BL Lac objects show a distinct tendency to lie either parallel or perpendicular to the direction of the inner VLBI jet (Gabuzda et al. 2000). The few sources whose core polarization angles clearly do *not* display one of these two orientations (about 15% of the total) can plausibly be understood as those in which the observed χ values are affected by non-uniform Faraday rotation in the core region, or in which the jets bend substantially on small scales (i.e., the polarization vectors may be well aligned with the jet on smaller scales). A correlation between the orientation of the core χ and the core degree of polarization was also found, with cores in which χ was aligned with the inner jet tending to be more highly polarized (Gabuzda et al. 1994). These results led to the suggestion that the polarization of the ground-based VLBI cores of BL Lac objects included a substantial contribution from highly polarized regions in the inner VLBI jet, and that the core

χ became aligned with the jet direction when the core was giving birth to new shock components (Gabuzda et al. 1994). This was clearly an idea that could be directly tested with VSOP polarization observations.

2. The Nature of the Ground-based "Cores"

2.1. Isolation of the 6 cm Cores with VSOP

VSOP polarization images are now available for the seven BL Lac objects 0735+178 (Gabuzda, Gómez, & Agudo 2001), OJ 287 (Gabuzda & Gómez 2001), 0954+658 (Gabuzda 2002; 2003a), 1334−127, 1418+546 (Gabuzda 2002; 2003a), 1749+096 (Gabuzda 2003a) and 1803+784 (Gabuzda 1999).

In virtually all cases, the VSOP images were able to separate the ground-based "core" into a relatively weakly polarized ($m \simeq 1.5 - 4\%$) feature at one end of the source structure and compact, much more highly polarized ($m \simeq 10 - 40\%$) emerging jet components. The only exception was 0735+178, for which both the ground-based and VSOP core polarizations agreed due to a lack of strong emerging jet features at our observing epoch. Four examples are shown in Figure 1. It is clear that polarization from the inner jet was dominating the total polarization observed for the ground-based "core." Thus, the VSOP polarization images of these seven BL Lac objects directly confirm the hypothesis that the ground-based core polarizations of these AGN include a substantial contribution from the polarization of optically thin regions in the inner jet.

The polarization angles χ for these better isolated 6 cm core components are all either roughly aligned with or perpendicular to the inner jet direction; there is no obvious correlation between the degree of polarization and the orientation of χ. In fact, there is no convincing evidence at present that there are a significant number of BL Lac objects whose intrinsic core polarization angles are neither aligned with nor perpendicular to the orientation of the inner jet [see the discussion of this issue by Gabuzda (2003b)]. Another striking result to come from VSOP polarization observations of OJ 287 may point toward optical depth effects as the origin of this bimodal behavior for both the ground-based and VSOP 6 cm cores, as is discussed below.

2.2. Optical Depths of the Core Components

Fortuitously, it turned out that 6 cm VSOP and 7 mm+1.3 cm VLBA observations of OJ 287 were obtained independently by two groups at epochs separated by only 16 days (April 4 and March 19, 1999, respectively). These groups learned of this happy coincidence and agreed to collaborate on a joint analysis of the 6 cm VSOP and 1.3 cm VLBA images, which had roughly the same resolution (Gabuzda & Gómez 2001).

Both the total intensity and polarization structures observed at the two wavelengths were strikingly similar, with one exception: the polarization angles for the core components were offset by about 73° – suspiciously close to the 90° offset expected for the transition from the optically thin regime (where χ is perpendicular to **B**) to the optically thick regime (where χ is parallel to **B**). Comparison of the total flux densities for the 6 cm and 1.3 cm cores yielded a

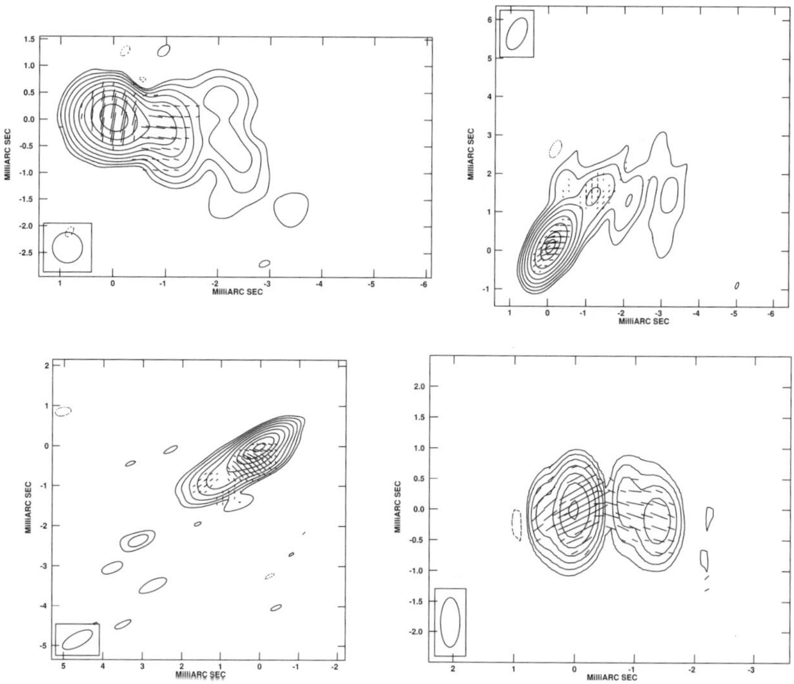

Figure 1. 6 cm VSOP images of OJ 287 (top left, peak 1.185 Jy/beam; *taken from Gabuzda & Gómez 2001, MNRAS, 320, L49*), 0954+658 (top right, peak 0.26 Jy/beam), 1418+546 (bottom left, peak 0.48 Jy/beam), and 1803+784 (bottom right, peak 1.07 Jy/beam; *taken from Gabuzda 1999, New Astron. Rev., 43, 695*) with P vectors superimposed. The contours increase in steps of two, and the bottom contours are $\pm0.25\%$ (OJ 287), $\pm0.50\%$ (0954+658), $\pm0.50\%$ (1418+546), and $\pm1.40\%$ (1803+784). In each case, what would be interpreted as the "core" in the ground-based image is comprised of emission from the modestly polarized VSOP "core" and highly polarized regions in the inner jet.

spectral index of $\alpha_{6-1.3} = +0.92$, demonstrating that the core was optically *thick* at 6 cm. The 7 mm VLBA image at the same epoch as the 1.3 cm image demonstrated, in turn, that the core was optically *thin* at 1.3 cm. This led Gabuzda & Gómez (2001) to interpret this roughly 90° offset between the core χ's at 6 cm and 1.3 cm as the first direct observation of the predicted 90° flip in the polarization electric vector in going from the optically thick to the optically thin regime. In this case, both the 6 cm and 1.3 cm core polarization angles indicate the core **B** field to be transverse to the jet.

Thus, we can understand the bimodal distribution for the 6 cm core polarization angles as a consequence of optical depth effects, if cores with χ *aligned* with the jet are those dominated by optically *thin* regions (**B** transverse), while those with χ *perpendicular* to the jet are those dominated by optically *thick* regions (likewise, **B** transverse). There is some evidence supporting this hypothesis from multi-wavelength VLBA polarization observations of sources in the 1-Jy complete sample of northern BL Lac objects defined by Kühr & Schmidt (1990): a preliminary analysis of the 2–4–6 cm VLBI core spectral indices for the sample sources suggests that cores in which χ is aligned with the jet have less flat (more "optically thin") spectra than cores in which χ is perpendicular to the jet (Pushkarev, Gabuzda, & Cawthorne 2001). A more detailed analysis of the data to more thoroughly test this hypothesis is underway.

There is another intriguing result that can easily be understood if the bimodal distribution of 6 cm VLBI core χ's reflects bimodality of the core optical depths. Simultaneous optical polarization and 6 cm VLBI polarization data are now available for 15 compact AGN, primarily BL Lac objects (Gabuzda, Sitko, & Smith 1996; Gabuzda, Garnich & Smith, in prep.). These data show clear evidence for a correlation between the optical polarization angle $\chi_{\rm opt}$ and the VLBI core polarization angle $\chi_{\rm core}$: $\chi_{\rm core}$ is essentially always either aligned with or perpendicular to $\chi_{\rm opt}$ (to within 20°, in 11 of 13 measurements). The bimodal behavior in the distribution of $|\chi_{\rm opt} - \chi_{\rm core}|$ is natural if (i) the optical and radio polarization are roughly co-spatial, and (ii) the VLBI core polarizations for which $\chi_{\rm core}$ is aligned with $\chi_{\rm opt}$ are dominated by optically *thin* regions, while those for which $\chi_{\rm core}$ is perpendicular to $\chi_{\rm opt}$ are dominated by optically *thick* regions. We are in the process of obtaining and analyzing new simultaneous optical and shorter-wavelength VLBA polarization data to test this hypothesis (Gabuzda, Rastorgueva, & Smith, in prep.): as we compare the optical measurements with VLBI measurements at shorter wavelengths, more $\chi_{\rm core}$ values should align with $\chi_{\rm opt}$ as more radio cores become dominated by optically thin regions; at short enough radio wavelengths, the optical and VLBI core polarization angles should virtually always be aligned.

Thus, there is evidence suggesting that the **B** fields in the core regions of BL Lac objects may essentially always be transverse to the inner jet, with the observed VLBI core polarization at 4–6 cm being determined by whether the "core" is dominated by the contribution of optically thin or optically thick regions. In this case, what we observe as the "core" polarization is nearly always a mixture of polarization from the actual core and from optically thin regions in the inner jet; this is a natural consequence of the fact that the resolution attained by ground VLBI arrays corresponds to scales larger than the intrinsic sizes of the VLBI cores at these frequencies.

If some VLBI cores are dominated by optically thick regions at 6 cm, their degree of polarization should be about a factor of six lower than in neighboring optically thin regions where the magnetic field has roughly the same degree of order (Pacholczyk 1970). The available VSOP polarization images of BL Lac objects at 6 cm (Gabuzda 1999; Gabuzda & Gómez 2001; Gabuzda 2002, 2003a) indicate the degree of polarization in the innermost 6 cm jets to be typically $\simeq 10 - 20\%$; accordingly, the expected degree of polarization for optically thick VSOP cores at 6 cm would be $\simeq 1.5 - 3.5\%$, consistent with the core polarizations observed in VSOP images.

3. The Nature of the Jet B Fields

It has been known for a long time that the **B** fields in the VLBI jets of BL Lac objects are most often perpendicular to the local jet direction. Usually, this has been interpreted as evidence for the presence of relativistic shocks that enhance the **B** field component transverse to the direction of compression (Laing 1980; Gabuzda et al. 1992). A "shock paradigm" has developed, in which essentially all of the compact features observed in the VLBI jets of AGN are thought to be related to shocks.

However, this view began to cardinally change with the analysis of the first VSOP polarization image, constructed from observations of 1803+784 obtained on July 21, 1998 (Gabuzda 1999; Fig. 1, lower right). The extra resolution provided by the space–ground baselines demonstrated that the inner few milliarcseconds of the VLBI jet were appreciably curved, and that the brightest component was not the easternmost detected feature, as might have been expected. The polarization vectors are everywhere aligned with (i.e., tangent to) the local jet direction, and remain aligned with the jet as it bends. Since the observed degrees of polarization in much of the jet were sufficiently high that these regions had to be optically thin, this indicated a **B** field that was everywhere transverse to the curved jet. The relatively modest degree of polarization of the easternmost feature ($m \simeq 3\%$) suggested that it was the core, and that the brightest feature was an emerging knot in the inner jet.

A "conventional" interpretation of this structure in the context of shock models would require the presence of a whole series of relativistic shocks along the jet, each of which enhanced the local transverse **B**-field component. Although theoretically possible, this picture seems contrived. In addition, the observed distribution of polarized flux in 1803+784 was fairly smooth, and the degree of polarization did not rise sharply in the brightest knots. This led Gabuzda (1999) to conclude that it was more likely that the observed transverse field corresponded to a dominant toroidal **B**-field component associated with the VLBI jet. The **B**-field structure of 1803+784 on larger scales revealed by analysis of the ground-based data shows that the **B** field remains transverse to the jet on all scales on which the jet is detected, from less than one milliarcsecond to tens of milliarcseconds (Gabuzda & Chernetskii 2003; Fig. 2).

Subsequently, a number of other sources in the Kühr & Schmidt (1990) sample of BL Lac objects were found to have similar **B**-field structures, which remained transverse all along the jets, even in the presence of appreciable bending (Gabuzda & Pushkarev 2002; Gabuzda 2003b,c). If these transverse **B** fields are associated with a dominant toroidal field component, maintenance of the transverse field orientation as the jets curve would be natural. It is also interesting that completely independent arguments have begun to point toward the presence of transverse **B** fields in the core regions of BL Lac objects (see Section 2).

If we are seeing a dominant toroidal **B**-field component, this has at least two very natural theoretical interpretations. This type of **B**-field structure is predicted by theoretical models for jet formation and propagation in which an initial "seed" field with a significant longitudinal component is "wound up" by the rotation of the central accreting object (e.g., Ustyugova et al. 2000; Nakamura, Uchida, & Hirose 2001). Another intriguing possibility is that the observed toroidal field components develop in association with currents flowing in the jet (Istomin & Pariev 1996; Blandford, these proceedings). Indeed, if

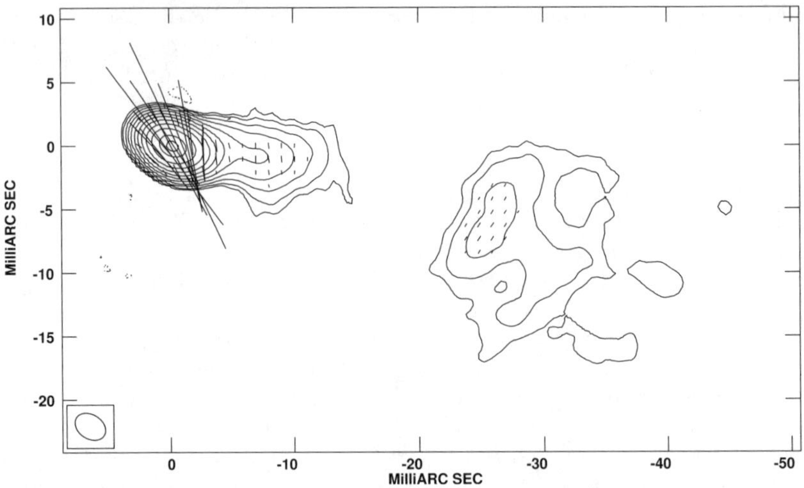

Figure 2. 6 cm ground-based image of 1803+784 for epoch July 21, 1998 (*taken from Gabuzda & Chernetskii 2003, MNRAS, 339, 669*) with **B** vectors superimposed. The bottom contours are $\pm 0.06\%$ of the peak of 2.02 Jy/beam, and the contours increase in steps of 2. The **B** field is transverse throughout the detected VLBI jet. The inner 2 mas of the jet can be seen in more detail in Figure 1 (bottom left).

we are seeing a dominant toroidal **B**-field component, one can argue that there *must* be currents flowing in the region enclosed by the field.

In reality, if the jet **B** fields of BL Lacs tend to be dominated by a toroidal component, we should still expect to see a fairly wide variety of **B**-field structures, since the observed structure will depend on the degree of dominance of the toroidal component and on the viewing angle. The viewing angle will play a role both geometrically and in terms of relativistic aberration effects. For example, if the observed **B** field is everywhere transverse, with little or no evidence for a longitudinal component near the edges, this may require that we are viewing the jet nearly side-on in its rest frame. Interaction with the surrounding medium could also affect the overall observed **B**-field structure.

In fact, a number of BL Lac objects and related AGN do show evidence for "spine+sheath" **B**-field structures, with a central region of transverse field and the longitudinal field component becoming dominant in some regions offset from the jet axis (e.g., 0745+241: Pushkarev & Gabuzda 2002; 0820+225: Gabuzda, Pushkarev, & Garnich 2001; 1055+018: Attridge, Roberts, & Wardle 1999; 1418+546: Gabuzda 2002, 2003a). One example is the VSOP image of 1418+546 in Figure 1 (bottom left), which shows a region of offset transverse χ (longitudinal **B**) in the inner 1 mas of the jet, with a region of well centered longitudinal χ (transverse **B**) 1.5–2 mas from the core. Thus, it may be that the variety of **B**-field structures observed in the BL Lac sample corresponds to jets dominated by toroidal **B**-field structures viewed at some range of angles and displaying varying degrees of interaction with the surrounding medium.

Although this paper has concentrated on the observational evidence for toroidal field structures in BL Lac objects, it may be that toroidal or helical jet **B** fields are characteristic of other types of AGN as well. One interesting possibility is that both BL Lac objects and quasars have helical jet **B** fields, but that slower flow speeds in BL Lac objects lead to an increased pitch angle for the helical field, increasing the dominance of the toroidal field component (Gabuzda et al. 2000; Asada et al. 2002).

It is clearly of interest to try to identify robust observational tests that can distinguish between transverse **B** fields due to the dominance of a toroidal field component and due to compression of an initial field by shocks. One possibility is to try to detect rotation-measure gradients across the jets: such gradients are expected in the case of a toroidal field structure due to the systematic change in the line-of-sight component of the **B** field across the jet. Asada et al. (2002) claim to have observed such a gradient across the VLBI jet of 3C 273; further work is needed to confirm this result and to search for rotation-measure gradients in other sources whose jets are well resolved in the transverse direction.

4. Conclusion

Our evolving understanding of the core region is bringing to our attention a fact that is simultaneously obvious and easy to forget: the observed "VLBI core" will correspond to the optically thick base of the jet only if the observations have resolution sufficient to enable meaningful and reliable estimation of the core parameters. If the size of the optically thick core is substantially smaller than the resolution provided by the observations, the observed "core" will actually correspond to a mixture of the actual core emission and emission from the inner jet. This appears to be the regime in which we have been working in ground-based VLBI observations at moderate wavelengths such as 4 and 6 cm. VSOP polarization observations have played a crucial role in the development of our understanding of the core region by directly demonstrating that the 6 cm ground-based "core" polarizations of BL Lac objects are virtually always dominated by the contribution of highly polarized regions in the inner VLBI jets.

Analyses based on quasi-simultaneous 6 cm VSOP and 1.3–2 cm VLBA observations indicate that we have begun to resolve the 6 cm cores with VSOP, enabling us to obtain reasonable estimates of the parameters of the 6 cm cores for the first time. The VSOP polarization observations of OJ 287 analyzed by Gabuzda & Gómez (2001) together with quasi-simultaneous 1.3 cm VLBA observations are of fundamental importance, since they have revealed for the first time the theoretically predicted 90° flip in the polarization electric vector in the transition from the optically thin to the optically thick regime.

The VSOP polarization image of 1803+784 (Gabuzda 1999) provided a beautiful example of a source with a transverse **B** field everywhere along a substantially curved VLBI jet. This led the way to the suggestion that this **B**-field structure was associated with the dominant toroidal component of the magnetic field of the jet itself, possibly due to the "winding up" of a seed field with a significant longitudinal component by the rotation of the central black hole and accretion disk. A number of other examples of this type of **B**-field structure have subsequently been found, showing that it is not unusual for BL Lac objects.

The realization that we are very likely seeing intrinsic "global" **B** fields associated with the jets themselves, rather than **B** fields produced purely by "local" phenomena, such as shocks and shear, fundamentally changes our view of the observed polarization structures. In addition, the fact that we are seeing dominant toroidal **B**-field components in many parsec-scale jets leads to the important conclusion that these jets are *Poynting-flux dominated*, which is of cardinal theoretical importance.

References

Asada, K., et al. 2002, PASJ, 54, L39

Attridge, J. A., Roberts D. H., & Wardle J. F. C. 1999, ApJ, 518, L87

Gabuzda, D. C. 1999, New Astron. Reviews, 43, 695

Gabuzda, D. C. 2002, in Proc. of the 6th European VLBI Network Symposium, ed. E. Ros, R. W. Porcas, A. P. Lobanov, & J. A. Zensus (Bonn: MPIfR), 83

Gabuzda, D. C. 2003a, in ASP Conf. Ser., High Energy Blazar Astronomy, ed. in press

Gabuzda, D. C. 2003b, in Ap&SS, Circular Polarisation from Relativistic Jet Sources, in press

Gabuzda, D. C. 2003c, in New Astron. Rev., in press

Gabuzda, D. C., et al. 1992, ApJ, 388, 40

Gabuzda, D. C., et al. 1994, ApJ, 435, 140

Gabuzda, D. C., & Chernetskii, V. A. 2003, MNRAS, 339, 669

Gabuzda, D. C., & Gómez, J. L. 2001, MNRAS, 320, L49

Gabuzda, D. C., Gómez, J. L., & Agudo, I. 2001, MNRAS, 328, 719

Gabuzda, D. C., & Pushkarev, A. B. 2002, in ASP Conf. Ser. Vol. 250, Particles and Fields in Radio Galaxies, ed. R. Laing & K. Blundell (San Francisco: ASP), 180

Gabuzda, D. C., Pushkarev, A. B., & Garnich, N. N. 2001, MNRAS, 327, 1

Gabuzda, D. C., Pushkarev, A. B., & Cawthorne, T. V. 2000, MNRAS, 319, 1109

Gabuzda, D. C., Sitko, M. L., & Smith, P. S. 1996, AJ, 112, 1877

Istomin Ya. N. & Pariev, V. I. 1996, MNRAS, 281, 1

Kühr, H., & Schmidt, G. D. 1990, AJ, 90, 1

Laing, R. 1980, MNRAS, 193, 439

Nakamura, M., Uchida, Y., & Hirose, S., 2001, New Astron., 6, 61

Pacholczyk 1970, in Radio Astrophysics (San Francisco: Freeman)

Pushkarev, A. B., & Gabuzda, D. C. 2002, in ASP Conf. Ser. Vol. 250, Particles and Fields in Radio Galaxies, ed. R. Laing & K. Blundell (San Francisco: ASP), 200

Pushkarev, A. B., Gabuzda, D. C., & Cawthorne, T. V. 2001, in IAU Symp. 205, Galaxies and Their Constituents at the Highest Angular Resolution, ed. R. T. Schilizzi et al. (San Francisco: ASP), 140

Ustyugova, G. V., et al. 2000, ApJ, 541, L21

Radio Astronomy at the Fringe
ASP Conference Series, Vol. 300, 2003
J.A. Zensus, M.H. Cohen, & E. Ros, eds.

Compact Structures in Flat-Spectrum Extragalactic Radio Sources

Y. Y. Kovalev
Astro Space Center of P.N.Lebedev Physical Institute, Profsoyuznaya 84/32, 117997 Moscow, Russia

Abstract. We have used visibility data obtained with the VLBA at 15 GHz to examine the most compact structures in 160 flat-spectrum extragalactic radio sources. With projected baselines out to 440 million wavelengths we are able to study source structure on angular scales as small as 0.12 mas. For about 40% of the sources, at least at one epoch, more than half of the flux density is unresolved on the longest baselines. These objects may have dimensions less than 0.12 mas. The most variable sources tend to have the most compact structure. Typical lower limits to the brightness temperature are in the range of 10^{11} to 10^{12} K but extend up to 10^{13} K; this being an apparent excess of the inverse Compton limit for stationary synchrotron sources. Longer baselines, such as shall be available in future space VLBI missions, are needed to study the most compact high brightness temperature regions in these sources. The fringe visibilities of the γ-ray loud AGN show, on average, higher values than for γ-ray quiet AGN and they are more variable. This supports a concept of a common emission mechanism at radio and γ-ray wavelengths. A visible correlation of VLBI–filled aperture monitoring data is also discussed.

1. Introduction

The structure and physics of the regions nearest to a super massive black hole where relativistic particles are accelerated can be studied with a linear resolution of a parsec or less by VLBI. The highest resolutions are achieved by earth to space interferometry. The first space VLBI missions (Levy et al. 1989, Hirabayashi et al. 1998) have already increased the baselines by a factor of about 3 in comparison with the maximum baselines of ground-based interferometers. By extending the baselines by a factor of ≥ 10 in expected space VLBI missions it will be possible to study very compact structures with high brightness temperature. E.g., for the planned *RadioAstron* mission the fringe size approaches the angular size of the gravitational radius of nearby super massive black holes such as in M 87. Such very compact regions with high surface brightness are surely present in pulsars and perhaps in the rapidly variable Galactic and extragalactic radio sources, as indicated by their time variability.

In this paper we analyze 15 GHz VLBA observations of 160 flat-spectrum compact extragalactic sources (sample selection procedure and first epoch maps are published by Kellermann et al. 1998; Zensus et al. 2002). These data are comparable with the space VLBI data from VSOP (Hirabayashi et al. 1998) obtained on longer baselines but at longer wavelengths and with lower accuracy. We use the data to study most compact structures and the way they change with time.

2. Observational Data in Use

In this analysis we have used visibility monitoring data obtained up to 1998. Three epochs of 15 GHz VLBA observations were available for each source, on average. For every source we have calculated the total flux density $S_{\rm tot}$ within a map and the flux density from the most compact component $S_{\rm comp}$. The "compact" flux density is defined by the data obtained at baseline projections near 400 Mλ. VLBA core parameters were derived for each source using DIFMAP.

A small subset of our sources are only slightly resolved even at the longest spacings. With the robust self calibration of the VLBA, we are able to state with some confidence when a source or component of a source has a fringe visibility greater than 0.9 at the longest spacings of about 400 million wavelengths, corresponding to an angular dimensions less than 0.12 mas.

We have also considered the total flux density $S_{\rm fa}$ at 15 GHz determined from observations with filled-aperture antennas. Data from the UMRAO and RATAN–600 monitoring programs (Aller et al. 1985; Kovalev et al. 1999) were used.

3. Discussion of General Characteristics

Figures 1 and 2 present the distributions of the flux densities and flux density ratios (or "indices of compactness"). We see that many sources have considerable correlated flux density at projected baselines more than 400 Mλ. More than a third of the sources have $S_{\rm comp}/S_{\rm tot} > 0.5$ and 90% of the sources have the unresolved flux density greater than 0.1 Jy. For 70 of these sources the flux density of the most compact component is greater than 0.5 Jy.

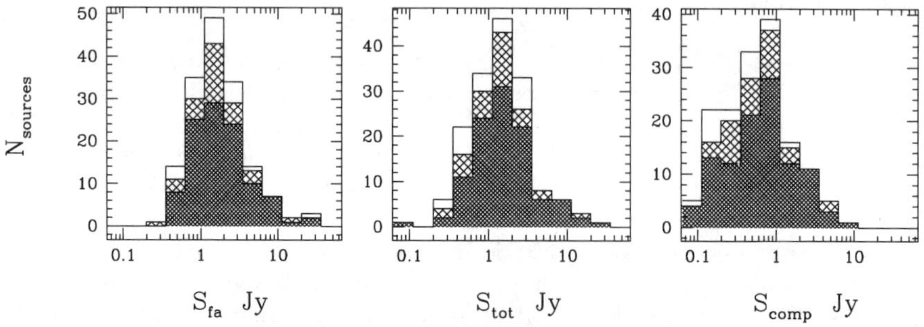

Figure 1. Distributions of flux densities $S_{\rm fa}$, $S_{\rm tot}$, $S_{\rm comp}$. We separate here and in Figure 2 quasars (most hatched area) from BL Lacs (medium hatched) and radio galaxies (empty).

The good correlation between $S_{\rm tot}$ and $S_{\rm fa}$ indicates that for most sources the VLBA sees nearly all of the flux density, and that there are no significant systematic errors in the flux density scale of the VLBA. Aside from the few sources that clearly have large scale structure, the mean compactness index $< S_{\rm tot}/S_{\rm fa} >$ for the sample is about 0.86. Several objects have a compactness index $S_{\rm tot}/S_{\rm fa} > 1$; most likely, this is due to the source variability and the non-simultaneity of the VLBA and filled aperture observations. Sources with the compactness index $S_{\rm tot}/S_{\rm fa}$ close to 1 at all epochs are suitable for calibration of other VLBI observations.

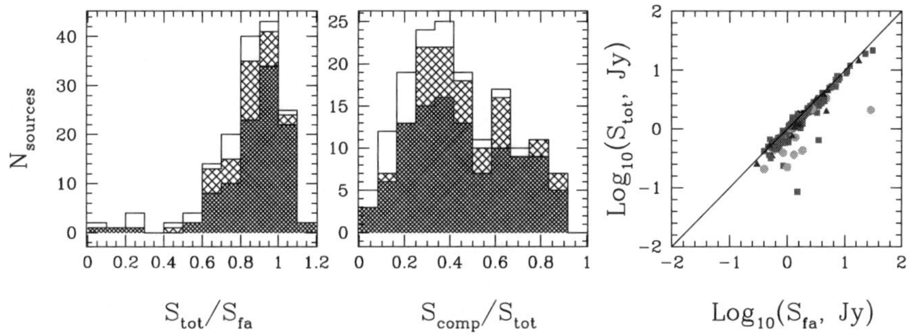

Figure 2. Source distributions on indexes of compactness $S_{\rm tot}/S_{\rm fa}$, $S_{\rm comp}/S_{\rm tot}$ and $S_{\rm tot}$–$S_{\rm fa}$ dependence (quasars are labeled by squares, BL Lacs—by triangles, radio galaxies—circles).

Generally, there are no major distinctions among sources included in Figures 1 and 2, although the radio galaxies have on average a smaller compactness index. This is easily interpreted in the context of the unified scheme (see, e.g., Urry & Padovani 1995). The radio galaxies probably have a greater angle between a jet and a line of view than quasars and BL Lacs do and have smaller beaming factor; thus, the radio galaxies are viewed as more extended and more symmetric objects.

The peak of the $S_{\rm fa}$ distribution corresponds to the selection limit of the sample. The peak position and the shape of the left wing reflect the discrepancy between the actual 15 GHz flux density and the value estimated from extrapolation by Kellermann et al. (1998) and Zensus et al. (2002), largely due to the variability of the sources.

The brightness temperature distribution of VLBA cores has a maximum near $3 \cdot 10^{11}$ K and extends up to 10^{13} K, which is close to the limit attainable by the VLBA for a source of a few Jansky. In many cases the measurement refers only to the upper limit of the angular sizes, corresponding to our limiting angular size of about 0.12 mas, defined by the beam size and the signal-to-noise ratio. Thus, the real brightness temperatures may extend to the higher values, beyond the synchro-Compton limit of 10^{12} K (Kellermann & Pauliny-Toth 1969). Such excessive brightness temperatures might be due to the Doppler boosting, transient events, coherent emission, emission by relativistic protons (see, e.g., Kardashev 2000, Kellermann 2002).

About one third of all VLBI cores involved are unresolved along one direction having a dimension less than 0.12 mas. For the nearest object, M 87, this corresponds to a linear size of $3 \cdot 10^{16}$ cm. Galaxies typically have larger VLBI cores. As many sources appear to have unresolved structures, we have examined a one dimensional brightness distribution along and perpendicular to the jet axis for each source. The CLEAN component model has been convolved with a knife-like beam (HPBW: 0.12×20 mas) which is very narrow along the direction of the scan and is very wide in the perpendicular direction. For many sources the fan beam distribution includes a considerable number of unresolved structures (often representing an unresolved width of a jet). If a direction of a jet is known, a (u, v)-coverage could be planned most effectively for a task of resolving a jet width. This task is attractive for planned space VLBI missions with stretched orbits (e.g., *RadioAstron*). In the modeling of annual changes of the timescale of scintillation variability (see, e.g., Jauncey & Macquart 2001; Rickett et al. 2001) it is also important to take into account an orientation to the Earth orbit of a VLBI core with unresolved width.

For the sources with many epochs of observations, VLBI core parameters (in particular, the brightness temperature) vary significantly with time. For 20% of the objects, the flux density of a core varies by a factor of two or more. The most variable sources tend to have the most compact structure.

On modeling the broad band radio spectra of compact extragalactic radio sources (Kovalev et al. 2001), we derived that a compact relativistic jet must be responsible for the centimeter and millimeter part of the spectrum. It is also supported by the compactness analysis done in the present paper (see Fig. 2). Consequently, correlation of a long term variability between $S_{\rm tot}$ and $S_{\rm fa}$ was anticipated. A check on its presence gave us a positive answer.

4. Radio Sources Detected by EGRET

The third catalog of high energy γ-ray sources detected by the EGRET telescope of the *CGRO* (Hartman et al. 1999) includes 66 high-confidence identifications of blazars. While the γ-ray sources were identified with flat-spectrum radio loud extragalactic radio objects ($\alpha > -0.5$, $F \sim \nu^{\alpha}$), not all radio sources with such characteristics have been detected as γ-ray sources. However, we note that the range of flux density detected by EGRET is very limited, so that the number of γ-ray detections may rise considerably with improved sensitivity, in contrast to the distribution of radio flux density where there are well defined radio loud and radio quiet objects.

Kellermann et al. (1998) did not find any clear differences in the milliarcsecond structure between the EGRET subsample (20% of radio sample) and the rest of the sources. We find no significant difference in the compactness parameter $S_{\rm comp}/S_{\rm tot}$ between EGRET detections and the sources of undefined γ-ray status (see Fig. 3, left). However, the EGRET non detected sources (von Montigny et al. 1995) show remarkably lower compactness at mas scale than the EGRET detections do. The situation changes if we use only the maximum values of the compactness index for each source (Fig. 3, right). For the EGRET detected sources the peak of the distribution becomes more pronounced: maximum compactness indexes of more than 75% of the EGRET detections are greater than 0.5. The distributions for all other sources become flatter. Notice also that two γ-bright objects with the lowest compactness were observed only once, and an object with the maximum compactness index 0.2 is the well known quasar 3C 273 with large resolved jet. Probably, sources are γ-ray strong only during the initial stages of strong radio flares (as it was proposed by Valtaoja & Teräsranta 1996) when the radio flux density is generally emitted by compact unresolved components. It is not clear if the relation between the observed γ-ray emission and radio compactness is the result of inverse Compton scattering mechanism, or the γ-ray emission is the synchrotron radiation of electrons in

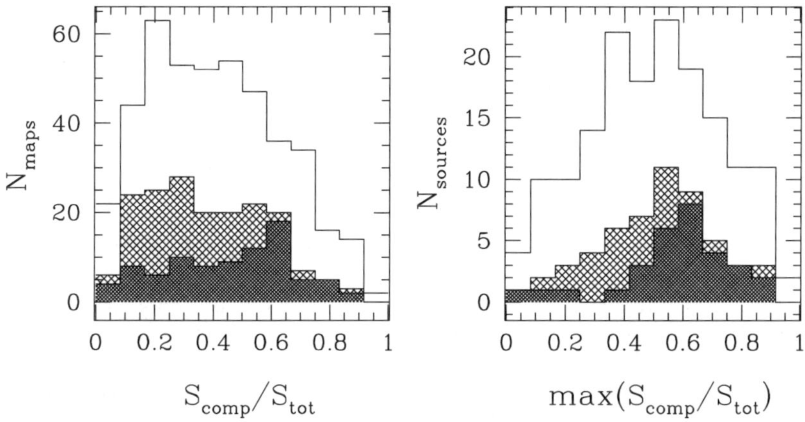

Figure 3. Distributions over the compactness index $S_{\rm comp}/S_{\rm tot}$: left—for all the maps analyzed; right—for all the sources (for each source a map with the maximum value of $S_{\rm comp}/S_{\rm tot}$ is chosen from the multi-epoch data). We separate flat-spectrum radio sources detected by EGRET with high confidence (most hatched area) from EGRET not detected sources (medium hatched) and the objects of undefined γ-ray status (empty).

the high energy tail of distribution of the electrons which are responsible for the radio radiation. With the next generation of γ-ray telescopes, such as *GLAST*, the sensitivity may be sufficient to actually separate the class of γ-ray loud and γ-ray quiet objects and better define the relation between radio and γ-ray emission of the sources.

5. Conclusion

We have carried out an analysis of the observational data of 160 extragalactic objects, obtained with the VLBA at 2 cm as a part of a program to monitor the structure of strong quasars and AGN (Kellermann et al. 1998, Zensus et al. 2002). We find that there are many sources with significant compact structure on sub-milliarcsecond scales. About 90% of the objects from the sample have unresolved flux density greater than 0.1 Jy. For 40% of the objects in the sample, more than half of the flux density comes from unresolved features. We have compiled a list of 70 sources with an unresolved structure stronger than 0.5 Jy. These sources will be good candidates for a study with space VLBI. A small subset of the sources are only slightly resolved even at the longest spacings and have angular size less than about 0.12 mas. About one third of all VLBI cores are unresolved along one direction with a corresponding dimension less than 0.12 mas. For the nearest object, M 87, this implies a linear size of $3 \cdot 10^{16}$ cm.

The distribution of brightness temperatures of the cores peaks at $3 \cdot 10^{11}$ K and extends up to 10^{13} K. For many sources the value of the brightness temperature is only a lower limit. Brightness temperatures in excess of the inverse Compton limit of 10^{12} K can be explained as the result of the Doppler boosting, transient phenomena, coherent emission, or synchrotron emission by relativistic protons.

EGRET detected radio sources show, on average, higher values of fringe visibility and more variability of their most compact structure than non EGRET sources, supporting models which relate the radio and γ-ray emission, such as inverse Compton scattering or synchrotron emission from a common electron energy distribution.

This paper is based in part on work done in collaboration with M. H. Cohen, K. I. Kellermann, E. Ros, R. C. Vermeulen, J. A. Zensus, N. S. Kardashev, M. F. Aller, and H. D. Aller. The NRAO is a facility of the NSF operated under cooperative agreement by Associated Universities, Inc. We have made use of the data base from the UMRAO which is supported by funds from the University of Michigan. This work has been partly supported by the NASA JURRISS program (W–19611), the RFBR (01–02–16812, 02–02–16305), and the Russian State Program "Astronomy".

References

Aller, H. D., Aller, M. F., Latimer, G. E., & Hodge, P. E. 1985, ApJS, 59, 513

Hartman, R. C., et al. 1999, ApJS, 123, 79

Hirabayashi, H., et al. 1998, Science, 281, 1825

Jauncey, D. L., & Macquart, J.-P. 2001, A&A, 370, L9

Kardashev, N. S. 2000, ARep, 44, 719

Kellermann, K. I., & Pauliny-Toth, I. I. K. 1969, ApJ., 155, L71

Kellermann, K. I., Vermeulen, R. C., Zensus, J. A., & Cohen, M. H. 1998, AJ, 115, 1295

Kellermann, K. I. 2002, PASA, 19, 77

Kovalev, Y. Y., et al. 1999, A&AS, 139, 545

Kovalev, Y. Y., Nizhelsky, N. A., & Bogdantsov, A.V., 2001, in Similarities and Universality in Relativistic Flows, ed. A. Guthmann et al., (Berlin: Logos Verlag), 74

Levy, G. S., et al. 1989, ApJ, 336, 1098

Rickett, B. J., et al. 2001, ApJ, 550, L11

Urry, C. M., & Padovani, P. 1995, PASP, 107, 803

Valtaoja, E., & Teräsranta, H. 1996, A&AS, 120, 491

von Montigny, C., et al. 1995, A&A, 299, 680

Zensus, J. A., et al. 2002, AJ, 124, 662

Radio Astronomy at the Fringe
ASP Conference Series, Vol. 300, 2003
J.A. Zensus, M.H. Cohen, & E. Ros, eds.

Gigahertz Peaked Spectrum Radio Sources

M. L. Lister
National Radio Astronomy Observatory, 520 Edgemont Road, Charlottesville, VA 22903, USA

Abstract. The study of compact AGN which possess convex radio spectra (the Gigahertz-peaked spectrum, or GPS sources) has generated a lot of interest over the last four decades, since these objects offer a unique opportunity to probe both the early evolutionary stages of relativistic AGN jets and their immediate nuclear environments. In this article I trace Ken Kellermann's early investigations of these sources, which played a major role in justifying the development of modern-day VLBI techniques. I describe how our understanding of these AGN has progressed since Kellermann's early discoveries, and discuss several ways in which the current classification scheme can be simplified to reflect intrinsic source characteristics, rather than observer-biased quantities. Finally, I discuss recent results from the VLBA 2 cm survey concerning the relativistic jet kinematics of the two-sided peaked-spectrum sources 4C +12.50 (PKS 1345+125) and OQ 208 (1404+286).

1. Initial discovery

The history of peaked-spectrum radio sources can be traced back to Ken Kellermann's graduate student days at the Owens Valley Radio Observatory in the early 1960s. At that time, the revised 3C catalog had just been published (Bennett 1962), and there was great interest in obtaining accurate broad-band radio spectra of these bright sources. For his Ph.D. thesis, Kellermann used the variable spacing interferometer at Owens Valley to obtain flux densities at four frequencies between 750 and 3200 MHz for 160 objects from the 3C list and other radio catalogs available at that time. Approximately half of the sample had straight, steep spectra consistent with incoherent synchrotron emission from a population of electrons having a power-law energy distribution.

The remaining radio sources had somewhat curved spectra, and two unusual objects: CTA 21 (0316+161) and CTA 102 (2230+114), had convex profiles that peaked around 900 MHz (Kellermann et al. 1962), hence the origin of the term "Gigahertz-peaked spectrum" (GPS) sources. These two CTA objects had been discovered serendipitously by Dan Harris at Owens Valley several years earlier, and subsequently generated a great deal of excitement in the Soviet Union as possible beacons from extraterrestrial civilizations. Kardashev (1964) had argued that considering the effects of quantum noise and competing background emission, the most efficient spectral profile for transmitting radio signals across the galaxy was remarkably similar to that of CTA 21 and CTA 102. This coincidence fueled intense speculation as to their origin. The situation grew more interesting when Sholomitskii (1965) reported highly unusual flux density variations in CTA 102 at 32 cm. This motivated Kellermann (1966) to look for "artificial" features as predicted by Kardashev (1964) in the spectrum of PKS 1934−638, but none were present.

Kellermann et al. (1962) had realized early on that the 3C sources with distinctly curved spectra had the highest brightness temperatures, but incorrectly attributed this to spectral ageing processes. Slysh (1963) was subsequently able to fit the curved spectra of 3C 48 (0134+329) and CTA 21 using formulae he derived for a synchrotron self-absorbed source, and calculated their angular sizes to be exceedingly small (140 and 10 milliarcsec, respectively). This led Shklovsky (1965) to reason, using his spherically expanding supernova model, that peaked-spectrum sources had to be quite young and should display substantial flux density variations over time.

At around the same time, Dent (1965) reported significant flux density changes in 3C 279, and Kellermann & Pauliny-Toth (1968) began to regularly monitor a sample of bright AGN at wavelengths ranging from 2 to 40 cm. They soon discovered that flux variability was quite common in flat-spectrum AGN, but virtually absent in the peaked-spectrum sources. This posed serious problems for Shklovsky's simple spherical expansion model for AGN variability, which was later dropped in favor of more complex shock-in-jet models (e.g., Aller et al. 1985). Several groups, including Maltby & Moffet (1965), failed to confirm Sholomitskii's discovery of variability in CTA 102, although later low-frequency observations of this source (e.g., Hunstead 1972) showed flux density variations that were probably associated with interplanetary scintillation.

1.1. Early VLBI observations

By 1966 there was overwhelming indirect evidence that many of the brightest radio-loud AGN were exceedingly compact. Both the high turnover frequencies of the peaked-spectrum sources and light travel time arguments applied to source variability suggested that some AGN could only be studied using very long interferometric baselines exceeding several million wavelengths. Kellermann (1964) had pointed out two years earlier that that the peaked-spectrum sources were all intrinsically very luminous, and therefore could provide a number of good targets for very long baseline interferometry (VLBI).

These findings provided the scientific motivation for several groups around the globe to independently pursue the development of VLBI techniques. Early VLBI observations between Green Bank and Haystack in 1967 at a wavelength of 18 cm (Clark et al. 1968) showed that many AGN were still unresolved, which underscored the need for even longer baselines at shorter wavelengths. The following year, Kellermann et al. (1968) obtained the first inter-continental fringes of several AGN at 6 cm, between Green Bank and Onsala, Sweden. These fringe detections gave the first direct measurement of sizes 1 mas or smaller for several AGN, including the peaked-spectrum source 2134+004.

The direct confirmation of small source sizes using VLBI renewed a debate launched two years earlier by Kellermann (1966) in his analysis of the spectrum of PKS 1934−638: was the low-frequency spectral turnover in peaked-spectrum sources due to free-free absorption (FFA) by thermal plasma, or synchrotron self-absorption (SSA) from relativistic electrons? Or was it related to the presence of cold plasma, as described by the Razin (1960) effect?

Surprisingly, this debate still continues nearly forty years later, with recent papers discussing either FFA (e.g., Bicknell, Dopita, & O'Dea 1997) or SSA models (e.g., Snellen et al. 1999). The main difficulty is that both models contain enough free parameters to fit virtually any convex spectrum, albeit with somewhat different constraints (see O'Dea 1998). Multi-frequency VLBI absorption studies (e.g., Marr, Taylor & Crawford 2001; Shaffer, Kellermann, & Cornwell 1999) may eventually settle this issue, but at present it remains plausible that both mechanisms play a role in creating the spectral turnovers in these sources.

2. Classification schemes

The standard view of peaked-spectrum sources today is that they are compact, powerful, young radio sources that reside in gas-rich environments at the centers of active galaxies (O'Dea 1998). Other typically quoted properties include low fractional polarization, low apparent jet speeds, and low flux density variability. However, in reality most individual peaked-spectrum sources display only some of these characteristics, and exceptions to the "classic" definition are unfortunately the norm.

Part of the problem is that the current definition is based only on the overall radio spectrum of an AGN, which is often the sum of many complex, time-variable emitting regions in a relativistic jet. The spectral shape can also be heavily influenced by the external environment, as in the case of FFA. It could be argued, therefore, that a peaked spectrum by itself tells us very little about an AGN, except for the obvious fact that it is dominated by an emitting region of small angular size. This suggests that the term "GPS" should either be abandoned entirely, or qualified in more detail so that it can be used as a true physical classification, and not solely a phenomenological one. Below I suggest some simple ways that the GPS classification could be modified so that it favors intrinsic rather than observer-biased properties.

2.1. "Masquerading" blazars

A first step toward a more definitive classification scheme would be to eliminate flat-spectrum, variable AGN (i.e., blazars) from the GPS class. What distinguishes the blazars from GPS sources is their bright, flat-spectrum cores, but in some blazars the signature of the core can be occasionally hidden in its integrated radio spectrum. Kovalev et al. (2002) have shown that the radio spectra of many blazars change dramatically with time, often showing remarkably GPS-like convex profiles during flux density outbursts. This can naturally be explained by the ejection of a new feature in the jet, whose individual self-absorbed spectrum briefly dominates the total radio flux. Many of the "high-frequency peakers" found by Torkinoski et al. (2001) appear to fall into this category.

A small fraction of blazars also display persistent features in their jets that often outshine the flat-spectrum core. These are likely associated with either stationary shocks or bends in the jet, and can give rise to a relatively stable, convex spectrum. In most cases, however, the signature of the flat-spectrum core can be found in another region of the spectrum, well away from the spectral peak of the stationary component. A well-known example is the blazar 4C 39.25 (0923+392), which has a convex spectrum above 1 GHz, but a distinctively flat spectrum over the region $0.1 < \nu < 1$ GHz. Ironically, CTA 102 also shows the distinct flat-spectrum and variability of a blazar at frequencies above 3 GHz, and should be excluded from the GPS category. As it turns out, the only reason CTA 102 was originally called a GPS source is that it was discovered during a relatively inactive state when its spectrum happened to be convex.

In order to eliminate these "masquerading" blazars from the peaked-spectrum class, two additional restrictions should be imposed. First, a *bona fide* peaked-spectrum source should display a canonical convex spectrum that a) peaks at *any* radio frequency, and b) maintains this spectrum consistently at all epochs. A canonical peaked spectrum has been constructed by de Vries et al. (1997), and has $\alpha > 0.5$ for $\nu < \nu_m$, where $S_\nu \propto \nu^\alpha$ and ν_m represents the spectral turnover frequency. In the high frequency part of the spectrum well past the turnover where the source is optically thin, the spectral index is steeper than -0.5.

2.2. Compact-steep spectrum sources

The current classification scheme is also complicated by the nebulous division between the GPS and compact steep-spectrum (CSS) sources. These are often grouped together based on their many shared characteristics (see the review by O'Dea 1998). For example, when plotted together on the intrinsic size vs. ν_m plane, they form a continuous trend having a single slope (size $\propto \nu_m^{-0.65}$; O'Dea & Baum 1997). The CSS sources originally got their name since they were unresolved with connected interferometers ($\theta < 1''$), and had overall steep spectra ($\alpha \simeq -0.7$). However, subsequent low-frequency observations of CSS revealed that many in fact had spectral turnovers at a few hundred MHz. The term GPS then became synonymous with sources with $\nu_m \simeq 1$ GHz, and those with lower frequency turnovers remained CSS sources.

This completely unphysical, arbitrary division generates additional redshift biases in "complete" samples of CSS and GPS sources, since it depends on the *observed* turnover frequency. Some authors (e.g., Fanti et al. 1985) have used an intrinsic size criterion of 1 kpc to divide CSS and GPS sources, but again this does not correspond to any physical quantity. Specifically, there is no evidence of a bi-modality in the observed size distribution. Instead, 1 kpc merely corresponds to the typical size of a source with a turnover frequency of 1 GHz. Furthermore, not all GPS/CSS have measured redshifts, making some size determinations impossible. To make matters worse, many intrinsically small GPS sources appear to have associated, large-scale emission, that is perhaps the result of past, episodic jet activity in the source (see review by Lara et al. 2002). It is unclear how to assign a characteristic size to these sources. Finally, there is again the issue of "masquerading" blazars, many of which tend to have small projected jet sizes due to foreshortening. These sources can display steep radio spectra if their flat-spectrum core is under-luminous compared to the jet. This appears to be the case for the superluminal CSS sources 3C 138 (0518+165), 3C 147 (0538+498), and 3C 216 (0906+430).

In light of these difficulties, the CSS classification should probably be reserved for physically small (< 20 kpc) sources that truly show no spectral turnover down to the lowest observed frequencies. Some CSS might very well be foreshortened, but these should be re-classified as blazars if their viewing angles can be constrained, for example, by measurements of superluminal motion. Furthermore, any CSS that shows a MHz-frequency spectral turnover and has a stable, canonical convex spectrum should be combined with existing GPS sources under the blanket classification "peaked-spectrum radio source".

2.3. Compact symmetric objects: a misnomer?

With the advent of hybrid imaging techniques for VLBI data in the early 1980's, it became possible to further classify GPS and CSS sources on the basis of their parsec-scale morphologies. Phillips and Mutel (1980) popularized the "compact double" class on the basis of VLBI observations of CTD 93 (1607+268) and 3C 395 (1901+319), which showed bright lobe-like features separated by a few hundred pc. As the sensitivity of VLBI arrays steadily improved, more and more "double" sources were seen to have faint emission that connected the lobes to a weak core component. The term "compact double" was soon abandoned in favor of "compact symmetric object" (CSO).

Although the term CSO is now widely used in the literature, it is a somewhat misleading description of these sources. First, the word "compact" can have different meanings depending on the context. VLBI researchers tend to use this term to describe a source with a large VLBI-to-single dish flux density ratio (i.e., brightness temperature), whereas most other astronomers would take it to mean a physically small object. Indeed, most AGN that are classified as CSOs do satisfy both of these definitions, but a growing number of CSOs (and other peaked-spectrum sources as well) appear to be associated with kpc-sized regions of weak, extended emission (Stanghellini et al. 1998).

The "symmetric" part of the class definition is also problematic. Wilkinson et al. (1994) first introduced the term CSO to describe sources such as 2352+495, which possess a fairly high degree of morphological symmetry between their jet and counter-jet. However, the term CSO is now widely applied to *any* GPS source that shows two-sided VLBI jets, regardless of apparent symmetry. This is illustrated by many CSO candidates in the COINS sample of Peck and Taylor (2000).

The current definition of a CSO, then, relies solely on two properties: a GPS spectrum, and a two-sided VLBI jet morphology, with source compactness and symmetry playing insignificant roles. For the remainder of this article, therefore, I will drop the term "CSO" in favor of "two-sided peaked-spectrum source".

3. Jet kinematics of two-sided peaked-spectrum sources

Since 1994, Kellermann and the VLBA 2 cm survey collaboration have been regularly monitoring a large sample of AGN in order to learn more about the kinematics and evolution of relativistic jets (Zensus et al. 2002). Although the goals of the program are mainly statistical, the sample also includes several peaked-spectrum sources which we are studying in closer detail. Here we report on two particularly interesting two-sided peaked-spectrum sources: 4C +12.50 (PKS 1345+125) and OQ 208 (1404+286). For the remainder of this paper we assume an accelerating cosmology with $\Omega_\lambda = 0.7$, $\Omega_\lambda = 0.3$, and $H_o = 70$ km s^{-1} Mpc^{-1}.

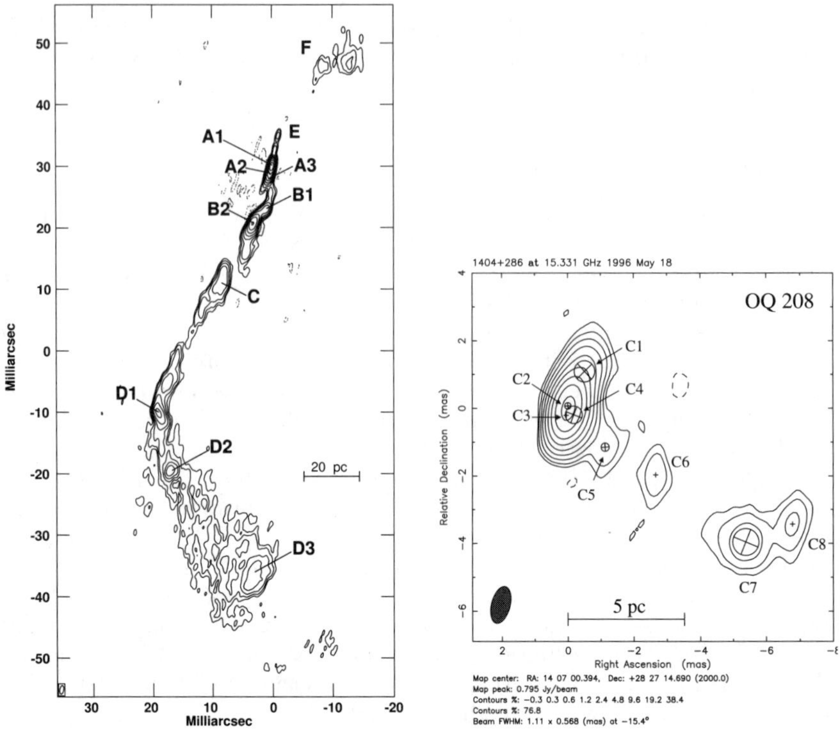

Figure 1. VLBA 2 cm images of 4C +12.50 (left panel; *taken from Lister et al. 2003, ApJ, 584, 135*) and OQ 208 (right panel).

3.1. 4C +12.50

The powerful radio source 4C +12.50 lies within an ultra luminous infrared galaxy at $z = 0.12$ that is rich in both molecular and atomic gas. The host contains two optical nuclei separated by ~ 3.5 kpc that appear to have undergone a recent merger, as evidenced by its tidal tail structure and distorted isophotes (Gilmore & Shaw 1986). Our VLBA observations (Lister et al. 2003) reveal a well-defined jet structure with a total extent of ~ 100 mas $= 220$ pc (Fig. 1). The brightest feature in our full-track VLBA image (A1) has an inverted spectrum and was identified as the core by Stanghellini et al. (2001). Like most peaked-spectrum sources, the integrated polarization of 4C +12.50 as measured by the VLA is very low ($m < 0.2\%$), however, our analysis has revealed small regions with extremely high fractional polarization in the southern jet. These range up to $m = 30\%$ at the major bend (D1) and $m = 60\%$ at the very tip of the jet (D3). The electric vectors in both regions are nearly perpendicular to the total intensity contours.

We have detected two components in the southern jet (A2 and A3) that display substantial apparent motion over our five year sampling interval. Their mildly superluminal speeds ($\simeq 1.2$ c) and the presence of the bright northern counter-jet suggest that the jet axes of 4C +12.50 lie fairly close to the plane of the sky. We obtained a good fit to the overall bent morphology of the jet and counter-jet using a simple constant-wavelength helix of opening angle 23° projected at 82° to the line of sight. Although some helical jets are the result of material being ejected ballistically from a precessing nozzle (e.g., SS 433; Hjellming and Johnston 1981), this is not the case for 4C +12.50, as there is ample evidence for streaming of material along the curved ridge line of the jet. These include the alignment of the polarization gradient and inferred magnetic field vectors along the jet ridge line, and the highly polarized flux at specific points in the jet (i.e., the major bend and southern tip). These features can naturally be explained as shocks in the flow that have ordered the magnetic field of the jet. The helical ridge line is likely the result of growing Kelvin-Helmholtz instabilities (e.g., Hardee 1987) that are driven by a small precession of the jet nozzle. Such a precession might be expected for the black hole of a galaxy such as this one that has undergone a recent merger event.

The relativistic speeds of the inner jets are significantly higher than the outer lobe expansion speeds measured for other two-sided peaked-spectrum sources ($v/c \simeq 0.3$; Owsianik, Conway & Polatidis 1999). The southern lobe is too diffuse to determine an accurate advance speed, but if it is typical of other sources, this implies that a significant amount of jet kinetic energy must be dissipated over a relatively small distance (~ 200 pc). This is perhaps being channeled into the large expansion of the jet at D1 or a terminal shock at D3.

3.2. OQ 208

The radio source OQ 208 is associated with the broad line Seyfert 1 radio galaxy Mrk 668, which like 4C +12.50, shows signs of tidal distortion (Stanghellini et al. 1993). Being one of the closest known peaked-spectrum sources ($z = 0.077$), we are able to study its jet kinematics with high spatial resolution (1 mas $\simeq$ 1.5 pc). Its radio structure is very compact, with $> 99\%$ of its cm flux originating from a region with a projected diameter of ~ 12 pc (Fig. 1). Aaron (1996) discovered a faint component ~ 33 mas SW of this region at position angle $-107°$. All of the parsec-scale structure lies within a large diffuse halo of diameter ~ 30 kpc (de Bruyn 1990)

We have assembled ten individual VLBA epochs at 2 cm from the VLBA 2 cm survey and other programs that span the time period 1995-2001. For the purposes of our proper motion analysis, we have modeled the inner jet structure using eight discrete circular Gaussian and delta-function components labeled C1-C8 in Figure 1. The precise location of the center of activity in this source is not well known. Our analysis shows the distance between C1 and C5 is shrinking at a rate of 54 μas y^{-1} ($v/c = 0.28$), which immediately rules out the possibility of the core being in the eastern region. Component C6 lies roughly halfway between the eastern and western "lobe" structures, and is a logical candidate for the core. However, the distance between C6 and the eastern structure is increasing at rate of ~ 180 μas y^{-1}, while the distance from C6 to the western structure is *decreasing* by 140 μas y^{-1}. If we assume there are no inward motions, the core must lie between C5 and C6, with C6 representing a jet feature that is moving to the SW.

We are unable to determine any individual component velocities without a stationary reference point, however, C3 and C7 are separating at a rate of 60

$\pm 4\ \mu$as y^{-1}. If OQ 208 has been steadily expanding at this rate, its parsec-scale emission is only $\sim 1100 \pm 73$ years old. This is typical of kinematic ages derived for other two-sided peaked-spectrum sources (Owsianik et al. 1999).

4. Summary

Since Ken Kellermann's initial discoveries regarding Gigahertz-peaked-spectrum sources nearly forty years ago, these powerful radio sources have greatly increased our understanding of active galaxies. They not only provided strong incentive for the development of modern-day VLBI techniques and synchrotron emission theory in compact radio sources, but are now being used to confirm the scenario by which AGN jet activity is triggered by galactic mergers. Unfortunately, the exposure of GPS sources to the larger astronomical community has been somewhat hindered by a complex and awkward classification scheme. However, with the small modifications discussed in this article, it should be possible to implement a scheme that distinguishes GPS sources on the basis of intrinsic, physical properties, and not observer-biased ones.

The author would like to thank the members of the VLBA 2 cm survey collaboration for helpful comments on this manuscript.

References

Aaron, S. E., 1996, Ph.D thesis, Brandeis University, Waltham, MA,USA.
Aller, H. D., Aller, M. F., & Hughes, P. A. 1985, ApJ, 298, 296
Bennett, A. S. 1962, MmRAS, 68, 163
Bicknell, G. V., Dopita, M. A., & O'Dea, C. P. O. 1997, ApJ, 485, 112
Clark, B. G., et al. 1968, AJ, 153, 705
Dent, W. A. 1965, Science, 148, 1458
de Bruyn, A. G. 1990, in Variability of Active Galaxies, Variability of Active Galaxies, ed. W. J. Duschl, S. J. Wagner & M. Camenzind, (Heidelberg, Germany: Springer), Lecture Notes in Physics 377, 105
de Vries, W. H., Barthel, P. D., & O'Dea, C. P. 1997, A&A, 321, 105
Fanti, C., et al. 1985, A&A, 143, 292
Gilmore, G. & Shaw, M. A. 1986, Nature, 321, 750
Hardee, P. E. 1987, ApJ, 318, 78
Hjellming, R. M., & Johnston, K. J. 1981, ApJ, 246, L141
Hunstead, R. W. 1972, Astroph. Lett., 12, 193
Kardashev, N. S. 1964, Soviet Ast., 8, 217
Kellermann, K. I. 1964, ApJ, 140, 969
Kellermann, K. I. 1966, Aus. J. Phys, 19, 195
Kellermann, K. I., et al. 1962, Nature, 195, 692
Kellermann, K. I. & Pauliny-Toth, I. I. K. 1968, ApJ, 152, 639
Kellermann, K. I. et al. 1968, ApJ, 153, L209
Kovalev, Y. Y., et al. 2002, PASA, 19, 83
Lara, L. et al. 2002, New Astron. Rev., 46, 89
Lister, M. L., et al. 2003, ApJ, 584, 135
Marr, J. M., Taylor, G. B., & Crawford, F. 2001, ApJ, 550, 160

Maltby, P. & Moffet, A. T. 1965, ApJ, 142, 409
O'Dea, C. P. 1998, PASP, 110, 493
O'Dea, C. P. & Baum, S. A. 1997, AJ, 113, 148
Owsianik, I., Conway, J. E., & Polatidis, A. G. 1999, New Ast. Rev., 43, 669
Peck, A. B. & Taylor, G. B. 2000, ApJ, 534, 90
Phillips, R. B. & Mutel, R. L. 1980, ApJ, 236, 89
Razin, V. A., 1960, Izvestia Vys. Ucheb. Zaved. Radiofiz., 3, 921
Slysh, V. I. 1963, Nature, 199, 682
Shaffer, D. B., Kellermann, K. I., & Cornwell, T. J. 1999, ApJ, 515, 558
Shklovsky, J. 1965, Nature, 206, 176
Sholomitskii, G. B. 1965, Inform. Bull. on Var. Stars, 83
Snellen, I. A. G., et al. 1999, New Astron. Rev., 43, 675
Stanghellini, C., et al. 1993, ApJ, 88, 1
Stanghellini, C., et al. 1998, A&AS, 131, 303
Stanghellini, C., et al. 2001, A&A, 377, 377
Tornikoski, M., et al. 2001, AJ, 121, 1306
Wilkinson, P. N., et al. 1994, ApJ, 432, L87
Zensus, J. A., et al. 2002, AJ, 124, 662

Radio Astronomy at the Fringe
ASP Conference Series, Vol. 300, 2003
J.A. Zensus, M.H. Cohen, & E. Ros, eds.

Radio Source Evolution on Small Scales

C. FANTI and R. FANTI
Dipartimento di Fisica, Università di Bologna, Via Irnerio 46, I-40126 Bologna, Italy
Istituto di Radioastronomia del CNR, Via Gobetti 101, I-40129 Bologna, Italy

Abstract. There is now a general consensus on a scenario for radio source evolution in which radio sources for most of their life dim in luminosity as they grow in size. We summarize our understanding of the arguments which have brought to such a scenario and comment on the extension of the evolution models to the very young phases, when the sources are expected to grow in an almost constant ambient density. We conclude that the models seem to fail at smaller linear sizes as they predict too few sources as compared to those which are found. We discuss some implications and alternatives.

1. Introduction: the evolution scenario

We discuss the radio source evolution from two complementary approaches:
a) observable statistical properties of the radio source population;
b) simple theoretical models based on rather general physical laws.

An important tool of the observational approach (Baldwin 1982) is the distribution of radio sources in the *radio power* (P_r)*– linear size (LS)* plane[1] which we parameterize as: $\frac{dN(LS)}{d(LS)} = N'(LS) \propto LS^{-m}$. This distribution gives valuable constraints on the source evolution. Assuming it to be stationary while the sources sweep across the plane with $P_r \propto LS^{-h}$ and $\frac{d(LS)}{dt} = v_a \propto LS^x$, it is possible to derive the following simple relation between m, x and h:

$$m = x + hb \tag{1}$$

The coefficient b is: *i*) the exponent of the source counts ($\log N - \log S$), if $N'(LS)$ is derived for all the sources in a flux density limited sample (Readhead et al. 1996 and Begelman 1996), or *ii*) the exponent of the (integral) Luminosity Function if the $N'(LS)$ is determined considering, in a flux density limited sample, only sources brighter than a minimum radio power, $P_{r,\rm min}$, (Fanti et al. 1995). We adopt here the second way of counting and take $b \approx 1.8$. Anyhow, in either case $b \gtrsim 1.5$.

Observationally, e.g., for the 3CR radio sources (Laing et al. 1983) with $LS \gtrsim 20\,\rm kpc$ and $P_{178\,\rm MHz} \gtrsim 10^{27}\,\rm W\,Hz^{-1}$ ($H_\circ = 100\,\rm km\,s^{-1}\,Mpc^{-1}$, $q_\circ = 0.5$)

[1] We exclude from our considerations the broad band radio flat spectrum sources as they are *core-jet Doppler boosted* objects.

one finds $m \approx 0.6 \pm 0.1$. From eq. (1) one gets immediately important constrains on h and x: *as the radio sources grow they have to either speed up or to dim in luminosity, or both.*

In order to proceed further one needs some theoretical modeling of x and h, such as, e.g., in Baldwin (1982), Begelman (1996), Kaiser & Alexander (1997), Snellen et al. (2000). In the following, for sake of simplicity, we will refer to the first two models only.

The simplest *x modeling* is based on ram-pressure arguments under the assumption of constant jet power (F_e). The advance velocity is given by

$$v_a \approx (F_e/c)^{0.5}(\rho_e A_h)^{0.5} \qquad (2)$$

where F_e is the jet power, $\rho_e \propto r^{-\beta}$ is the external density and A_h is the *working area* where the jet thrust is discharged, identified observationally with the *hot spot*. It is assumed that $\beta \approx 1.5 \div 2$ on scales larger than several kiloparsec (kpc). Observationally (see Jeyakumar & Saikia 2000) A_h increases almost quadratically, or slightly less, with LS. Baldwin (1982) assumed $A_h \propto (LS)^2$ (*B conical* model). Begelman (1996, *Be* model) assumes self–similarity in the source growth which, very roughly speaking, requires taking $A_h \propto F_e^{1/3} LS^{\frac{\beta+4}{3}}$.

For $\beta \lesssim 2$, one expects: $x \lesssim 0$ and, therefore, $h \gtrsim 0$

For *h modeling* one assumes equipartition and has to model the growth of the lobe volume with time (or LS). As a first approximation one takes into account expansion losses only, neglecting radiation losses. The luminosity is assumed to be mostly that of the lobes, although the hot spot contribution might become dominant at small sizes (Begelman 1996). A summary of the *B conical* and the *Be* models as a function of β is given in Table 1.

Table 1. Parameters of Evolution Models

Model	x	$h_{\rm lobe}$	h_{hs}
B conical	$\frac{\beta-2}{2}$	$\frac{11\beta-14}{16}$	0.5
Be	$\frac{\beta-2}{3}$	$\frac{7\beta-8}{12}$	$\frac{\beta+4}{12}$

$$v_a \propto F_e^{1/2}(LS)^x, \quad P_{r,\rm lobe} \propto F_e\,(LS)^{-h_{\rm lobe}}, \quad P_{r,hs} \propto F_e^{1/2}(LS)^{-h_{hs}} \qquad (3)$$

The models show that: $h \gtrsim 0$ for $\beta \gtrsim 1.2$

Combining the h and x models, m depends on β only.

The analysis performed on the 3CR catalogue, for $LS \gtrsim 20$ kpc and $P_{178\,\rm MHz} \gtrsim 10^{27}$ W Hz^{-1} ($m \approx 0.6$), gives: $\beta \approx 1.8$ $h \approx 0.4$ $x \gtrsim -0.1$.

In conclusion, in that range of LS we have a consistent picture for source evolution, in which the radio luminosity decreases significantly as the source size increases, while the growth velocity v_a remains roughly constant.[2]

2. Discovery of the young radio source population?

According to the evolution scenario *young radio sources are expected to be small and quite luminous.* Where are they?

In the early eighties Kapahi (1981) and Peacock & Wall (1982) pointed out that, besides the classical populations of "flat compact" and "steep extended" radio sources there appeared to be a third population: the "Compact Steep–spectrum Sources" (*CSS*s). These objects were characterized by a high luminosity and a low frequency spectral curvature, bearing some similarity with the "flat spectrum" population. However, on the basis of rather poor structure information and on the similarity of the source counts, Peacock & Wall concluded: "*... there are grounds for regarding the Compact Steep–spectrum Sources as being more closely related to the extended sources than to the compact flat–spectrum sources...*".

Those *CSS*s whose spectrum peaks at about 1 GHz were named **G**Hz **P**eaked-spectrum **S**ources, *GPS*s (see e.g., O'Dea, Baum, & Stanghellini 1991).

Actually, the existence of *CSS*s and *GPS*s has been known long before. Examples of compact ($\lesssim 1''$) 3C sources with steep radio spectra flattening at $\lesssim$ few hundred MHz, were discovered in the early sixties (see Conway, Kellermann & Long 1963) and the archetype *GPS* source 1934–63 was discussed in detail by Kellermann (1965).

Fanti et al. (1990) and O'Dea & Baum (1997) pointed out a good anti-correlation between source rest-frame turnover frequency and linear size LS. This relation has been discussed alternatively in terms of free-free absorption (see e.g., Bicknell et al. 1997) and synchrotron self-absorption (e.g., Snellen et al. 2000). Fanti & Fanti (2001) have shown that the synchrotron self–absorption explanation works well if the magnetic field is close to equipartition. Independent of the interpretation, the $\nu_{\rm turn}^{\rm rest} - LS$ relationship shows clearly that *CSS*s and *GPS*s *are not distinct classes* but rather form a continuous sequence.

Phillips & Mutel (1982) discussed the properties of a sub–class, which appeared to have double, < 1 kpc, structure and suggested that it represents the young phase of the evolution of powerful radio sources. Subsequently this has been called the *youth scenario.*

According to the $\nu_{\rm turn}^{\rm rest} - LS$ relationship, the "young radio sources" are expected to have their radio spectrum peaking at high frequencies. Radio source samples selected at low frequency largely miss them. In order to pick out sources down to a given LS one has to use surveys at frequencies in the spectral transparent regime of the searched objects.

[2] We should not expect to see a clear correlation between P_r and LS in samples of radio sources, either flux density selected or volume selected, as we expect a very broad range in *jet power* which would largely wash out the correlation implied by the evolution. Furthermore, in flux density selected samples other additional effects, related to the visibility volume, are present.

For a long time a pending problem has been that (with the subliminal assumption of constant luminosity and constant velocity growth during the source lifetime) the fraction of young sources should correspond to the ratio of the time spent in the young phase to the total source lifetime, which is of the order of the ratio of *CSS* sizes over the sizes of the larger sources: a rather small number, against the fraction of *CSS*s found in radio catalogs ($\approx 15-20\%$). This *population problem* has contributed somewhat to support an alternative scenario, *the frustration scenario*, which assumes that the environment of *CSS*s is so dense that it inhibits the growth of these objects, keeping them small for their whole lifetime.

3. *CSS*s/*GPS*s: a homogeneous class?

In the majority of *CSS*s ($LS \gtrsim 0.5$ kpc) the radio morphology consists of *two radio lobes* on either side of a (sometimes un–seen) radio core. At the appropriate resolution most of them appear as scaled down versions of powerful extended radio sources. The two–sidedness is often referred to as "symmetry", though the characteristics of the two mini-lobes may be quite different. In a minority of cases more complex morphologies, including very distorted and one–sided ones, are found.

For *GPS*s there is a larger variety of structures. Besides the sub-population of sources with double-lobed structure, there is also a significant fraction of *core-jet type* sources (Stanghellini 2003; Lister, these proceedings, page 71), very likely with Doppler boosting effects. Actually, the fraction of core-jet sources among *GPS*s seems to increase with the peak frequency and is rather prominent in quasars.

Terms like *CSO*, *MSO* have been introduced (Wilkinson et al. 1994; Fanti et al. 1995) to indicate different sub–classes of those **O**bjects displaying the two sided (or **S**ymmetric) emission and are either **C**ompact (<1 kpc, *CSO*s) or **M**edium sized (*MSO*s), while *LSO*s was adopted to indicate the powerful **L**arge–sized **S**ymmetric **O**bjects, typically FRII. It is these *symmetric objects* which are assumed to be *the young parent population* of the large size radio sources. As several authors have pointed out, it is not easy to distinguish the double symmetric structure objects from the core-jet ones on the basis of the spectral properties alone and high resolution observations at several frequencies are required.

4. Evidence for youth in *CSS*s/*GPS*s

Lobe proper motions. So far lobe proper motions have been measured in a handful of the smallest *CSO*s ($10 \lesssim LS_{\rm pc} \lesssim 150$) by several authors (Owsianik & Conway 1998; Polatidis & Conway 2002 and references therein). Velocities between the edges of the two mini–lobes up to $0.4\,c$ have been found. Polatidis & Conway (2002) quote a mean value $\approx 0.18\,c$. The corresponding *kinematic ages* are in the range of few hundred to few thousand years. No measurements of this kind exist for larger sources (*MSO*), due to the difficulty of mapping hot-spots with the required resolution ($\lesssim 1$ mas), typically some tens of mas in size.

Radiative ages. Murgia et al. (1999) have fitted the integrated spectra of a number of *CSO*s and *MSO*s with a continuous injection synchrotron model. The model fits the data remarkably well over three orders of magnitude in frequency ($\sim 0.1 - 200\,\mathrm{GHz}$). Adopting the equipartition magnetic field, they derived radiative ages ($\tau_{\rm syn}$) of these radio sources, which turned out to be in the range $\approx (1 - 100) \times 10^3$ years. The average expansion speed, derived as $LS/\tau_{\rm syn}$, is $(\approx 0.3 \pm 0.1)\,c$, close to what found from the proper motions of *CSO*s. In the remarkable case of 1943+546, $\tau_{\rm syn}$ (Murgia 2003) is amazingly close to the kinematic age obtained by Polatidis et al. (2002). An interesting byproduct of this analysis is that the source magnetic field seems to be within a factor ≈ 2 of the equipartition value.

It is worth comparing the expansion speed of *CSO*s and *MSO*s with what is known for the large powerful sources *LSO*s. Arshakian & Longair (2000), from a study of the asymmetries in the arm lengths of the radio lobes, estimated an average lobe speed $v_{LSO} = (0.11 \pm 0.04)\,c$ and further pointed out that these "kinematic ages" are consistent with those obtained from radio spectral studies ("radiative ages").

It is remarkable that the speeds obtained for *LSO*s are in the same range as those for *CSO*s and *MSO*s.

5. The evolution on small scales

a) What we expect. At small linear scales the density of the ambient medium, assumed to be described by a *King-like* law, is expected to have $\beta \approx 0$ for $r < r_c$, and $\lesssim 2$ for $r > r_c$, were r_c is the *core radius*, estimated to be $\lesssim 1\,\mathrm{kpc}$. The evolutionary models summarized in Table 1 predict the following.

- $LS > r_c$ The evolution of P_r and v_a should be as found for the *LSO*s ($LS > 20\,\mathrm{kpc}$) and the $N'(LS)$ should remain a power law with unchanged m (≈ 0.6) down to about r_c.

- $LS \lesssim r_c$ According to the models summarized in Table 1, we should have:

  $x \approx -(0.66 - 1.0) \qquad h_{lobe} \approx -(0.75 - 0.9) \qquad h_{hs} \approx (0.3 - 0.5).$

In other words, *within* r_c we expect the following:

i) The lobe luminosity increases with the source size, while the hot spot luminosity decreases. The ratio of hot-spot to lobe luminosities is expected to change as $LS^{-(1-1.4)}$. Very likely the hot spots are dominant at the earlier phases.

ii) The advance velocity of the lobes is strongly decelerated.

iii) From equation (1) we should have $-2 \lesssim m \lesssim -0.1$, depending on whether the source luminosity is dominated by the lobes or by the hot spots. Therefore $N'(LS)$ should increase with LS or be roughly constant.

b) What the observations show.

i) *The N'(LS) counts.* The P&W catalog (Peacock and Wall 1982), owing to its high frequency (2.7 GHz) selection, includes small sources down to a few tens of parsecs. Other source samples, such as that by Stanghellini et al. (1998) selected at 5 GHz, allow us to pick out even smaller sources. They allow the $N'(LS)$ relationship as compared to the 3CR catalog, to be extended to significantly small LS.

Down to $\lesssim 1$ kpc, the supposed *core radius*, $N'(LS)$ remains a power law with unchanged m (≈ 0.6), as expected from the models (Fanti et al. 1995).

Surprisingly this relationship still holds for $LS < 1$ kpc, down to a few tens of parsec (Readhead et al. 1996), where we expected m to become negative (point *iii)* above).

ii) *Hot-spot versus lobe luminosity.* Observationally this ratio does not seem to change much with LS and the lobes are already well formed and dominant at $\lesssim 100$ pc. According to equation (1) we should have $m \approx -2$.

iii) *Advance speed of the lobes.* The advance velocity of the lobes is expected to be strongly decelerated. The available data, either from measured proper motions (Polatidis & Conway 2002) or from radiative ages (Murgia 2003) do not show any significant correlation between v_a and LS. For sources with the same F_e we should have $v_a \propto LS^x \approx LS^{-(.67-1)}$. On the other hand for sources with roughly the same total (lobe) luminosity and therefore with $F_e \propto LS^{h_{lobe}}$, from equations (3) we would expect a correlation between v_a and LS such as $v_a \propto LS^{0.5h_{lobe}+x} \approx LS^{-(1\div1.4)}$, which is inconsistent with the observations. The actual situation may be in between. In either case the lack of correlation between v_a and LS is incompatible with a flat gas atmosphere.

6. Discussion

In the previous section we have pointed out some significant discrepancies between what was expected and what was found on very small scales ($10 \lesssim LS(\text{pc}) \lesssim 1000$). We examine, in the following, two possible explanations.

An intruder population? An intruder population has already been mentioned in section 3, particularly amongst *GPS*s: it is a *core-jet boosted population*. The counts at small sizes referred to in the previous section do contain this intruder population, mainly composed of quasars. To test the evolutionary models we should consider only the *CSO*, *MSO* and *LSO* population, but in a number of cases the structure information is not enough to select the right sources.

Nevertheless, we suspect that this explanation is not working for the following two reasons. First, we may skip most of the core-jet sources simply by rejecting *quasars* and using *radio galaxies* only, for which the contamination from boosted sources is expected to be much less. We find is that the $N'(LS)$ still keeps rising at decreasing LS. Second, even restricting the $N'(LS)$ to those sources which are well known *CSOs* (Readhead et al. 1996), and so with the possibility of missing some *bona fide ones*, still $N'(LS)$ keeps raising at decreasing LS.

An alternative possibility is that a significant fraction of the small sources is made up either by *frustrated* or by *short lived* (Readhead et al. 1994) sources.

It seems to us that none of these possibilities is tenable, as both the *frustrated* and the *dying* sources are expected to grow in the constant density atmosphere and therefore their $N'(LS)$ as well should increase with LS, with $m \approx -2$. In addition a significant population of frustrated sources of small linear sizes is ruled out by the observed fast motions and the spectral ages. Finally only a minority ($\lesssim 15\%$) of sources show radio spectra such as expected for a dying source (Murgia 2003) and anyhow the alternative interpretations of *intermittent activity* is also possible.

A non flattened gas atmosphere? As seen in the previous section, the lack of correlation between v_a and LS gives strong constraints on x. Combining these constraints with equation (1) one finds that $h \lesssim 0.45$ and $x \gtrsim -0.2$. According to the models in Table 1, this corresponds to $\beta \approx 1.8$.

In other words, if the gas distribution never flattens when approaching the nucleus all problems disappear. Of course an explanation is required for a *never flattening atmosphere.*

7. Conclusions.

At present it seems that simple evolutionary models, coupled with a power law density distribution of the ambient gas, justify pretty well the distribution of radio sources in the $P_r - LS$ plane for $LS \gtrsim 0.5 - 1$ kpc.

However on smaller scales, where the gas density is expected to be roughly constant, the models over-predict the source number.

We tend to exclude that this source excess is totally justified by intruder populations, although there is evidence for them.

We make, instead, the non-orthodox suggestion that the source distribution in the $P_r - LS$ plane probes an ambient gas whose density decreases with distance from the nucleus roughly as a power law with $\beta \approx 1.8$, at least from a few tens of parsec on. Confirmation or rejection of this suggestion requires direct studies of the ambient gas at linear resolutions from few tens to several hundred parsec.

References

Arshakian, T. G., & Longair, M. S. 2000, MNRAS, 311, 846

Baldwin, J. 1982, in Extragalactic Radio Sources, in IAU Symp. 97, ed. D. S. Heeschen & C. M. Wade (Dordrecht: Reidel), 21

Begelman, M. C. 1996, in Cygnus A: Study of a Radio Galaxy, ed. C. Carilli & D. Harris (Cambridge: CUP), 209

Bicknell, G., Dopita, M. A., & O'Dea, C. P. 1997, ApJ, 485, 112

Conway, R.G., Kellermann, K.I., & Long, R.J. 1963, MNRAS, 125, 261

Fanti, C., et al. 1990, A&A, 231, 346

Fanti, C., et al. R.E., & Stanghellini, C. 1995, A&A, 302, 317

Fanti, C., & Fanti, R. 2001, in ASP Conf. Ser. Vol. 258, Issues in Unification of Active Galactic Nuclei, ed. Maiolino, R., Marconi, A., & Nagar, N., (San Francisco: ASP), 261

Kapahi, J. K. 1981, A&A, 43, 181

Kaiser, C. R., & Alexander, P. 1997, MNRAS, 286, 215

Kellermann K.I. 1965, AJ, 70, 681

Jeyakumar, S., & Saikia, D.J. 2000, MNRAS, 311, 397

Laing, R. A., Riley, J. M, & Longair, M. S., 1983, MNRAS, 204, 151

Murgia, M., et al. 1999, A&A, 345, 769

Murgia M. 2003, PASA, 20, 19

O'Dea, C. P., Baum, S. A., & Stanghellini, C. 1991, ApJ, 380, 66

O'Dea, C. P., & Baum, S. A. 1997, AJ, 113, 148

Owsianik, I., & Conway, J. E. 1998, A&A, 337,690

Peacock, J. A., & Wall J. V. 1982, MNRAS, 198, 843

Phillips, R. B., & Mutel, R. L. 1982, A&A, 106, 21

Polatidis, A. G., & Conway, J. E. 2003, PASA, 20, 69

Readhead, A. C. S., et al. 1994, in Compact Extragalactic Radio Sources, ed. J. A. Zensus, K. I. Kellermann, (Socorro, NM: NRAO), 17

Readhead, A. C. S., et al. 1996, ApJ, 460, 612

Snellen, I. A. G., et al. 2000, MNRAS, 319, 445

Stanghellini, C., et al. 1998, A&AS, 131, 303

Stanghellini, C. 2003, in Third International Workshop on Compact Steep Spectrum and Gigahertz-Peaked Spectrum Radio Sources, ed. A. Koekemoer, I. Snellen, A. Tzioumis, & W. de Vries, PASA, 20, 118

Wilkinson, P. N., et al. 1994, ApJ, 432, L87

Radio Astronomy at the Fringe
ASP Conference Series, Vol. 300, 2003
J.A. Zensus, M.H. Cohen, & E. Ros, eds.

VLBI Properties of Gamma-Ray Blazars

S. G. JORSTAD
Institute for Astrophysical Research, Boston University, 725 Commonwealth Ave., Boston, MA 02215-1401, USA
Astronomical Institute of St. Petersburg State University, Universitetskij pr. 28, 198504 St. Petersburg, Russia

A. P. MARSCHER
Institute for Astrophysical Research, Boston University, 725 Commonwealth Ave., Boston, MA 02215-1401, USA

Abstract. We have completed a program in which we monitored a sample of γ-ray blazars with the VLBA at 22 and 43 GHz. The distribution of apparent speeds, determined for 23 quasars and 10 BL Lac objects, reveals that the majority of superluminal components have velocities exceeding 10c in both γ-ray quasars and BL Lac objects. On average, EGRET-detected blazars are characterized by highly relativistic jets with Lorentz factor $\Gamma \sim 20$, viewing angle $\Theta \sim 1^\circ - 2^\circ$, opening half-angle $\alpha \sim 1^\circ$, and Doppler factor $\delta \sim 25 - 30$. However, more intensive and sensitive surveys are needed to determine whether the Lorentz factor, and viewing and opening angles are constant parameters of the jet or, if not, the ranges of the variability of these parameters for a typical source.

1. Introduction

We have observed a sample of 31 quasars and 11 BL Lac objects, roughly 60% of known γ-ray blazars (Hartmann et al. 1999), with the VLBA from November 1993 to August 1997 (Jorstad et al. 2001a). The main goal of our program was to establish whether the VLBI properties of γ-ray blazars are distinct from those of other subclasses of flat radio spectrum, non-γ-ray active galactic nuclei (AGNs). Our main findings can be summarized as follows:

- In our sample of 42 γ-ray bright blazars, apparent superluminal jet velocities are measured in 33 sources. The distribution of apparent velocities of jet components peaks at $12 - 14\,c$, which is significantly higher than the average superluminal speed of jet components in the general population of strong compact radio sources. For example, according to Kellermann et al. (1999), the mean jet velocity for 48 non-γ-ray sources with comparable radio luminosity and structure is equal to $(3.9\pm0.4)c$. This result strongly supports the thesis that γ-ray emission is enhanced by extreme relativistic beaming.

- We have detected multiple moving components in 16 quasars and 8 BL Lac objects. The existence of a number of sources with multiple moving knots with different proper motions in the same source indicates that either the

Lorentz factor of the jet flow or the pattern speed of bright features (or both) is variable.

- In 27 of the 42 γ-ray blazars in our sample, we find a total of 45 components that show no motion with respect to the core over the duration of our VLBA observations. Out of these 27 sources, 21 are observed to contain moving components as well, some of which propagate at very high apparent speeds, $> 12c$. Hence, the existence of standing features does not correspond to very slow flow speeds in the jet.

- The majority of jets of γ-ray blazars bend significantly on parsec scales, although no more so than do jets of blazars not detected by EGRET.

- We have obtained 36 images of γ-ray bright blazars with polarization vectors and conclude that EGRET blazars possess no observable signature of jet polarization structure.

- We have identified 23 γ-ray flares (in 18 sources) that are contemporaneous with our VLBA observations. In these 23 cases, 10 of the flares (in 8 objects) fall within 1σ uncertainties of the extrapolated epoch of zero separation from the core of a superluminal radio component. This establishes a statistical association between the ejection of superluminal radio knots and high states of γ-ray emission in blazars.

Our primary results have already been published (Marscher 2002; Mattox et al. 2001; Jorstad et al. 2001a; 2001b) and images obtained at all epochs may be found at our web site `http://www.bu.edu/blazars`. Here we present some analysis which was not performed in the aforementioned papers. A standard Friedman cosmology with $H_\circ$=65 km s^{-1} Mpc^{-1} and $q_\circ = 0.1$ is used to calculate the apparent transverse velocities.

2. Superluminal Motion

We have obtained the jet velocities in 23 quasars and 10 BL Lac objects. Figure 1a presents the distribution of apparent speeds for quasars (21 knots, light shade) and BL Lac objects (10 knots, dark shade) when only a single well-determined apparent speed is included for each source. This means that at least 3 epochs within 1 year of observation have been used to determine an apparent speed. For sources with multiple moving components, the speed with the smallest uncertainty is included. For quasars there is no significant difference between this distribution and that obtained when all detected moving components (57 knots) are included (Jorstad et al. 2001a): the mean velocities for 21 knots $(13.1\pm7.1)c$ and 57 knots $(16.3\pm9.5)c$ are comparable. The situation changes for BL Lac objects. The distribution of apparent speeds for all detected moving components in BL Lac objects (28 knots) peaks at low apparent speed of about 3–4c (Jorstad et al. 2001a), which agrees with the previous finding by Gabuzda et al. (1994) that typical VLBI component speeds in BL Lac objects are systematically lower than those in quasars. However, if each source is represented by a single component (Fig. 1a), the apparent speeds of BL Lacs are distributed over a wide range of velocities from 2 to 30c and there is no

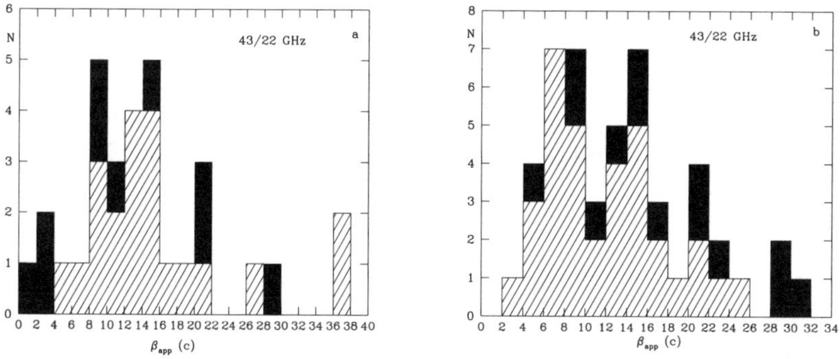

Figure 1. Distribution of jet velocities for γ-ray blazars a)with a single well-determined apparent speed for each source; b)for all detected moving knots in sources with $0.1 < z < 1.0$. The dark shade corresponds to BL Lac objects.

significant difference between jet velocities of quasars and BL Lac objects. Any difference disappears completely if we compare the distributions of the jet velocities for quasars (34 knots) and BL Lac objects (14 knots) with $0.1 < z < 1.0$, including all detected components (Fig. 1b). In this case the mean transverse apparent speed for BL Lac objects $(12.4\pm7.1)c$ is even slightly higher than for quasars $(10.0\pm5.4)c$. It is possible, however, that this conclusion is valid only for γ ray blazars.

A popular model for the γ-ray production in AGNs involves inverse Compton scattering of either synchrotron photons from the jet (SSC) or photons external to the jet (ECS). If this is the case, one can expect a connection between γ-ray flares and ejections of VLBI components. We have identified 23 γ-ray flares (in 16 quasars and 2 BL Lac objects) that are contemporaneous with our VLBA observations (Jorstad et al. 2001b). Note that the BL Lac objects are Mrk 421 and ON 231, with $z \leq 0.1$, and are characterized by VLBI components with low apparent speeds. We have found 10 of 23 γ-ray flares where the epochs of zero separation coincide, to within 1σ uncertainty, with the time of the γ-ray flares. The list consists only of the quasars 0336−019 (CTA 26), 0528+134, 0836+710, 1222+216, 1226+023 (3C 273), 1622−253, 1611+343 (DA 406), and 1730−130 (NRAO 530). The distributions of apparent speeds for these 10 cases peaks at $12c$, which indicates high superluminal speed for VLBI components associated with the γ-ray flares. Therefore, for low redshift γ-ray bright BL Lac objects, which belong to the subclass of TeV blazars, the mechanism of γ-ray flares might be different and could correspond, for example, to an increase in the upper-energy cutoff to the relativistic electron energy distribution without a change in the normalization or slope (Marscher & Travis 1996).

3. Physical Parameters of the Gamma-Ray Blazar Jets

Lister & Marscher (1999) created simulated flux-limited samples of γ-ray blazars based on the ECS and SSC models. They found that both models predict a high median Doppler factor and small median viewing angle. Here we estimate these parameters based on our observational results. We have sufficient resolution to measure projected opening angles of the jets for half of the blazar sample with well-determined jet velocities. Figure 2 and 3 show images of some sources used to measure the projected opening half-angle of the jet ϕ by deconvolution of the size of the jet component a (a might also be the sum of sizes of components lined up across the jet width) with the size of the CLEAN beam b transverse to the jet direction: $\phi = \arctan(\frac{f\sqrt{a^2-b^2}}{2R})$, where R is the distance from the VLBI core to the jet component and f is equal to 1.6 (the transverse cross-section of the jet should be similar to a disk). We assume that the VLBI core is close enough to the jet vertex that R is roughly the distance from the jet base. We use the *well-determined* apparent speeds to estimate the Lorentz factors Γ of the jet flows. Vermeulen & Cohen (1994) and Lister & Marscher (1997) found that the most likely viewing angle Θ is $(0.5-0.6)\Theta_m$, where Θ_m is the angle at which a source's apparent velocity is maximized. This value of Θ corresponds to the following relation between Γ and apparent velocity: $\beta_{app} \sim 0.88\Gamma$. According to Daly & Marscher (1988), in the gas dynamical model of compact jets the intrinsic opening half-angle of the jet α depends upon only two parameters: Γ and ξ, where ξ is the ratio of external pressure to the initial pressure of the gas in the jet: $\alpha = \alpha_\circ - \alpha_e$, where

$$\alpha_\circ = \sqrt{3}\ \arctan\sqrt{\frac{M_\circ^2-1}{3}} - \arctan\sqrt{M_\circ^2-1}$$

$$\alpha_e = \sqrt{3}\ \arctan\sqrt{\frac{M_e^2-1}{3}} - \arctan\sqrt{M_e^2-1},$$

and $M_\circ^2 = 2(\Gamma^2-1)\xi^{\frac{1}{2}}$, $\quad M_e^2 = 2(\Gamma^2\xi^{-\frac{1}{2}}-1)$. Here $M_\circ$ and M_e are the Mach numbers of the flow inside the jet (at the injection point) and along the boundary of the jet, respectively. Comparison of the projected opening half-angle ϕ and intrinsic opening half-angle α allows us to estimate the angle Θ between the jet axis and line of sight: $\tan\Theta = \frac{\sin\alpha}{\sin\phi}$. We used ξ as a free parameter to get the best agreement between our observational results and the numerical simulations obtained by Lister & Marscher (1999). The best fit occurs when the ratio between the external and internal pressure equals 0.7. The derived values of the parameters are given in Table 1. The last column of Table 1 is the Doppler factor δ obtained from the values of Γ and Θ.

Figure 4 shows the distribution of the viewing angles and Doppler factors. Under the adopted approximations and assumptions the typical parameters of γ-ray blazar jets can be described as follows: the average value of the intrinsic opening half-angle α=0.9° $\pm$ 0.3°; the distribution of the viewing angles has a global maximum at 1° $-$ 2°, and the distribution of Doppler factors peaks in the range $\delta \sim 25-30$. Although Lister & Marscher (1999) conclude that the SSC and ECS models make similar predictions for a small γ-ray selected sample, the

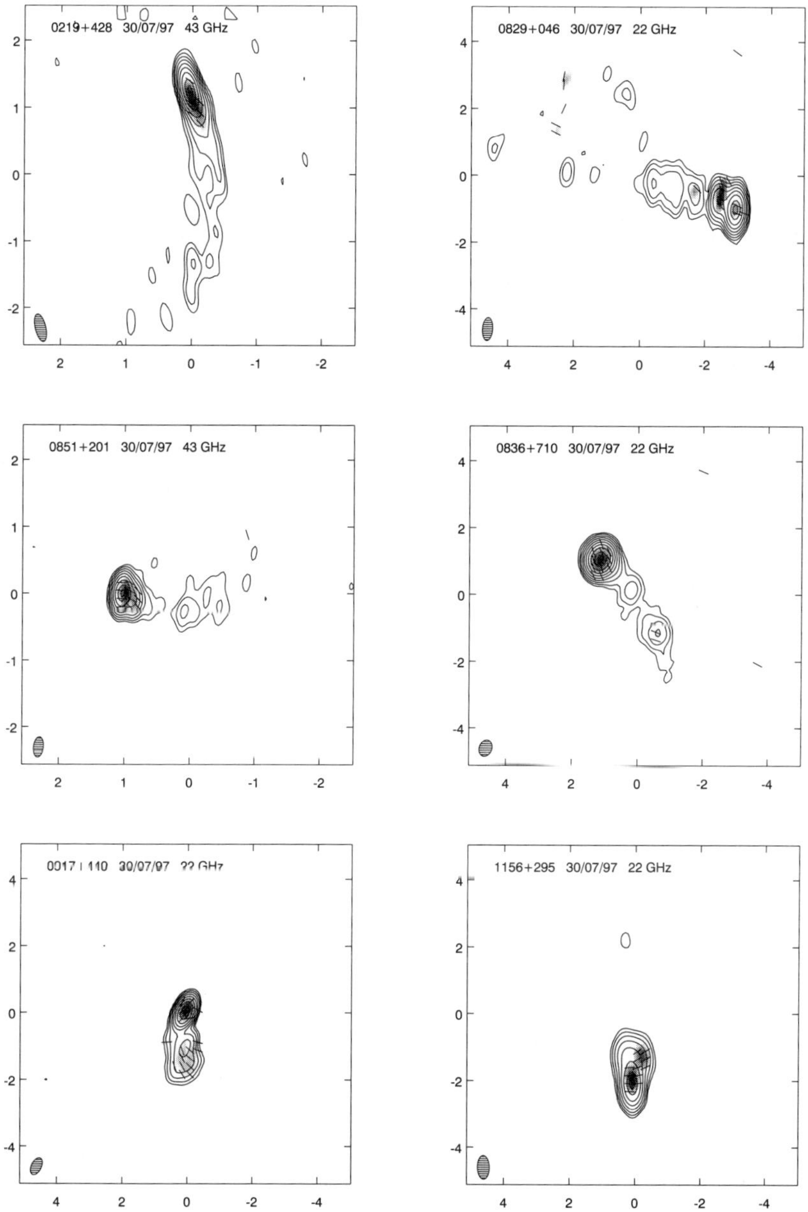

Figure 2. Sample of hybrid maps of γ-ray blazars whose jets are resolved across their widths.

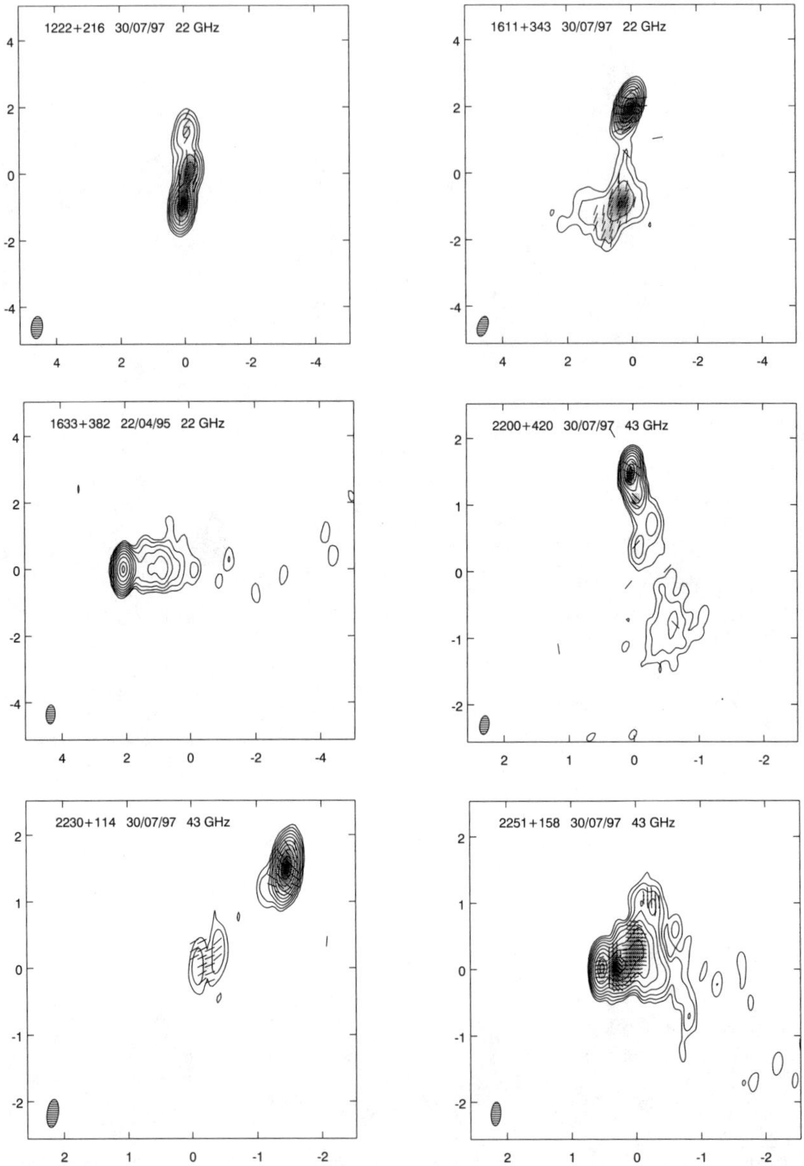

Figure 3. Sample of hybrid maps of γ-ray blazars whose jets are resolved across their widths (continued).

Table 1. Parameters of Jets

Source	β_{app} [c]	Γ	b [mas]	a [mas]	R	ϕ [°]	α [°]	Θ [°]	δ
0219+428	21.8±0.8	24.8	0.16	0.25	0.50	17.1	0.7	2.4	29
0336−019	19.1±1.8	21.7	0.46	0.57	0.65	22.5	0.7	1.7	30
0829+046	8.3±1.1	9.5	0.69	0.76	0.53	25.7	1.5	3.5	14
0836+710	21.7±1.8	24.7	0.39	0.53	1.05	15.3	0.6	2.2	26
0851+202	10.5±1.1	11.9	0.43	0.46	0.23	29.6	1.2	2.4	19
0917+499	13.5±1.8	15.4	0.30	0.59	1.00	22.1	0.9	2.1	21
1127−145	15.8±2.9	18.0	0.81	1.65	4.40	14.6	0.8	3.2	18
1156+295	21.7±3.6	24.7	0.30	0.61	0.80	28.0	0.6	1.2	38
1222+216	23.7±4.6	26.9	0.34	0.47	1.05	13.9	0.5	2.2	26
1226+023	9.1±1.8	10.4	0.35	0.49	0.34	38.9	1.4	2.2	18
1611+343	40±14	42	0.30	0.82	2.35	14.6	0.3	1.4	42
1633+382	16.0±1.5	18.2	0.57	0.88	1.30	22.4	0.8	2.1	26
2200+420	0.2±0.5	10.6	0.16	0.43	1.14	15.6	1.4	5.1	11
2230+114	17.4±0.5	19.8	0.35	0.86	1.94	18.0	0.7	2.4	24
2251+158	15.5±1.8	17.7	0.33	0.85	0.70	41.8	0.8	1.2	31

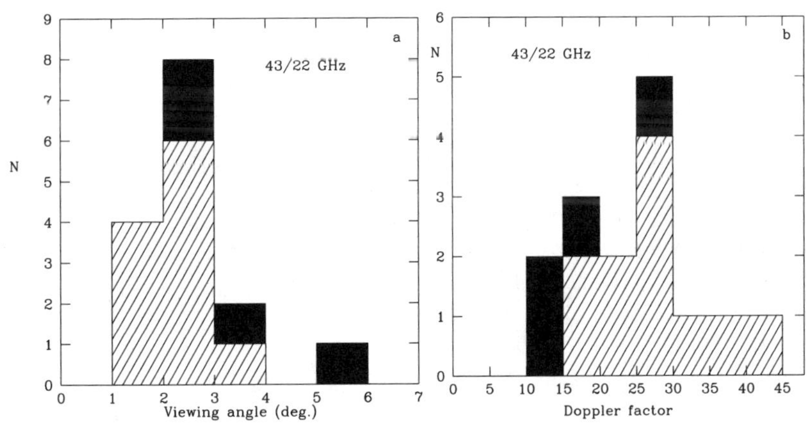

Figure 4. Distributions of jet parameters: a) angle between the jet axis and line of sight; b) Doppler factors. The intensive shade corresponds to BL Lac objects.

parameters listed in Table 1 are closer to the parameters obtained in the case of the SSC model, which favors a somewhat lower Doppler factor. However, the parameters depend strongly on the adopted ratio between the internal and external pressure of the jet. An increase in the external pressure leads to smaller opening and viewing angles. For example, for $\xi = 0.8$ the average value of α is about 0.5° and the maximum of the Doppler factor distribution shifts to higher values, 30–35. A small viewing angle restricts the possible range of apparent speeds seen close to the VLBI core (before the first bending of the jet) if the Lorentz factor of the jet flow is constant for a given source. Therefore, one of the possibilities to distinguish between the two models is to determine for a number of sources the distributions of apparent speeds for different superluminal components close to the VLBI core within each object, to find the range of variability of Γ and Θ in a typical source.

4. Conclusion

We have determined the average parameters of the jet in γ-ray blazars. The Doppler boosting factors are closer to those expected for the SSC model of γ-ray production in blazars than for the ECS model, which is favored for higher Doppler factors. However, there are still too few constraints to allow one to distinguish between the models reliably. Nevertheless, more intensive VLBI observations of a larger sample of γ-ray sources will significantly decrease the range of uncertainty. This should be possible with the VLBA when GLAST is operating.

This research was funded in part by US National Science Foundation grants AST-9802941 and AST-0098579 and NASA grants NAG5-7323 and NAG5-2508.

References

Daly, R.A., & Marscher, A.P. 1988, ApJ, 334, 539
Gabuzda, D. C., et al. 1994, ApJ, 435, 140
Hartman, R. C., et al. 1999, ApJS, 123, 79
Jorstad, S. G., et al. 2001a, ApJS, 134, 181
Jorstad, S. G., et al. 2001b, ApJ, 556, 738
Kellermann, K. I., et al. 1999, New Astron. Reviews, 43, 757
Lister, M. L., & Marscher, A. P. 1999, Astroparticle Phys., 11, 65
Lister, M. L., & Marscher, A. P. 1997, ApJ, 476, 572
Marscher, A. P., et al. 2002, ApJ, 577, 85
Marscher, A. P., & Travis, J. P. 1996, A&AS, 120, 537
Mattox J. R., et al. 2001, ApJ, 549, 906
Vermeulen, R. C., & Cohen, M. H. 1994, ApJ, 430, 467

Radio Astronomy at the Fringe
ASP Conference Series, Vol. 300, 2003
J.A. Zensus, M.H. Cohen, & E. Ros, eds.

VLBI Imaging of Seyfert Galaxies

J. S. ULVESTAD
National Radio Astronomy Observatory, P.O. Box O, Socorro, NM 87801, USA

Abstract. The advent of the Very Long Baseline Array and its phase-referencing capability have enabled milliarcsecond-scale imaging of radio sources at sub-milli-Jansky sensitivity, opening the door to high-resolution imaging of Seyfert galaxies. Over the last few years, this has led to a number of interesting new results that shed light on the characteristics of the inner cores of Seyfert galaxies. These include the following: (1) classical Seyfert galaxies with steep radio spectra have a strong tendency to exhibit radio jets at milliarcsecond scales, sometimes with considerable curvature in the jets; (2) apparent speeds of jets imaged at multiple epochs generally are considerably less than c, often in the vicinity of $0.1c$ or less, although there sometimes are faster motions seen after strong outbursts; and (3) lower luminosity Seyfert galaxies have a much stronger tendency to show flat or somewhat inverted radio spectra, usually are unresolved on milliarcsecond scales, and probably are dominated by a combination of low-radiative-efficiency accretion flows and very compact radio jets. This paper discusses some examples that illustrate these properties, and summarizes some of what we have learned about weak active galaxies on milliarcsecond scales.

1. Introduction

Seyfert galaxies were first identified as a class by Seyfert (1943), now 60 years ago. They are cousins of the more luminous quasars, having the advantage of being much more numerous, so that they include many nearby objects that can be studied at high linear resolution. Seyfert galaxies exhibit forbidden optical emission lines of moderate width (a few hundred $\mathrm{km\,s^{-1}}$), while their permitted emission lines may be very broad (Seyfert 1 galaxies, with line widths of a few thousand $\mathrm{km\,s^{-1}}$) or of widths similar to the forbidden lines (Seyfert 2 galaxies). This has been explained generally by a unified scheme in which Seyfert galaxies contain an obscuring nuclear torus or disk that is seen pole-on in Seyfert 1 galaxies or edge-on in Seyfert 2 galaxies (e.g., Antonucci 1993). Most of the observational manifestations of Seyfert galaxies, certainly in the inner few milliarcseconds, are thought to be powered by central supermassive black holes that typically have masses in the range of $\sim 10^7 M_\odot$ to $\sim 10^8 M_\odot$ (e.g., Wandel, Peterson, & Malkan 1999).

2. Scientific Goals and Context

There are a number of scientific problems that can be addressed by imaging Seyfert galaxies on milliarcsecond scales, a realm that is just beginning to be explored in these objects. Among the key questions are the following:

1. Why is there such a wide range of spectral properties in active galactic nuclei (AGNs), with radio/optical luminosity ratios spanning many orders of magnitude?

2. What do Seyfert galaxies reveal about the formation and fueling of supermassive black holes and radio jets?

3. What are the observational manifestations of supermassive black holes in nearby galaxies?

The above questions can be asked in the context of several currently unsolved problems in AGN physics. Specifically, one current question of interest is the assessment of the differences between radio-quiet and radio-loud AGNs. Are there two populations or a continuous distribution, and what are the parameters that control the relative strengths in different spectral regimes? Another question is that of the general nature and physics of the low-luminosity AGNs (LLAGNs). Here, we would like to know how LLAGN accretion and jet-formation processes differ from radio galaxies and quasars.

The general areas of interest lead to a few specific observations of Seyfert galaxies that may be made with VLBI arrays such as the Very Long Baseline Array (VLBA):

1. *Jet properties and speeds:* Are Seyfert and LLAGN jets intrinsically weak and slow, or quenched by gas in their nuclei? Do VLA and VLBI-scale jets have the same symmetry axes?

2. *Compact radio-source spectra and luminosities:* Are these quantities consistent with various flavors of accretion models, or jet models, or is a combination required?

3. *Radio core sizes:* For nearby galaxies at distances of tens of megaparsecs, classical models of advection-dominated accretion flows, or ADAFs (e.g., Narayan, Mahadevan, & Quataert 1998) predict sizes of tens of microarcseconds. Models in which the radio emission is dominated by compact jets (e.g., Falcke 1999) predict sizes of hundreds to thousands of microarcseconds. Can the two be distinguished by means of VLBI imaging?

4. *Location of gas relative to the AGNs:* The location of various parcels of gas can be studied by imaging H_2O mega-maser emission (Greenhill et al. 2003), free-free absorption (Walker et al. 2000), or H I absorption (Mundell et al. 2003). Discussion of this issue is beyond the scope of the current contribution, and the reader is referred to the cited references instead.

3. Jet Properties and Speeds

A few Seyfert galaxy jets have been imaged at multiple epochs, so that their jet speeds can be measured on parsec scales. Examples are shown in Figure 1, taken from Ulvestad et al. (1999). Here, the cores of the Seyfert galaxies Mrk 231 and Mrk 348 are shown at two epochs from early VLBA imaging; the respective relative component speeds are $\beta_{\rm app} = 0.14 \pm 0.052$ and 0.074 ± 0.035, where $\beta_{\rm app} \equiv v_{\rm app}/c$. These values appear fairly typical of most Seyfert galaxies measured to date, although objects in outburst, such as III Zw 2 (Brunthaler et al. 2000) and Mrk 348 (Peck et al. 2003) may display apparent speeds on sub-parsec scales that are (at least) a substantial fraction of the speed of light.

A summary of published measurements of Seyfert jet speeds is given in Table 1. In contrast to the Seyfert galaxies, radio galaxies and radio-loud quasars generally have highly relativistic jets on scales of a few parsecs to hundreds of parsecs. Since the properties of nuclear gas in Seyfert galaxies and the more powerful objects are fairly similar (as indicated by emission-line measurements), it seems likely that Seyfert galaxy jets initially are much less powerful relative to the nuclear environment, and may even have different acceleration mechanisms.

Table 1. Published Jet Speeds in Seyfert Galaxies

Galaxy	$\beta_{\rm app}$	Scale	Comment	Reference
III Zw 2	< 0.04	0.2 pc	Quiescent	Brunthaler et al. (2000)
	> 1.25	0.4 pc	Outburst	Brunthaler et al. (2000)
Mrk 348	0.07	0.5 pc	Start of outburst	Ulvestad et al. (1999)
	~ 0.4	0.5 pc	Outburst	Peck et al. (2003)
NGC 1068	< 0.08	21 pc	Quiescent	Roy et al. (2000)
NGC 4151	< 0.14	7 pc	Quiescent	Ulvestad et al. (1998)
	< 0.05	7 pc	Quiescent	Ulvestad et al., in prep.
Mrk 231	0.14	1 pc	Moderate outburst	Ulvestad et al. (1999)
NGC 5506	< 0.25	3.4 pc	Quiescent?	Roy et al. (2001)

At present, there is no obvious systematic difference between objects of different Seyfert classes in the few objects studied. In a large sample, we might expect that Seyfert 2 galaxies would have larger apparent jet speeds as a class; their more edge-on tori should translate into jets that have a larger projection in the plane of the sky than for Seyfert 1 galaxies. However, this would be true only if the small-scale jet axes truly indicate the orientation of the nuclear torus.

Recently, it has been shown conclusively that VLA-scale jets in Seyfert galaxies essentially are randomly oriented with respect to their host galaxy major axes (Kinney et al. 2000). This generally is thought to indicate that the central tori are misaligned with respect to the galaxy disks, and therefore had a different origin. However, it is not necessarily the case that the VLA-scale jet must indicate the direction of the torus axis, since there often is significant curvature between milliarcsecond and arcsecond scales. For example, Mrk 231 shows a position angle difference of 65° between scales of 1 pc and tens of parsecs (Ulvestad, Wrobel, & Carilli 1999) and NGC 5506 has a jet bend of 90° at a few parsecs from the nucleus (Roy et al. 2001). The beautiful 1.4 GHz contin-

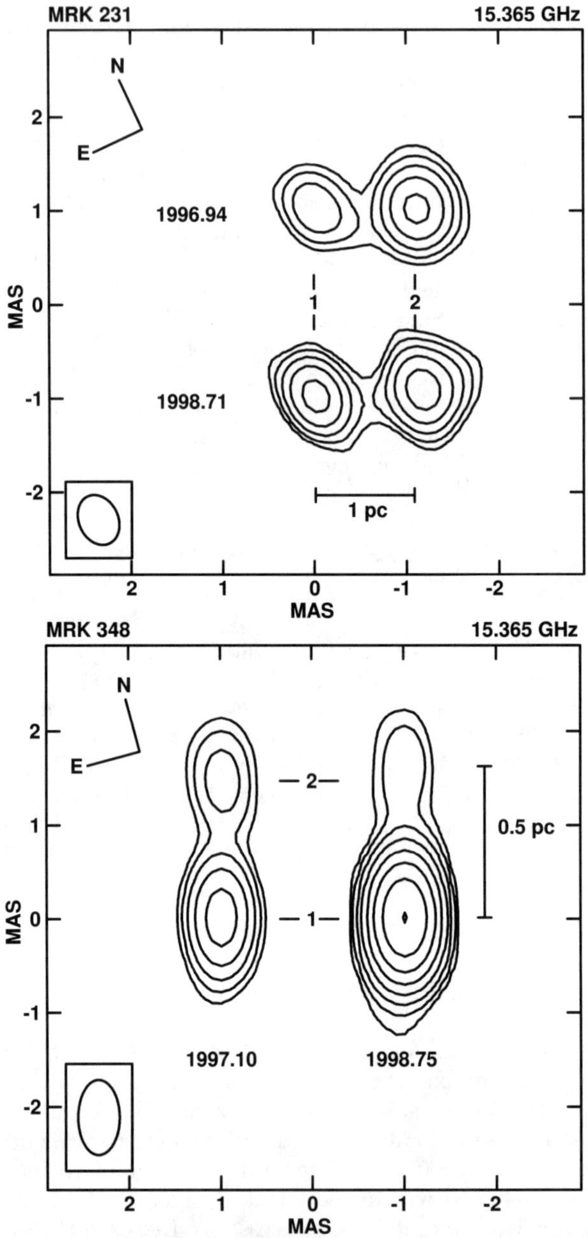

Figure 1. Two-epoch VLBI images of the Seyfert galaxies Mrk 231 (*top*) and Mrk 348 (*bottom*) are shown, *(taken from Ulvestad et al., 1999, 517, L81)*. The apparent speeds corresponding to the small relative motions of the two components are $\sim 0.1c$ within the central parsec of each galaxy.

uum image of NGC 4151 by Mundell et al. (2003) demonstrates that a jet which looks straight in VLA and MERLIN images (e.g., Mundell et al. 1994) actually contains a number of remarkable changes in jet direction when viewed with milliarcsecond resolution, possibly due to interactions with narrow-line gas clouds. Given that nuclear tori can be strongly warped by radiation on sub-parsec scales (Pringle 1997), we must be wary of inferring the directions of the central tori even from milliarcsecond-scale jet directions.

4. Compact Radio-source Spectra and Luminosities

Significant samples of LLAGNs, including both Seyfert and LINER galaxies, have been imaged with the VLBA in order to determine whether they contain milliarcsecond-scale cores that must be powered by supermassive black holes (Falcke et al. 2000; Nagar et al. 2002). In virtually all objects that have unresolved cores on arcsecond scales, a compact radio source with a brightness temperature $T_B \geq 10^8$ K has been detected, a sure indicator of a nucleus powered by a massive black hole.

Classical Seyfert galaxies typically have steep radio spectra, indicative of optically thin synchrotron radiation; no more than $\sim 10\%$ of such objects appear to have flat or inverted spectra (Ulvestad & Wilson 1989). In the classical Seyfert galaxy NGC 1068, a weak flat-spectrum radio core has been resolved by the VLBA. This core has multiple components with brightness temperatures of $\sim 10^6$ K, interpreted as direct or reflected emission from the nuclear torus itself (Gallimore, Baum, & O'Dea 1997). However, efforts to find similar objects by VLBA imaging of flat-spectrum cores in other classical Seyfert galaxies have failed thus far (Wilson et al. 1998; Mundell et al. 2000), indicating that most of these objects are likely to have flat spectra due to synchrotron self-absorption. In the classical Seyfert NGC 4151, a high-sensitivity VLBI observation in 2002, using the VLBA together with several 100 m class telescopes, has revealed a flat-spectrum 2 mJy radio core at 15 GHz. That core is shown here in Figure 2; it appears to be the AGN and the origin of the two-sided jet seen on larger scales, but shows no evidence of emission from a nuclear torus (Ulvestad et al., in preparation). Comparison of this image with an 8.4 GHz image made from data taken in 1998 is under way, in order to measure the apparent speed on a 0.1-pc scale.

Surprisingly, within the lower-luminosity Palomar Seyfert sample (Ho, Filippenko, & Sargent 1997), it appears that roughly half of the objects in an optically selected sample have flat or inverted radio spectra (Ulvestad & Ho 2001a). In these objects, which appear to be radiating far below their Eddington luminosities, the accretion flows almost certainly have low enough densities so that they have low radiative efficiencies, Then, synchrotron radiation from hot electrons in the accreting gas can be self-absorbed, yielding a slightly inverted spectrum at radio frequencies in the Gigahertz range (Narayan et al. 1998). Ulvestad & Ho (2001b) used the VLBA to image three of the stronger LLAGNs with inverted spectra at radio frequencies ranging from 1.6 GHz to 8.4 GHz. In all three cases, they found unresolved sources with radio spectra that increase gradually with frequency, similar to the spectrum of the radio source Sgr A* at the center of our own galaxy.

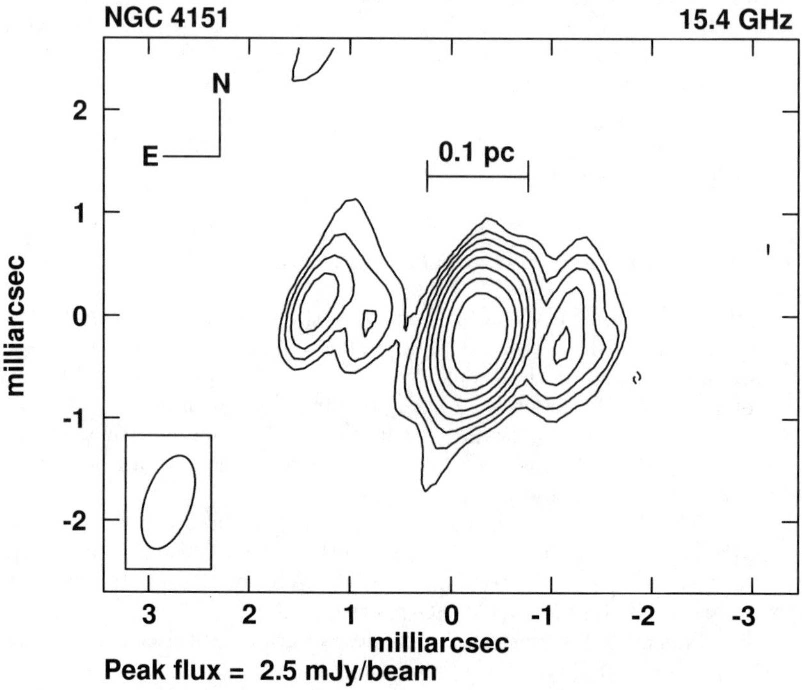

Figure 2. VLBI image of the AGN in NGC 4151, made using the VLBA, phased VLA, Green Bank Telescope, and Effelsberg at 15 GHz (Ulvestad et al., in preparation). Contour levels are incremented by factors of $2^{1/2}$, beginning at 3 times the root-mean-square noise of 45 μJy beam^{-1}.

One of the well-known problems of classical ADAF models is that the radio emission in LLAGNs and Sgr A* is observed to be far stronger relative to the X-ray emission than is predicted by the models. This problem can be "fixed" if a large fraction of the observed radio emission comes from a compact radio jet rather than from the accretion flow itself (Falcke & Markoff 2000; Falcke 2001). The jet models seem to predict that the radio emission should become optically thin and turn over above a few Gigahertz. However, Anderson & Ulvestad (in preparation) have used the VLBA to image the LLAGNs from Ulvestad & Ho (2001b) at frequencies up to 43 GHz, and found that the radio spectra continue to be flat or rising with frequency, in contradiction to the current jet models.

Ho & Peng (2001) and Terashima & Wilson (2003) have used high-resolution optical and X-ray imaging to attempt to isolate the active nuclei of Seyfert galaxies and LLAGNs, and to divide them into radio-quiet and radio-loud objects. Terashima & Wilson studied the radio/X-ray ratio using the quantity $R_X \equiv \nu L_\nu(5\,\mathrm{GHz})/L_X(2-10\,\mathrm{keV})$, useful even for heavily obscured nuclei. They found that $\log R_X = -4.5$ is the rough dividing line between radio-quiet and radio-loud objects. When compact radio flux is used, most of their Seyfert galaxies have $\log R_X \sim -5$ and are in the radio-quiet category, whereas using the total radio flux would make more of these objects radio loud. The general trend found by Terashima & Wilson (2003), using the VLBI core powers from some of the references cited above, is that most LLAGNs are radio loud. These LLAGNs almost certainly have accretion flows with low radiative efficiency. Therefore, the implication is that the central power sources in such objects put a significantly larger fraction of their total energy output into radio emission than in galaxies with classical accretion flows.

5. Radio Core Sizes

Radio emission from low-efficiency accretion flows is thought to originate very close to the central black holes, perhaps within a few to a few tens of Schwarzschild radii (Mahadevan 1997). In contrast, emission from compact jets connected to these flows is thought to come from scales a factor of ~30–40 times larger, up to as much as 1000 Schwarzschild radii (Falcke & Biermann 1999). Ulvestad & Ho (2001b) found upper limits of 10^3–10^4 Schwarzschild radii from their VLBA imaging of three LLAGNs, and these limits have been extended downward by factors of ~ 5 by 43 GHz VLBA imaging of the same objects (Anderson & Ulvestad, in preparation). The upper limits now are getting uncomfortably small for the compact jet models. Putting this together with the problem mentioned above, that the radio spectra do not turn over above ~ 10 GHz, it is becoming clear that the accretion/jet models used to date may need more work in order to account for the compact radio emission from LLAGNs.

Since the Earth is not going to get any bigger, what can we do in order to attempt to measure true sizes of the radio sources in low-luminosity Seyferts and AGNs? Space VLBI is one option, but the relative weakness of the radio cores in most Seyfert galaxies makes this only a fond hope for the relatively distant future. But a more promising technique may be to use the interstellar medium in our Galaxy to "image" the radio cores. It recently has become clear that intraday variability in strong extragalactic radio sources is caused by scintillation in the interstellar medium (Rickett et al. 2001; Jauncey & Macquart 2001; Dennett-Thorpe & de Bruyn 2002), providing a means of estimating radio core sizes in the vicinity of a few microarcseconds. Although measuring intraday variability in milli-Jansky-strength sources will be a real challenge to modern instrumentation, it may be possible with the current VLA, and certainly is well within the reach of the more sensitive Expanded VLA.

The National Radio Astronomy Observatory is a facility of the National Science Foundation operated under cooperative agreement by Associated Universities, Inc. I thank the many collaborators whose work has been summarized in this contribution.

References

Antonucci, R. 1993, ARA&A, 31, 473

Brunthaler, A., et al. 2000, A&A, 357, L45

Dennett-Thorpe, J., & de Bruyn, A. G. 2002, Nature, 415, 57

Falcke, H. 1999, in ASP Conf. Ser. Vol. 186, The Central Parsecs of the Galaxy, ed. H. Falcke, A. Cotera, W. J. Duschl, F. Melia, & M. J. Rieke (San Francisco: ASP), 148

Falcke, H. 2001, Reviews in Modern Astronomy, 14, 15
Falcke, H., & Biermann, P. L. 1999, A&A, 342, 49
Falcke, H., & Markoff, S. 2000, A&A, 362, 113
Falcke, H., et al. 2000, ApJ, 542, 197
Gallimore, J. F., Baum, S. A., & O'Dea, C. P. 1997, Nature, 388, 852
Greenhill, L. J., et al. 2003, ApJ, 582, L11
Ho, L. C., Filippenko, A. V., & Sargent, W. L. W. 1997, ApJS, 112, 315
Ho, L. C., & Peng, C, Y. 2001, ApJ, 555, 650
Jauncey, D. L., & Macquart, J.-P. 2001, A&A, 370, L9
Kinney, A. L., et al. 2000, ApJ, 537, 152
Mahadevan, R. 1997, ApJ, 477, 585
Mundell, C. G., et al. 1994, MNRAS, 272, 355
Mundell, C. G., et al. 2000, ApJ, 529, 816
Mundell, C. G., et al. 2003, ApJ, 583, 192
Nagar, N. M., et al. 2002, A&A, 392, 53
Narayan, R., Mahadevan, R., & Quataert, E. 1998, in Theory of Black Hole Accretion Disks, ed. M. A. Abramowicz, G. Björnsson, & J. E. Pringle (Cambridge: Cambridge University Press), 148
Peck, A. B., et al. 2003, ApJ, 590, in press (astro-ph/0303423)
Pringle, J. E. 1997, MNRAS, 292, 136
Rickett, B. J., et al. 2001, ApJ, 550, L11
Roy, A. L., et al. 2000, in EVN Symposium 2000, Proceedings of the 5th European VLBI Network Symposium, ed. J. E. Conway, A. G. Polatidis, R. S. Booth, & Y. M. Pihlström (Onsala: Onsala Space Observatory), 7
Roy, A. L., et al. 2001, in IAU Symposium 205, Galaxies and Their Constituents at the Highest Angular Resolutions, ed. R. T. Schilizzi, S. N. Vogel, F. Paresce, & M. S. Elvis (San Francisco: ASP), 70
Seyfert, C. K. 1943, ApJ, 97, 28
Terashima, Y., & Wilson, A. S. 2003, ApJ, 583, 145
Ulvestad, J. S., et al. 1998, ApJ, 496, 196
Ulvestad, J. S., et al. 1999, ApJ, 517, L81
Ulvestad, J. S., & Ho, L. C. 2001a, ApJ, 558, 561
Ulvestad, J. S., & Ho, L. C. 2001b, ApJ, 562, L133
Ulvestad, J. S., & Wilson, A. S. 1989, ApJ, 343, 659
Ulvestad, J. S., Wrobel, J. M., & Carilli, C. L. 1999, ApJ, 516, 127
Walker, R. C., et al. 2000, ApJ, 530, 233
Wandel, A., Peterson, B. M., & Malkan, M. 1999, ApJ, 526, 579
Wilson, A. S., et al. 1998, ApJ, 505, 587

Radio Astronomy at the Fringe
ASP Conference Series, Vol. 300, 2003
J.A. Zensus, M.H. Cohen, & E. Ros, eds.

Microquasars

V. Dhawan
National Radio Astronomy Observatory, P.O. Box O, Socorro, NM 87801, USA

Abstract. Microquasars are here taken to be accreting stars with relativistic jets. The accretion disks emit thermal X-rays of a few keV, often with a power-law tail to higher energies, attributed to Compton up-scattering by relativistic electrons. These electrons in turn may produce synchrotron emission - radio, IR, even X-rays. About a dozen jet sources are now established in the Galaxy, with some 30 more neutron stars and black holes detected in sporadic radio monitoring, out of ~300 X-ray binaries. Many are found by high-energy satellites, and are essentially transients, though timescales range from days to years.

Selected recent VLBA observations are reviewed. The VLBA provides rapid response to transients, by dynamic scheduling; multi-frequency coverage; and easy phase-referencing, with ~1 mJy detectability and ~1 mas astrometry. For a few objects, very rich data sets have resulted from the confluence of radio imaging, X-ray light curves from RXTE, and spectroscopy from Chandra. The time variations of disk, corona, and radio jets show the interplay of accretion and ejection in these systems. Such 'laboratory quality' data is more readily had for galactic objects, compared to distant, slowly varying AGN. The problems of imaging time-variable, moving sources await the sensitivity and snapshot coverage of the EVLA, and the New Mexico Array, respectively. SKA users can look forward to the whole iceberg.

1. Introduction

The first microquasar, SS 433, stood alone for over a decade, though several objects were known with variable X-ray and radio emission. In the early and mid 90's, X-ray satellites supported by radio light curves and imaging, identified a distinct subset of X-ray binaries with relativistic ejecta (Mirabel & Rodriguez 1999). Accreting compact stars are now proven owners of strong magnetic fields, rapid rotation, strong gravity, high radiation pressure and gas pressure, and relativistic flow. Insights from this area are (bravely?) being applied to the ultra-luminous X-ray objects in external galaxies (Georganopoulos, Aharonian, & Kirk 2002); to jets in AGN and quasars (Marscher et al. 2002); and gamma ray bursts (Paczynski 1998). I focus on a few observational, partly contrasting cases. Theory and much detail is available in the references and in proceedings of the Hjellming Memorial Conference (ed. Castro-Tirado, Greiner, & Paredes 2001). Exciting prospects abound: Corbel et al. (2002) recently reported superluminal synchrotron X-ray jets in XTE J1550−564, decelerated by interaction with the ISM. Very fast jets exist, e.g., in V4641 Sgr (Hjellming et al. 2000). And radio emission may be detectable with the SKA, from thousands of the

$\sim 10^9$ isolated NS and BH adrift in the galaxy, accreting from molecular clouds (Agol & Kamionkowski, 2002; Fender & Kuulkers, 2001)

2. Anatomy of X-ray binary/ microquasar

To set the stage, the colour graphic (from Rob Hynes, see `http://www-xray.ast.cam.ac.uk/~sav/blackholes/stellar.html`) ties the broadband spectra to the physical components of an X-ray binary. Thermal X-rays of $\sim$1keV come from the inner accretion disk; some of these photons are Compton up-scattered to give a power-law tail out to $\sim$100 keV. The location and geometry of the hot electron corona are debated, but could be the base of the radio synchrotron jet. A variety of X-ray modulations were discovered by RXTE, including kHz fluctuations from Keplerian orbits in the inner accretion disk (van der Klis, 2000) and $\sim$1 Hz oscillations from the disk/corona. Several mechanisms are proposed for the latter, but they are clearly related to the radio jet (Muno et al. 2001).

In LMXB the low mass companion fills its Roche lobe, transferring matter into an accretion disk around the compact star, e.g., Cyg X$-$1 & GRS 1915+105, (section 4.1). In the HMXB, the accretion is via capture of the wind from the massive donor, e.g., Cyg X$-$3 & SS 433, (section 4.2). Garcia et al. (2001) report under-luminous X-rays from BH accretors, compared to NS at the same $\dot{M}$, ascribed to an event horizon which swallows the influx, versus a solid re-radiating surface, respectively.

A further variable is the magnetic field—above 10^7G or so, it disrupts the inner accretion disk by causing the accreting plasma to flow along the field lines to the poles, instead of in the equatorial plane (Li & Wilson 1999). High-field NS, or pulsars, are not expected (nor observed) to produce radio jets, while weak-B NS do, e.g., Sco X$-$1, (Fomalont, Geldzahler, & Bradshaw 2001). A magnetic switch (Meier et al. 1997) may modulate the jet speed.

3. X-ray States and State Transitions

States are defined by the broadband X-ray spectra, characterised at the simplest level by increasing $\dot{M}$, illustrated by GX 339$-$4 (Fig. 15 in Corbel et al. 2000). Contrary cases exist (McCollough et al. 1999)

At very low $\dot{M}$, models for under-luminous black holes (Yuan, Markoff, & Falcke 2002) employ the ADAF and related schemes, augmented by a synchrotron jet to fit the radio emission. At modest $\dot{M}$ the 'low/hard' state is attributed to a truncated disk (innermost part missing) with a strong hot-electron corona (Esin et al. 2001; Fender 2001). This state (low X-ray counts, with prominent hard power-law) most often exhibits steady radio emission, presumably tied to the the relativistic electrons. At very high $\dot{M}$, the disk radiates strongly in the soft X-rays (Gierliński & Done 2002), near the Eddington limit of $10^{38}\,\mathrm{erg\,s^{-1}}$ for a NS; and extends in to the innermost stable orbit ($6R_g$ for non-rotating). Radio emission is not usually detected in this high/soft state, though Ballantyne, Ross, & Fabian (2002) argue for high $\dot{M}$ in radio loud AGN. Intermediate, unstable, or oscillatory states may exist at transitions between these regimes, where gas pressure, radiation pressure, and magnetic field vie for dominance (e.g., Nayakshin, Rappaport, & Melia 2000)

4. Examples

4.1. GRS 1915+105

GRS 1915+105 is a 14 $M_\odot$ black hole (Greiner, Cuby, & McCaughrean 2001), with a low-mass primary, and unstable accretion disk. Approximately once per year, (Fig. 1A) the source emerges from an extended low/hard state with X-ray and radio flares, when superluminal 'cannonballs' are ejected and decay over 3 days. Though the ejecta are hard to image with the current VLBA due to the lack of short baselines, the superluminal motion is measurable in just hours, (Fig. 1B). The bulk and pattern speeds coincide in this case. In Sco X−1 however, (Fomalont, Geldzahler, & Bradshaw 2001) the hot-spot and flow speeds differ, familiar ghosts from the extragalactic case.

Following the flares, the disk is often wildly unstable, with oscillating X-ray and radio emission (Fig. 1C,D). VLBA imaging shows a compact jet of length ∼20–50 AU, and brightness temperature $T_B \sim 10^9$ K on scales of 0.2–0.6 mas. The jet varies in ∼10–60 min, as the relativistic plasma is ejected from the inner disk and flows through it, radiating with a flat spectrum from radio to IR (Mirabel et al. 1998; Fender & Pooley 1998; Falcke & Biermann 1999).

Some other black hole systems (e.g., Cyg X−1, 1E 1740−2942, GRS 1758−258, GX 339−4, XTE J1118+48, Sgr A*) have compact radio cores in the low/hard state, without flares and transient ejecta. GRS 1915+105 is the only source known to exhibit both states. The most recent RXTE/VLBA observations capture the state transition and jet birth with unprecedented clarity, (Dhawan et al., in preparation)

4.2. SS 433

SS 433 has an unknown primary, with a high mass, possibly WR companion, and massive wind outflow. It has famously persistent, precessing, bi-directional radio jets moving at 0.26c, but is only a modest X-ray source. Hjellming & Johnston (1988) from VLA images, fit the decay of jet flux vs. distance with slowed expansion followed by free expansion $\sim 10^4$ AU downstream. Vermeulen et al. (1993) followed the moving ejecta with an image sequence from a VLB campaign unprecedented till then.

Recent MERLIN, VLA, VLBA images illustrate the spectacular range of scales present in the jets with their ∼163 day precession period. The VLA image of W50 on degree scales (Dubner et al. 1998) shows the circular supernova remnant that presumably produced the collapsed object, with 'ears' blown out by the impact of the jet plasma. The VLA+MERLIN image (Fig. 2A, from Blundell et al. 2001) shows the jets on arcsec scales, while the VLBA image (Fig. 2B, one frame of a movie) reveals, in addition to the jets, an equatorial ruff of emission from a wind with outflow velocity 300 km s^{-1} and estimated mass loss $10^{-5}\,M_\odot\,yr^{-1}$. The equatorial outflow is poorly imaged due to the lack of short VLBA baselines, but is also seen in independent VLBA images (Paragi et al. 1999). The wind was predicted by the evolutionary scenario of King, Taam,

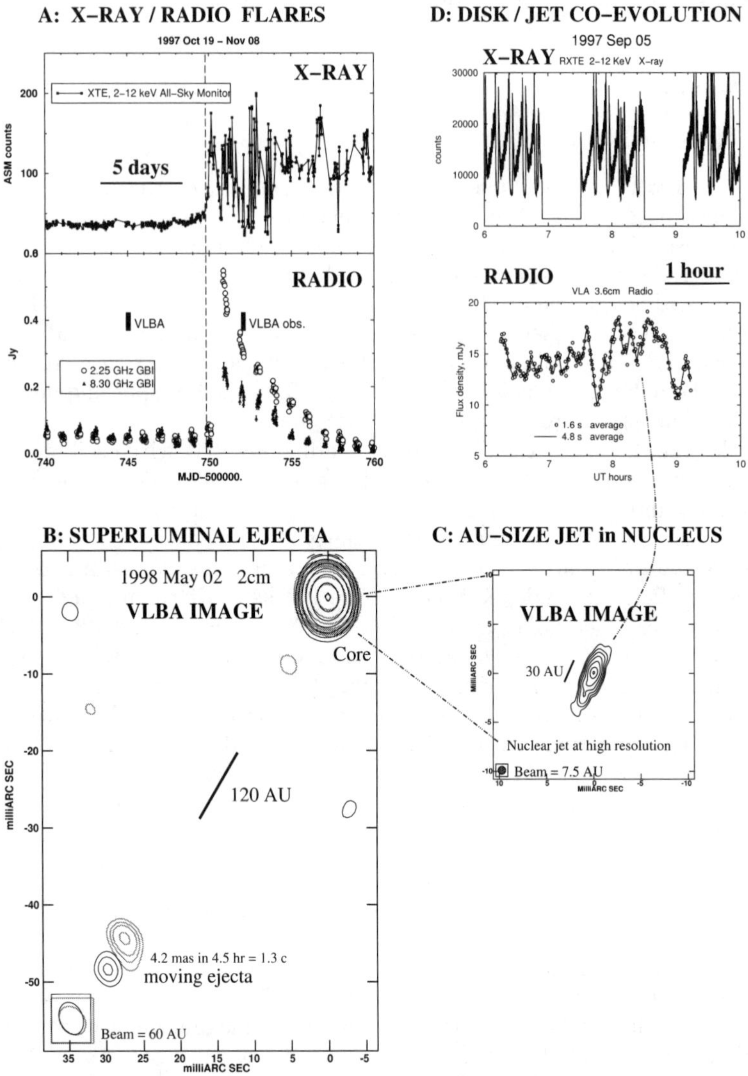

Figure 1. The disk/jet connection illustrated by GRS 1915+105.
A: X-ray state transition from low/hard to unstable disk, accompanied by a radio flare, imaged in a similar but later event, B.
B: Phase-referenced image showing the steep-spectrum 'cannonball' ejecta ~67 h after the flare, and stationary flat spectrum core.
C: High resolution image of the nucleus, showing the AU-sized jet that varies on timescales of ~10 min, in response to the injection and flow-through of bulk relativistic plasma.
D: The plasma originates with the ejection of matter from the inner disk, seen as soft X-ray dips.

& Begelman (2000), with the weak X-rays explained by the mass diversion to outflow instead of accretion.

X-ray data tightly constrain the conditions in the jets. Chandra detection of Doppler-shifted lines from many ionised atomic species, (Marshall, Canizares, & Schulz 2002), indicates that the jets are accelerated to terminal velocity of $0.26c$, within 0.01 AU from the star, where the temperature is 10^{7-8}K. The electron density and density gradient in the expanding jet are measured, and the mechanical luminosity of the jet is $\sim 10^{38}\,\mathrm{erg\,s^{-1}}$, some 10^3 times the X-ray emission. Next, the limit on the time delay between the red and blue-shifted optical lines, restricts the size of the optical emission region with $T \sim 10^4$ K to < 100 AU—the plasma has cooled at constant bulk velocity. Further, spatially resolved Fe spectra (Migliari, Fender, & Méndez 2002) indicate reheating of plasma to $T \sim 10^7$ K, 10^4 AU downstream, presumably by conversion of the abundant jet mechanical energy. Finally, note that there is measured circular polarisation in this jet of baryonic, *not leptonic* plasma (Fender et al. 2000; Ruszkowski & Begelman 2002)

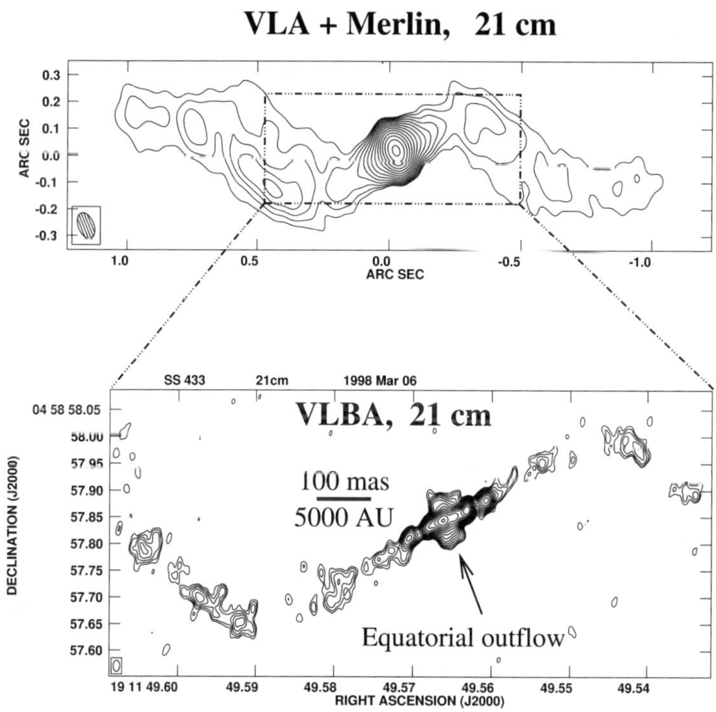

Figure 2. **A:** Precessing Jets; **B:** Equatorial Outflow in SS 433: both features have low surface brightness, and would benefit from a larger UV range than can be mustered at present. (Contours are 1, 1.4, 2%... with total flux ~500mJy. *Courtesy A.J. Mioduszewski (from Blundell et al. 2001, ApJ, 562, L79).*

5. VLBA Astrometry

Though the Hipparcos tie to VLBA/ICRF included some persistent stellar sources (Lestrade et al. 1999), X-ray transients have not been well served by astrometry, but for a few cases. Sco X−1 is one famous source with a VLBA parallax distance of 2.8±0.3 kpc (Bradshaw, Fomalont, & Geldzahler 1999). HST astrometry was used to infer the Galactocentric orbit of the XRT GRO J1655−40 (Mirabel et al. 2002). A more systematic attack is in progress, as was done for pulsars by Brisken et al. (2002). Two preliminary examples are below.

5.1. GRS 1915+105

The black hole is located to ∼1 mas. After accounting for the secular parallax due to Galactic rotation, a velocity of 20–40 km s^{-1} is found for its motion with respect to its neighbourhood, implying that this 14 $M_\odot$ black hole has received a modest kick at birth (Dhawan et al., in preparation). A more precise but finite velocity might rule out evolutionary schemes in which the massive star collapses directly to the black hole with no supernova and hence no birth kick. (Belczinski & Bulick 2002)

5.2. XTE J1118+480

This unusual X-ray nova at high Galactic latitude was observable in all wave bands. Though the radio emission lasted only 3 months, significant proper motion was measurable with the VLBA. With the optical radial velocity, the 3-dimensional orbit was found to be similar to that of halo objects (Mirabel et al. 2001). A disk origin for this binary (7 $M_\odot$ primary) is not excluded, but it would require an extreme kick in the originating supernova, and is inconsistent with the age of the system. It thus may be a wanderer ejected from a globular cluster, one of many dark halo objects that happened to brighten in a mass transfer episode.

6. Conclusion

Microquasars complement distant AGN:

- They offer a speedup, proportional to mass, of $>10^6$, comparable to the advantage to biology of *e.coli* vs. humans. However, the timescales demand new ways of scheduling and coordinating multiple observatories; and the spatial scales demand wider (u, v) range, and better snapshot coverage for radio imaging.

- Galactic objects are close, and brighter by up to 10^3 in X-rays, allowing detailed spectroscopy of the disk and jet contents. Radio imaging has milliarcsec resolution, but needs sensitivity to thermal material. The advantage in linear resolution in units of R_g per beam lies with big AGN, the best case being Sgr A*.

- For Sgr A* and many extragalactic black holes, the $\dot{M}$ is low. Accreting binaries yield more examples at high $\dot{M}$. Conditions in the accretion region are more directly measurable, unlike in most extragalactic cases—see next item.

- Unification faces more diversity: NS, accreting millisecond pulsars, magnetars, white dwarfs, in addition to BH. On the other hand, this variety may be key to controlled experiments, by comparing the effects of event horizons and and magnetic fields. Also, the physical quantities are more likely to be understood in galactic objects, e.g., M, (Bailyn et al. 1998; Galloway et al. 2002), $\dot{M}$, evolutionary history, metalicity, age, binary orbit, inclination, distance (by VLBA parallax). One can hope to measure B directly as in pulsars and CV, from cyclotron lines, (Coburn et al. 2002) or from the period derivative $\dot{P}$ (Cheng 1989). The spin period P of the central mass is measured from radio or X-ray pulses, while Fe line profiles are diagnostic of rotation and GR effects in the inner disk (Martocchia, Karas, & Matt 2000).

References

Agol, E., & Kamionkowski, M. 2002, MNRAS, 334, 553
Bailyn, C. D., et al. 1998, ApJ, 499, 367
Ballantyne, D. R., Ross, R. R., & Fabian, A. C. 2002, MNRAS, 332, L45
Belczynski, K., & Bulick, T. 2002, ApJ, 574, L147
Blundell, K. M., et al. 2001, ApJ, 562, L79
Bradshaw, C. F., Fomalont, E. B., & Geldzahler, B. J. 1999, ApJ, 512, L121
Brisken, W. F., et al. 2002, ApJ, 571, 906
Castro-Tirado, A. J., Greiner, J., & Paredes, J. M. (ed.) 2001. Proceedings of the third Microquasar Workshop, Ap&SS, 276
Cheng, A. F. 1989, ApJ, 337, 803
Coburn, W., et al. 2002, ApJ, 580, 394
Corbel, S., et al. 2000, A&A, 359, 251
Corbel, S., et al. 2002, Science, 298, 196
Dubner, G. M., et al. 1998, AJ, 116, 1842
Esin, A. A., et al. 2001, ApJ, 555, 483
Falcke, H., & Biermann, P. L. 1999, A&A, 342, 49
Fender, R. P. 2001 MNRAS, 322, 31
Fender, R. P., et al. 2000, ApJ, 530, L29
Fender, R. P., & Kuulkers, E. 2001, MNRAS, 324, 923
Fender, R. P., & Pooley, G. G. 1998, MNRAS, 300, 573
Fomalont, E. B., Geldzahler, B. J., & Bradshaw, C. F. 2001, ApJ, 558, 283
Galloway, D. K., et al. 2002, ApJ, 576, L137
Garcia, M. R., et al. 2001, ApJ, 553, L47
Georganopoulos, M., Aharonian, F. A., & Kirk, J. G. 2002, ApJ, 388, L25
Gierliński, M., & Done, C. 2002, MNRAS, 331, L47
Greiner, J., Cuby, J. G., & McCaughrean, M. J. 2001, Nature, 414, 522
Hjellming, R. M., et al. 2000, ApJ, 544, 977
Hjellming, R. M., & Johnston, K. J. 1988, 328, 600
King, A. R., Taam, R. E., & Begelman, M. C. 2000, ApJ, 530, L25
Lestrade, J.-F., et al. 1999, A&A, 344, 1014

Li, J., & Wilson, G. 1999, ApJ, 527, 910
Marscher, A. P., et al. 2002, Nature, 417, 625
Marshall, H. L., Canizares, C. R., & Schulz, N. S. 2002, ApJ, 564, 941
Martocchia, A., Karas, V., & Matt, G. 2000, MNRAS312, 817
McCollough, M. L., et al. 1999, ApJ, 517, 951
Meier, D. L., et al. 1997, Nature, 388, 350
Migliari, S., Fender, R. P., & Méndez, M. 2002, Science, 297, 1673
Mirabel, I. F., et al. 1998, A&A, 330, L9
Mirabel, I. F., et al. 2001, Nature, 413, 139
Mirabel, I. F., et al. 2002, A&A, 395, 595
Mirabel, I. F., & Rodríguez, L. F. 1999, ARA&A, 37, 409
Muno, M. P., et al. 2001, ApJ, 556, 515
Nayakshin, S., Rappaport, S., & Melia, F. 2000, ApJ, 535, 798
Paczynski, B. 1998, ApJ, 494, L45
Paragi, Z., et al. 1999, A&A, 348, 910
Ruszkowski, M. & Begelman, M. C. 2002, ApJ, 573, 485
van der Klis, M. 2000, ARA&A, 38, 717
Vermeulen, R. C., et al. 1993, A&A, 270, 177
Yuan, F., Markoff, S., & Falcke, H. 2002, A&A, 383, 854

Radio Astronomy at the Fringe
ASP Conference Series, Vol. 300, 2003
J.A. Zensus, M.H. Cohen, & E. Ros, eds.

3C 84

R. C. Walker
National Radio Astronomy Observatory, P.O. Box O, Socorro, NM, USA

Abstract. VLBI studies of 3C 84 show a bright, complex structure that is expanding sub-luminally and whose origin extrapolates back to the major increase in flux density that occurred around 1960. At high frequencies, the structure is two sided. The relative brightness and size of the northern and southern features can be explained in terms of a twin jet inclined about 44° to the line-of-sight and moving at about $0.37c$. The smaller and weaker feature to the north is free-free absorbed at low frequencies while the southern feature is not. The free-free absorption constrains the properties of ionized gas in the region of the accretion disk feeding the black hole that is thought to power the source.

1. Introduction

The radio source 3C 84 is in the center of NGC 1275, the dominant galaxy of the Perseus cluster. It has a very bright central radio source with most of the emission coming from a core and jets on scales of about 15 milliarcseconds (mas), which is about 5 pc for the Hubble constant of $75\,\mathrm{km\,s^{-1}\,Mpc^{-1}}$ used in this contribution. There are also larger scale jets and diffuse emission extending out into the cluster (Pedlar et al. 1990). These large scale structures are in regions of low X-ray emission. It seems that the hot, X-ray emitting gas has been pushed away by the radio lobes (Boehringer et al. 1993). The systemic velocity of 3C 84 is $5260\,\mathrm{km\,s^{-1}}$. Overlayed on the elliptical galaxy is what appears to be a spiral or remnants thereof. That system has a velocity of $8200\ \mathrm{km\,s^{-1}}$ and is seen in absorption against the lower velocity emission. Whatever that system is, it must be in the foreground and is falling toward the main system at about $3000\,\mathrm{km\,s^{-1}}$.

As one of the brightest compact sources in the sky, 3C 84 has been a target of VLBI observations since very early in the history of VLBI. But it is also one of the most complex of the bright sources, so it has always proven difficult to image, even with modern arrays. The structure is well represented in the 23 GHz image of Figure 1 (from Walker & Anantharamaiah 2003). That structure is dominated by a core that is slightly resolved and a bright southern structure that has been expanding over the years. If the size of the southern feature is extrapolated back in time, it appears to have been formed at the core at about the time that the source started to increase in flux density from below 10 Jy, as seen in the earliest cm radio measurements before 1960, to the 50 to 60 Jy eventually seen. It is now decreasing again, but is still between 10 and 20 Jy.

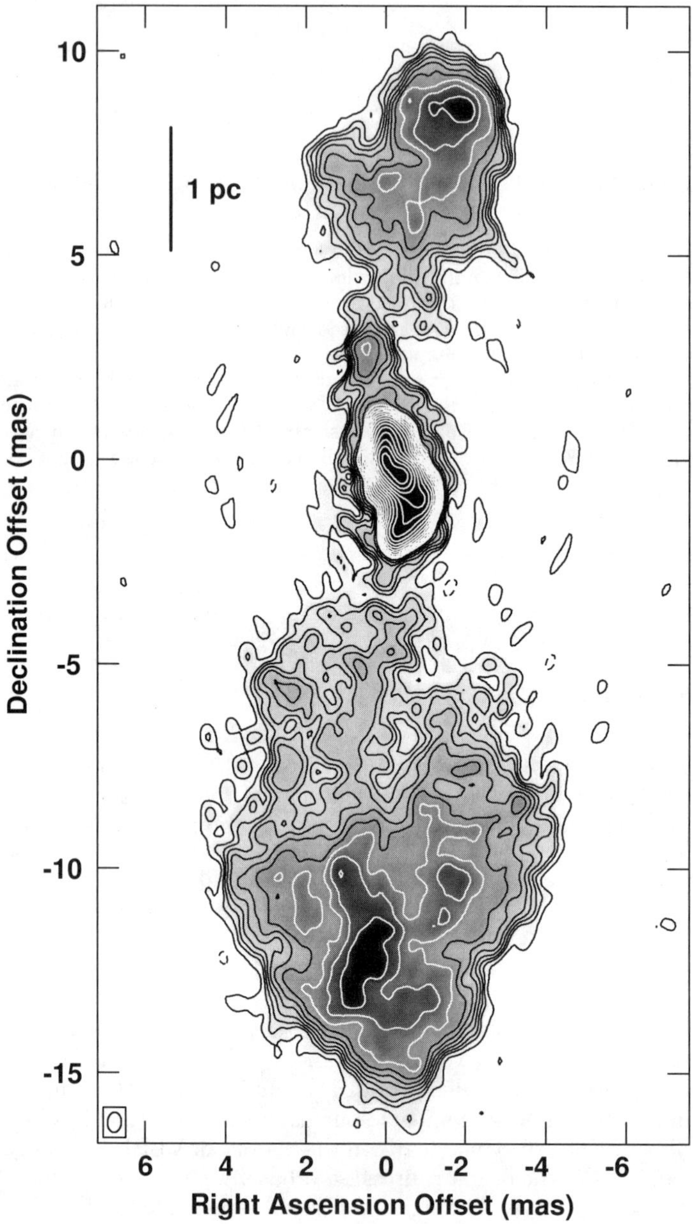

Figure 1. An image of 3C 84 at 23 GHz showing the bright core region and the structures north and south of the core thought to be related to the far and near side jets respectively. The contour levels start with -1, 1, 2, 2.83, 4, 5.66 mJy beam^{-1} and increase from there by factors of $\sqrt{2}$. The peak flux density is 1.9 Jy beam^{-1}. The resolution is 0.49×0.35 mas elongated in position angle $-4.3°$.

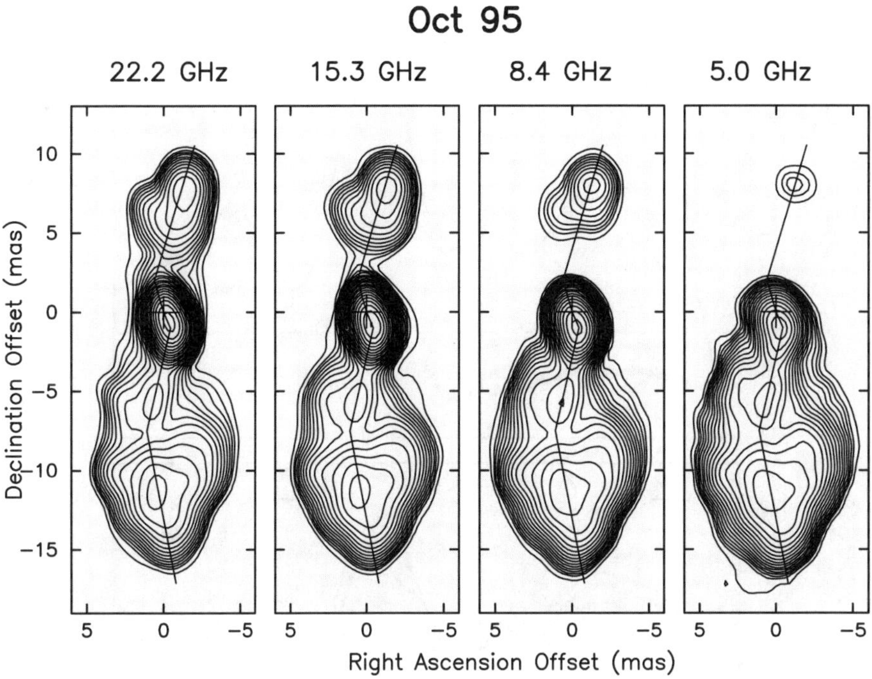

Figure 2. Images of 3C 84 at 22.2, 15.3, 8.4, and 5.0 GHz in 1995 October. All images are shown at the same resolution, 1.6 × 1.2 mas, elongated north-south. The contour levels start with 5, 10, 14, 20 mJy beam^{-1} and increase from there by factors of $2^{1/2}$. The southern feature has a uniform spectral index of about −0.7. In contrast, the northern feature has a strongly inverted spectrum indicating free-free absorption. At 2.3 GHz, at slightly lower resolution, the northern feature is completely missing. The segmented lines show where the profiles in Figure 3 were made.

2. Free-free Absorption

The VLBA first science observations, made at 8.4 GHz, revealed the presence of a northern feature that looks a lot like the southern feature, except that it is smaller and less bright (Walker, Romney, & Benson 1994). This presumed counter-jet was also seen in 22 GHz Global Network data from a couple of years earlier, but reduced at the time of the VLBA first science (Vermeulen, Readhead, & Backer 1994). When the 8.4 and 22 GHz images were compared in detail, it became clear that something interesting was happening. The southern feature was stronger at 8.4 GHz, as is typical for jets. But the northern feature was much stronger at 22 GHz. This was modeled as being the result of free-free absorption of the counter-jet by ionized material associated with the accretion disk.

In January and October of 1995, both groups that had done the earlier observations combined to make observations at all bands from 2.3 GHz to 22 GHz (Walker et al. 2000). The October images at 5.0, 8.4, 15.3, and 22.2 GHz are

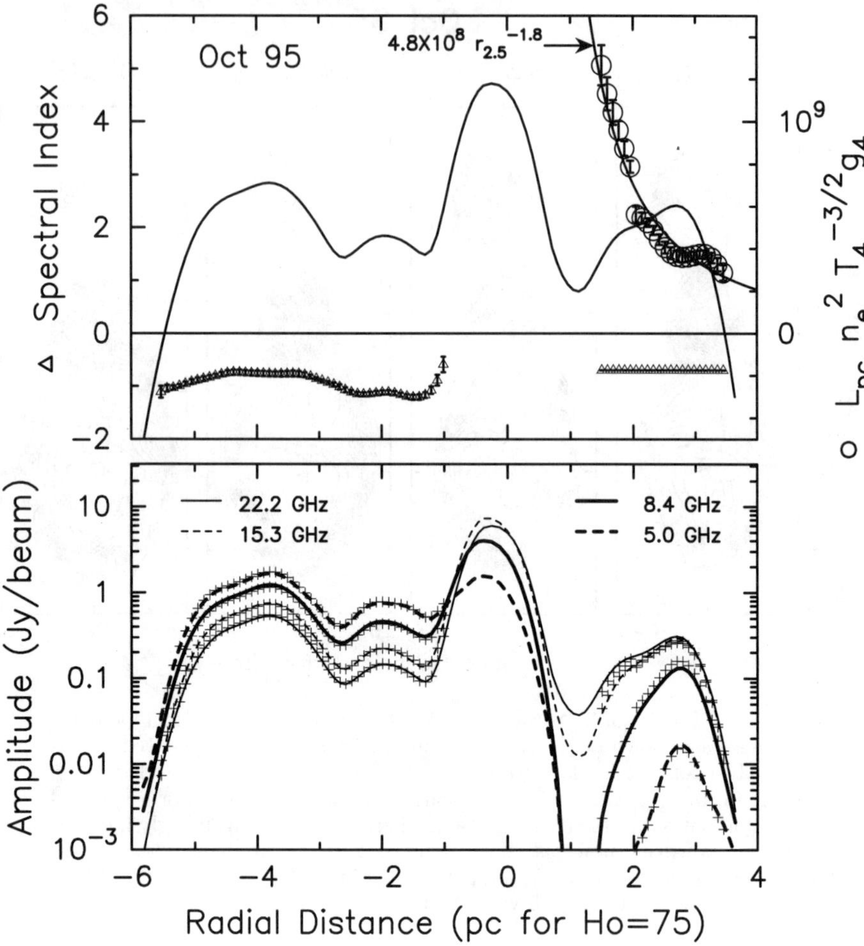

Figure 3. In the bottom panel, the lines are profiles along the ridge line of the northern and southern jets. The crosses are the results of a least squares fit. For the southern jet, the fit is for spectral index, with the results shown as triangles in the upper panel. For the northern jet, a fixed spectral index of −0.7 was assumed, and the fit is for the free-free absorption, with the results shown as circles in the top panel. Note that $L_{\rm pc}$ is the path length through the absorbing medium in parsecs, n_e is the electron density, T_4 is the temperature in units of 10^4 K and g_4 is the Gaunt factor for a temperature of 10^4 K. The profile used for orientation in the top panel is for 22 GHz. The top panel also shows the result of a power law fit to the absorption as a function of core distance.

shown in Figure 2. The free-free absorption idea nicely explains the results. The 1995 observations showed a spectral index of 4 between the 5 and 8.4 GHz images of the counter-jet, leaving free-free absorption as the only viable explanation. Also, the expected spectrum from free-free absorption fits all bands well as shown in Figure 3, taken from Walker et al. (2000). The northern feature is well resolved so the observations allow the absorption, which effectively gives the Emission Measure (EM), to be imaged over a two dimensional area. The main spatial variation is in the form of a gradient that is radial relative to the core. Thus the amount of absorbing material appears to decrease with core distance. Observations, also made in 1995 October at lower frequencies, show that the absorption continues to scales of at least 80 mas (Silver, Taylor, & Vermeulen 1998).

The situation where the northern feature, presumably the counter-jet, is absorbed while the southern feature, presumably the near-side jet, is not is easily explained if the absorbing gas is associated with the accretion disk feeding the central black hole. In the most simple models, the disk is not expected to be ionized at the radii of a few parsecs where the absorption is occuring. But clearly there is ionized gas present. It could be in a corona or wind above the disk, but the geometry is not established.

It appears that the northern feature is not absorbed much at 22 GHz. As a result, this source makes a very good case in which simple beaming arguments can be used to derive the actual jet speed and angle to the line-of-sight. This is aided very much by the drastic change in the jet brightness at the end of the main southern feature—the feature that extrapolates back to the big brightness increase in about 1960. It is reasonable to assume that the northern feature is also the result of that same activity. The situation is simplified by the fact that the measured velocities are relatively modest—a few tenths of the speed of light. Unlike in the superluminal sources, the counter-jet is not expected to be orders of magnitude weaker than the front-side jet. With an apparent speed of the southern jet of $0.34c$ and a length ratio of 1.69 for the two jets, we derive an angle to the line-of-sight of 44° and a true jet speed of $0.37c$. The brightness ratio of the northern and southern features, which is about 4, gives a fairly similar result.

In January 2000, we made another set of multi-frequency VLBI observations of 3C 84 to look for temporal changes in the free-free absorption. Those data are still being reduced, but the initial results show that any changes are small. The overall expansion of the source is seen, as expected. But at 8.4 GHz, the region that is strongly absorbed has not changed. More of the northern feature is seen, consistent with the idea that the lengthening jet is coming out from under the absorbing screen.

The ionized gas responsible for the free-free absorption could, under some conditions, be a source of stimulated recombination line emission. An attempt to detect such a line was made by Walker & Anantharamaiah (2003). An upper limit of about 15% of the continuum was placed on any such line emission. A non-detection is not surprising because the intense radio radiation field near the jets is expected to cause both saturation and line broadening by radiation damping (Shaver 1978). The absorbing region would have to be at the very least twice as far from the counter-jet as where the line-of-sight is expected to encounter the accretion disk, and probably much farther, to avoid these effects. If it were not for saturation and radiation damping, the non-detection would imply that the gas has a wide velocity spread or high offset velocity, or is in a very dense region.

3. Summary

The compact radio source 3C 84 has a complex, two-sided structure. The bright features appear to be the result of a period of enhanced activity that started shortly before 1960. The observed jet speed can be used, along with the relative lengths and brightnesses of the the two jets, to derive the geometry of the system. The jets appear to be at about 44 degrees to the line of sight and have a true jet velocity of about $0.37c$. The northern feature is strongly free-free absorbed while the southern feature is not. This situation can be explained if the absorbing material is associated with the accretion disk that is expected to be feeding the central black hole. Thus the free-free absorption places interesting constraints on the material in the accretion region. An attempt to detect a radio recombination line from the absorbing material was not successful.

The National Radio Astronomy Observatory is a facility of the National Science Foundation, operated under cooperative agreement by Associated Universities, Inc.

References

Boehringer, H., et al. 1993, MNRAS, 264, L25
Pedlar, A., et al. 1990, MNRAS, 246, 477
Shaver, P. A. 1978, A&A, 68, 97
Silver, C. S., Taylor, G. B., & Vermeulen, R. C. 1998, ApJ, 502, 229
Vermeulen, R. C., Readhead, A. C. S., & Backer, D. C. 1994, ApJ, 430, L41
Walker, R. C., Romney, J. D., & Benson, J. M. 1994, ApJ, 430, L45
Walker, R. C., et al. 2000, ApJ, 530, 233
Walker, R. C., & Anantharamaiah, K. R. 2003, AJ, 125, 1756

Radio Astronomy at the Fringe
ASP Conference Series, Vol. 300, 2003
J.A. Zensus, M.H. Cohen, & E. Ros, eds.

Ionised, Atomic, and Molecular Gas around the Twin Radio Jets of NGC 1052

R. C. VERMEULEN
Astron, P.O. Box 2, NL–7990 AA Dwingeloo, The Netherlands

E. ROS, J. A. ZENSUS
Max-Planck-Institut für Radioastronomie, Auf dem Hügel 69, D-53121 Bonn, Germany

K. I. KELLERMANN
National Radio Astronomy Observatory, 520 Edgemont Road, Charlottesville, VA 22903, USA

M. H. COHEN
California Institute of Technology, Pasadena, CA 91125, USA

H. J. VAN LANGEVELDE
Joint Institute for VLBI in Europe, P.O. Box 2, NL–7990 AA Dwingeloo, The Netherlands

Abstract. The bright radio structure of the LINER elliptical galaxy NGC 1052 is dominated by bi-symmetric jets on parsec scales. Features move outward on both sides of the core at $v_{\rm app} \sim 0.26c$. We have established the occurrence of free-free absorption, and suggest the presence of a patchy, geometrically thick region oriented roughly orthogonal to the jets; components have a wide range of spectral shapes and brightness temperatures. We distinguish three velocity systems in H I absorption. The "high velocity" system is the most prominent of these; it is redshifted from systemic by about 150 km s^{-1}. In H I VLBI it is seen towards both jets, but appears to be restricted to a shell 1 to 2 pc away from the core. The central hole might be largely ionised, and could be connected to the free-free absorption. WSRT spectroscopy shows 1667 and 1665 MHz OH main line absorption over at least the full ~250 km s^{-1} velocity range seen in H I. In the "high velocity" system, the profiles of the OH main lines and H I are similar, which suggests co-location of molecular and atomic gas. The OH satellite lines are also detected in the "high velocity" system: 1612 MHz in absorption and 1720 MHz in emission, with complementary strength. But we have no satisfactory model to explain all properties; the connection to H_2O masing gas at the same velocity but apparently a different location is also unclear.

NGC 1052, an elliptical LINER galaxy (e.g., Gabel et al. 2000), is at a distance of only 22 Mpc (assuming $H_\circ = 65\,{\rm km\,s^{-1}\,Mpc^{-1}}$ and using $cz = 1474\,{\rm km\,s^{-1}}$ from Sargent et al. 1977), so that detailed sub-parsec scale scrutiny is possible with VLBI (1 mas $\approx$ 0.1 pc). The unusually bright (1 to 2 Jy) radio structure is core-dominated, and has two lobes with a span of only about 3 kpc (Wrobel 1984). The overall spectrum is fairly flat and has sometimes been classified as Gigahertz peaked (e.g., de Vries, Barthel, & O'Dea 1997).

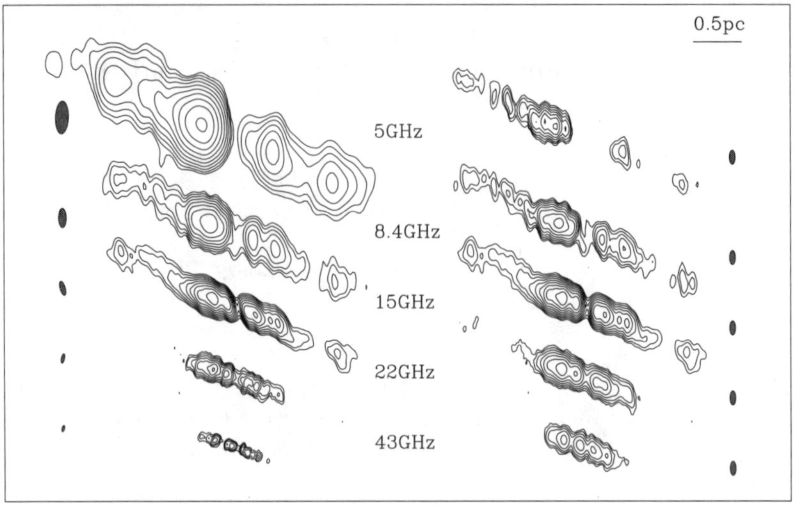

Figure 1. Contour images of NGC 1052 at 43, 22, 15, 8, and 5 GHz, observed at epoch 1997.52. The images on the left-hand side are at a resolution appropriate to the frequency (see restoring beam shown). To facilitate intercomparison, the images in the right-hand column have a fixed restoring beam of 0.14×0.06 pc. The lowest contour levels are 0.83, 1.21, 1.57, 3.44, and 1.04 mJy beam^{-1}, and in all cases the contours increase by factors of $2^{1/2}$. *(Taken from Vermeulen et al. 2003, A&A, 401, 113.)*

Fortunately, given all of the valuable information which has now emerged about it, NGC 1052 was bright enough to be included in the 15 GHz VLBA survey (Kellermann et al. 1998), which was initiated by Ken when it became possible to undertake such a substantial effort at a relatively high frequency. The first 15 GHz VLBA observation, in July 1995, revealed that, in contrast to most of the sources in the sample, NGC 1052 has a bi-symmetric morphology (see also Fig. 1). We therefore not only included NGC 1052 in many of our regular VLBA 15 GHz monitoring observations, but also initiated dedicated observations of this source. The first report on those was given by Ken, and included our discovery of two-sided motions as well as of free-free absorption (Kellermann et al. 1999). We have since then written a comprehensive paper on all of our findings (Vermeulen et al. 2003). The main results can be summarised as follows:

1. In the slightly curved bi-symmetric jets (e.g., Fig. 1), multiple sub-parsec scale features have been traced in ten 15 GHz VLBA epochs over five years. They move outward, reasonably linearly, at typically $v_{\rm app} \sim 0.26c$ on both sides. Given their symmetry, the jets must be oriented near the plane of the sky.

2. Nearly simultaneous VLBA observations were made at up to seven frequencies; some of the images are displayed in Figure 1. They show a central gap in the radio emission, which widens at lower radio frequencies. Detailed spectra demonstrate that it is due to free-free absorption. The western jet is covered more extensively, showing partial obscuration along at least 1 pc; this must be the receding jet. But the eastern jet is also partially covered, along approximately 0.3 pc, and so the free-free absorber is probably geometrically thick, and oriented roughly orthogonal to the jets. The ionised gas near the core, if distributed uniformly along a path-length of 0.5 pc, would have a volume density of $n_e \sim 10^5\,\mathrm{cm}^{-3}$. The opacity globally decreases away from the centre, but the variability of moving components, acting as "probes behind the screen", suggests considerable patchiness, and the wide range of component spectral shapes suggests that synchrotron self-absorption and free-free absorption both play a role in the inner parsecs of the jets.

3. VLBI observations show that H I absorbing atomic gas is distributed in front of the approaching as well as the receding jet, and has sub-pc scale structure. This is illustrated in Figure 2. We distinguish three absorption systems with different characteristics. From their substructure, at least two of them are situated close to the AGN rather than in the galaxy as a whole. The most prominent absorption system, with a peak optical depth of 20–25%, is at "high velocity", receding by 125–200 $\mathrm{km\,s^{-1}}$ with respect to the systemic velocity. Under the uncertain assumptions that this atomic gas has a spin temperature $T_{sp} = 100\,\mathrm{K}$, and uniformly fills a path-length of 0.5 pc, we derive a column depth of $N_H \sim 10^{21}\,\mathrm{cm}^{-2}$, and a density of $n_H \sim 1000\,\mathrm{cm}^{-3}$. This absorber may have a continuous velocity gradient of some $10\,\mathrm{km\,s^{-1}\,pc^{-1}}$ across the nucleus, but while it is seen at a distance of 1–1.5 pc on both sides of the centre, there is a deficit in the innermost parsec; this "central hole" in atomic gas may be largely ionised.

4. WSRT observations show, from 1667 and 1665 MHz OH main lines seen in absorption, that molecular gas exists along the full velocity span found also in atomic gas, roughly −30 to 200 $\mathrm{km\,s^{-1}}$. The peak absorption depth, about 0.4%, occurs in the "high velocity" system, and indicates a column depth of order $10^{14}\,\mathrm{cm}^{-2}$ The ratio between the OH 1667 and 1665 MHz lines is about 2 in the "high velocity" system, but closer to 1 at lower velocities. In the "high velocity" system, the 1612 and 1720 MHz OH satellite lines have also been detected, both with a peak strength near 0.25%, but with conjugate profiles, in absorption and emission, respectively. The profiles of the OH main lines and the integrated H I line are remarkably similar in the "high velocity" system, which suggests co-location of the atomic and molecular gas at these velocities. But H_2O masers, which cover the same velocity range, were found very close to the nucleus, at 0.1–0.2 pc along the receding jet (Claussen et al. 1998) We have no satisfactory model to explain this, in light of the central hole in H I, the OH and H I profile similarity, the velocity gradient across the core (suggestive of rotation), and the substantial overall velocity offset compared to systemic (suggestive of infall).

References

Claussen, M. J., et al. 1998, ApJ, 500, L129
de Vries, W.H., Barthel, P.D., & O'Dea, C.P. 1997, A&A, 321, 105
Gabel, J. R., et al. 2000, ApJ, 532, 883

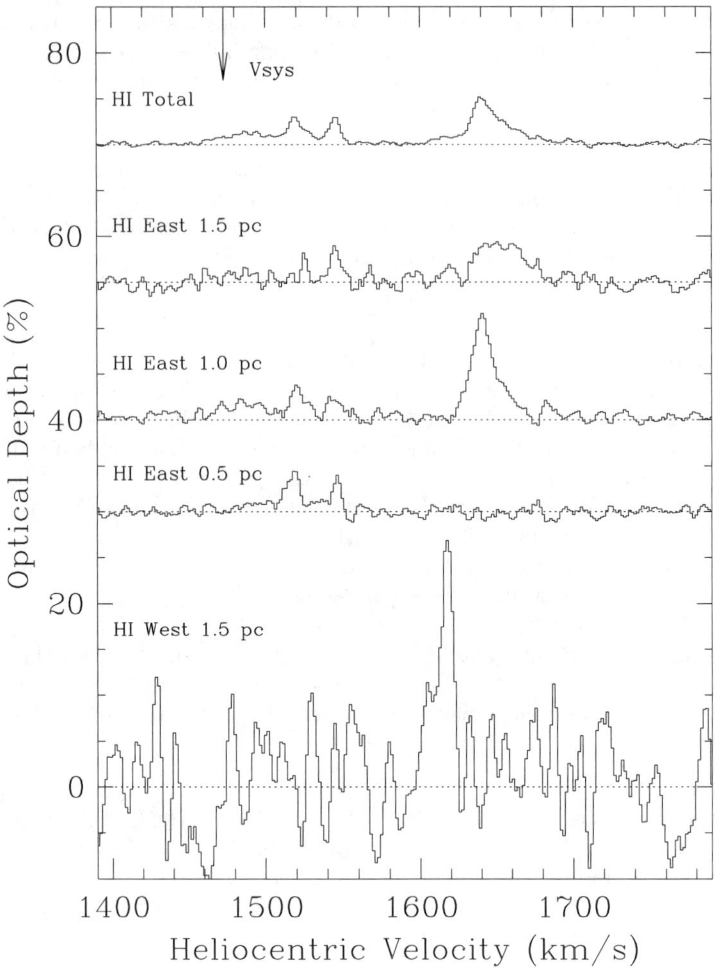

Figure 2. H I optical depths as function of velocity, at various locations along the jets of NGC 1052, as observed with VLBI in July 1998. Offsets between the spectra are used for clarity; dotted lines show the zero levels. *Taken from Vermeulen et al. (2003, A&A, 401, 113).*

Kellermann, K. I., Vermeulen, R. C., Zensus, J. A., & Cohen, M. H. 1998, AJ, 115, 1295

Kellermann, K. I., Vermeulen, R. C., Cohen, M. H., & Zensus, J. A. 1999, BAAS, 31, 856

Sargent, W. L. W., Schechter, P. L., Boksenberg, A., & Shortridge, K. 1977, ApJ, 212, 326

Vermeulen, R. C., et al. 2003, A&A, 401, 113

Wrobel, J. M. 1984, ApJ, 284, 531

Radio Astronomy at the Fringe
ASP Conference Series, Vol. 300, 2003
J.A. Zensus, M.H. Cohen, & E. Ros, eds.

Sagittarius A* as an AGN

W. M. Goss
National Radio Astronomy Observatory, Array Operations Center, P.O. Box O Socorro, NM, USA

Abstract. I present a concise and incomplete history of the discovery of Sgr A* almost 30 years ago (1974). The early VLBI attempts and successes are discussed. A more complete historical overview will be presented by Goss, Brown, & Lo (2003). A number of recent radio and mm observations are summarized: circular and linear polarization detections, time variations and possible quasi-periodic behavior, and determinations of the motions of Sgr A*.

1. Introduction

The discovery of Sgr A* by Balick & Brown in 1974 led to a renewed interest in the galactic center in the following 30 years. The realization that this source is related to a 2.6×10^6 $M_\odot$ black hole at the center of the Milky Way (Ghez et al. 2000; Eckart et al. 2002; Schödel et al. 2002) is one of the more significant advances in galactic research made in the last decades. The early VLBI investigations of Sgr A* were begun by Ken Kellermann and K. Y. Lo (plus many collaborators). Both groups realized that Sgr A* is a low luminosity AGN. Due to the improved linear resolution for an object at a distance of 8 kpc (where 1 mas = 8 AU), studies of this radio source at the center of the Milky Way led to a more complete understanding of the AGN phenomenon, even though scattering by the interstellar electrons hides many of the details of the structure of Sgr A*.

A detailed history of the discovery and early years of Sgr A* research will be included in a contribution by Goss, Brown, & Lo (2003). In this communication, I provide a short summary of the discovery process and the naming of Sgr A* (section 2). In section 3, I present some new results on the radio and mm properties of Sgr A*.

A basic radio property of Sgr A* is the inferred low luminosity compared to other radio AGNs in the nearby universe. Thus Sgr A* is a good laboratory to study AGNs, but the low luminosity implies that the underlying emission mechanisms may well be quite different. As an example, if Sgr A* were at a distance of 10 Mpc, the observed flux density in the cm range would be $\lesssim 1\,\mu$Jy (or even at the distance of M 31 only ~ 0.3 mJy). As has been pointed out by many authors, Sgr A* is an AGN with a low Eddington luminosity; the $L_{\rm bol}$ is only $\sim 10^{-8.5}\,L_{\rm Edd}$. The inferred accretion rate in a number of models is 10^{-3} to 10^{-11} $M_\odot$ yr^{-1}. Polarization results have led to a restriction of this wide range in a substantial manner. A number of ADAF, CDAF, and jet models have been proposed by several authors.

2. A selected historical summary 1974–2001

The earliest suggestion that the center of the galaxy might contain a compact object ($\sim 10^7$ M$_\odot$) was in an extensive paper in Nature entitled "Galactic Nuclei as Collapsed Old Quasars"(Lynden-Bell 1969). In subsequent years, this credit is often attributed to a later paper by Lynden-Bell & Rees (1971): "On Quasars, Dust and the Galactic Centre." This later paper suggests four specific tests. As a second test, they suggest: ". . . very long baseline interferometry may soon be possible with broad enough bandwidths to measure sources as weak as 0.5 f.u. (presently Jy) and diameters of 10^{-3} $''$."

The breakthrough in the detection of Sgr A* occurred in February 1974 by Bruce Balick and Robert (Bob) L. Brown and their results were published later that year (Balick & Brown 1974). The first detection was made using the NRAO Green Bank radio link interferometer at 11 and 3.7 cm with a baseline of 35 km. The beam sizes in the direction of maximum resolution were $\sim 0\rlap{.}''7$ and $\sim 0\rlap{.}''2$. The distant radio telescope is shown in Figure 1; the 45 foot telescope is still a prominent part of the Green Bank landscape (in 2003). This telescope played a major role in the tracking of the Japanese VLBI spacecraft, *HALCA*, in the years 1997 to 2002. (The VLBI satellite was tracked almost every day for five years and participated in more than 200 VLBI experiments.) The reason for the success of the Balick & Brown observation is that it provided a "matched filter" for the detection of Sgr A*. For the detection of the compact source, the extended structure of Sgr A West was resolved out (see Downes & Martin 1971) and yet the over-resolution of early VLBI observations was avoided. In October of 1973, a VLBI experiment at 6 cm was carried out by K. Y. Lo and collaborators. This observation is described by Lo (1974) in his Ph.D. thesis at MIT. The baseline was 228 km between the NRAO 140 foot telescope and the 85 foot Maryland Point telescope of the Naval Research Laboratory. The fringe spacing was 50 mas with a sensitivity of 0.1 Jy. As we now know, the size of Sgr A* at 6 cm is broadened by interstellar scattering to a size of 51 × 27 mas with a position angle (P.A.) of 81° (Lo et al. 1998). The orientation of the interferometer was quite unfavorable for a positive detection; if the baseline had been oriented in a more N-S direction, the source might have been detected.

The discovery of Sgr A* was made on 13 and 15 February, 1974. Balick and Brown write: " The unusual nature of this sub-arcsecond structure and its positional coincidence with the inner 1-pc core of the galactic nucleus strongly suggests that this structure is physically associated with the galactic center (in fact defines the galactic nucleus)." Balick and Brown compare the compact source with energetic nuclei of other galaxies and even suggest that variations in the radio flux density and structure might be observed.

The first VLBI detection was made the following year (19 May 1975) by Lo et al. (1975) using the OVRO 40 m (Caltech) and the NASA Goldstone 64 m at 3.7 cm. The inferred size of Sgr A* was 20 mas. In the period 10 June 1974 to 10 September 1975, Sgr A* was observed with the early MERLIN array at 0.408, 0.96 and 1.66 GHz with baselines of 24 and 127 km. The detections at the latter two frequencies suggested the angular size scales as λ^2, originating in a turbulent electron distribution along the line of sight (Davies, Walsh, & Booth 1974).

Figure 1. The 45 foot telescope used at Huntersville as the 35 km outstation for the NRAO Green Bank radio link interferometer. The discovery of Sgr A* was made in February 1974 by Bruce Balick and Bob Brown. The dual frequency instrument operated at 11 and 3.7 cm and was a prototype for the VLA during the early 1970s (see Fomalont 2000). The 45 foot telescope was the fourth element of the interferometer; correlations were performed with all 3 of the 85 foot antennas at Green Bank. This smaller antenna is now at the Green Bank site and has had an illustrious career as a component of the tracking stations for the *HALCA* VLBI spacecraft. The two antennas for telemetry are pointing to the Green Bank site to the NE; the local oscillator signal was transmitted by a two way link at 1.3 GHz while the 18 GHz link was used for telemetry and IF transmission. During the summer there was no clear line of sight to the main interferometer site due to leaves in intervening trees. A passive reflector was used on a hill behind the main site to overcome this problem. Much of the development work for the radio link was done by N. G. V. Sarma who was on a sabbatical from the Tata Institute for Fundamental Research (Ooty) in India.

Ken Kellermann and colleagues began VLBI observations in May and November of 1974, using the Haystack 37 m, the Green Bank 140 foot and the NASA Goldstone 64 m at 7.8 GHz (Kellermann et al. 1977). The baseline range was 15 to $105 \times 10^6\,\lambda$. The compact source of size 10 to 20 mas was in good agreement with the Lo et al. (1975) result; the mas core with a flux density of ~ 0.16 Jy has not been confirmed in subsequent observations. The final paragraph of this publication is quite prescient: "The possible existence of a black hole at the center of the galaxy has been suggested to explain the various phenomena observed at the Galactic Center (e.g., Lynden-Bell 1969) and the detection of a radio source with milli-arcsec dimensions has been suggested as a possible test of the presence of such an object (Lynden-Bell & Rees 1971). The observed size of the radio source corresponds to the Schwarzschild radius of a black hole of mass $\sim 10^8\,M_\odot$." Kellermann et al. (1977) also discuss the low luminosity of Sgr A* : ". . . . 4 to 13 orders of magnitude less than the radio nuclei in other spiral and elliptical galaxies or in radio galaxies and quasars." Kellermann and his group continued VLBI observations of Sgr A* at 8.45 GHz in June, 1977 (Geldzahler, Kellermann, & Shaffer 1979).

Ken Kellermann was involved in an important new observation carried out in 1983 (May and June) at 3.6 and 1.35 cm. The Mark III recording system with an order of magnitude increase in bandwidth was used. The phased VLA was used for the 1.35 cm observations (Lo et al. 1985). A clear asymmetry in the broadening size was observed at 3.6 cm with an axial ratio of 0.55. Also, the name Sgr A* was used for the first time after it had been introduced by Brown (1982). The symmetrical elongated structure likely arises in a confined layer of anisotropic turbulence. In 1991, Ken Kellermann was involved in one of the first VLBA observations (plus VLA, Goldstone, Haystack and the Green Bank 140 foot) (Lo et al. 1993). With the improved coverage of the (u, v)-plane, an image of Sgr A* was made with a beam of 12.4×3.6 mas with a P.A. = 7°; the deconvolved source size was 17.5×8.5 mas at P.A. = 87°. No evidence of small scale structure was observed; in addition the closure phases were inconsistent with any asymmetric small-scale structure. These results imply that the scattered size is most likely due to electron density fluctuations along the line of sight; the previous suggestions that the size might be due to the effects of either synchrotron or free-free opacity were thus unlikely.

Lo et al. (1998) have carried out simultaneous VLBA observations at 6, 3.6, 2, 1.35, and 0.7 cm of Sgr A*. They suggest that the axial ratio of the observed source at 7 mm is ~ 0.83 compared to ~ 0.55 at longer wavelengths. If this important result is confirmed, the instrinsic size along the minor axis is ~ 0.44 mas or 3.5 AU (~ 72 Schwarzschild radii).

VLBI observations at higher frequencies at 3.5 mm (86 GHz) are summarized by Doeleman et al. (2001), following the pioneering earlier results of Rogers et al. (1994) and Krichbaum et al. (1998). Based on the Doeleman et al. data, the best fit is a circular source with size 0.18 ± 0.02 mas, consistent with the scattering law derived at lower frequencies. In addition, there is a VLBI detection of Sgr A* at 1.4 mm by Krichbaum et al. (1998). The goal of the current mm-VLBI observations is to determine the intrinsic size of the emission region around the black hole and to use this size to restrict emission models.

3. A few recent results

3.1. Polarization

The detection of polarization in the source Sgr A* can lead to important information about the emission process, the surrounding plasma (rotation measure), the accretion rate, the magnetic field and interstellar propagation. As Bower, Falcke, & Backer (1999) point out, interstellar scattering at cm and even mm wavelengths dominates the structure of Sgr A*, preventing VLBI observations of details of the source structure which might have lead to an "important diagnostic of physical processes". In the 1999 paper, Bower, Falcke, & Backer present convincing detections of circular polarization in Sgr A* using the VLA at 4.8 and 8.4 GHz. The level is about −0.36% at the lower frequency. These results were confirmed with more extensive observations using both the VLA and ATCA (Australia Telescope Compact Array) (Bower et al. 2002). Over the period 1981 to 1998, the total intensity of Sgr A* at 4.8 GHz decreased by a factor of two while the fractional circular polarization was more or less steady at a level of −0.31% with a root-mean-square (rms) scatter of 0.13%. The 1999 VLA and ATCA observations showed that the spectrum of the circular polarization fraction is variable and can be inverted, with a spectral index of ~ 0.5 over the frequency range 1.4 to 15 GHz. During the same period any linear polarization was less than 0.1 % at these frequencies. The polarization results for Sgr A* are thus fundamentally different from those of other AGN; the galactic center source has a high circular polarization content and low linear polarization content at cm wavelengths.

The initial detection of linear polarization at 150, 220, 375, and 400 GHz has been reported by Aitken et al. (2000) using the JCMT with an angular resolution of 22″ at 220 GHz. In March–May 2002, Bower and colleagues (Bower et al. 2003) confirmed the linear polarization in Sgr A* at 230 GHz using the BIMA array. A fractional linear polarization in Sgr A* of 7.2 ± 0.6 % with a beam of $3.6 \times 0.9''$ was determined. Based on the new BIMA data, the favored rotation measure (RM) is $-4.3 \times 10^5\,\mathrm{rad\,m^{-2}}$. If unusual geometries or numerous magnetic field reversals can be ignored, this RM suggests that the accretion rate has an upper limit of $\sim 10^{-7}\,M_{\odot}\,\mathrm{yr}^{-1}$. This value is not consistent with the higher accretion rate used in the standard ADAF (accretion dominated accretion flow) models but is consistent with both CDAF (convection) (Quataert & Gruzinov 2000) and jet models (Falcke & Markoff 2000).

3.2. Time variations of Sgr A*

Time variations in the compact source at the Galactic Center were first detected by Brown & Lo (1982) at 11 and 3.6 cm using the Green Bank radio link interferometer. The time variations were followed up using VLA data by Zhao et al. (1989) and Zhao et al. (1992). The latter paper described the detailed campaign of Sgr A* monitoring in 1990 to 1993, and suggested that the amplitude of the variations increased to higher frequencies and that the rate of "outbursts" was about three per year. In addition, pronounced mm variations were observed by Wright & Backer (1993) and Tsuboi, Miyazaki, & Tsutsumi (1999). Falcke (1999) had also reported prominent time variations in the Green Bank interferometer data at 11 and 3.6 cm.

In a a paper by Zhao, Bower, & Goss (2001), two decades of VLA high resolution observations of Sgr A* at 20, 6, 3.6, 2 and 1.3 cm were used to study the variability of Sgr A*. This project illustrates the power of the historical VLA archive. Only data with baselines longer than 80 kλ (corresponding to a resolution of $\sim 3''$) were used in order to eliminate confusion from the strong H II region, Sgr A West. (At mm wavelengths this restriction can be relaxed due to the steeply inverted spectral index of Sgr A*. At 1.3 mm the limit is about 20 kλ.) The final light curve at 1.3 cm shown by Zhao, Bower, & Goss (2001) has 92 observations with a median sampling of 39 days in the period March

1990 to November 1999. At the highest frequencies of 1.3 and 2 cm, there is strong evidence for a quasi-periodic signal with a period of 106 ± 10 days; at 3.6 cm there is only marginal evidence for any periodic signals and at the lower frequencies there is none.

In mid 2000, an intense VLA campaign was initiated to monitor Sgr A* with a frequency of about four times per month at 2, 1.3 and 0.7 mm. All VLA arrays were used. The leaders of this campaign have been Robin M. Herrnstein and Jun-Hui Zhao, joined in early 2002 by a Harvard undergraduate Jenny Pegg (Herrnstein et al., in preparation). The light curve is shown in Figure 2 at these three frequencies. At 1.3 cm there are 114 data points with a median sampling interval of seven days.

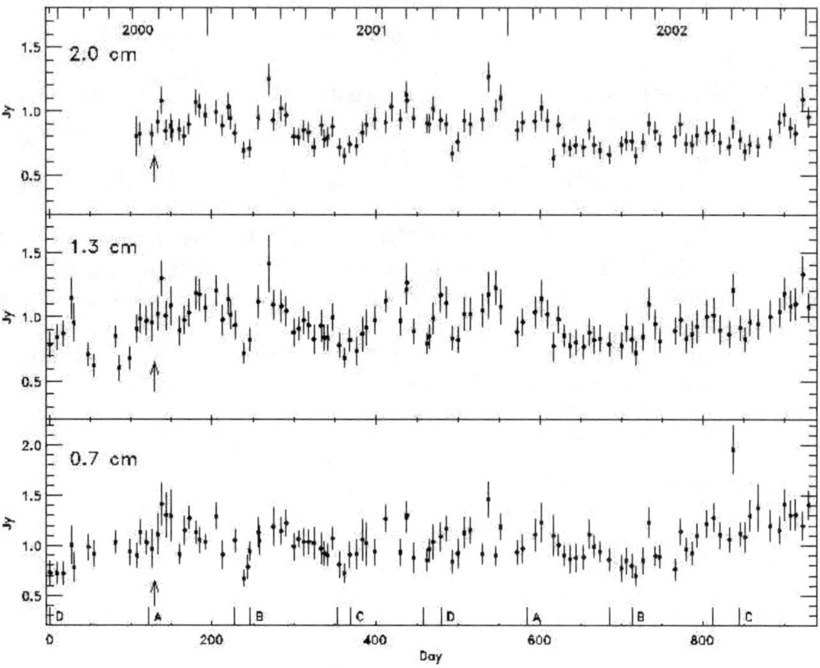

Figure 2. The results of the VLA campaign to monitor Sgr A* from 21 June 2000 to the beginning of 2003 (Herrnstein et al., in preparation) at 2,1.3 and 0.7 cm. The intrinsic variability of Sgr A* is apparent. The arrow marks the epoch of the prominent X-ray flare described by Baganoff et al. (2001) near day 120—27 October 2000. The light curve (in Jy) is shown with respect to the relative day number since 21 June 2000. The moves to the various VLA arrays (A, B, C, & D) are shown; note that the interval between moves is variable. The adjacent double lines indicate the move to the relevant hybrid array (e.g., A to B near day 225 to 240).

Based on a power spectral density (PSD) analysis of the VLA archival data at 1.3 cm from 1990 to 1999, the presence of a $\sim$ 100 day periodic signal has been confirmed with an improved confidence level ($> 99.5\%$) using the Lomb method of analysis of a non-uniformly sampled time series. In addition to the

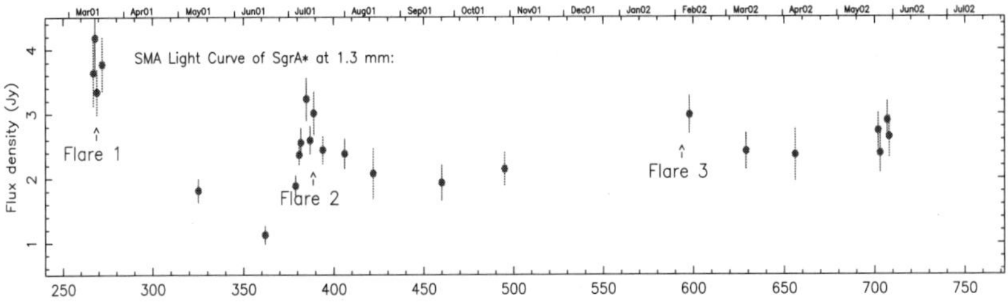

Figure 3. The 1.3 mm light curve obtained with the SMA from March 2001 to May 2002. Sgr A* was calibrated with respect to the compact extragalactic sources OV −236 and NRAO 530. *Taken from Zhao et al. (2003, ApJ, 586, L29).*

strong periodic feature near $0.01\,\mathrm{d}^{-1}$ ($\sim$ 100 days), there is a weak periodic feature at $0.004\,\mathrm{d}^{-1}$ (240 days), suggesting the presence of a double oscillation signal in the fluctuations of flux density at 1.3 cm. The double oscillation signals are also confirmed in the densely sampled light curves observed at 1.3 cm from 2000 June to the present (beginning of 2003). All the PSDs (derived from the different observing periods during the course of this recent monitoring program) are dominated by features at $0.0075\,\mathrm{d}^{-1}$ (133 days) and $\sim 0.003\,\mathrm{d}^{-1}$ (333 days), while the power of the shorter period is decreasing in magnitude. With nearly 200 epochs over the past 13 years, the variability of Sgr A* at 1.3 cm is dominated by two oscillation modes with periods of $P1 = 100\,\mathrm{d} - 133\,\mathrm{d}$ and $P2 = 250\,\mathrm{d} - 333\,\mathrm{d}$ (based on a preliminary analysis of the data using both white noise and $1/f$ noise analyses). Both the intensity and the period of the oscillation drift in time, but the ratio of the periods maintains a nearly constant ratio of 2.4–2.5. The DQPO (double quasi-periodic oscillation) may well provide crucial information about activity near the black hole ; this activity may be driven by an orbital resonance occurring in the central few thousand R_s (Schwarzschild radius) (Zhao et al., in preparation).

A recent publication (Zhao et al. 2003) summarizes 25 epochs of observations of Sgr A* at 1.3 mm (232 GHz) using the Sub-Millimeter Array (SMA) of SAO and ASIAA on Mauna Kea, which will be completed within the next year. Over the time period March 2001 to May 2002, three possible "flares" were observed. Figure 3 shows the observations at 1.3 mm. In addition during the flare of 22 March 2001, Sgr A* was also detected at 0.87 mm (345 GHz), implying a steeply inverted spectrum with $\alpha \sim 1.5$ between 232 and 345 GHz. These new observations suggest that Sgr A* shows a pronounced sub-mm increase during a flare with a break in the spectral index near 3 mm. These flares may well represent the collective mass ejection associated with the recently discovered X-ray flares (Baganoff et al. 2001) that may arise from the inner region of the accretion disk near the black hole.

3.3. Motions of Sgr A*

In 1976, Don Backer and Dick Sramek began a Green Bank radio link interferometer project (with the 35 km baseline) to determine the proper motion of Sgr A*. As Backer & Sramek (1999) point out, Rick Fisher played a major role in suggesting this observation in circa 1975. Backer & Sramek (1982) observed a proper motion consistent with the expected secular parallax; thus, it was clear that Sgr A* is galactic and not a chance superposition of an extragalactic source. In 1981, a VLA 6 cm program was begun to determine the secular motion using the VLA in the A array (resolution $\sim 0.3''$). Eight epochs were observed during the 16 year interval. The results are summarized in Backer & Sramek (1999). The longitudinal motion ($-6.18 \pm 0.19\,\mathrm{mas\,yr^{-1}}$) is consistent with the secular parallax that results from Galactic rotation. The latitude motion ($-0.65 \pm 0.17\,\mathrm{mas\,yr^{-1}}$) can be used to conclude that Sgr A* must be "attached" to a black hole if the center of the Milky Way does, in fact, contain a compact object. Equipartition arguments based on the observed motions of nearby IR stars indicate that Sgr A* must have a mass $\gtrsim 2 \times 10^4\,\mathrm{M_\odot}$.

In a companion paper to Backer & Sramek (1999), Reid et al. (1999) describe the results of a two year VLBA 7 mm program that consists of 5 separate observations. The relative positions of Sgr A* with respect to two nearby extragalactic radio sources has been determined with a precision of ~ 0.1 mas. As Reid et al. state: "It is interesting to note that, while it takes ~ 220 Myr for the Sun to complete an orbit around the galactic center, the east-west component of the parallax from only 10 days motion can be detected with the VLBA!" Reid et al. estimate that the circular rotation speed at the position of the Sun is $219 \pm 20\,\mathrm{km\,s^{-1}}$ (scaled by $R_\odot = 8$ kpc). The slow speed in latitude yields a lower limit to the mass of $\sim 10^3\,\mathrm{M_\odot}$. Reid et al. suggest that future observations of the latitude motion should improve by at least two orders of magnitude. An important future goal being pursued by this group is the determination of the parallax distance to the galactic center.

4. Summary

In summary, Sgr A*'s discovery almost 30 years ago has had far reaching consequences. The subsequent radio, IR, and X-ray observations have elucidated many properties of this dim AGN in the Milky Way. The observational properties and summaries of current theoretical models are summarized in the 2002 Annual Reviews of Astronomy and Astrophysics summary by Melia & Falcke (2001).

The National Radio Astronomy Observatory is a facility of the National Science Foundation operated under cooperative agreement by Associated Universities, Inc. I want to thank my long term collaborator, Jun-Hui Zhao, with whom I have worked on Galactic Center problems since 1986. I thank Geoff Bower, Robin Herrnstein and Ashley Zauderer for helpful comments and Pat Smiley for assistance with the photo archive of the Green Bank interferometer.

References

Aitken, D. K., et al. 2000, ApJ, 534, L173

Backer, D. C., & Sramek, R. A. 1982, ApJ, 260, 512

Backer, D. C., & Sramek, R. A. 1999, ApJ, 524, 805

Baganoff, F. K., et al. 2001, Nature, 412, 45

Balick, B., & Brown, R. L. 1974, ApJ, 194, 265

Bower, G. C., Falcke, H., & Backer, D. C. 1999, ApJ, 523, L29

Bower, G. C., Falcke, H., Sault, R. J., & Backer, D. C. 2002, ApJ, 571, 843

Bower, G. C., Wright, M. C. H., Falcke, H., & Backer, D. C. 2003, ApJ, in press (astro-ph/0302227)

Brown, R. L. 1982, ApJ, 262, 110

Brown, R. L., & Lo, K. Y. 1982, ApJ, 253, 108

Doeleman, S. S., et al. 2001, AJ, 121, 2610

Downes, D., & Martin, A. H. M. 1971, Nature, 233, 112

Davies, R. D., Walsh, D., & Booth, R. S. 1976, MNRAS, 177, 319

Eckart, A., Genzel, R., Ott, T., & Schödel, R. 2002, MNRAS, 331, 917

Falcke, H. 1999, in ASP Conf. Ser. Vol. 186, The Central Parsecs of the Galaxy, ed. H. Falcke, A. Cotera, W. T.Duschl, F. Melia, & M. J. Ricke (San Francisco: ASP), 113

Falcke, H., & Markoff, S. 2000, A&A, 362, 113

Fomalont, E. B. 2000, in NRAO Symp. 27, Radio Interferometry: The Saga and the Science, ed. D. Finley, & W. M. Goss (Socorro, NM: NRAO), 41

Geldzahler, B. J., Kellermann, K. I., & Shaffer, D. B. 1979, High resolution observations of the compact radio sources CL4, Cygnus X-3, and the Galactic Center", AJ, 84, 186

Ghez, A. M., Morris, M., Becklin, E. E., Tanner, A., & Kremenek, T. 2000, Nature, 407, 349

Goss, W. M., Brown, R. L., & Lo, K. Y. 2003, in The Central 300 parsecs of the Milky Way, ed. A. Cotera, H. Falcke, T. R. Geballe, & S. Markoff, Astronomische Nachrichten, in press

Kellermann, K. I., Shaffer, D. B., Clark, B. G., & Geldzahler, B. J. 1977, ApJ, 214, L61

Krichbaum, T. P., et al. 1998, A&A, 335, L106

Lynden-Bell, D. 1969, Nature, 223, 690

Lynden-Bell, D., & Rees, M. J. 1971, MNRAS, 152, 461

Lo, K. Y. 1974, Ph.D. thesis, (Cambridge, MA: MIT)

Lo, K. Y., Schilizzi, R. T., Cohen, M. H., & Ross, H. N. 1975, ApJ, 202, L63

Lo, K. Y., et al. 1985, Nature, 315, 124

Lo, K. Y., et al. 1993, Nature, 362, 38

Lo, K. Y., Shen, Z. Q., Zhao, J.-H., & Ho, P. T. P. 1998, ApJ, 508, L61

Melia, F., & Falcke, H. 2001, ARA&A, 39, 309

Quataert, E., & Gruzinov, A. 2000, ApJ, 545, 842

Reid, M. J., Readhead, A. C. S., Vermeulen, R. C., & Treuhaft, R. N. 1999, ApJ, 524, 816

Rogers, A. E. E., et al. 1994, ApJ, 434, L59

Schödel, R., et al. 2002, Nature, 419, 694

Tsuboi, M., Miyazaki, A., & Tsutsumi, T. 1999, in ASP Conf. Ser. Vol. 186, The Central Parsecs of the Galaxy, ed. H. Falcke, A. Cotera, W. T. Duschl, F. Melia, & Rieke, M. (San Francisco: ASP), 105

Wright, M. C. H., & Backer, D. C. 1993, ApJ, 417, 560

Zhao, J.-H., Ekers, R. D., Goss, W. M., Lo, K. Y., & Narayan, R. 1989, in IAU Symp. 136, The Center of the Galaxy, ed. M. Morris (Dordrecht: Kluwer), 535

Zhao, J.-H., Goss, W. M., Lo, K. Y., & Ekers, R. D. 1992, in ASP Conf. Ser. Vol. 31, Relationship between Active Galactic Nuclei and Starburst Galaxies, ed. A. V. Filippenko, (San Francisco: ASP), 295

Zhao, J.-H., Bower, G. C. & Goss, W. M. 2001, ApJ, 547, L29
Zhao, J.-H., et al. 2003, ApJ, 586, L29

Radio Astronomy at the Fringe
ASP Conference Series, Vol. 300, 2003
J.A. Zensus, M.H. Cohen, & E. Ros, eds.

Relativistic Jets in Blazars: Where the Action Is

A. P. Marscher
Institute for Astrophysical Research, Boston University, 725 Commonwealth Ave., Boston, MA 02215, USA

Abstract. I discuss detailed multi-waveband monitoring of several compact extragalactic radio sources. X-ray dips followed by superluminal ejections in 3C 120 support the super-massive black hole accretion model and place the core of the radio jet at least 0.3 pc from the central engine. The relative timing of radio, optical, X-ray, and γ-ray events in blazars implies that the high-frequency emission occurs at or downstream of the radio core rather than closer to the central engine. Apparent speeds of radio knots are as high as $\sim 40c$. I relate these observations to the brightness temperature problem and conclude that relativistic beaming—sometimes extreme—is the most likely solution for the majority of blazars. In general, incoherent synchrotron and inverse Compton radiation in relativistic jets explains the observations well.

1. Introduction

As Kellermann & Pauliny-Toth (1969) realized, the brightness temperature is an elegant diagnostic of the radiative processes that occur in compact non-thermal sources. If all objects obeyed the $\sim 10^{12}(\delta/1+z)^{1.2}$ K inverse Compton limit (where δ is the Doppler factor and z is the redshift), our field would have been much duller over the past 35 years. To the contrary, many of the bright sources—the ones we now call "blazars"—vary so rapidly that their implied angular sizes give brightness temperatures far in excess of the "KP limit" (e.g., Jones & Burbidge 1973).

It was recognized early-on that there is a loophole to the inverse Compton limit: the brightness temperature can exceed 10^{12} K if the X-ray or γ-ray luminosity is much higher than that of the radio–infrared synchrotron radiation. (The Klein-Nishina limit suppresses the so-called "catastrophe" of higher-order scattering.) I have spent much of my career attempting to determine the relationship between the observed high-energy and synchrotron (radio to optical) emission. The results of the programs in which I have been involved support the conclusion (now considered conservative!) that relativistic beaming shortens the timescale of variability and increases the apparent brightness temperature. This gives the impression that blazars routinely violate the 10^{12} K limit. In fact, save for a few possible exceptions (e.g., Kedziora-Chudczer et al. 1997), current observations are consistent with general obedience to the KP limit (e.g., Tingay et al. 2001), although we need to extend space-VLBI to longer baselines to verify this.

While the results discussed here support what has now become the accepted paradigm—relativistic plasma jets spurting out from disks of gas accreting onto super-massive black holes—they suggest some rather uncomfortable features of the model. Apparent velocities range up to $\sim 40c$, well above Phinney's (1987) limit ~ 10 for propagation out of the nuclear region. This implies that acceleration of the jet continues well outside the immediate environs of the central engine (Li, Begelman, & Chiueh 1992). Perhaps then the radio "core" is the site where the bulk Lorentz factor reaches its asymptotic value (Marscher 1980). Secondly, the time lags of X-ray and γ-ray flares relative to superluminal ejections and radio-optical flares indicate that even the high-energy photons come from the radio-emitting portion of a blazar jet. In fact, it is not clear whether any of the observed emission arises from the section of the jet between the radio core and the central engine. This leaves the question of how the relativistic electrons are energized parsecs downstream of the accretion disk.

2. Observations

Since 1996, a host of collaborators and I have been observing several X-ray bright blazars, as well as the radio galaxy 3C 120, with the Rossi X-ray Timing Explorer (RXTE) (Marscher et al. 2002; 2003). We support the RXTE observations with monitoring of (1) the structure and polarization of the radio jet at a resolution of about 0.1 milliarcseconds (mas) with the Very Long Baseline Array, (2) the total radio-mm flux density at various frequencies, and (3) the optical and/or infrared (IR) brightness. From these data, we have determined that the X-ray variations in blazars occur in the jet, in contrast to lower-luminosity AGN (including 3C 120), whose X-rays appear to be emitted from the central engine. Presumably, this non-jet emission is swamped by beamed radiation in blazars.

I summarize below the results on the radio galaxy 3C 120, the quasars PKS 1510−089 and 3C 279, and BL Lacertae. Each object exhibits a distinct pattern of multi-waveband behavior, although for some the pattern changes with time. The sequences of VLBA images and light curves can be viewed at the web site of the Boston Univ. blazar research group (`http://www.bu.edu/blazars`).

3. The Radio Galaxy 3C 120

The active galactic nucleus 3C 120 is a schizophrenic object with three personalities: radio galaxy, Seyfert galaxy, and blazar. It sports a prominent relativistic radio jet with apparent superluminal motion at $\sim 5c$ (Gómez et al. 2000; 2001). In the X-ray, it has Seyfert-like characteristics, complete with an iron emission line (e.g., Grandi et al. 1997) and a continuum spectrum that is flatter at lower flux levels. Since the X-ray emission of black-hole binary systems ("microquasars") in our Galaxy also possesses these properties, one might expect 3C 120 to share the behavior of the microquasar GRS 1915+105: X-ray dips followed by ejections of superluminal knots (Mirabel & Rodríguez 1998). As reported in Marscher et al. (2002), this is indeed the case: there are four significant X-ray dips that precede the "time of ejection" (coincidence of the moving feature with the core at the narrow end of the radio jet) by ~ 30 days. The observed time lag between the minimum in the X-ray light curve and the appearance of the moving knot at the site of the radio core implies that the latter is at least 0.3 pc downstream of the black hole. On a radio image, the angular distance between the two ~ 0.15 mas, projected on the plane of the sky.

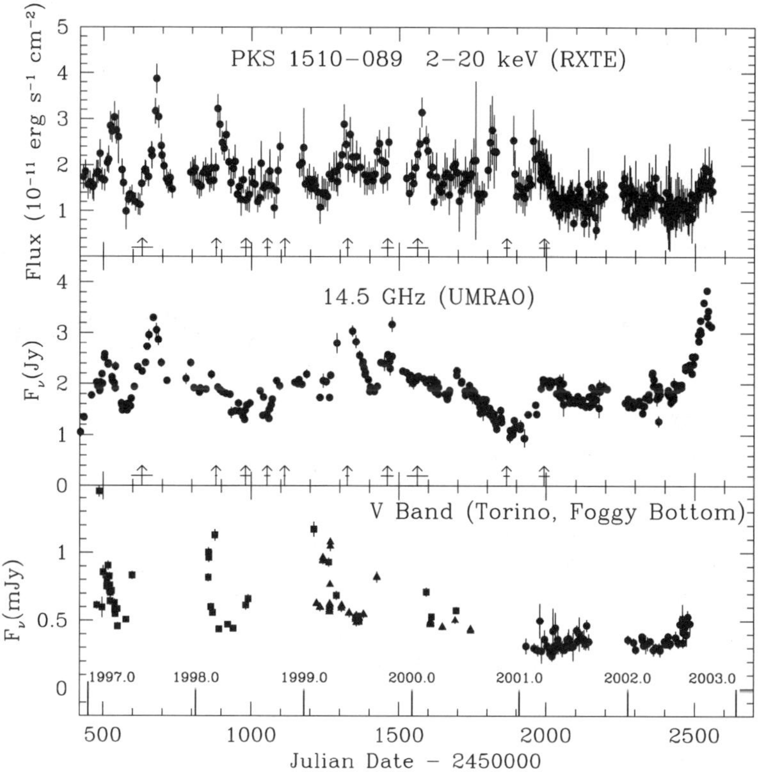

Figure 1. Multi-frequency light curves of PKS 1510−089. The vertical arrows denote epochs of ejection of superluminal radio knots, with the crossing horizontal lines indicating the uncertainties in the dates. (Data from private communications: 14.5 GHz—M. Aller, optical—M. Villata and T. Balonek.)

4. The Quasar PKS 1510−089

The $z = 0.361$ quasar PKS 1510−089 is an extreme member of the already extreme class of blazars. Jorstad et al. (in preparation) have clocked its apparent speed at $40c$ (for $H_\circ = 70\,\mathrm{km\,s^{-1}\,Mpc^{-1}}$ and $q_\circ = 0.1$).

Figure 1 displays the X-ray, radio, and optical light curves, with the dates of superluminal ejections marked, while Figure 2 gives the discrete cross-correlation function (DCCF; Edelson & Krolik 1988) of the X-ray and radio variations for each year and pairs of years. The correlation was very good during the first two years, with a *reverse* (radio leads the X-ray) time lag of 15–20 days. Such a reverse delay is expected if the X-rays are synchrotron self-Compton (SSC) emission and light-travel effects are important (Marscher 2001; Sokolov & Marscher 2003). In order for this to work, the X-ray emitting region must be *at or downstream of the radio core*, far from the accretion disk. This is similar to the conclusion drawn by Lähteenmäki & Valtaoja (2003) and Jorstad et al.

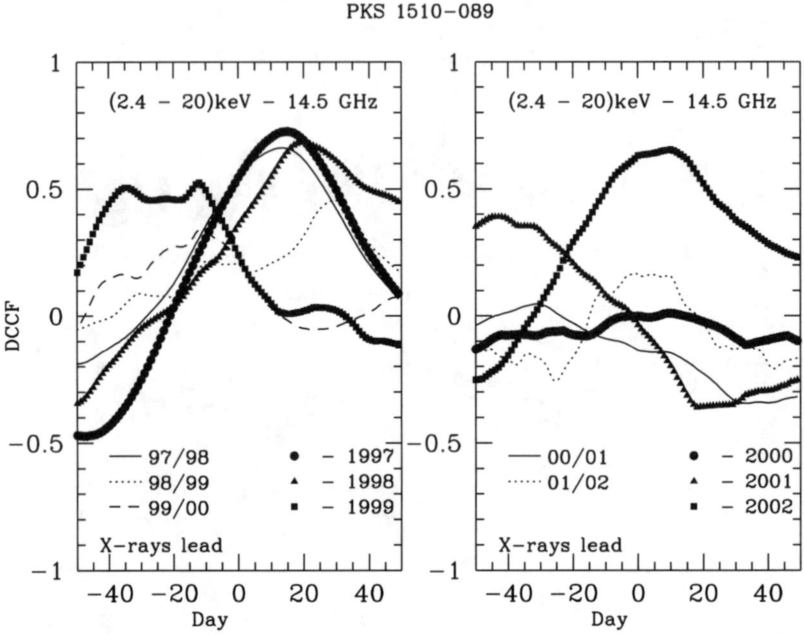

Figure 2. Discrete cross-correlation function for the radio and X-ray light curves of PKS 1510−089. Positive time lag corresponds to radio leading the X-ray.

(2001) from comparison of episodes of high γ-ray flux with either radio flares or ejections of superluminal radio knots.

The X-ray–radio correlation became weaker in 1999, with the X-ray leading. In 2000 and 2001, the correlation essentially vanished, although so did the connection between radio outbursts at 14.5 GHz and superluminal ejections at 43 GHz. We conclude that the opacity of the jet increased by late 1999, causing time delays of variations at 14.5 GHz relative to $>$ 40 GHz. This could have been caused by a slight change in the direction of the jet or a decrease in the bulk Lorentz factor. Indeed, the observed direction of the jet within 0.5 mas of the core differed by $\sim 40°$ from that seen previously. Furthermore, the apparent speed of the superluminal knot ejected in early 2000 appears to be more than a factor of 2 slower than those appearing in 1999, although there are considerable uncertainties given the time coverage and variable image quality.

Ejections of superluminal knots tend to occur near the epochs of X-ray flares. However, the X-ray light curve is so complex that it is difficult to specify the beginning and end of a flare, and periods when there is no active X-ray flare tend to be brief. The problem is exacerbated by the presence of 8-week annual RXTE sun-avoidance gaps.

5. The Quasar 3C 279

Figure 3 shows our X-ray and optical light curves of the quasar 3C 279. As is demonstrated in Figure 4, there is a strong correlation between the X-ray and optical variations. Most noteworthy is the behavior in 2001, when both the optical and X-ray varied wildly, although not in complete synchronization. The discrete cross-correlation function over the entire dataset has two peaks: at 0–3 days and 15–20 days (optical leading the X-ray in each case). The lag indicates that the X-rays are emitted downstream of the site where the optical photons arise, but the double peak implies that the geometry is complex.

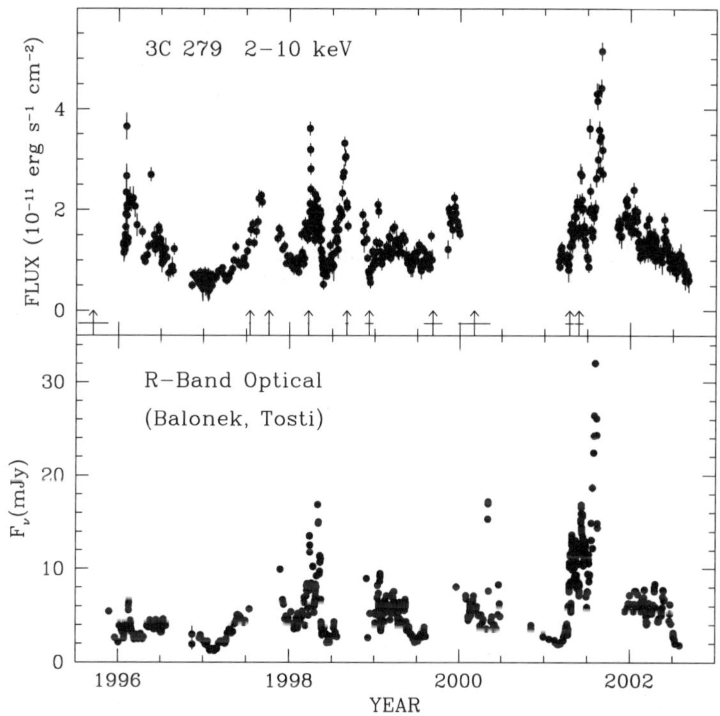

Figure 3. X-ray and optical light curves of 3C 279. The vertical arrows have the same meaning as in Figure 1. (The optical data were communicated by T. Balonek and G. Tosti.)

The increased level of variability in 1998 followed a sudden increase in apparent speed of radio knots from $5c$ to $13c$ within 1 mas of the core. The violent activity of 2001 followed a rather sudden (over a several-month period) shift in the direction of the jet by 10° to the south. Such changes in the jet should generate strong shock waves, which we speculate to be the cause of the enhanced variability. As with PKS 1510−089, there appears to be some correspondence between X-ray flares and superluminal ejections, but this requires statistical verification with a longer data train.

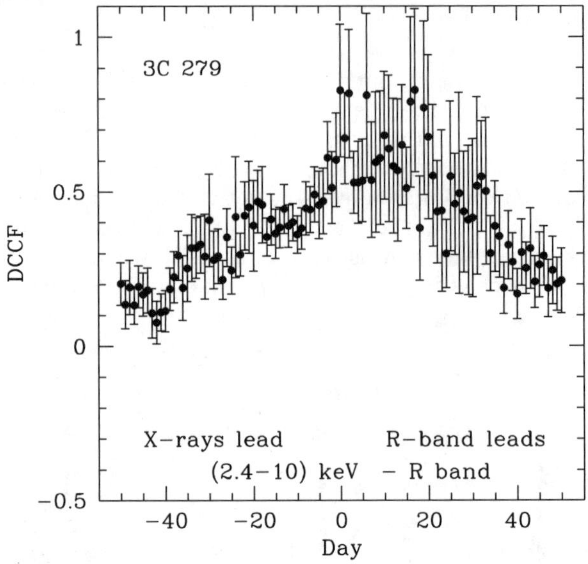

Figure 4. Discrete cross-correlation function for the optical and X-ray light curves of 3C 279. Positive time lag corresponds to radio leading the X-ray.

6. BL Lacertae

We only monitored BL Lac with RXTE for about two years, and managed to leave a gap during which a pronounced optical and radio outburst occurred, accompanied by the emergence of a superluminal radio knot. However, we did catch a prolonged optical outburst in late 2000 that also coincided with a superluminal ejection. The correspondence between the optical and X-ray activity is excellent, and the time lag is $\sim$ zero. Interestingly, the 2–10 keV X-ray spectral index was uncharacteristically steep during the event, which suggests that the X-ray flare may have been synchrotron rather than inverse Compton radiation.

7. Theoretical Considerations

The close relationship between the X-ray and radio (PKS 1510−089) or optical (3C 279, BL Lac) emission implies that nearly all the radiation observed in these blazars is relativistically beamed and produced in the jet. The site of activity is at, or slightly downstream of, the radio core, not near the black hole. We still need to determine the physical processes that cause the jet to light up at the core. If the core is a system of standing shocks (Daly & Marscher 1988), what causes the change in pressure at the boundary to sustain the shock? If it is the position where the jet energy density becomes rest-mass dominated and the Lorentz factor of the flow reaches its asymptotic value (Marscher 1980), are the most energetic disturbances ever seen upstream of this point?

The relationship between synchrotron self-Compton (SSC) X-ray emission and the radio–optical synchrotron radiation in blazars is not as obvious as we

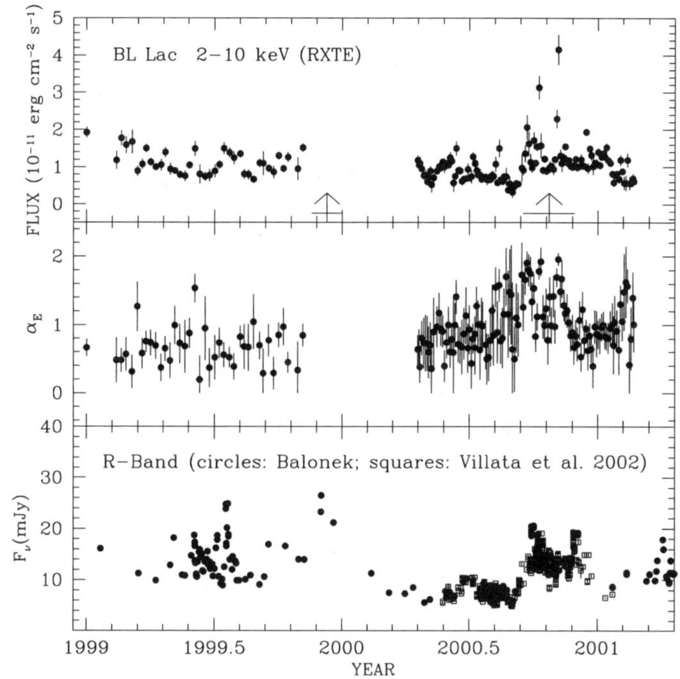

Figure 5. Light curves of BL Lacertae; α_E is the X-ray spectral index. The vertical arrows have the same meaning as in Figure 1.

had naïvely thought. As mentioned above with regard to the reverse time lag in PKS 1510−089, internal light-travel delays can cause the X-ray changes to lag behind the radio–optical variations. Furthermore, as shown by McHardy et al. (1999), several decades of synchrotron seed photon frequencies are scattered up to a given X-ray energy if the electron energy distribution is a power law. Finally, gradients in the magnetic field and/or density and maximum energy of relativistic electrons can cause different synchrotron frequencies to be emitted from different volumes (Marscher 1980; Marscher & Gear 1985; Ghisellini et al. 1985; Georganopoulos & Marscher 1998). All this complicates the expected multi-frequency behavior of flares.

More sophisticated models are needed that include the factors listed above as well as other effects that might be important. On the observational side, GLAST and AGILE promise to produce γ-ray light curves of many blazars that may be as well sampled as our RXTE light curves. New, high-frequency space VLBI antennas should be high priority, since we predict that they would resolve the X-ray and γ-ray emitting regions.

8. Conclusions

There is considerable evidence for complex geometries and frequency stratification of the emission regions in blazars, but overall the relativistic jet-incoherent synchrotron and inverse Compton model appears capable of explaining the general characteristics of non-thermal emission from the objects we have studied. It is true that very high Doppler factors are needed in some sources, but our proper motions indicate that this is indeed the case. An apparent speed of $40c$ requires a Lorentz factor of at least 40, which allows a Doppler factor (if the jet is viewed exactly head-on) of 80. Of course, it is a challenge to understand how the jet flow can be accelerated to such a high Lorentz factor, especially if this mainly occurs far from the central engine.

There are still some extreme objects, such as 0405−305, that may have coherent emission, but at this point they seem to be exceptions rather than the norm. Of course, it is often the aberrant sources that teach us the most!

This material is based upon work supported by the National Aeronautics and Space Administration under grant no. NAG5-11811 issued through the Office of Space Science, and by the National Science Foundation under grant AST-0098579. I am indebted to many collaborators who contributed to the projects reported here.

References

Daly, R. A., & Marscher, A. P. 1988, ApJ, 334, 539

Edelson, R. A., & Krolik, J. H. 1988, ApJ, 333, 646

Georganopoulos, M., & Marscher, A. P. 1998, ApJ, 506, L11

Ghisellini, G., Maraschi, L., & Treves, A. 1985, A&A, 146, 204

Gómez, J. L., et al. 2000, Science, 289, 2317

Gómez, J. L., et al. 2001, ApJ, 561, L161

Grandi, P., et al. 1997, ApJ, 487, 636

Jorstad, S. G., et al. 2001, ApJ, 556, 738

Kedziora-Chudczer, L., et al. 1997, ApJ, 490, L9

Kellermann, K. I., & Pauliny-Toth, I. I. K. 1969, ApJ, 155, L71

Jones, T. W., & Burbidge, G. R. 1973, ApJ, 186, 791

Lähteenmäki, A., & Valtaoja, E. 2003, ApJ, in press

Li, Z., Begelman, M. C., & Chiueh, T. 1992, ApJ, 384, 567

Marscher, A. P. 1980, ApJ, 235, 386

Marscher, A. P. 2001, in ASP Conf. Ser. Vol. 224, Probing the Physics of Active Galactic Nuclei by Multiwavelength Monitoring, ed. B. M. Peterson, R. S. Polidan, & R. E. Pogge (San Francisco: ASP), 23

Marscher, A. P., & Gear, W. K. 1985, ApJ, 298, 114

Marscher, A. P., et al. 2002, Nature, 417, 625

Marscher, A. P., et al. 2003. In ASP Conf. Ser. Vol. 299, High Energy Blazar Astronomy, ed. L. O. Takalo, & E. Valtaoja (San Francisco: ASP), in press

McHardy, I. M., et al. 1999, MNRAS, 310, 571

Mirabel, I. F., & Rodríguez, L. F. 1998, Nature, 392, 673

Phinney, E. S. 1987, in Superluminal Radio Sources, ed. J. A. Zensus & T. J. Pearson (Cambridge: CUP), 301

Sokolov, A., & Marscher, A. P. 2003, in ASP Conf. Ser. 299, High Energy Blazar Astronomy, ed. L. O. Takalo, & E. Valtaoja (San Francisco: ASP), in press

Tingay, S. J., et al. 2001, ApJ, 549, L55

Villata, M., et al. 2002, A&A, 390, 407

Radio Astronomy at the Fringe
ASP Conference Series, Vol. 300, 2003
J.A. Zensus, M.H. Cohen, & E. Ros, eds.

Thermal Electrons in the Helical Magnetic Field of 3C 273

M. Inoue
National Astronomical Observatory, Osawa 2-21-1, Mitaka, Tokyo 181-8588, Japan

K. Asada
Grad. Univ. for Advanced Studies, Osawa 2-21-1, Mitaka, Tokyo, 181-8588, Japan

Y. Uchida
Dept. of Physics, Science University of Tokyo, Kagurazaka 1-3, Shinjyuku, Tokyo, 162-8601, Japan

Abstract. The location of thermal electrons is discussed under the situation of a helical magnetic field in the jet of 3C 273. The helical field was recently suggested by a gradient of Faraday Rotation using VLBI polarimetry, and it is suggested that the thermal electrons are distributed outside of the emitting region, like a sheath.

Introduction High resolution, sensitive VLBI polarimetry has been revealing structures of magnetic field in jets near the core. Asada et al. (2002) showed a transversal gradient of Faraday Rotation Measure (RM) across a jet component in 3C 273. They interpreted this transversal gradient as a helical magnetic field in a jet, and explained successfully the abrupt change of the observed intrinsic field direction from parallel to perpendicular to the jet axis. This explanation was generally extended to the bimodal nature of the observed intrinsic field direction, parallel or perpendicular to the jet axis, by means of the pitch angle of the helical magnetic field. They defined the orientation of the helical magnetic field, and further suggested the rotational direction of the possible accretion disk around the supermassive black hole, following the MHD model of the helical magnetic field.

It is, however, not clear how and where the thermal electrons which are responsible for the RM are located in terms of the jet structure. It has been thought that RM is due to a Faraday Screen in front of the jet, and independent of the jet. Asada et al. (2002) claimed that the thermal electrons are closely associated with the jet, and gave a successful explanation in terms of a helical magnetic field. Here, we discuss the possible location of the thermal electrons from the observational point of view.

Distribution of thermal electrons First, for simplicity we assume a helical magnetic field and axisymmetric structure of the jet. The relativistic electrons that are responsible for synchrotron radiation lie between radius r_1 and r_2 ($r_1 < r_2$), so that we observe a jet size of r_2, although the helical field extends over more than r_2. When $r_1 \ll r_2$ or $r_1 \approx 0$, we will see a bright central ridge along jet, while when $r_1 \approx r_2$, rim brightening would be observed. Although we do not have enough resolution for pc scale jets, it seems that observations of M 87

(Junor, Biretta, & Livio 1999; Reid et al. 1989), Mrk 501 (Edwards et al. 2000), and 3C 273 (Lobanov & Zensus 2001) suggest the latter case, r_1/r_2 being ≈ 0.2. If this is the case, there is a cavity at the center of jet, to some extent. Further, although relativistic electrons themselves might cause Faraday Rotation, we do not consider this possibility.

Faraday Depolarization When Faraday Rotation occurs within a synchrotron emitting region, the observed polarization is reduced even if the magnetic field is uniform. This is due to a vector summation of different polarization angle components along the line of sight, and we call this Faraday depolarization. This is a critical test to identify the source of Faraday Rotation. In our case of 3C 273, $RM \approx 300\,\mathrm{rad\,m^{-2}}$ and we should observe some amount of depolarization at 6 cm, if the thermal electrons are distributed between r_1 and r_2. We did not see such a clear depolarization, and we rule out this possibility. Thus, we discuss the case that the thermal electrons are in the cavity ($< r_1$), or outside of the emitting region ($> r_2$). It is, however, difficult to identify the case when the region of thermal electrons is partially mixed up with that of relativistic electron. Below we discuss cases hwen the two regions of thermal electrons and relativistic electrons are clearly separated.

Thermal electrons in the central cavity

The distribution of RM might show various features depending on r_1 and r_2, as only the polarization from the far side of the jet suffers from Faraday Rotation, and the line of sight component of the magnetic field and path length change across the jet in a complex manner. Then, we rule out the case.

Thermal electrons outside the emitting region

As the line of sight component of the helical magnetic field simulates well the observed RM distribution (see Figs. 5 and 4 of Asada et al. 2000), it is likely that the thermal electrons are distributed around the emitting region, or $> r_2$, like a sheath around the emitting region of relativistic electrons. Although it is not easy to estimate the extent of the distribution, it should be within a structure of the helical magnetic field. A "spine-shear layer" seen in blazars 1055+018 (Attridge, Roberts, & Wardle 1999) and Mrk 501 (Edwards et al. 2000) would give a hint of the sheath.

Summary Recent VLBI polarimetry of 3C 273 (Asada et al. 2002) that made use of Faraday Rotation to derive a helical magnetic structure, is studied. As no significant depolarization is seen, it is suggested that thermal electrons which are responsible for the gradient in RM distribution, are located outside the emitting region where the helical magnetic field is still prevailing.

References

Asada, K., et al. 2002, PASJ, 54, L39

Attridge, J. M., Roberts, D. H., & Wardle, J. F. C. 1999, ApJ, 518, L87

Edwards, P. G., et al. 2000, PASJ, 52, 1015

Junor, W., Biretta, J. A., & Livio, M. 1999, Nature, 401, 891

Lobanov, A. P., & Zensus, J. A. 2001, Science, 294, 128

Reid, M. J., et al. 1989, ApJ, 336, 112

Radio Astronomy at the Fringe
ASP Conference Series, Vol. 300, 2003
J.A. Zensus, M.H. Cohen, & E. Ros, eds.

Magnetic Field Structures in Parsec Scale Jets: Oblique Shocks

H. D. ALLER, M. F. ALLER, P. A. HUGHES
University of Michigan, Ann Arbor, MI, 48109–1090, USA

Abstract. A selected group of ten highly variable extragalactic sources was observed with the VLBA at 15 and 43 GHz, on nine epochs spanning a 30-month period; five of these objects were also observed at 8.0 and 22 GHz. All objects exhibited variability; and we present here the evolution of the observed polarization features. We interpret the temporal changes of the polarization in terms of oblique shocks developing in relativistic jets.

1. Introduction

The early success of fitting quantitative transverse shock models to the University of Michigan (hereafter UMRAO) variability data for selected radio outbursts in extragalactic sources (Hughes, Aller, & Aller 1989), gave strong support to the hypothesis that the moving source components observed by VLBI experiments in parsec-scale jets were shocks propagating outwards in relativistic flows. However, many sources have exhibited bursts where the radio polarization is at an arbitrary angle to the jet flow direction, and we suspect that in these cases, the linear polarization is produced by *oblique* shocks. The removal of the constraint that the shocks must be transverse to the flow direction, when combined with the effects of relativistic aberration, permit polarization structures at arbitrary angles to the flow direction; but it also adds two additional free model parameters (see Fig. 5) which makes it difficult to derive a unique set of parameters for an observed polarized feature. To better understand the evolution of oblique polarization structures, we made multi-epoch observations of ten highly active extragalactic sources selected from the UMRAO variability program: DA 55, 0607−157, OJ 287, 1055+018, 3C 279, 1510−089, OT 081, 1928+738, BL Lacertae, and 3C 446; light curves of these sources over the time of this study are shown in Figure 1 and Figure 2. We report on preliminary results for one program source, BL Lacertae, based on 15.5 GHz measurements with the VLBA at five epochs and the monitoring results from the UMRAO 26-meter telescope. The long term behavior of BL Lac is described in Aller, Aller, & Hughes (these proceedings, page 159).

2. Results

The integrated flux and polarization variability and VLBA images of OT 081 at two sample epochs are shown in Figure 3. This source was used as our VLBA polarization calibrator: the VLBA instrumental polarization (the "D terms") was determined from observations over a wide range of plano-centric angle. The

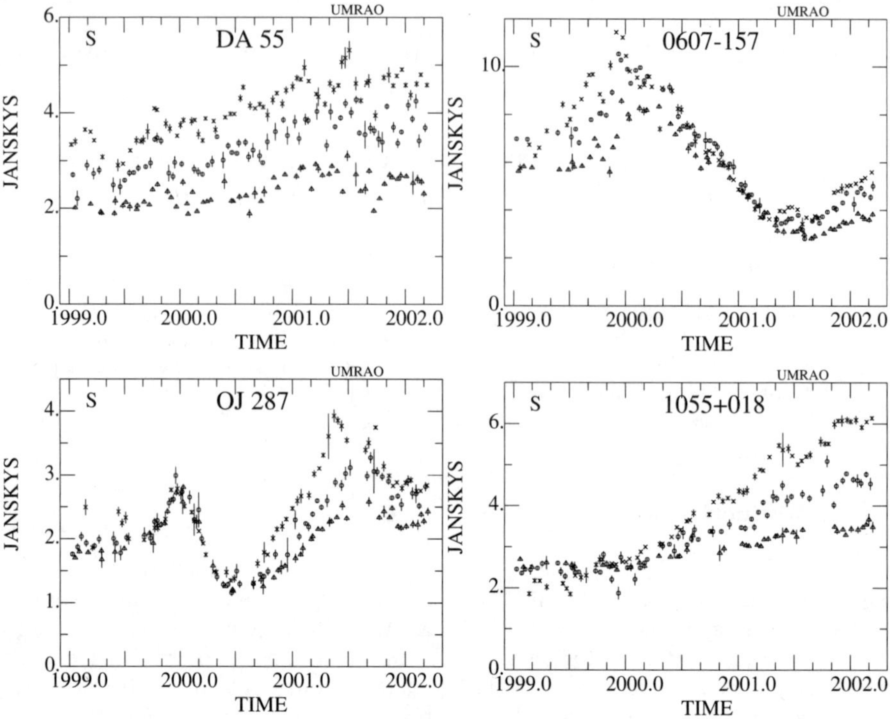

Figure 1. Two week averages of the UMRAO single antenna flux density measurements of four of the ten program sources at 4.8, 8.0, and 14.5 GHz, shown as triangles, circles, and crosses respectively.

EVPA calibration was obtained by a comparison with the UMRAO data which are in turn calibrated using an absolute reference. Figure 4 is a similar display for BL Lac with VLBA images at five selected epochs, with the addition of images of the polarized emission. The contour levels (chosen to emphasize the diffuse emission) are the same in all images, but the noise levels vary due to weather conditions or temporary equipment problems. The basic morphology of the source remained unchanged over the two-year observing period, and is very similar to the structures observed by Denn, Mutel, & Marscher (2000). A 'hot spot' evident in the jet appears to change position in an erratic manner; this is likely the result of motion of the effective position of the core as new components appear. Ultimately, however, the outer bright component appears to move outward along a fixed direction (arrow A in Fig. 6). The majority of the variability (and the polarized flux) originates in the unresolved (and, part of the time, partially opaque) 'core'. The most highly polarized component, however, is often the 'hot spot' located two or more mas from the core, where the degree of polarization is as high as 25%. We believe that this is the site of a shock in the jet flow. The striking feature is that this region maintains a relatively fixed polarization position angle (in the vicinity of 45°) from epoch

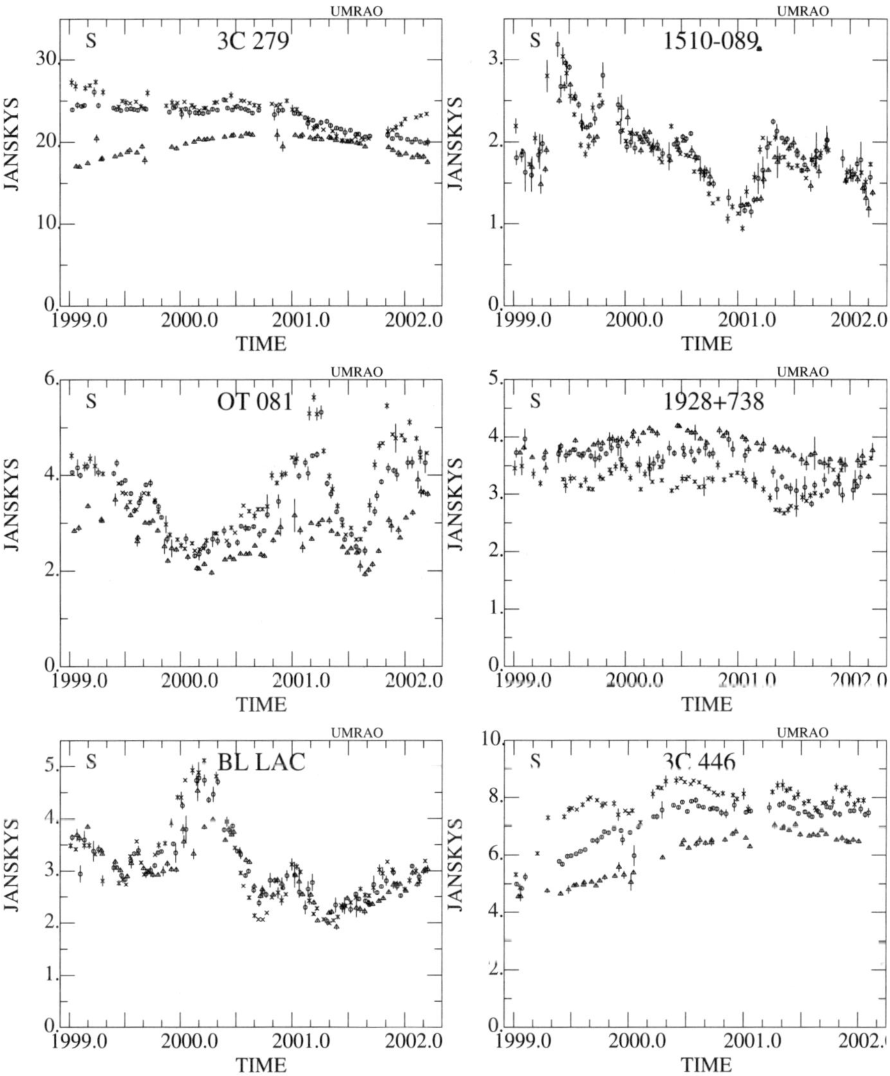

Figure 2. Two week averages of the UMRAO single antenna flux density measurements of the remaining six sources at 4.8, 8.0, and 14.5 GHz, shown as triangles, circles, and crosses respectively.

to epoch. The observed orientation is consistent with an oblique shock which could be associated with the deflection or bending of the jet flow towards the south east.

Figure 6 shows, on an expanded scale, the total flux density contours and polarized flux at the July 2000 epoch. The apparent orientation of the shock

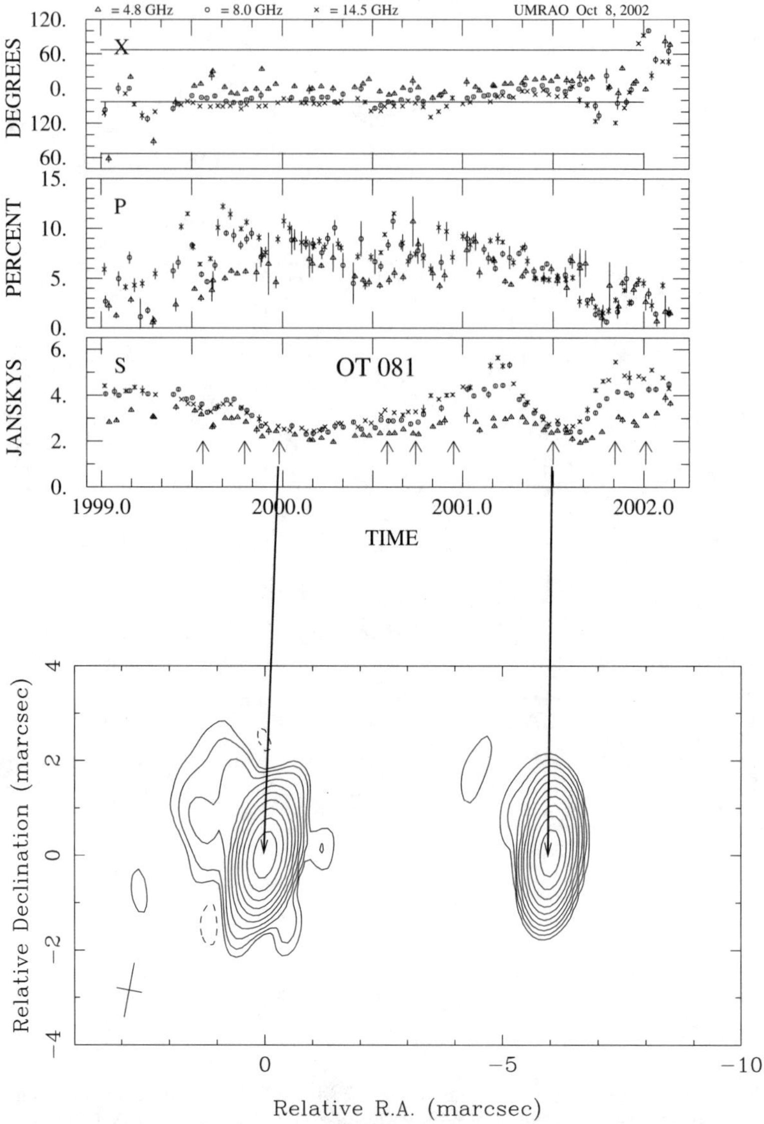

Figure 3. The top panel shows two-week averages of the UMRAO total flux density, degree of polarization and position angle of the electric vector of the polarized emission (EVPA) of OT 081. The horizontal lines at an EVPA of 63° represent the average orientation of extended structure in the source; the observed EVPA remains nearly orthogonal to this orientation. The bottom panel shows total flux maps of the source at two of the observed epoches using the same contour levels. The extended structure observed in the first displayed epoch is approximately 1% of the peak flux.

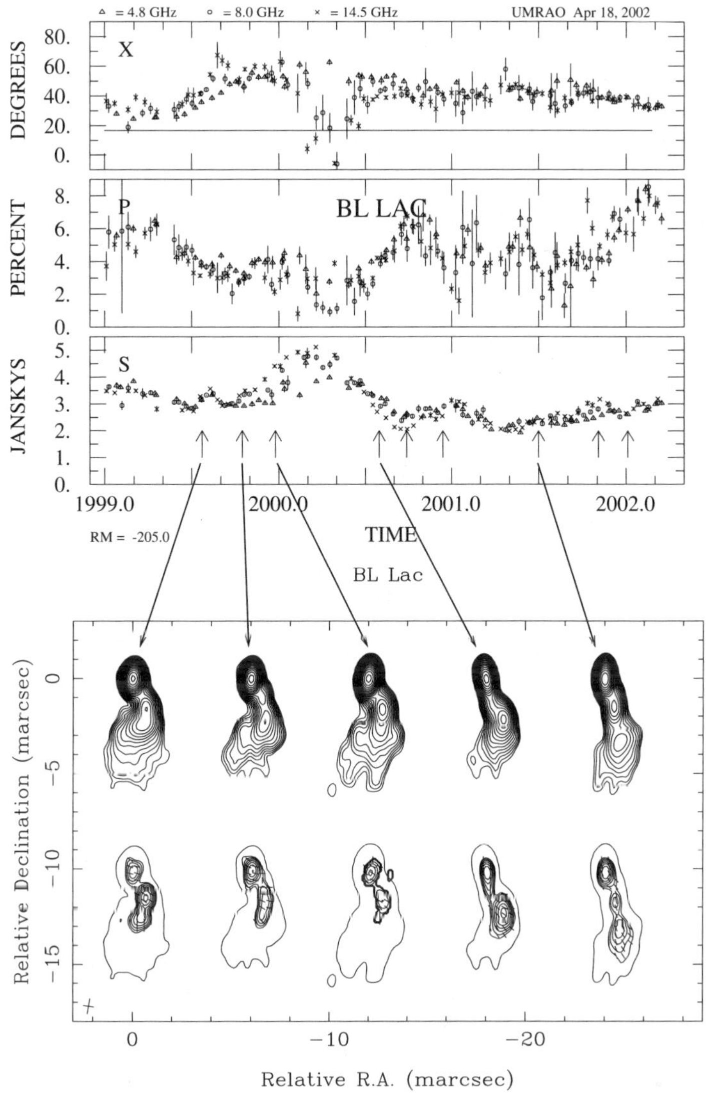

Figure 4. The top panel shows from bottom to top two-week averages of the UMRAO total flux density, degree of polarization, and position angle of the electric vector of the polarized emission (EVPA) of BL Lac. The horizontal line at an EVPA of $17° = 197°$ (there is a $180°$ ambiguity) represents the average orientation of the extended structures observed in the VLBI maps. The lower panel show the 15.4 GHz structures observed at five epochs. The top set of images are the total flux density contours, the bottom images show the outer contour of the total flux density plus contours of linearly polarized emission. The small lines show the local EVPA of the polarized emission.

plane (line C in Fig. 6) is perpendicular to the EVPA, and is inclined by an angle of approximately 30° to the apparent direction of the motion of the bright component (line A) that is evident in the multi-epoch plot (Fig. 4). Although the apparent motion of the bright component is along line A, material appears to be deflected along the line D. The relevant parameters of a shock (orientation and strength) are illustrated in Figure 5. Figure 7a shows the allowed range of deflection angles (χ in Fig. 4) of the flow through a shock (e.g. Lind & Blandford 1985) for a range of upstream Lorentz factors versus the shock orientation. Figure 7b shows the range of possible co-η values on a ϕ versus θ plot which could produce the observed polarized component (the range of upstream Lorentz value has been limited to less than 5.0). There is no difficulty producing the observed emission with an oblique shock.

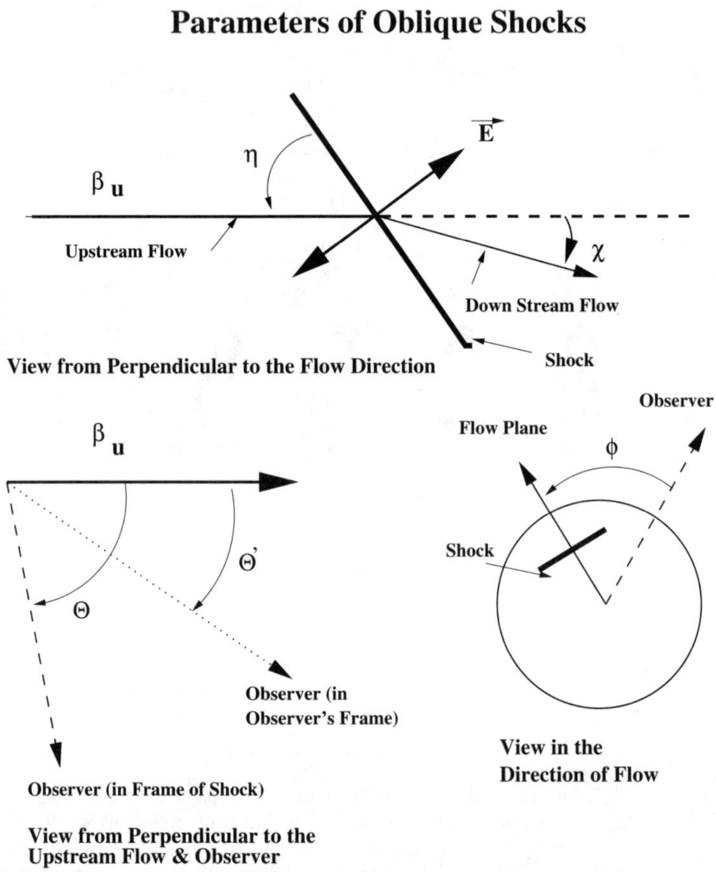

Figure 5. Parameters needed to describe an oblique shock in a relativistic flow (following Cawthorne & Cobb 2000). The strength of the shock (which sets the degree of compression and hence the degree of polarization) is set by β_u and η. The observed polarization (orientation and degree of polarization) also depend on the velocity of the shock, the angle to the observer, θ', and the angle of the flow plane to the observer, ϕ.

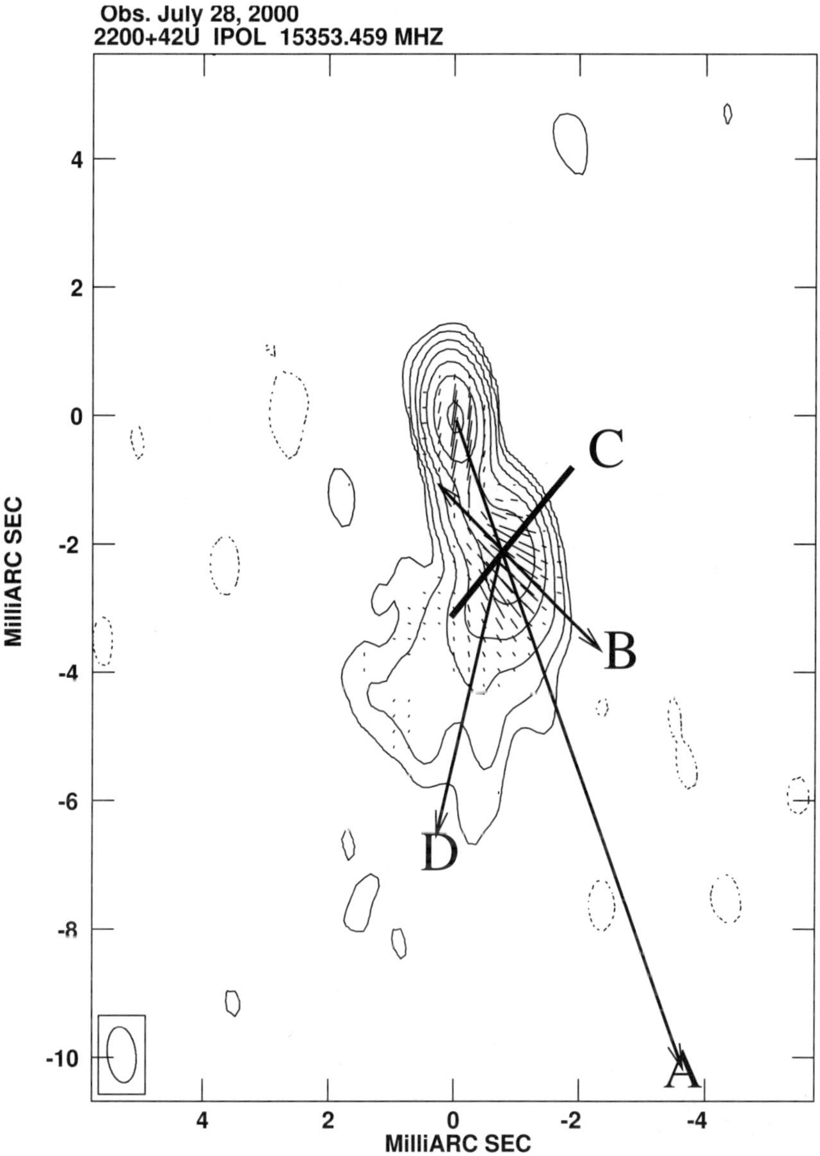

Figure 6. The total flux contour map of the July 2000 image of BL Lac with superimposed lines representing the amplitude and orientation of the polarized emission. Arrow A represents the direction of motion of the bright component away from the core, B represents the orientation of the electric vector in the bright component, C represents the inferred apparent orientation of the shock front in the bright component, and D represents the apparent direction of the motion away from the shock front.

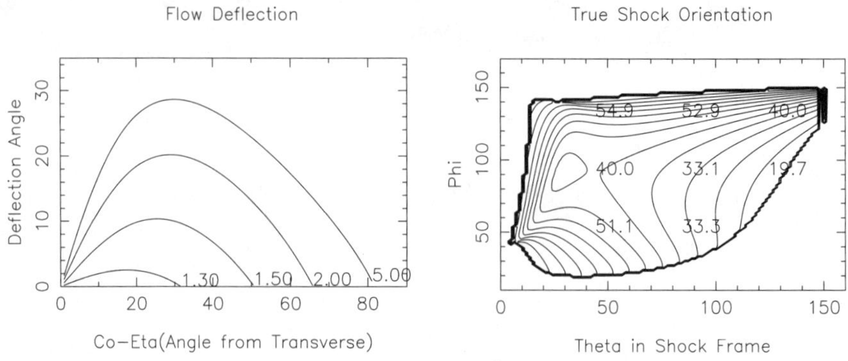

a. b.

Figure 7. **a.** The angle through which the flow is deflected at the shock front, χ, versus the strength of the shock (parameterized by the Lorentz factor of the upstream flow) and the orientation of the shock plane relative to the transverse direction. **b.** Possible solution angles for the true angle of the shock front relative to the initial flow direction for an apparent polarization of 25% at an EVPA (relative to the initial flow) of 30°. The contours are at 5° intervals.

3. Conclusions

The polarized structures observed in the images that we have analyzed to date are all consistent with oblique shocks which have similar strengths to those found in previous studies of transverse shocks. Although we can not define *unique* model parameters (shock orientation, shock strength, and viewing angle) from single images, weak limits can be placed on these parameters. Quite reasonable model parameters can produce polarization structures with the observed degree of polarization that are highly oblique to the apparent jet flow direction when aberration effects are included, and we conjecture that apparent magnetic field directions that are essentially parallel to the jet flow direction may also be produced by this mechanism.

This research has been supported in part by grant AST-9900723 from the National Science Foundation. UMRAO is operated with funds from the Department of Astronomy of the University of Michigan. The National Radio Astronomy Observatory is a facility of the National Science Foundation operated under cooperative agreement by Associated Universities, Inc.

References

Cawthorne, T. V., & Cobb, W. K. 1990, ApJ, 350, 536
Denn, G. R., Mutel, R. L., & Marscher, A.P. 2000, ApJS, 129, 61
Hughes, P. A., Aller, H. D., & Aller, M. F. 1989, ApJ, 341, 68
Lind, K. R., & Blandford, R. D. 1985, ApJ, 295, 358

Radio Astronomy at the Fringe
ASP Conference Series, Vol. 300, 2003
J.A. Zensus, M.H. Cohen, & E. Ros, eds.

What Can We Learn about Extragalactic Radio Jets from X-Ray Data?

D. E. Harris
SAO, 60 Garden St., Cambridge, MA 02138, USA

Abstract. We review the current status of resolved X-ray emission associated with extragalactic radio jets and hot-spots. The primary question for any particular jet is to decide if the X-rays come from the synchrotron process or from inverse Compton scattering. There is considerable evidence supporting synchrotron emission for knots in the jets of FRI galaxies. For FRII terminal hot-spots detected in the X-ray band, synchrotron self-Compton emission continues to provide viable models with one possible exception (so far). Inverse Compton scattering on photons of the cosmic microwave background is indicated for a few powerful jets, and is expected to be an important contributor if not the dominating mechanism for higher redshift objects. The application of a model generally yields physical parameters and in many cases, these include the Doppler boosting factor.

1. Introduction

The study of relativistic jets via X-ray emission provides certain key advantages unobtainable at lower frequencies. If the X-rays come from synchrotron emission, then we are dealing with Lorentz energies $\gamma \approx 10^7$. This fact leads to two important features: E^2 half-lives are very short which means that observed emission regions correspond to sites of injection or acceleration of particles, and that variability timescales are of order a year or less. For the most part, X-ray variability has been limited to unresolved cores; whereas now we can observe variability for knots which are well separated from the nucleus (Harris et al. 2003).

If the X-ray emission arises from the inverse Compton (IC) process, then we are dealing with electrons for which $\gamma \leq 1000$ and thus we can obtain amplitudes of the electron spectra at energies not available to ground based radio observations.

For the case of synchrotron self-Compton emission (SSC), we are able to estimate the average magnetic field strength and for the case of equipartition fields, constrain the filling factor and the contributions to the total particle energy density from low energy electrons and from protons.

In this overview which is limited to jet knots and hot-spots clearly distinct from emission associated with the nucleus, we discuss the emission processes and summarize what we have learned so far. We end with a discussion of some of the current problems and goals. We concentrate on aspects for which new information has become available subsequent to a previous short review prepared for a meeting in 2000 August (Harris 2002) and discussions on the enhancement of IC scattering on the cosmic microwave background (CMB) experienced by relativistic jets (Celotti, Ghisellini, & Chiaberge 2001; Tavechhio et al. 2000; Harris & Krawczynski 2002).

We maintain a web-site (`http://hea-www.harvard.edu/XJET/`) which currently lists 37 radio galaxies and quasars with known X-ray emission from jets or hot-spots. Before the launch of the *CHANDRA* X-ray Observatory, this table had only 7 entries: 3C 273, Cen A, M 87, Cyg A, Pic A, 3C 120, and 3C 390.3.

2. X-ray Emission Processes and what they reveal

For any given feature, it is necessary to identify the emission process before we can estimate physical parameters. For knots and hot-spots there is no longer any serious consideration given to thermal emission, although X-ray emission from hot gas may be closely associated with jets in some cases. The main problem is to decide between synchrotron and IC emissions (the latter comes of course in many different flavors depending on the nature of the target photons). Every synchrotron source must also produce IC emission from at least the CMB and the synchrotron photons; our question is, which process dominates the X-ray emission.

2.1. Synchrotron Self-Compton Emission

SSC emission models have been successfully applied to several terminal hot-spots of FRII radio galaxies and the resulting magnetic field strengths are close to equipartition for filling factors close to 1 and little or no contribution to the particle energy density from protons (Harris 2002). Hardcastle et al. (2002) have recently added to the small number of FRII hot-spots with sufficiently large photon energy densities to provide detectable SSC X-rays. In Figure 1 we show one of their sources, the double hot-spot in the quasar 3C 351. This is not a typical SSC hot-spot, nor is it a typical FRII radio structure since the hot-spot on the other side of the source is very weak. It is one of the few where the X-ray emission is substantially greater than predicted by SSC with equipartition fields. Here, and in the western hot-spot of Pic A, Doppler boosting, a significant departure from equipartition, and/or a substantial contribution from synchrotron emission are possible explanations for the excess emission.

2.2. IC/CMB

The notion that kpc scale jets of FRII sources are still moving with significant bulk relativistic velocity was argued by Bridle (1996) and others. Celotti et al. (2001) and Tavecchio, Maraschi, & Sambruna (2000) suggested that in the jet frame, the photon energy density, $u(\nu)$, of the CMB would be augmented by Γ^2 (Γ is the jet's Lorentz factor), and this could shift the primary energy loss mechanism from the normally dominate synchrotron channel to the IC channel, thus increasing the ratio of the IC to synchrotron luminosities. These authors successfully applied this idea to the jet of PKS 0637−752 and derived $\Gamma \approx 10$ for equipartition fields. However, when the model is applied to low redshift FRI radio galaxy jets, it fails since extremely large values of Γ, coupled with unreasonably small angles of the jet to the line of sight, are required (Harris & Krawczynski 2002).

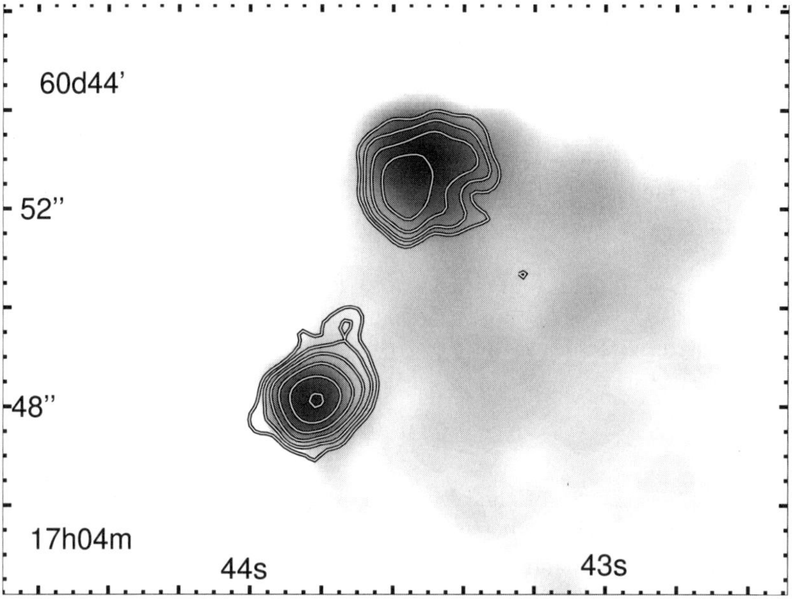

Figure 1. The double hot-spot system in 3C 351. The grey-scale shows an 8GHz VLA map with a beam size of 0.86″, comparable to the *CHANDRA* resolution. The contours are from the *CHANDRA* X-ray image smoothed with a Gaussian of FWHM=0.5″. Contour levels increase by factors of two. Note that the northern hot-spot is well resolved in the X-rays, and that there is a significant mismatch between the X-ray and radio morphologies. For most SSC hot-spots, the X-ray morphology closely mimics the radio structure.

Once the redshift becomes significant, the $(1+z)^4$ term in the energy density of the CMB becomes increasingly important, and for such sources, the 'heavy lifting' to increase $u(\nu)$ comes from the redshift rather than the Γ^2 term (see e.g., Siemiginowska et al. 2002).

2.3. Synchrotron Emission

Synchrotron X-ray emission is normally modeled by extending the power-law distribution of electron energy (with or without breaks) to Lorentz factors, γ, of 10^7 to 10^8. This sort of model can be applied to most or all X-ray knots in the jets of FRI sources and does not require excessive amounts of energy. It is true however that the E^2 lifetimes are short, and hence observed X-ray emission must demarcate injection/acceleration regions.

When constructing synchrotron (or IC) models, we often obtain estimates of the beaming parameters as a by product. In the case of the knot HST-1 in the M 87 jet, we were able to demonstrate that the variability characteristics were consistent with a synchrotron model, but not with IC emission (Harris et al. 2003). Moreover, by solving for synchrotron parameters for a range of Doppler factors δ, it was found that the observed decay rates in the light curves (harder

X-rays decay faster than softer X-rays) were consistent with synchrotron losses in a field of order 1 mG with $\delta \approx 3$.

To relate the decay of the light curve to a synchrotron loss model, we start with the normal synchrotron equations (Pacholczyk 1970) and make the simplifying assumption that if the emission in a small band drops by a certain amount, then there will be a proportionally smaller number of electrons after the drop. Although this approach assumes that the magnetic field strength remains constant, it demonstrates the method of obtaining parameters.

The electron spectrum is the usual power law.

$$N(E) = k_e \, E^{-p} . \tag{1}$$

Synchrotron losses go as E^2 (c_1 & c_2 are constants):

$$\frac{dE}{dt} = c_2 \, B^2 \, E^2 \text{ erg s}^{-1} . \tag{2}$$

Consider a segment of the electron energy distribution responsible for a given observed X-ray band between ν_1 and ν_2. Let $h \equiv \frac{\nu_2}{\nu_1}$. The number of electrons entering the band (per second) at E_2 will be $N(E_2) \times \frac{dE_2}{dt}$ and the number leaving at E_1 will be $N(E_1) \times \frac{dE_1}{dt}$.

The net change will be

$$\Delta \, N = N(E_1) \, \frac{dE_1}{dt} \, [h^{1-\frac{p}{2}} \; - \; 1] \text{ s}^{-1} , \tag{3}$$

and for a drop in flux, $f(t)$, over some given time, $t' = t'_2 \; - \; t'_1$ (primes denote the jet frame):

$$\frac{f(t'_2)}{f(t'_1)} = \frac{\int \; N(E, t'_1) \; + \; t' \times \Delta \; N}{\int \; N(E, t'_1)} . \tag{4}$$

With $t' = \delta \; t_o$ and $E' = \sqrt{\frac{\nu}{\delta \; c_1 \; B}}$, we find:

$$\delta \; B^3 \; = \; \frac{c_1}{\nu_1} \, \frac{1}{t_o^2 \; c_2^2} \left[1 \; - \; \frac{f(t_2)}{f(t_1)} \right]^2 \; \frac{1}{(p-1)^2} \left[\frac{h^{(1-p)/2} - 1}{h^{(1-p/2)} - 1} \right]^2 . \tag{5}$$

All the terms on the far right depend only on the spectral index and the bandwidth. In the center we have the ratio of fluxes at two times, and the time between intensity observations is given by t_o. Thus on the right are observables, providing a measure of $\delta \; B^3$ which can be compared with this product from trial δs and resulting equipartition fields (see Harris, Biretta, & Junor 2003 for the relevant table). For the 3 energy bands used in the M 87 analysis of HST-1, we find $\delta = 3.3 \pm 1$ (slightly different values are obtained if $\alpha \neq 1.0$)

This 'modest beaming' model is consistent with beaming parameters deduced from proper motions of substructures in the same knot at radio wavelengths (Biretta, Sparks, & Macchetto 1999).

3. Problems

3.1. High frequency, flat spectrum components - the 'bowtie' problem: Some knots have flatter X-ray spectra than they 'should'.

It has become common to differentiate IC from synchrotron emissions on the basis of the X-ray spectral index. If $\alpha_x \leq \alpha_r$, then IC emission is indicated since we expect the low frequency radio spectrum to extrapolate to unobservable frequencies with α_r or less. If $\alpha_x \geq \alpha_r$, then losses or cutoffs in the synchrotron spectrum are invoked. We believe that this approach is reasonable, but should not be considered as definitive. Fresh injection may be manifest by flat spectra at high energies, and at the other end of the electron spectrum, we don't actually know what slope the power law distribution (if indeed it is still a power law) might have at energies below those responsible for synchrotron radiation observable from the ground.

The 'bowtie problem' arises from the expectation that synchrotron spectra ($\log S_\nu$ vs. $\log \nu$) are generally concave downwards and such a spectrum extrapolated from the optical data cannot accommodate the 'bowtie' delineating the X-ray intensity and spectral index (with uncertainties) for some jet knots and hot-spots; see e.g., Pic A (Wilson, Young, & Shopbell 2001) or M 87 (Wilson & Yang 2002). Although we have argued in the past for 'concave downwards' sort of spectral fits, we no longer believe that this argument is germane to knots in jets. Rather it is a characteristic of the continuous injection (CI) model for stationary sources. For relativistic jets, the electrons resulting from a power law generation at a shock are convected downstream instead of accumulating in the local emitting volume as assumed for CI. Our view is that while synchrotron spectra of jet knots may be a single or broken power law over a wide frequency range, at acceleration sites (wherever X-rays are observed), it is expected that there may be a flat spectrum component visible only at the highest frequencies because the lower segment of such a power law is lost in comparison to the much higher intensity component which has arrived from upstream and no longer contains electrons with $\gamma \approx 10^7$. The only non-standard ingredient of this scenario is that the local acceleration site should not intercept all of the flow from upstream, thus allowing, in a sense, two spectral components: the upstream contribution which is cutoff at a lower energy and the locally (re)accelerated flat spectrum component.

3.2. Offsets in peak brightnesses between bands & progression of relative intensities in different bands

Even when care has been exercised to ensure that the effective beam sizes are close to being the same, relative offsets are sometimes observed between peak emissions at different bands. In Figure 2, we show a comparison of radio and X-ray images of M 87. Offsets can be seen for knots D, E, and F (2.5″ to 9″ from the core). Other cases are the nearest X-ray jet (Cen A, Kraft et al. 2002) and one of the more distant jets, the quasar PKS 1127−148 ($z = 1.18$; Siemiginowska, Bechtold, & Aldcroft 2002).

For synchrotron models of FRI radio galaxies (i.e., with physical offsets of order parsecs, not kpc), a reasonable explanation of the offsets would be that the acceleration region is distributed along the jet and the magnetic field strength

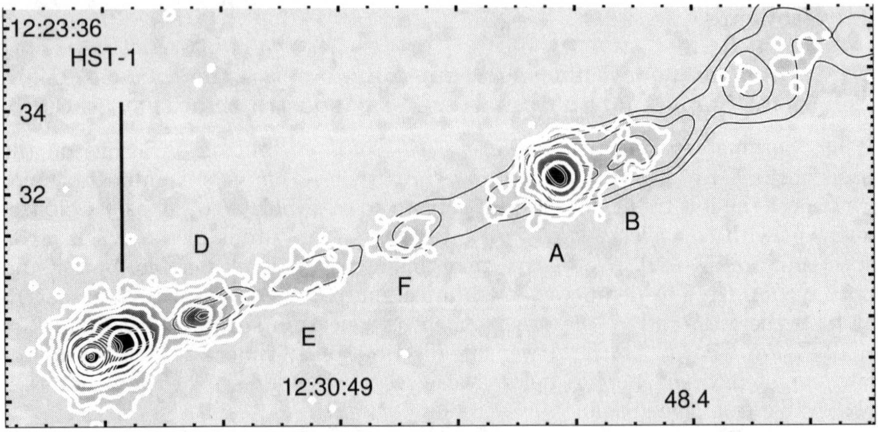

Figure 2. Radio and X-ray images of M 87. The grey-scale and fat white contours are the *CHANDRA* data with 0.5″ resolution (FWHM). Contour levels increase by factors of two and the lowest contour is $1.0\times10^{-16}\,\mathrm{erg\,cm^{-2}\,s^{-1}}$ per 0.049″ pixel in the 0.2 to 6 keV band. The thinner contours (combination of black and white) show the 8GHz VLA map with a beam size of 0.3″, comparable to the *CHANDRA* resolution. Note the offsets between the radio and X-ray brightnesses for knots D, E, & F.

is increasing after the initial shock front. Moving downstream, the increasing field strength will lower the maximum energy attainable in subsequent shocks and increase the emissivity. Since the half-lives of X-ray emitting electrons are of order a year, whereas the optical and radio emitting electrons would endure ten and 10,000 times longer (respectively), such an acceleration region would naturally produce the observed offsets. No offsets would be expected for a single shock, although the lower frequency emissions could extend further down the jet than the X-ray knot.

A somewhat similar situation exists for the overall structure of a few jets: the brightest X-ray features are close to the nucleus and the knots get fainter moving out the jet. In the case of 3C 273, the optical features have similar intensities but the radio intensities increase as one moves further downstream (the final radio 'knot' is weak or undetected at optical and X-ray energies). If one were to observe the 3C 273 jet from a great distance with only a few resolution elements for the jet, one would see an offset between the radio and X-ray peaks. Put another way, each small segment of some jets mimics the gross characteristics of the whole (observable) jet. Once again, a natural explanation would be an increasing average field strength as one moves out along the jet.

3.3. Other items needing attention

- Is there any method to check on the extrapolation to lower γ in the electron spectra which is required for IC/CMB models?

- Are there serious departures from $B_{\rm eq}$?

- Is complex jet structure required? Celotti et al. (2001) suggested the possibility of a fast spine with large Γ, surrounded by a sheath with low Γ. This obviously provides more latitude to explain observed fluxes in different bands, but is it necessary?

- How much beaming is there, and where (i.e., in FRI jets? in hot-spots?)? Currently we think Γ is of order a few in FRI jets, but we don't really have convincing limits for FRII jets.

- X-ray emission between discrete knots. If, as seems likely in several jets, there is quasi continuous X-ray emission along the jets, then the standard synchrotron model of acceleration at discrete locations (shocks=knots) fails to explain the emission between knots. Two possibilities are IC/CMB emission from cold pairs (Harris & Krawczynski 2002) and synchrotron emission from distributed acceleration such as might occur from magnetic reconnection along a magnetically dominated jet (Blandford, private communication).

4. Summary

4.1. Goals

It seems reasonable that we may expect progress in understanding a number of jet properties in the not too distant future.

- Composition: is the major carrier of momentum Poynting flux, normal plasma, or pair plasma? NB: large Γs at large distances (kpc scales) effectively kills hot pairs.

- Distribution of Γ: refining models, we should be able to obtain reasonable estimates of Γ for various jets and for features within jets.

- Jet Structure: does the fluid follow a helical path or is there a spine plus sheath structure?

4.2. What have we learned so far?

Except for terminal hot-spots, all detected X-ray jets are one sided. Therefore, whatever the emission process, Doppler favoritism is operating and we need to include beaming effects. With increased sensitivity, examples of bona fide two sided jets will most likely be found, but this is expected for the small values of the bulk relativistic velocity currently estimated for many FRI radio galaxies.

As reviewed in Harris (2002), X-ray emission from terminal hot-spots provides the following conclusions. If the X-rays are synchrotron emission from high energy electrons which are generated by Proton Induced Cascades (rather than the usual high energy tail of a power law distribution), the average magnetic field strength will be large (of order a mG) and hot-spots would be good candidates for the origin of UHE cosmic rays.

If the hot-spot X-rays come from SSC emission as is commonly accepted, then the fact that the average magnetic field required for SSC is consistent with the conventional equipartition field strength, can be taken as circumstantial evidence that relativistic protons do not contribute to the total particle energy density, and thus are most likely absent.

For the handful of hot-spots successfully modeled with SSC, we find $\alpha_x \approx \alpha_r$, as expected, and the ratio of photon to magnetic energy densities $\leq$ 0.1

From X-ray synchrotron emission from knots in jets, $\gamma \approx 10^7$ and the half-life of the highest energy electrons, $\tau_{\frac{1}{2}} \leq$ a few years. Therefore, X-rays mark the spot of acceleration sites.

From models involving IC/CMB emission, we find the following.

- Since $\nu_{IC/CMB} = 3 \times 10^{11} \times \Gamma \times (1+z) \times \gamma^2$, the electrons responsible for the observed X-rays have energies γ=30–300 and we may obtain an estimate of the amplitude of the electron spectrum at very low energies.
- Some jets are relativistic on kpc scales.
- $\alpha_x \approx \alpha_r$ (or flatter).

The radio map of M 87 was kindly supplied by F. Owen. The work at SAO was supported by NASA contract NAS8-39073 and grant GO2-3144X.

References

Biretta, J. A., Sparks, W. B., & Macchetto, F. 1999, ApJ, 520, 621

Bridle, A. H. 1996, in ASP Conf. Ser. Vol. 100, Energy Transport in Radio Galaxies, ed. P. E. Hardee, A. H. Bridle, & J. A. Zensus, (San Francisco: ASP), 383 eds.

Celotti, A., Ghisellini, G., & Chiaberge, M. 2001, MNRAS, 321, L1

Hardcastle, M. J., Birkinshaw, M., & Cameron, R. A., 2002, ApJ, 581, 948

Harris, D. E., Biretta, J. A., Junor, W., 2003, ApJ, 586, L41

Harris, D. E. & Krawczynski, H. 2002, ApJ, 565, 244

Harris, D. E. 2002, in ASP Conf. Ser. Vol. 250, Particles and Fields in Radio Galaxies, ed. R. A. Laing & K. M. Blundell, (San Francisco: ASP), 204

Kraft, R. P., Forman, W. R., Jones, C., 2002, ApJ, 569, 54

Pacholczyk, A. G. 1970, Radio Astrophysics, (San Francisco: W. H. Freeman)

Siemiginowska, A., Bechtold, J., & Aldcroft, T. L., 2002, ApJ, 570, 543

Tavecchio, F., Maraschi, L., & Sambruna, R. M., 2000, ApJ, 544, 23

Wilson, A. S., Young, A. J., & Shopbell, P. L. 2001, ApJ, 547, 740

Wilson, A. S. & Yang, Y. 2002, ApJ, 568,133

Radio Astronomy at the Fringe
ASP Conference Series, Vol. 300, 2003
J.A. Zensus, M.H. Cohen, & E. Ros, eds.

Survey of Variability

M. F. Aller, H. D. Aller, P. A. Hughes
University of Michigan, Ann Arbor, MI 48109–1090, USA

Abstract. Centimeter-band total flux density and linear polarization variability is a common property of extragalactic objects, including many with a steep spectrum: these variations occur on a variety of timescales and, in several cases, are clearly associated with the development of VLBI-scale hot spots. Within the framework of a hydrodynamical relativistic flow model we attribute these variations to the development of weak shocks in the parsec-scale flows; these arise from naturally-developing Kelvin-Helmholtz instabilities. The observed changes in electric vector position angle (EVPA) and degree of polarization during well-resolved events are consistent with a scenario in which an initially-tangled magnetic field becomes more ordered with the passage of a shock. Differences in magnetic field orientation relative to flow direction as a function of optical classification are discussed. Dual-wavelength observations of circular polarization suggest that variations in circular polarization (CP) are produced by linear-to-circular mode conversion in a partially opaque region of the source, while well-documented changes in polarity in the Stokes parameter, V, demonstrate that the magnetic field does not retain its handedness over time. Combined X-ray/radio-band monitoring of 3C 120 provides evidence for causally-related activity, plausibly yielding the first connection between activity in the radio jet and the central engine.

1. Introduction

In the 1960s, there were few multi-frequency, total flux density observations of variable sources delineating the spectral evolution during outbursts; a noteworthy exception was the double-peaked event in the radio galaxy 3C 120 in which the variations were well-fit by a simple, spherically-expanding source model (Pauliny-Toth & Kellermann 1968). We now know of hundreds of sources which exhibit this kind of variability, on a variety of timescales, and that such variations are an underlying property of the radio jets of extragalactic objects. In some cases the onset of the outbursts is clearly associated with the development of hot spots in the parsec-scale flows revealed by temporally well-sampled, coeval VLBA observations at centimeter-to-millimeter wavelengths. The spectral evolution of the total flux density, coupled with information on the linear polarization variability, continue to support an interpretation in terms of shocks which propagate within the relativistic flows of the parsec-scale jets of these sources, and which develop naturally within such flows. Both linear and circular polarization observations provide important information on the orientation, and degree of stability of the magnetic field within these flows on timescales of decades. The status of these data is reviewed here.

2. Variability in total flux: amplitude, duty cycle, & timescale

Several groups are currently engaged in variability monitoring. We concentrate here on results from the University of Michigan (UMRAO) program; these include observations of total flux and linear polarization at three primary frequencies, 14.5, 8.0 and 4.8 GHz, and of circular polarization at 8.0 and 4.8 GHz. The time-span of these data (up to 3 decades), combined with the sampling rate (semi-weekly to trimonthly depending on activity level), allow us to probe timescales of order a few weeks to decades.

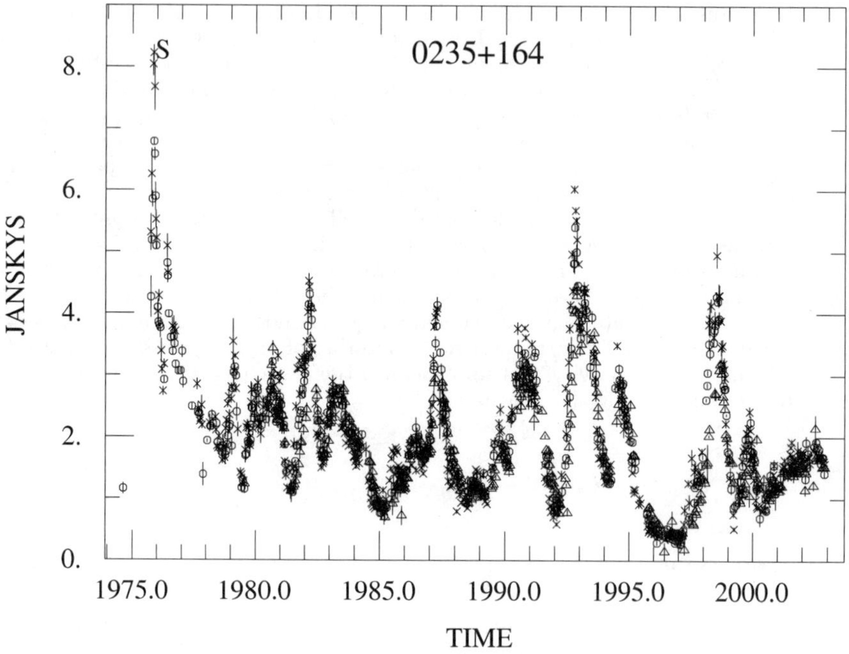

Figure 1. Two-week averages of the total flux density observations for 0235+164. The data at 14.5, 8.0 and 4.8 GHz are denoted by crosses, circles, and triangles respectively; this convention is adopted in all of the UMRAO light curves shown. Note that the light curves at the three frequencies track, indicating that the spectrum remains flat throughout the activity. Such spectral behavior is characteristic of BL Lac objects.

Figure 1 shows the total flux density data for the highly-active BL Lac object 0235+164; these illustrate the behavior that we find in the light curves for many of our program sources: nearly continuous activity in the form of discrete events which differ in amplitude and time-spacing from event to event. Embedded within these variations, which are generally accepted to be intrinsic in origin, can be intraday-to-week variations (e.g., Kraus et al. 1999) which in several cases are plausibly attributed to scattering in an external medium (Jauncey et al., these proceedings, page 199).

To obtain a characteristic timescale, we have carried out first-order structure function analyses for many of our program sources (Hughes, Aller, & Aller 1992); the turnover in plots of log(structure function) versus log(lag) gives the maximum correlation timescale; the slope of the linear portion gives an indication of the noise process. In general, these timescales are of order 2 yr with some spread; in the case of the very rapid variable 0235+164, it is 0.87 yr (0.45 yr in the rest frame). The slopes of the structure functions are almost always near 1, indicative of shot noise; this result suggests that a common process produces the variability.

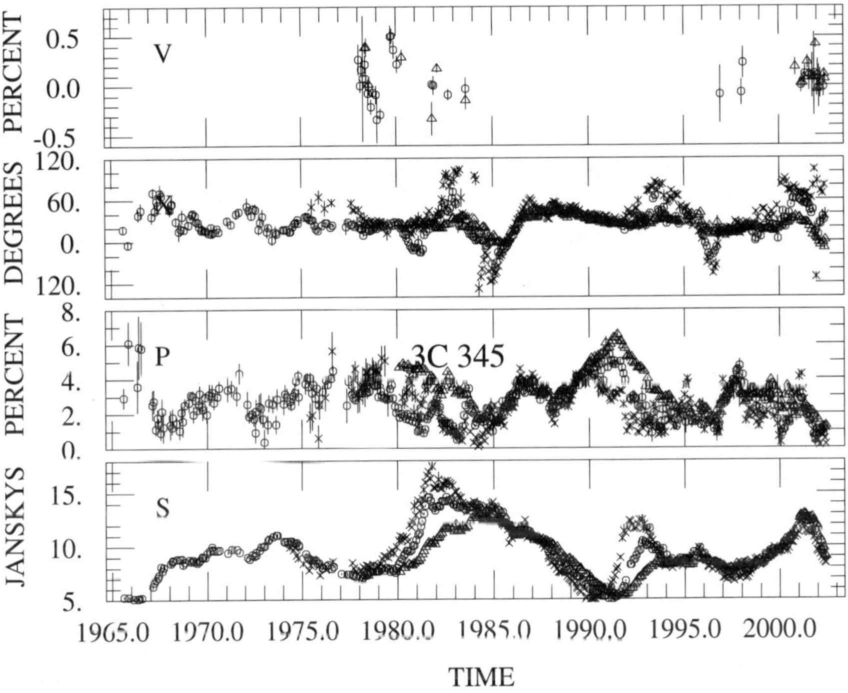

Figure 2. From bottom to top (panels 1 to 4): (1) Monthly averages of the total flux density, (2) fractional linear polarization, (3) EVPA, and (4) fractional circular polarization for 3C 345. The polarization has been corrected for Faraday rotation assuming a value of 29 rad m^{-2} (Rudnick & Jones 1983). Note the stability of the EVPA over extended time periods.

We have searched for evidence of periodicity in our program sources using both Scargle periodograms and cross wavelets, and find no strong evidence for periodic behavior, with the exception of OJ 287. However, periodogram analyses can be affected by windowing and aliasing, hampering the identification of significant peaks. A recent cross-wavelet analysis, a technique which can emphasize a quasi-periodic component in the signal, has been systematically applied to our data for the Pearson-Readhead (1988) flux and position limited sample, and has identified quasi-periodicity in several sample members (Kelly et al. 2003). In 0235+164 this technique identifies a periodicity of $\sim$ 3.7 yr. We note that these are NOT true periodicities such as might be expected from a binary star, but rather quasi-periodicities which may be associated with perturbations arising from the excitation of naturally-developing modes of oscillation of the flow (Hardee et al. 2001).

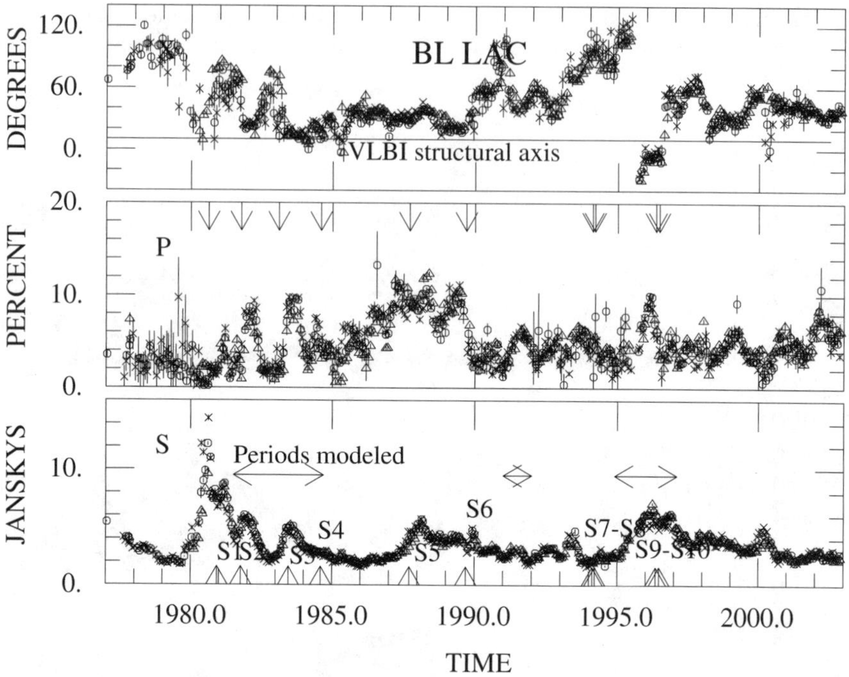

Figure 3. From bottom to top: Monthly averages of the total flux density, fractional linear polarization, and EVPA as functions of time for BL Lacertae. The polarization has been corrected for Faraday rotation assuming a value of $-205\,\mathrm{rad\,m^{-2}}$ (Rudnick & Jones 1983). The vertical arrows mark the onset of VLBI component ejections (R. Mutel, private communication). The line in the upper panel marks the flow direction indicated by VLBI morphology. Horizontal arrows in the lower panel mark time periods modeled.

In addition to the well-known flat spectrum objects, some steep spectrum objects also vary in total flux density; such behavior is found both in sources which change in character from steep to flat spectrum over time periods of order a decade (such as 3C 111), and in sources such as 3C 147 which maintain their steep spectrum character. In 3C 147, and in other similar sources, the contribution from the variable, but weak, core component is dominated by the contribution from extended structure in our integrated flux density measurements, in all but the strongest events (Aller, Aller, & Hughes 2003).

Sources also show variability in linear polarization, as is illustrated in Figure 2 which shows both the total flux density and polarization data for the QSO 3C 345. Note in panel 1 that there is overlapping of events in the individual total flux density curves, and evidence in the multi-frequency data for opacity; the latter suppresses both the amplitude of the variations and produces a time-delay with decreasing frequency. The events are more clearly resolved in polarized flux, as can be seen in panel 2. The third panel shows the position angle of the electric vector of the polarized emission, which is orthogonal to the magnetic field orientation in a transparent source. The circular polarization, shown in the top panel, is discussed in Section 4.

In several sources, there is a clear association between the development of hot-spots observed with VLBI and the onset of outbursts in total flux and linear polarization. The best longterm example is BL Lac, shown in Figure 3, but such behavior is now well-documented in other sources as well (Homan et al. 2002).

3. Interpretation/origin of variability: the shock-in-jet model

The fractional polarizations in the radio-emitting regions of extragalactic objects are typically on the order of a few percent. This can be seen in Figure 4 which shows results based on time-averaged Q and U Stokes parameters corrected for bias for two large samples: the Pearson-Readhead sample and the UMRAO BL Lac sample; these sources have been systematically observed since 1984 and 1979 respectively. During a few isolated events we have observed high values of fractional linear polarization near 20%, but high values have been short-lived, and we have *never* observed values of the order of 65-70% as expected for a canonical synchrotron emitter, nor the high values of tens of percent observed in the optical band for some BL Lac objects. The low values observed have provided convincing evidence that the magnetic fields are tangled or turbulent in the radio-emitting regions (Jones et al. 1985). However, at times of activity onset, we can sometimes identify an increase in fractional polarization coupled with a swing in the EVPA direction: these changes are plausibly attributed to a compression associated with the passage of a shock. In some sources with highly resolved activity, including BL Lac (see Fig. 3), we have been able to successfully fit the multi-wavelength flux and linear polarization light curves with radiative transfer models incorporating weak transverse shocks (Hughes, Aller, & Aller 1985a,b; 1989).

Table 1. BL Lacs versus QSOs

BL Lacs	QSOs
$\bar{P} \leq 10\%$; $\mathbf{B}_q$ turbulent	$\bar{P} \leq 10\%$; $\mathbf{B}_q$ turbulent
EVPA stable$\geq$ 1/3 of time	EVPA stable $\geq$ 3/4 of time
No preferred $\lvert$EVPA-$\Theta\rvert$	$\lvert$EVPA-$\Theta\rvert$ near 90° (4.8 GHz)
$\Gamma \sim$ few	$\Gamma \gg 1$

In spite of the high degrees of variability, in many of our sources we find a longterm preferred orientation of the EVPA relative to the flow direction, with a difference in behavior in BL Lacs and QSOs. The preferred EVPAs are identified from statistically significant peaks in histograms of the monthly averages of the EVPA, corrected for Faraday rotation, following the procedure described in

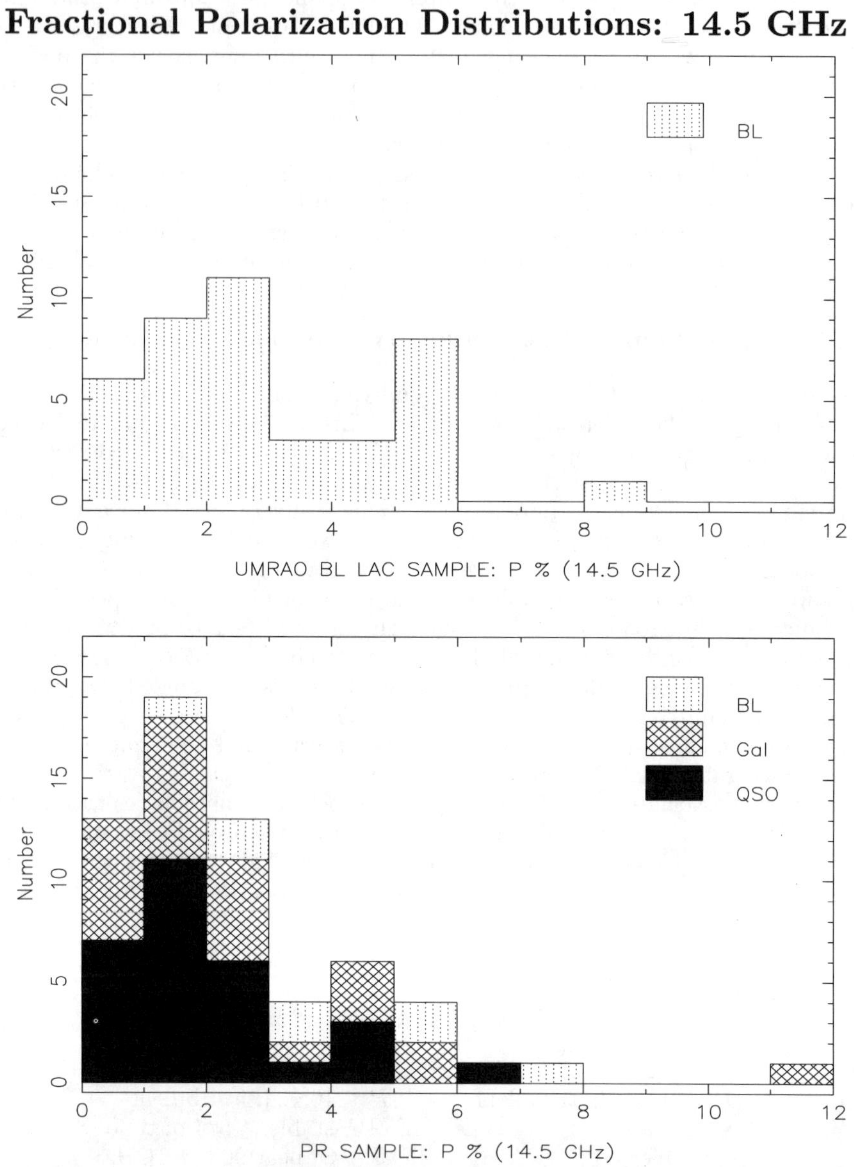

Figure 4. Histograms of time-averaged fractional polarization based on observations of the UMRAO BL Lac sample (Aller et al. 1999) and of the Pearson-Readhead (1988) sample.

Aller, Aller, Hughes, & Latimer (1999). The fraction of 30 day periods when the EVPA measurements were within ±20° of the most common value is also identified. As an example of such behavior, we show in Figure 5 the data for

1928+738, a source found from VLBI monitoring to exhibit structural variations consistent with the presence of a helically twisted magnetic field (Hummel et al. 1992). The magnetic field direction is preferentially aligned nearly parallel to the flow direction, Θ, identified from VLBI morphology. In QSOs the offset, |EVPA-Θ|, is preferentially near 90°; in BL Lacs there is a larger range of values, in general, and a shorter fractional time maintained. The fact that the orientation offsets lie within a band defining a preferred range including consecutive events suggests that there is a memory of the magnetic field from event to event and that new components do not emerge with random orientations. A comparison of the BL Lac and QSO variability and linear polarization properties is given in Table 1.

A plausible explanation for the class-dependent differences in variability coming from hydrodynamical simulations (and consistent with results from some statistical VLBI studies of component motions) is that the flows in the jets of BL Lacs are slower, giving rise more readily to the development of Kelvin-Helmholtz instabilities. In the case of the QSOs, developing shocks may be at highly oblique angles to the flow direction.

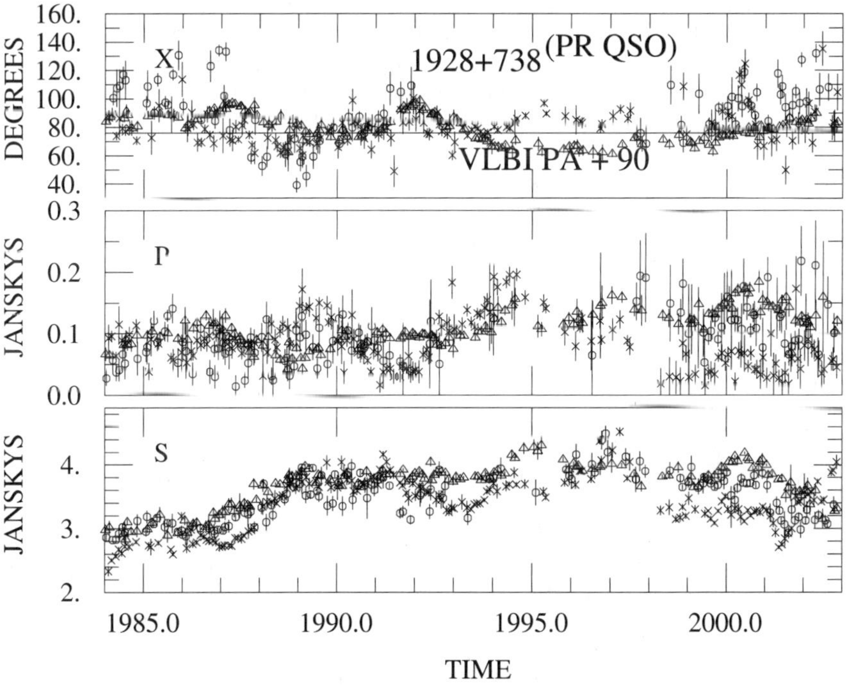

Figure 5. From bottom to top: Monthly averages of the total flux density, fractional linear polarization, and EVPA as functions of time for 1928+738. The polarization has been corrected for Faraday rotation assuming a value of 33 rad m^{-2} (Wrobel 1993). The line in the upper panel marks the flow direction +90°.

4. Circular Polarization: statistics, origin, & polarity

Circular polarization observations are exceedingly difficult because of the low levels of the emission, which are typically on the order of a few tenths of a percent. Nevertheless, there has been renewed interest in this area in the past few years because of new results from ATCA, an instrument designed with linearly polarized feeds amenable to such measurements, and because of the development of a new technique permitting the extraction of CP from VLBA measurements (Homan & Wardle 1999). In the case of linear polarization, the behavior of the variability is well-documented, as discussed above. In contrast, much of the work on CP has been devoted to initial explorations of detectability. Searches for CP in flat spectrum samples using the VLBA (Homan & Wardle 1999; Homan, Attridge, & Wardle 2001), ATCA (Rayner, Norris, & Sault 2000) and UMRAO (Aller, Aller, & Plotkin, in preparation) at centimeter wavelengths find CP in 1/4 to 1/2 of the sources surveyed at a level $\geq 3\sigma$; these results indicate that CP is relatively common in AGNs. Monitoring of CP with UMRAO at good signal-to-noise levels can only be carried out for the stronger sources, or for sources with unusually high levels of CP. Previously, it has been suggested that polarity is stable over timescales of years, suggesting that the sign is determined by the net magnetic flux carried from the central engine to the jet (Wardle & Homan 2002), but long term UMRAO observations clearly show that polarity changes are common on timescales of a few years and, further, that increased levels of CP occur during times when opacity effects are present as identified by the spectral behavior of the total and linearly polarized flux.

These characteristic behaviors are illustrated in Figure 2: 3C 345 exhibits large-amplitude variability in all four Stokes parameters, and changes in polarity of the CP at both 4.8 and 8.0 GHz which is temporally associated with the onset of a large outburst in total flux in early 1980. Both the total flux and linear polarization indicate that there is opacity in the radio-emitting region, in support of our belief that the circularly polarized emission is produced by linear-to-circular mode conversion in a partially opaque region of the radio source. We find evidence of variability and polarity changes in several additional well-observed sources, including 3C 84, 3C 273, and 3C 279.

5. Relation of radio/X-ray variability: jet-accretion disk connection

Recent RXTE monitoring in the medium soft X-ray band for the Seyfert galaxy 3C 120 has revealed unusual X-ray dips in the keV band light curve, reminiscent of the behavior seen in galactic superluminals; these are followed by the ejection of superluminal components, identified from VLBA monitoring observations at 43 GHz, with a mean time delay of 0.10 ± 0.03 yr relative to the X-ray dips (Marscher et al. 2002; see Fig. 6). Statistical analysis of several events now included in the available time window suggests that there is a causal relationship between the activity in the two bands. A plausible explanation is that the X-ray dips correspond to times when a portion of the inner accretion disk has broken off. Part of the material falls into the central black hole, while part is ejected into the jet - a scenario providing a direct link between the black hole/accretion disk and the radio jet. Further X-ray and radio monitoring is in progress to

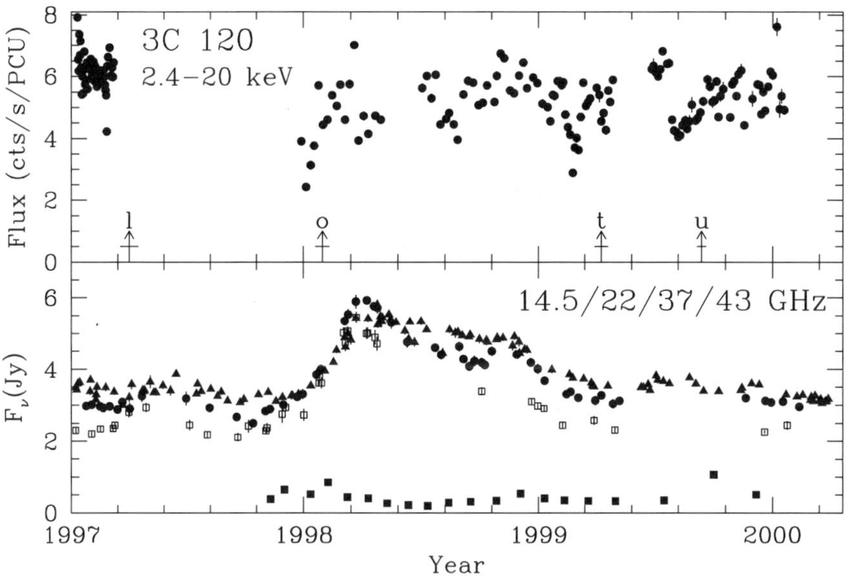

Figure 6. Radio and RXTE light curves for 3C 120. The vertical arrows in the upper panel mark the ejection times of superluminal components from 43 GHz VLBA monitoring. The radio data are from UMRAO (14.5 GHz: black triangles), Metsähovi (22 and 37 GHz: black circles and unfilled squares respectively), and the VLBA (43 GHz core only: black squares).

determine whether this correlated behavior continues. This result illustrates that variability continues to provide new and unexpected information on AGNs, giving important clues for understanding the origin and fueling of the radio jets.

This research has been supported in part by grants from the NSF and from NASA. UMRAO is operated with funds from the University of Michigan Department of Astronomy.

References

Aller, M. F., Aller, H. D., & Hughes, P. A. 2003, ApJ, 586, 33

Aller, M. F., et al. 1999, ApJ, 512, 601

Hardee, P. E., et al. 2001, ApJ, 555, 744

Homan, D. C., Attridge, J. M., & Wardle J. F. C. 2001, ApJ, 556, 113

Homan, D. C., et al. 2002, ApJ, 568, 99

Homan, D. C., & Wardle, J. F. C. 1999, ApJ, 118, 1942

Hughes, P. A., Aller, H. D., & Aller, M. F. 1985a, ApJ, 298, 296

Hughes, P. A., Aller, H. D., & Aller, M. F. 1985b, ApJ, 298, 301

Hughes, P. A., Aller, H. D., & Aller, M. F. 1989, ApJ, 341, 68

Hughes, P. A., Aller, H. D., & Aller, M. F. 1992, ApJ, 396, 469

Hummel, C. A., et al. 1992, A&A, 257, 489

Jones, T. W., et al. 1985, ApJ, 290, 627

Kelly, B., et al. 2003, ApJ, in press (astro-ph/0301002)

Kraus, A., et al. 1999, A&A, 344, 807

Marscher, A. P., et al. 2002, Nature, 417, 625

Pauliny-Toth, I. I. K., & Kellermann, K. I. 1968, ApJ, 152, L169

Pearson, T. J., & Readhead, A. C. S. 1988, ApJ, 328, 114

Rayner, D. P., Norris, R. P., & Sault, R. J. 2000, MNRAS, 319, 484

Rudnick, L., & Jones, T. W. 1983, AJ, 88, 518

Wardle, J. F. C., & Homan, D. C. 2002, in ASP Conf. Ser. Vol 250, Particles and Fields in Radio Galaxies, eds. R. A. Laing & K. M. Blundell (San Francisco: ASP), 152

Wrobel, J. M. 1993, AJ, 106, 444

Radio Astronomy at the Fringe
ASP Conference Series, Vol. 300, 2003
J.A. Zensus, M.H. Cohen, & E. Ros, eds.

Variability and Brightness Temperature

E. Valtaoja, T. Savolainen, K. Wiik
Tuorla observatory, Väisälä Institute for Space Physics and Astronomy, University of Turku, Väisäläntie 20, FIN-21500 Piikkiö, Finland

A. Lähteenmäki
Metsähovi Radio Observatory, Helsinki University of Technology, FIN-02540 Kylmälä, Finland

Abstract. We compare two methods of estimating the brightness temperatures of synchrotron sources, VLBI and flux variability. The latter method is not observationally limited, and shows that $T_{b,\mathrm{obs}}(\mathrm{var}) = T_{b,\mathrm{int}} \cdot D^3 < 10^{15}$ K. Comparing VLBI and variability data, we find that $T_{b,\mathrm{int}} \approx 5 \times 10^{10}$ K, closer to the equipartition limit than to the inverse Compton limit.

1. Introduction

It is a sign of the vigorous development of astronomy, in particular of extragalactic astrophysics, that new publications only rarely include references to papers from the 1980s, still less from the Seventies. There is, however, one persistent anomaly: the seminal paper by Ken Kellermann and Ivan Pauliny-Toth on the upper brightness temperature limit of synchrotron sources, published in 1969 - to most of today's younger astronomers, during later Middle Ages (Kellermann & Pauliny-Toth 1969). In that paper the authors established that inverse Compton losses limit the brightness temperature to about 10^{12} K, often called the KelPau limit. Strictly speaking, this limit applies only to stationary sources, and may be exceeded for a considerable time in non-stationary processes (Slysh 1992), and, of course, in coherent processes.

Although the theory of synchrotron radiation is quite well understood, the theory of real synchrotron sources is not. Thus, it is possible to argue that in realistic physical circumstances the KelPau limit is rarely, or never, reached, and the practical upper limit is determined by other processes. Readhead (1994) and others have argued that a better estimate is the equipartition brightness temperature, which for most sources is around, or below, 10^{11} K.

Why is the exact value of the upper limit important? One of the first, and still strongest, arguments for relativistic boosting in extragalactic radio sources is that the observed brightness temperatures (in the source frame) commonly exceed the synchrotron limit. It is then possible to estimate one of the most important source parameters, the Doppler boosting factor D, from simple observations, if the intrinsic limit $T_{b,\mathrm{int}}$ is known. For VLBI observations,

$$T_{b,\mathrm{obs}}(\mathrm{VLBI}) = T_{b,\mathrm{int}} \cdot D, \tag{1}$$

and for total flux density observations (where, in essence, the observed VLBI angular size θ_{VLBI} is replaced by θ_{var} estimated from the variability timescale τ_{var}):

$$T_{b,\mathrm{obs}}(\mathrm{var}) = T_{b,\mathrm{int}} \cdot D^3. \tag{2}$$

From D and the apparent VLBI expansion speed β_{app} it is further possible to determine both the viewing angle and the flow Lorentz factor Γ (e.g., Lähteenmäki & Valtaoja 1999). An uncertainty of an order of magnitude or even more in $T_{b,\mathrm{int}}$ translates into a corresponding uncertainty in these basic source parameters, fatal for arguments about unification schemes, differences between various classes of AGN, the need for relativistic boosting, etc. The possible existence of ultra-high brightness temperatures is also important for planners of future space VLBI missions.

2. Is there an observational upper limit to brightness temperature?

It is sometimes stated that due to observational limitations we do not know how high the brightness temperatures can go in radio sources. In fact, the highest-resolution VLBI can just about reach brightness temperatures around the KelPau limit (e.g., Tingay et al. (2001) for 5 GHz space VLBI with VSOP), and the upper cutoff in $T_{b,\mathrm{obs}}$ thus cannot be determined from the data. However, this argument neglects the other method for determining the brightness temperature: variability observations, which are not similarly limited.

The 22 and 37 GHz total flux density variations of over 100 brightest Northern hemisphere AGN have been monitored at Metsähovi for over twenty years (Teräsranta et al. 1998). During this period, virtually all sources have had a number of major flares, from which one can estimate the variability timescale $\tau_{\mathrm{var}} = \frac{dt}{d(\ln S)}$. A rather surprising (and still theoretically unexplained) fact was noted by Teräsranta and Valtaoja in the early 1990s: all the variations appear to have an exponential rise, a sharp cutoff, and an exponential decay with a timescale consistently about 1.3 times that of the rise (Teräsranta & Valtaoja 1994; Valtaoja et al. 1999; Fig. 1 shows an example). The only uncertainty in measuring τ_{var} is in estimating the flare flux (removing the non-varying part of the total flux density before determining the slope of the logarithmic flux density). The variability brightness temperature (in the source frame) can then be calculated with

$$T_{b,\mathrm{obs}}(\mathrm{var}) \propto h^{-2} \frac{\lambda^2 S_{\mathrm{var}}}{\tau_{\mathrm{var}}^2} (\sqrt{1+z} - 1)^2. \tag{3}$$

These data have been used for a number of purposes in a series of papers (e.g., Valtaoja et al. 1999; Lähteenmäki, Valtaoja, & Wiik 1999; Lähteenmäki & Valtaoja 1999, ; Lähteenmäki & Valtaoja 2003). The most basic result, the distribution of the observed variability brightness temperatures, is shown in Figure 2. The upper diagram shows the distribution at a single epoch (corresponding to the VSOP pre-launch survey epoch of Moellenbrock et al. (1996)). The highest variability brightness temperatures are slightly below 10^{14} K. A number of sources have very low brightness temperatures: they were not flaring

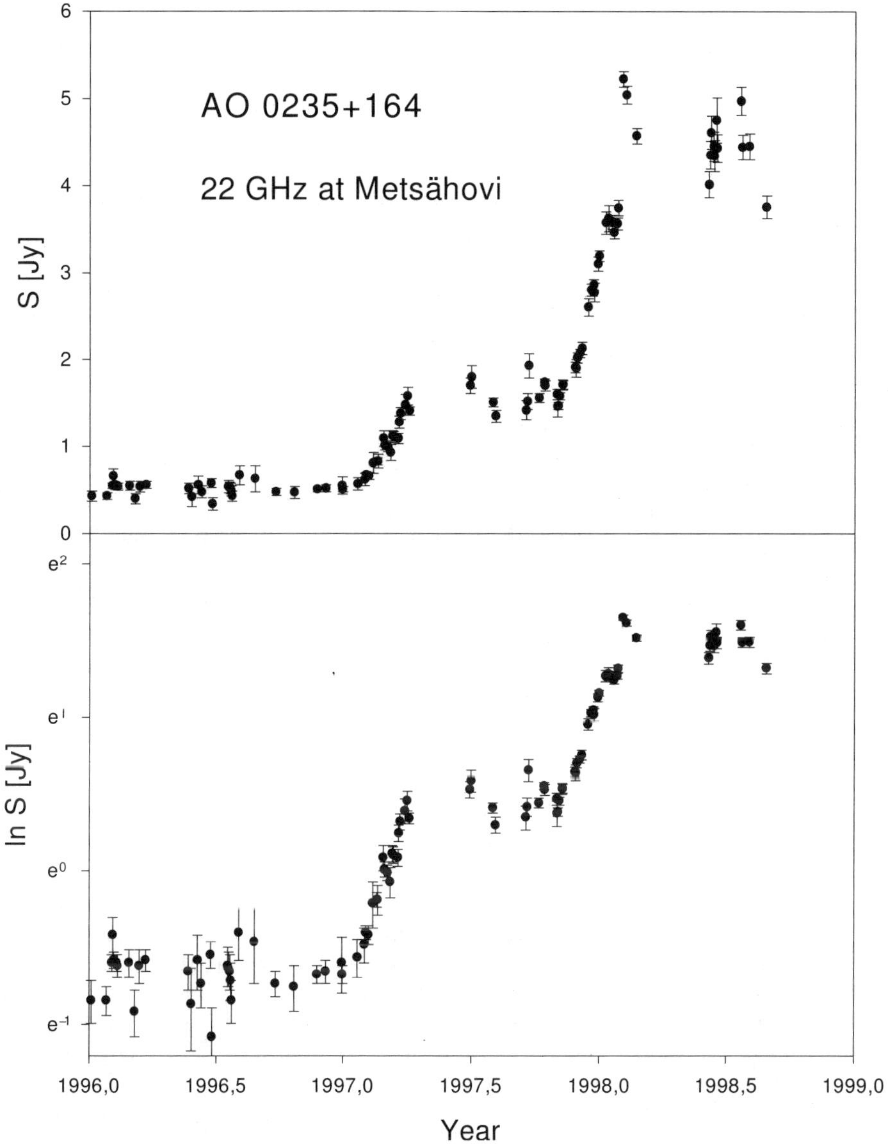

Figure 1. Metsähovi 22 GHz monitoring data for AO 0235+164, plotted both in linear and in ($\ln S$) scales. Note how closely exponential the two major flares are. Both have also same ($\ln S$)-slopes corresponding to $\tau_{\mathrm{var}} = 64$ days. From Eq. 3, with assumptions as in Lähteenmäki & Valtaoja (1999), we get $T_{b,\mathrm{obs}}(\mathrm{var}) > 4 \times 10^{13}$ K for the first and $T_{b,\mathrm{obs}}(\mathrm{var}) = 1.5 \times 10^{14}$ for the second flare.

at the time. Excluding these, the distribution of $T_{b,\mathrm{obs}}(\mathrm{var})$ is quite similar to the $T_{b,\mathrm{obs}}(\mathrm{VLBI})$ of Moellenbrock et al. (or, to take two other examples, to that obtained by Tingay et al. from VSOP observations, or the one corresponding to the flattest visibility plots in the latest installment of the 15 GHz survey; see Tingay et al. (2001) and Zensus et al. (2002)).

There is one difference: the variability values peak around 10^{13} K and the VLBI values around $10^{11.5}$ K. This difference is easily understood as a relativistic boosting effect, since $T_{b,\mathrm{obs}}(\mathrm{VLBI})$ depends on the first and $T_{b,\mathrm{obs}}(\mathrm{var})$ on the third power of D. If average $D \approx 5$, a value commonly thought to be reasonable for AGN, the distributions are in good agreement.

The lower diagram in Figure 2 shows the all-time highest observed variability brightness temperatures for each source in the sample: values derived from the "brightest" $(\tau, \Delta S)$ flares observed during all the years of monitoring at Metsähovi. The distribution has two peaks, the lower one mainly due to the BL Lacs and radio galaxies and the higher one to quasars. However, what is important is that the cutoff around 10^{15} K is *not* observational: given the average Metsähovi sampling and sensitivity, we could easily measure variability timescales an order of magnitude shorter than actually seen (the shortest observed being around a month).

This is a crucial result for discussions of brightness temperature limits: from total flux density variability monitoring we *do* know that there is an upper cutoff, around 10^{15} K, and that the highest ever observed values of $T_{b,\mathrm{obs}}(\mathrm{var})$ peak around $10^{13.5}$ K for quasars and around $10^{11.5}$ K for BL Lacs and radio galaxies. Moreover, these values are totally consistent with the VLBI results, assuming that the sources are moderately Doppler boosted, as independent other arguments suggest.

We can therefore conclude that the cutoff seen in VLBI measurements of $T_{b,\mathrm{obs}}$ is real, reflecting a true limit for the observed brightness temperature. Future higher-resolution VLBI observations will not be detecting new 'ultra-small, ultra-bright' components with $T_{b,\mathrm{obs}}(\mathrm{VLBI})$ in excess of $10^{14} - 10^{15}$ K - an important consideration for planned new space VLBI missions.

This discussion excludes the intraday variability (IDV), for which brightness temperatures up to 10^{20} K have been claimed. As Figure 2 shows, IDV — apparently seen in most of the sources at times— cannot be an extension of the major flaring events (i.e., shocks) towards smaller, more rapid variations with higher T_b's: there are no 'intermediate' variations with timescales of, say, a week, and brightness temperatures of $10^{15} - 10^{16}$ K (cf. Fig. 1). There also seems to be an increasing consensus, boosted by new observations, that IDV is not intrinsic to the sources; see, e.g., the contributions of Jauncey et al. (these proceedings, page 199) and Kedziora-Chudczer et al. (these proceedings, page 211).

3. Observational determination of the limiting brightness temperature

In addition to establishing the upper limits for the observed brightness temperatures, we can also —at least in principle— determine the intrinsic brightness temperature from the observations. As Eqs. 1 and 2 show, by measuring simultaneously both $T_{b,\mathrm{obs}}(\mathrm{VLBI})$ and $T_{b,\mathrm{obs}}(\mathrm{var})$, we can calculate both $T_{b,\mathrm{int}}$ and D

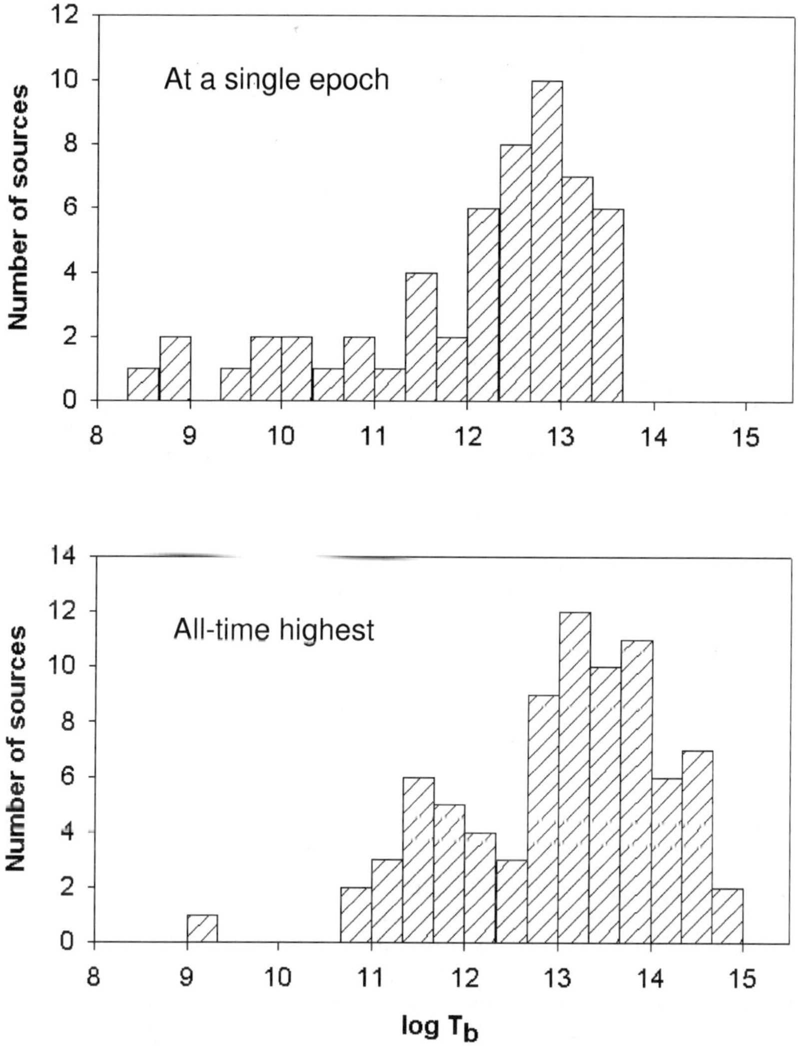

Figure 2. Distribution of the observed variability brightness temperatures (in the source frame) for the Metsähovi sample of bright, compact AGN. Upper figure: at a single epoch ("snapshot" distribution). Lower figure: all-time (1980 – 1999) highest values for each source.

for the ongoing flare (evolving shock) in the source. Such a procedure is possible, since the total flux density flares correspond to new VLBI components ejected from the core; cf. Savolainen et al. 2002). This was done for about 50 sources using three sets of VLBI data and simultaneous Metsähovi observations by Lähteenmäki, Valtaoja, & Wiik (1999). The results strongly suggested that only a very few sources seemed to have intrinsic brightness temperatures close to the KelPau limit of 10^{12} K; for most, the calculated $T_{b,\mathrm{int}}$ was around 10^{11} K, and for the data judged most reliable by the authors, $T_{b,\mathrm{int}} \approx 5 \times 10^{10}$ K. (See Fig. 1 in Lähteenmäki, Valtaoja, & Wiik (1999) showing the $T_{b,\mathrm{obs}}$(VLBI) versus $T_{b,\mathrm{obs}}$(var) for the sources.)

The error analysis of this result is very difficult. As anybody who has ever made a VLBI map knows, there are always a number of reasonable choices which lead to different final component parameters. In contrast, the determination of the variability brightness temperatures is much more straightforward - the fluxes and the timescales can be measured to good accuracy from the flux curves - but there are a number of 'hidden' assumptions such as the relation between true source size and τ_{var} (see also Section 4 below). Lähteenmäki, Valtaoja and Wiik concluded that their result is consistent with the assumption that all synchrotron sources reach the same intrinsic brightness temperature of 5×10^{10} K during the strongest flares; it is therefore reasonable to assume that this corresponds to the limiting intrinsic brightness temperature. In that case, the synchrotron sources are limited by equipartition, not by the inverse Compton cooling.

Both the Doppler boosting values for individual sources and the distributions for different types of AGN, obtained by the same method and using $T_{b,\mathrm{int}} = 5 \times 10^{10}$ K for all sources, are also consistent with those derived by other, independent methods, e.g., from X-ray and gamma-ray data or from VLBI modeling (Lähteenmäki & Valtaoja 1999). Similarly, the derived viewing angles and the Lorentz factors are also in general agreement with, e.g., unification schemes. These results serve to strengthen our confidence in the derived intrinsic brightness temperature limit also.

4. Detailed VLBI and total flux density comparisons of shocks

There are a number of questions which need to be addressed before one can decisively conclude that the equipartition limit of $\approx 5 \times 10^{10}$ K should be preferred over the KelPau limit of $\approx 10^{12}$ K. First of all, we need to be sure that the total flux density flares really correspond to the new VLBI components. This has been a tacit assumption since the advent of shocked jet models, but only recently has the correspondence been established using a larger data set (Savolainen et al. 2002).

Other uncertainties remain, however. Does τ_{var} give a reliable estimate of the source linear diameter, and if it does, what is the coefficient K in the relation $\theta_{\mathrm{VLBI}} \propto Kc\tau_{\mathrm{var}}$, used to calculate $T_{b,\mathrm{obs}}$(var) through Equation (3)? Are the VLBI component sizes systematically over/underestimated? What about the VLBI and total flux density fluxes? Is the intrinsic limiting brightness temperature, whatever its true value, same for all the sources?

These questions can be addressed by using simultaneous high quality millimeter VLBI and total flux density data. High frequencies are needed, since the typical angular sizes of 'new' (say, less than a year old) VLBI components are of the order of 0.1 mas or less, and because the individual flares at cm-wavelengths develop so slowly that they usually blend together.

The method we are presently applying is straightforward, at least (once more!) in principle. One can derive four different expressions for the Doppler boosting factor, each depending on a different combination of the observed parameters $S_{\rm var}$, $\tau_{\rm var}$, $S_{\rm VLBI}$, $\theta_{\rm VLBI}$ and the assumed values of H_0 and $T_{b,{\rm int}}$:

$$D_1 = \left(\frac{T_{b,{\rm obs}}({\rm var})}{T_{b,{\rm int}}}\right)^{1/3} \quad (4) \qquad D_3 = \left(\frac{T_{b,{\rm obs}}({\rm var})}{T_{b,{\rm obs}}({\rm VLBI})}\right)^{1/2} \quad (6)$$

$$D_2 = \left(\frac{\theta_{\rm VLBI}}{\theta_{\rm var}}\right) \quad (5) \qquad D_4 = \left(\frac{T_{b,{\rm obs}}({\rm VLBI})}{T_{b,{\rm int}}}\right). \quad (7)$$

If everything is in order (and the four expressions are all measuring the same physical factor), we should have $D_1 = D_2 = D_3 = D_4$. If this is not the case, a comparison can reveal which of the six factors on the right side of the equations is the culprit. For example, if the VLBI angular sizes are systematically overestimated, D_2 and D_3 ($\propto \theta_{\rm VLBI}$) should be larger and D_4($\propto \theta_{\rm VLBI}^{-2}$) much smaller.

A preliminary comparison, with 14 events in 9 sources, has been made using 22 and 43 GHz VLBI data kindly provided by a number of colleagues. The most obvious trend is that the 'VLBI only' Doppler boosting factor D_4 is significantly smaller than the other three, except in the two nearest sources in the small sample, 3C 120 (at $z = 0.033$) and 3C 273 (at $z = 0.158$). The four values are in general agreement, if the VLBI angular sizes are, indeed, too large: the true component sizes are smaller than average VLBA beam sizes by a factor of 1.5–2. This is in agreement with recent investigations of Savolainen et al. (2002), who made a careful comparison between the Metsähovi monitoring data and the 43 GHz VLBA monitoring data of Jorstad et al. (2001). They found that in many cases the so-called "VLBI core" consists of two components, the true core and the newest VLBI component ejected less than a year ago from the core. The new shock component is still hidden within the VLBA beam size, but spectacularly visible as a new large total flux density flare (and as a "VLBI core flare", although it is probably not the core that is doing the flaring).

5. Conclusions

We summarize our conclusions as follows:

1) Highest $T_{b,{\rm obs}}({\rm var}) < 10^{15}$ K in large total flux density flares. This is not an observational cutoff, but corresponds to a real upper limit. There is not much hope of finding ultra-small, ultra-bright components suitable for high-resolution, low sensitivity space VLBI.

2) The individual $T_{b,{\rm obs}}({\rm var})$ -values and the derived source parameters as well as distributions for different types of sources are in accordance with expectations. (For example, average Γ(BL Lac) $<$ Γ(QSO)).

3) There is a general agreement between variability and VLBI data: we are measuring the same thing!

4) $T_{b,\rm obs}$(VLBI) is usually underestimated, mainly due to insufficient resolution. At present, one should probably trust variability brightness temperatures more than VLBI brightness temperatures.

5) 43 GHz VLBA is getting to within a factor of two or so from resolving the new shock components at their brightest. (Super-resolution is not a dirty word.)

6) And, finally, the best present data favors the equipartition brightness temperature limit of about 5×10^{10} K over the $\approx 10^{12}$ K KelPau limit. However, not all the possible hidden systematic errors have been eliminated.

References

Jorstad, S. G., et al. 2001, ApJS, 134, 181
Kellermann, K. I., & Pauliny-Toth, I.I.K. 1969, ApJ, 155, L71
Lähteenmäki, A., & Valtaoja, E. 1999, ApJ, 521, 493
Lähteenmäki, A., & Valtaoja, E. 2003, ApJ, in press
Lähteenmäki, A., Valtaoja, E., & Wiik, K. 1999, ApJ, 511, 112
Moellenbrock, G. A., et al. 1996, AJ, 111, 2174
Readhead, A. C. S. 1994, ApJ, 426, 51
Savolainen, T., et al. 2002, A&A, 394, 851
Slysh, V. I. 1992, ApJ, 391, 453
Tingay, S. J., et al. 2001, ApJ, 549, L55
Teräsranta, H., et al. 1998, A&AS, 132, 305
Teräsranta, H., & Valtaoja E. 1994, A&A, 283, 51
Valtaoja, E., et al. 1999, ApJS, 120, 95
Zensus, J. A. et al. 2002, AJ, 124, 662

Radio Astronomy at the Fringe
ASP Conference Series, Vol. 300, 2003
J.A. Zensus, M.H. Cohen, & E. Ros, eds.

Variability and Velocity of Superluminal Sources

M. H. Cohen, M. A. Russo
Department of Astronomy, MS 105-24, California Institute of Technology, Pasadena, CA 91125, USA

D. C. Homan, K. I. Kellermann, M. L. Lister
National Radio Astronomy Observatory, 520 Edgemont Road, Charlottesville, VA 22903, USA

R. C. Vermeulen
Netherlands Foundation for Research in Astronomy, P.O. Box 2, NL-7990 AA Dwingeloo, The Netherlands

E. Ros, J. A. Zensus
Max-Planck-Institüt für Radioastronomie, Auf dem Hügel 69, D-53121 Bonn, Germany

Abstract. We investigate the relation between the Doppler factor determined from variations in total flux at 22 and 37 GHz, and the apparent transverse velocity determined from VLBA observations at 2 cm. The data are consistent with the relativistic beaming theory for compact radio sources, in that the distribution of $\beta_{\rm app}/\delta_{\rm var}$, for 30 quasars, is roughly consistent with a Monte Carlo simulation. The intrinsic temperature appears to be $\sim 2 \times 10^{10}$ K, close to the "equipartition value" calculated by Readhead (1994). We deduce the distribution of Lorentz factors for a group of 48 sources; the values range up to about $\gamma = 40$.

1. Introduction

Relativistic effects are commonly invoked to explain the superluminal motion and high brightness temperature ($T_{\rm b}$) seen in compact radio sources. $T_{\rm b}$ can be found from VLBI measurements or from variations in flux density; but these two methods depend differently on the Doppler factor. By measuring $T_{\rm b}$ in both ways, the Doppler factor and the Lorentz factor can be deduced, as well as $T_{\rm b}$(int), the intrinsic temperature in the synchrotron source. Lähteenmäki et al. (1999a) have done this for several sets of sources, and found that most of the $T_{\rm b}$(int) have a range around 10^{11} K. This is close to the "equipartition" value $\sim 5 \times 10^{10}$ K suggested by Readhead (1994), and the diamagnetic limit $\sim 3 \times 10^{11}$ K calculated by Singal (1986). It is at the bottom end of the range suggested by Kellermann & Pauliny-Toth for the limit due to the inverse-Compton catastrophe, $1-10\times10^{11}$ K. (See also Kellermann, these proceedings, page 185.)

In this paper we compare superluminal velocities, $\beta_{\rm app}$, with $\delta_{\rm var}$, the Doppler factor derived from variability. If a simple relativistic beaming theory is correct, there will be a close connection between these two quantities. The new large and reliable sample of superluminal sources found in the VLBA 2 cm survey (Zensus et al., these proceedings, page 27) allows this to be done with some confidence, but see Section 6 for a discussion of the $\delta_{\rm var}$. We work with velocities from the VLBA 2 cm survey and Doppler factors from the Metsähovi flux density monitoring program at 22 and 37 GHz (Lähteenmäki & Valtaoja 1999b, hereafter LV99). We convert the Metsähovi values to our assumed cosmology, $H_\circ = 65\,{\rm km\,s^{-1}\,Mpc^{-1}}$, $\Omega_{\rm m} = 0.3$, $\Omega_\Lambda = 0.7$, but otherwise use the values found in their paper.

2. The Velocities

We only use velocities from the 2 cm survey which have three or more epochs of observation with an appropriately small error to the fit, and which satisfy several criteria for morphology. (Kellermann et al. 2003, in preparation.) The velocities are determined by a linear regression through the locations of the components being followed, and the errors shown in Figure 1 are the curve-fitting errors. We are using the *fastest* component for each source, on the grounds that their velocities should be representative of the true flow velocities. Slower-moving components, especially those at a bend in the jet, may be dominated by backward shock waves. Forward shock waves might also exist, and trying to understand their role is also a goal of the survey. Other geometries have been suggested for the jet, including a fast "spine" which we would preferentially see, surrounded by a slower shell. In this case, we think, the spine would also control the flux variations, so that using the fastest (spine) velocity for $\beta_{\rm app}$ is appropriate. In the next sections we assume that the pattern velocities are identical to the flow velocities; we comment on this assumption in Section 7.

Five sources in common to the $\delta_{\rm var}$ and $\beta_{\rm app}$ lists were excluded on the grounds that they only had components situated at a bend in the jet. We believe that in these cases we see a standing shock wave, or perhaps a stationary location in a helical jet where the flow is closest to the LOS, and hence boosted most strongly. In either case the measured velocity is a poor indicator of the flow velocity, and not useful in looking for relativistic effects. The final sample we use contains 48 sources: 4 galaxies, 14 BL Lacs, and 30 quasars.

3. The Doppler Factors

The 22 and 37 GHz light curves give the flux density in an outburst and the time constant Δt. As a model we take $c\Delta t$ to be the radius of an optically thick sphere. A solid angle is calculated and then the brightness temperature $T_{\rm b}$. (See Lähteenmäki et al. 1999a for details.) A further assumption is now needed, that the measured brightness temperature may be relativistically boosted above an intrinsic temperature $T_{\rm b}({\rm int})$. LV99 choose $T_{\rm b}({\rm int}) = 5 \times 10^{10}$ K, from the estimates of equilibrium temperature for a self-absorbed synchrotron source calculated by Readhead (1994). The variability Doppler factor is calculated as $\delta_{\rm var} = [T_{\rm b}({\rm var})/T_{\rm b}({\rm int})]^{1/3}$.

A more direct way to get the brightness temperature of a compact source is to measure its diameter and flux density with VLBI. The Doppler factor can then be calculated by reference to an assumed $T_b(\text{int})$. In most cases of interest this yields a lower limit, because the cores of the sources are only slightly resolved with terrestrial interferometer baselines (Kellermann, these proceedings, page 185).

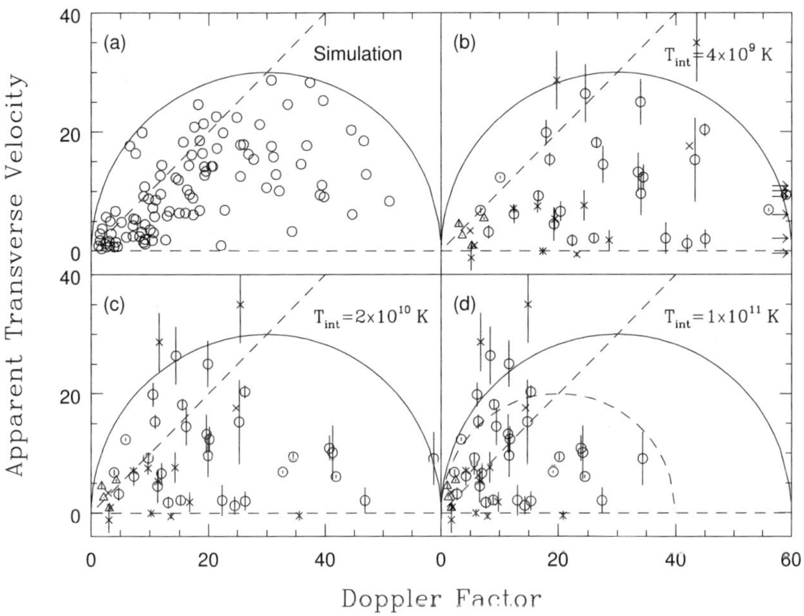

Figure 1. (a) Monte Carlo simulation of $(\beta_{\text{app}}, \delta)$ for a flux-limited sample with N=100, $a = -1.25$. The diagonal line represents the "$1/\gamma$ cone", where the apparent velocity is maximized for a fixed γ. The semi-circle is the locus of points for $\gamma = 30$. (b) As in (a) with δ_{var} calculated for $T_{\text{int}} = 4 \times 10^9$ K. Circles: Quasars; Crosses: BL Lacs; Triangles: Galaxies. N=48 (c) As in (b) with $T_{\text{int}} = 2 \times 10^{10}$ K. (d) As in (b) with $T_{\text{int}} = 1 \times 10^{11}$ K. $\gamma = 20$ is shown as the dashed curve.

An alternative method of finding a Doppler factor involves the inverse Compton effect and the ratio of X-ray to radio flux. This method depends on knowing the diameter from VLBI, but is only weakly dependent on the X-ray flux, and so is close to the pure radio VLBI method described in the previous paragraph. A few cases have been studied in detail and give good results (e.g., Unwin et al. 1994).

4. The $(\beta_{\text{app}}, \delta)$ Relation

Lister & Marscher (1997, hereafter LM97) have made Monte Carlo simulations of a flux-limited sample of compact sources drawn from a population that has a distribution of Lorentz factor $N(\gamma) \sim \gamma^a$. They assume flux boosting of the form S$\sim \delta^2$, which is appropriate for the flat-spectrum core. Figure 1(a) shows a realization for $a = -1.25$. (See LM97 for a discussion of the exponents.) In this case the Lorentz factors are set such that $\gamma = 30$ is the maximum value in the sample, so the points all fall under the $\gamma = 30$ curve. Note that a majority of the sources lie under the diagonal line $\beta_{\rm app} = \delta$, or $\theta \approx 1/\gamma$, as they should, since the probability density for selecting a source peaks near $\theta = 1/2\gamma$ (Cohen 1989; Vermeulen & Cohen 1994, hereafter VC94; LM97). For a fixed γ, the apparent velocity is a maximum on this line.

In Figures 1(b), (c) and (d) we show values of $(\beta_{\rm app}, \delta_{\rm var})$ for our sample, for three values of $T_{\rm b}$(int). We believe that our data are representative of a flux-limited sample (Zensus et al., these proceedings, page 27), and LM97 show that $a = -1.25$ gives a reasonable fit to the apparent–velocity distribution of the Caltech-Jodrell flat spectrum survey (Taylor et al 1996). Hence our points should lie within a region bounded by a γ = const curve, as in Figure 1(a). In Figure 1(b) 6 points are off scale on the x-axis, and the values of $\delta_{\rm var}$ go to 100. In Figure 1(d) we show two γ = const curves, and it appears that, as in Figure 1(b), the points will not properly fill any γ=const curve. Figure 1(c) comes closest to the simulation, and, given the small number of points (30), it perhaps provides an adequate match. These comments about Figure 1 are confirmed by calculating the fraction f of sources inside the $1/\gamma$ cone. For the Monte Carlo simulation $f = 0.80$, and for the observations the values for the 30 quasars are (for increasing temperature) 0.87, 0.77, and 0.50. On this basis 1×10^{11} can be excluded, and 2×10^{10} is somewhat better than 4×10^9.

5. The distributions of $\beta_{\rm app}/\delta_{\rm var}$ and γ

The product $\gamma\theta$ is a useful quantity because the probability density function for θ depends only on $\gamma\theta$ for $\gamma^2 \gg 1$ (see VC94, Fig. 7). Hence the measured distribution of $\gamma\theta$ can be compared with a theoretical distribution, or one generated by a simulation, to see how closely the observations match the standard theory. In the previous section we found f, the fraction of sources with $\gamma\theta < 1$, and we picked $T_{\rm b}(\rm int) = 2 \times 10^{10}$ K because its value of f is close to the expected value. However, it is clear that looking at many values of $\gamma\theta$ would be more powerful than looking at f alone, as this would test the general distribution throughout the region below the γ = const curves.

We actually use the ratio $\beta_{\rm app}/\delta_{\rm var}$, which, for $\gamma^2 \gg 1$, equals $\gamma\theta$. (A general result for a relativistic beam is $\beta_{\rm app} = \beta\gamma\delta \sin\theta$, where $\beta = v/c$.) We form the histogram of $\beta_{\rm app}/\delta_{\rm var}$, which counts the points between radii from the origin in Figure 1. In principle we can mix galaxies, BL Lacs, and quasars even though they have different parent populations, if we assume that a, and the slope of the luminosity function, are the same for all the populations. These assumptions seem somewhat dubious, however, and we only use the quasars.

Figure 2(a) shows the distribution of $\beta_{\rm app}/\delta_{\rm var}$ for the 100 sources in the simulation. Figures 2(b), (c), and (d) are for the 30 quasars, for the three values of $T_{\rm b}$(int). The histogram in Figure 2(d) is substantially flatter than that in 2(a) and supports our statement that $T_{\rm b}(\rm int) = 1 \times 10^{11}$ K is too high. In Figure 2(b) the distribution is too narrow, with a large spike in the first bin. This corresponds to the high values of $\delta_{\rm var}$ for this temperature. In Figure 2(c)

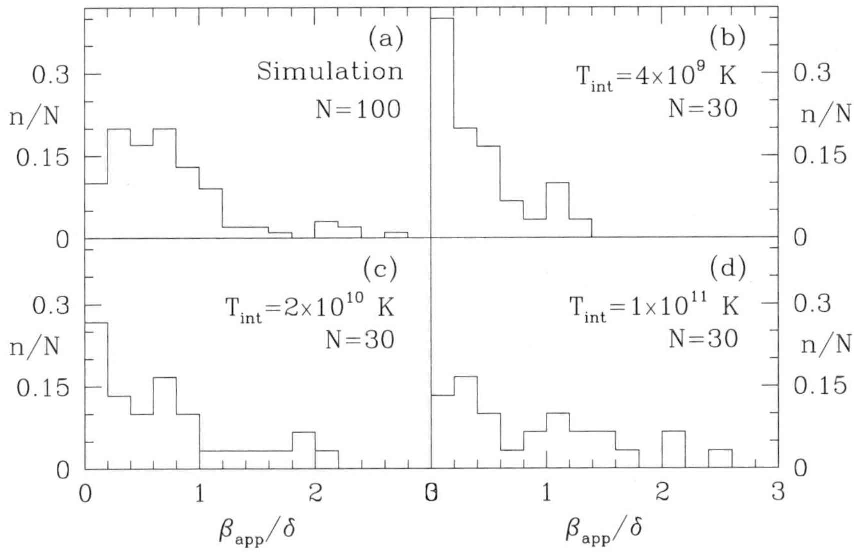

Figure 2. Histograms of $\beta_{app}/\delta = \gamma\theta$. (a) Results from a Monte Carlo simulation, $N = 100$. (b), (c), and (d) Calculated for 30 quasars, with three different values of T_b(int) used to calculate δ_{var}.

the histogram is a better match than the others, but it still has an excess in the first bin. We originally discarded 5 quasars because they had slow or stationary components located at a bend, and we believed that in those cases the measured velocity was not a good measure of the flow velocity. If we had kept those sources, they would have been in the first bin and the fit would be worse. In fact we discarded only the most egregious cases, and it is likely that there are other sources where we have measured a slow pattern rather than the flow velocity. This, we believe, explains at least part of the excess in the first bin.

The number of sources is too small to allow us to pick a particular "best" value for T_b(int). However, it seems likely that for many quasars the intrinsic temperature in the compact synchrotron source is $1 - 3 \times 10^{10}$ K. For illustrative purposes below we use $T_b(\text{int}) = 2 \times 10^{10}$ K, and we apply it to galaxies and BL Lacs as well as quasars.

We now calculate values of γ, using $T_b(\text{int}) = 2 \times 10^{10}$ K. However, four BL Lacs have negative β_{app}, with values within 1.5σ of zero. We assume that $\beta_{app} + 2\sigma$ is an upper limit to the velocity, and calculate the resulting upper limit to the Lorentz factor. The values of γ found in this way are nearly independent of how we pick β_{app}, as can be seen from Figure 1; the γ = const curves are vertical at $\beta_{app} = 0$, where $\gamma = \delta_{var}/2$. We could instead adopt a lower limit and assume that the negative sources are in the forward jet, but moving towards the center. In this case their flux would be de-boosted, whereas in fact they are strong sources. Alternately, we could assume that they are inward-moving in the back jet. We reject this because there is plentiful evidence that in these compact sources we see only forward jets.

Figure 3 shows the distribution of Lorentz factor for the 48 sources, with upper limits denoted by arrows. The quasars have a broad distribution between 5 and 25, with a few quasars both higher and lower. The galaxies and most of the BL Lacs have $\gamma < 10$. However, the two highest γ have large error bars (the two BL Lacs above the curves in Fig. 1) and they are not reliable. The quasars and BL Lacs have a wide distribution of γ, but there are insufficient data to see if they might match any of the distributions found in the Monte Carlo simulations by LM97. We note that some of their distributions for γ have a minimum in the first bin, as in Figure 3. It is interesting to note that Homan (these proceedings, page 35) has estimated γ for the source 3C 279, by analyzing a change in velocity, and assuming that it is due to a change in θ, not γ. He finds $\gamma \gtrsim 15$. Our calculation using $\delta_{\rm var}$ and $\beta_{\rm app}$ gives $\gamma = 19$, for $T_{\rm b}({\rm int}) = 2 \times 10^{10}$ K. These methods of finding γ are largely independent. Their good agreement adds to our conclusion that for many sources the intrinsic temperature is $\sim 1 - 3 \times 10^{10}$ K.

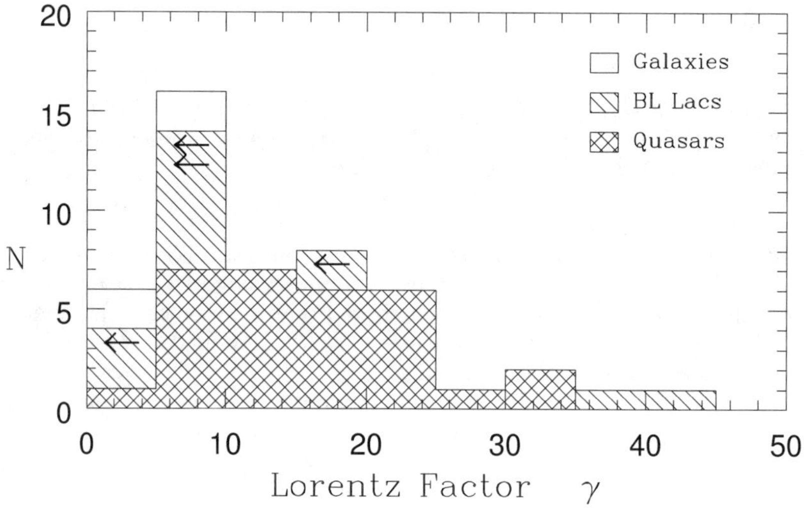

Figure 3. Histogram of Lorentz factors, calculated from $\beta_{\rm app}$ and $\delta_{\rm var}$ based on $T_{\rm int} = 2 \times 10^{10}$ K. Arrows indicate upper limits. $N = 48$

6. Reliability of $\delta_{\rm var}$

We have several concerns over the reliability of the $\delta_{\rm var}$ values. The first involves the procedure used to calculate $\delta_{\rm var}$ from a light curve. To partially address this we made some independent determinations of $\delta_{\rm var}$ using the same procedures as Lähteenmäki et al. (1999), and Valtaoja, et al. (1999). We used the 8 and 15 GHz variability curves from the University of Michigan web site, and the 22 and 37 GHz data from the Metsähovi web site. We found that the higher-frequencies gave better results, because there was less blending between outbursts. We did not try to investigate the effect of non-overlap of the epochs of the $\delta_{\rm var}$ and $\beta_{\rm app}$

measurements; but we have seen that for most sources successsive outbursts have a similar velocity. Also, Valtaoja et al. (1999) state that the apparent brightness temperature of well-defined flares does not vary much in any one source. However, the most important factor appears to be the identification of individual outbursts, even at the highest frequencies. This identification is somewhat subjective, and apparently we were more conservative and our values for $\delta_{\rm var}$ were consistently about 30% lower than the values published by LV99. In a few cases there were larger discrepancies, apparently caused by our missing some large flares not yet posted on the Metsähovi web site (Valtaoja, private communication). Thus we think that this systematic effect might limit the accuracy of $\delta_{\rm var}$ to about 30%.

The second general concern involves the model used for the source. The calculation simply assumes that the time constant of the flare is related to a radius by $r = c\Delta t$, and then that the solid angle of the source is $\Omega = \pi(r/D)^2$, where D is the metric distance to the source. If we were to assume instead that the diameter, not the radius, is given by $c\Delta t$, then the $T_{\rm b}$(int) would have to be increased by a factor of 4 to reproduce the simulation in Figure 1(a). Actually, $c\Delta t$ should be regarded as an upper limit to r; this might further increase $T_{\rm b}$(int). Furthermore, to be more realistic, the sphere should be replaced by a shock wave. See e.g., Marscher & Gear (1985).

7. Pattern and Flow Velocities

In this work we have assumed that the pattern velocity we measure with the VLBA is the same as the flow velocity of the beam. However it is clear that this is not always so. For example, we eliminated several cases where the pattern is stationary at a bend in the jet, on the grounds that the bright spot is due to a standing shock wave, and its zero motion is not representative of the flow velocity.

VC94 made a simple model to study differing pattern and flow velocities; namely, that the flow and the pattern have different γ, and their ratio $r = \gamma_{\rm p}/\gamma_{\rm b}$ has a characteristic value. From the early VLBI data they showed that if $r = 1$ in all sources, then γ cannot be constant, and vice versa. From Figure 3 it is quite clear that γ is not constant in the compact sources. That does not mean that r is constant; indeed, since we count stationary sources as having $r = 0$, then r has some distribution also. If the distribution of r mostly has values below unity, then in Figure 1 $\beta_{\rm app}$ must be increased. The distribution of γ will shift to higher values, and its shape may change. It is difficult to predict whether the best-matching $T_{\rm b}$(int) will increase or decrease, as that depends on the details of the r distribution.

8. Conclusions

We have combined variability Doppler factors with superluminal motions, for a sample of about 50 radio sources. The distribution of $\beta_{\rm app}/\delta_{\rm var}$ is roughly similar to that found in a Monte Carlo simulation, provided the intrinsic temperature in the synchrotron-emitting medium is $T_{\rm b}$(int)$\sim 2 \times 10^{10}$ K. This is near the "equipartition temperature", $\sim 5 \times 10^{10}$ K suggested by Readhead (1994). It is below the upper limit based on the inverse-Compton effect, $1 - 10 \times 10^{11}$ (Kellermann & Pauliny-Toth 1969; Kellermann, these proceedings, page 185), and the diamagnetic limit calculated by Singal (1986), $\sim 3 \times 10^{11}$ K.

The galaxies, and most of the BL Lacs, have $\gamma < 10$, when calculated with $T_b(\mathrm{int}) = 2 \times 10^{10}$ K. The quasars have a distribution which is flat between $\gamma = 5$ and $\gamma = 25$, with only a few quasars above and below these limits.

We are indebted to H. Aller, M. Aller, M. Kadler, and E. Valtaoja for helpful discussions. The NRAO and the VLBA are operated by AUI, under cooperative agreement with the NSF. This research has made use of data from the University of Michigan Radio Astronomy Observatory which is supported by funds from the University of Michigan.

References

Cohen, M. H. 1989, in BL Lac Objects, ed. L. Maraschi, T. Maccacaro, & M.-H. Ulrich, Lecture Notes in Phys. Vol. 334, (Berlin: Springer), 13
Kellermann, K. I., et al. 1998, ApJ, 115, 1295
Kellermann, K. I., & Pauliny-Toth, I. I. K. 1969, ApJ, 155, L71
Lähteenmäki, A., Valtaoja, E., & Wiik, K. 1999a, ApJ, 511, 112
Lähteenmäki, A., & Valtaoja, E. 1999b, ApJ, 521, 493 (LV99)
Lister, M. L., & Marscher, A. P. 1997, ApJ, 476, 572 (LM97)
Marscher, A. P. & Gear, W. K. 1985, ApJ, 298, 114
Readhead, A. C. S. 1994, ApJ, 426, 51
Singal, A. K. 1986, A&A, 155, 242
Unwin, S. C., et al. 1994, ApJ, 432, 103
Taylor, G. B., et al. 1996, ApJS, 107, 37
Valtaoja, E., et al. 1999, ApJS, 120, 95
Vermeulen, R. C., & Cohen, M. H. 1994, ApJ, 430, 467 (VC94)
Zensus, J. A., et al. 2002, AJ, 124, 662

Radio Astronomy at the Fringe
ASP Conference Series, Vol. 300, 2003
J.A. Zensus, M.H. Cohen, & E. Ros, eds.

Variability, Brightness Temperature, Superluminal Motion, Doppler Boosting, and Related Issues

K. I. KELLERMANN
National Radio Astronomy Observatory, 520 Edgemont Road, Charlottesville, VA, 22903, USA

Abstract. We review the observations of rapid flux density variations in compact radio sources, and discuss the inverse Compton limit to the maximum brightness temperature of incoherent synchrotron sources in comparison with recent VLBA observations. The apparent agreement of the theoretical brightness temperature limit due to inverse Compton cooling and the brightness temperatures observed by early VLBI observations appears to have been fortuitous. VLBA observations have greatly improved the quality of the data, but many of the early issues remain unresolved.

1. Variability Issues

Starting in the mid 1960s, it was becoming increasingly clear that many flat spectrum radio sources exhibited very rapid flux density variations (e.g., Sholomitskii 1965; Dent 1965; Pauliny-Toth & Kellermann 1968). This presented serious problems for conventional synchrotron models as it appeared to require a huge energy content and an excessive amount of inverse Compton cooling.

1.1. The Saga of 3C 120

One of most variable sources observed in the early years was PKS 0430+05 (Day et al. 1966) also known as NRAO 182 (Pauliny-Toth et al. 1966), which is now widely, but incorrectly, known as 3C 120. The misunderstanding about the nomenclature of this source goes back to the program of flux density and variability observations made with the newly completed 140-ft radio telescope at 10, 6, and 2 cm. The observing list for these programs was created by combining the source lists which Ivan Pauliny-Toth and I had separately made from our previous work at the 300 ft (mostly norther sources) and Parkes telescopes (mostly southern sources) respectively. One source was common to our two lists, but was called NRAO 182 by Pauliny-Toth and PKS 0430+05 by myself. Thinking that such a strong centimeter source should have been included in the better known low frequency catalogs, we searched in vain for a source at the same or nearby position in the 3CR catalog (Bennett et al. 1962)

Our inspection of the original confusion limited 3C Catalog was, at best, confusing. The position (B1950) given in the NRAO VLA Calibrator Manual is $RA = 04^{\rm h}30^{\rm m}31\overset{\rm s}{.}6$ and $Dec = +05°14'59\overset{''}{.}6$. The position of 3C 120 in the original 3C Catalog is $RA = 04^{\rm h}29^{\rm m}32^{\rm s} \pm 4^{\rm s}$ and $Dec = +01°55' \pm 8'$ which is more than 3 degrees away in declination. However, due to noise and confusion, the positions of some sources in the 3C Catalog can be in error by one or more full lobe shifts of the interference pattern. At +5 degrees declination, $\Delta RA = 57^s$ and $\Delta Dec = 2°13'59''$. Applying a correction of one lobe shift in each coordinate and allowing for an additional 2 sigma error in declination (16′) gives $RA = 04^h30^m29^s \pm 4^s$ and $Dec = +05°12' \pm 8'$. Although this is in reasonable agreement with the modern VLA position, after allowing for a lobe shift in both coordinates as well as a 2 sigma shift in declination, a 3C source can be found at almost any position in the sky. So, there is probably no connection between the original 3C 120 and the variable superluminal source which has received so much attention over the past three decades. Nevertheless, just for our own bookkeeping, somewhat frivolously we called our source "3C 120" in order not to favor either one of us with the NRAO or Parkes designation.

Subsequently, we discovered remarkably rapid flux density outbursts in "3C 120" with observed increases of as much as 1 Jy per week at 2 cm (Pauliny-Toth & Kellermann 1968). We were able to fit the change in peak flux density with frequency and time with a simple expanding source model (Shklovskii 1965; van der Laan 1966; Pauliny-Toth & Kellermann 1966) with multiple outbursts each separated by about a year. We were excited by these observations, and rushed into publication, having forgotten that we had essentially *invented* the name 3C 120 for this source. Since then, numerous authors have used this nomenclature in hundreds of subsequent publications (see NED). Apparently no one ever went back to look for 3C 120 in the original 3C or 3CR catalogs. This is a good opportunity to set the record straight. There is no 3C 120!

Curiously, no radio source since, has shown the same simple behavior. Rather most sources have multiple outbursts that overlap in frequency and time so that there is no unique decomposition into individual outbursts. In spite of the early success in interpreting 3C 120 and 1934–63 (Shklovskii 1965), and other sources, it isn't clear how to accommodate the expanding source model within currently popular shock in jet models (e.g., Marscher & Gear 1985). Unfortunately, there were no VLBI observations at that time to study how the structure changed with time. Were there really multiple expanding components? If so, were they coincident in space, or did they propagate along a jet-like path? What was the size and expansion velocity of these transient outbursts? These questions partially motivated the development of the first NRAO-Cornell VLBI system. But, by the time the MkI VLBI system became operational in 1967, the isolated outbursts in 3C 120 had ceased. In subsequent years the outbursts have been weaker, and like other quasars and AGN individual outbursts overlapped in both frequency and time.

2. The Inverse Compton Catastrophe and Superluminal Motion

As 3C 120 is relatively nearby, it did not present any particular energetic problems. Other sources, such as 3C 279 appeared to require enormous energy, with up to 10^{60}erg in the form of relativistic particles generated in less than one year (Pauliny-Toth & Kellermann 1966). Moreover, causality arguments placed limits on the size and corresponding radiation density of these transient sources. The amplitude and variability time scales of the observed outbursts appeared to be inconsistent with the calculated energy losses due to inverse Compton scattering (Hoyle, Burbidge, & Sargent 1966). It was particularly difficult to understand the variations that we and others observed at the longer wavelengths (Pauliny-Toth & Kellermann 1966; Fanti 1979; Hunstead 1972). These issues were exacerbated with the discovery of even more rapid intraday variability (IDV) (see Jauncey et al., these proceedings, page 199). Although now both the low frequency variablity and IDV are thought to arise from scattering in the ISM, the required dimensions still strain conventional theory.

The discovery of superluminal motion (Cohen et al. 1971, Whitney et al. 1971) offered a simple solution in terms of relativistic beaming (Cohen et al. 1977) following ideas suggested earlier by Rees (1966; 1967) and Woltjer (1966). However, recently, more extensive observations show little or no direct relation between variability time scales and observed velocity for individual sources; but, there remain ambiguities in the definition of the variability time scales. It is encouraging, however, that the distribution of observed velocities and variability time scales is consistent with a realistic distributions of Lorentz factors (Cohen et al., these proceedings, page 177).

3. T_b(max) and Inverse Compton Cooling

Kellermann & Pauliny-Toth (1969) expressed the inverse Compton cooling arguments in terms that could be directly related to observations. We showed that for any realistic magnetic field strength that the ratio of inverse Compton radiation to synchrotron radiation is simply a function of the peak brightness temperature, T_b. We pointed out that if $T_b > 10^{12}$ K, the relativistic electron population would be quickly quenched by inverse Compton cooling. Early VLBI observations appeared to indicate that for the compact flat spectrum radio sources, $T_b \sim 10^{11-12}$ K. However, with the benefit of 30 years of hindsight, both the derivation and the apparent agreement with VLBI observations were due to a series of fortuitous accidents.

Ivan and I did not intend to discuss the brightness temperature of synchrotron sources. Rather, we wanted to show from our multi-frequency flux density observations, that the flat spectra observed for many sources actually consisted of multiple peaks and valleys; and that this was due to the superposition of multiple components each of which becomes optically thick at different wavelengths.

This was in the very early days of VLBI, but multi-baseline multi frequency observations were already showing that, as expected from synchrotron theory, that components with the highest self absorption cutoff frequency, had the smallest angular diameter (Kellermann et al. 1969). Partly as a test of synchrotron radiation models, and partly to see if we could determine the magnetic field strength in the individual components, we plotted the spectral cutoff frequency, ν_m vs. S/θ^2. As shown in Figure 1, we also plotted lines of constant brightness temperature. Using the relation between angular size, surface brightness, and magnetic field, we showed lines of constant magnetic field strength, and we noted that the observed angular size and self absorption cutoff frequencies were consistent with $B \sim 10^{-5}$ to 10^{-4} Gauss. We were surprised to see that the observed values of T_b were essentially all in the narrow range 10^{11} to 10^{12} K.

We spent several weeks trying to understand what was special about $T_b \sim 3 \times 10^{11}$ K. Finally, we realized that with a simple substitution of variables

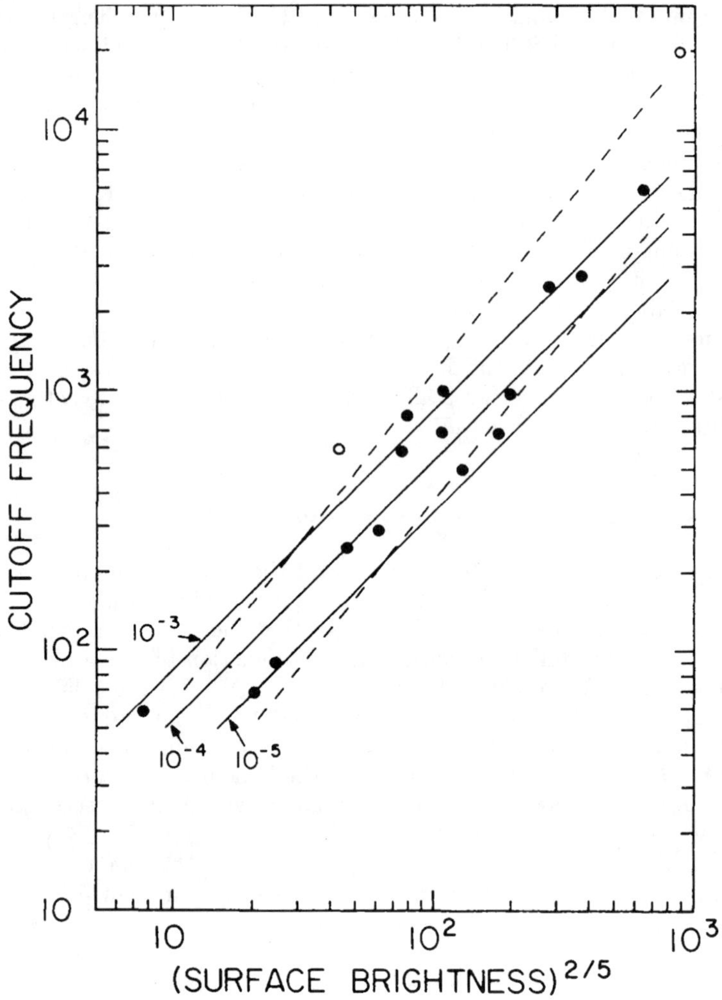

Figure 1. Spectral cutoff frequency (MHz) vs. (surface brightness)$^{2/5}$ for a number of opaque sources or components of sources. The solid parallel lines show the expected dependence for different values of the magnetic field. The upper and lower dashed lines represent constant brightness temperatures of 10^{11} and 10^{12} K respectively. *Taken from Kellermann & Pauliny-Toth 1969, ApJ, 155, L71.)*

the ratio of inverse Compton scattering to synchrotron radiation, L_c/L_s can be expressed simply in terms of the observed brightness temperature. Hoyle, Burbidge, & Sargent (1966) had noted that if $L_c/L_s > 1$, then the effect of inverse Compton cooling diverges, so we realized that values of $T_b > 10^{12}$ K cannot be sustained.

With the benefit of hindsight, the agreement between our derivation of $T_b(\max)=10^{12}$ K and the apparent maximum observed brightness temperature was fortuitous and was the result of the following over-simplifications and coincidences.

1) The limiting brightness temperature depends on the electron energy distribution, the upper energy limit, and the detailed geometry. We made our calculations for a uniform slab geometry as that was the only geometry for which we could handle the math, and reported that $T_b(\max)$ is in the range 10^{11-12} K, not 10^{12} K as is often stated in the literature. Based on more detailed calculations, Readhead (1994) estimates 3×10^{11} K as the limiting brightness temperature due to inverse Compton cooling.

2) There is no reason to assume that the VLBI observations were made at the peak of the spectrum. Below the peak, the flux density falls off rapidly as $\nu^{2.5}$. Above, the cutoff frequency, the brightness temperature also falls off rapidly as $\nu^{-2.7}$. Thus the observed brightness temperature is likely to be less than the actual peak value.

3) The early VLBI observations used in our study had only a limited number of visibility measurements which we interpreted in terms of simple circular Gaussian components. We now know, of course, that the true structure is more complex, usually with multiple separated components. Thus, the apparent sizes that we used to calculate T_b reflected the component separation rather than the size of any individual component, and so it should not be used to calculate T_b.

4) The inverse Compton limit to the maximum observed brightness temperature refers only to a source at rest. Doppler boosting increases the observed brightness temperature by the Doppler factor, δ. Most of the sources which we included in our 1969 analysis are now known to show superluminal motion, (see Zensus et al., these proceedings, page 27). These sources have significant Doppler factors, so their brightness temperature can significantly exceed the inverse Compton limit.

5) There is growing evidence that that in some sources, at least, the maximum brightness temperature is limited by free-free absorption from an intervening ionized medium rather than by synchrotron self absorption, (e.g., Vermeulen et al. 2003; Shaffer, Kellermann, & Cornwell 1999; Lister, these proceedings, page 71). However, these sources also have measured angular sizes which are consistent with self absorption, so that there appears to be a mixture of SSA and FFA taking place. It is curious however, that in these sources, the FFA and SSA opacities appear to be comparable.

6) Coherent processes, which can increase the observed brightness temperature above 10^{12} K may be important (e.g., Melrose 2002).

7) Synchrotron radiation from ultra relativistic protons can reach a peak brightness temperature which is about a factor of $(m_p/m_e)^{9/7} \sim 10^4$ higher than from electrons (Kardashev 2000). However, in this case, the required magnetic field is much larger by a factor of $(m_p/m_e)^2$ which is probably unrealistic.

8) Pauliny-Toth & Kellermann (1966) noted that for a source at rest, causality arguments indicated that the energy in relativistic electrons greatly exceeds that in the magnetic field and would be unrealistically large. The effect of Doppler boosting is to increase the apparent angular size and magnetic field strength, and so lower the relativistic particle energy requirements by $\delta^{-7/2}$. Doppler factors $\delta \sim 10$ will typically bring the particle and magnetic energies

into equilibrium. In this case the the peak brightness temperature is reduced by about a order of magnitude over the inverse Compton limit (Readhead 1994).

9) As we pointed out in our 1969 paper (see also Slysh (1992), the inverse Compton limit only applies after the source has reached equilibrium. For a limited period following the injection of a new supply of relativistic electrons, or if there is continuous reacceleration, brightness temperatures may be observed which are in excess of the inverse Compton limit. In this case, copious X-ray emission should be observed. For example first order Compton scattering will allow T_b to exceed 10^{12} K for about one year following the injection of relativistic electrons. Higher order scattering will reduce the lifetime further, but the scattering is ultimately limited by quantum effects.

10) Finally, we point out that the measured values of brightness temperature which prompted our discussion of $T_b(\max)$ and inverse Compton cooling was merely the natural consequence of the size of the Earth and the limited range of flux density observed, and had nothing to do with radio source physics. This is because the resolution of any interferometer is given by $\theta = \lambda/D$ where λ is the wavelength of observation, and D is the baseline length. For a radio source of flux density S and angular size, θ, the brightness temperature $T_b = 2k\lambda^2 S/\theta^2$. Thus $T_b \propto SD^2$, and is independent of angular size and observing wavelength. For $1 < S_{\mathrm{Jy}} < 10$, and $D \sim 5000$ km, $T_b \sim 10^{11-12}$ K.

4. Current Issues

More than three decades have passed since Ivan and I considered the question of a limiting brightness temperature. In many ways we are still asking the same questions about the same issues. However, due to tremendous advances in both the observational data and theoretical models, our questions are at a much higher level than before. Although the observations sometimes strain conventional synchrotron models and relativistic boosting, there does not appear to be any show stoppers. However, the number of free parameters needed to accommodate the observations has become uncomfortably large, and it has proven difficult to confirm by observation the predicted effects of relativistic motion. The early models make simple predictions. But, today, consideration of jet instead of ballistic flows, possible differences between the bulk velocity and the pattern velocity, combined with a distribution of intrinsic Lorentz factors has removed the predictability, and thus the simplicity, if not the attractiveness of these models.

It is perhaps sobering to recall that the concept of relativistic beaming was first introduced in the mid 1960's to explain the rapid variability that was observed in quasars, and which appeared to be dramatically confirmed by the discovery of superluminal motion in 1971. But, predictions of the distribution of observed velocity distribution can only be reconciled with observations by introducing modifications to the simple theory which involves new free parameters. Meanwhile simple non-Doppler boosting models such as suggested by Ekers and Liang (1989) are surprisingly consistent with the observed velocity distribution. The main evidence for Doppler boosting remains, as in the 1960's, the rapid flux density variations, not superluminal motion.

As discussed by Cohen et al. (these proceedings, page 177) application of the standard causality argument combined with the measured velocities can be used to estimate the intrinsic brightness temperature and Doppler factor. Although there is no simple direct relationship between the brightness temperatures calculated from variability time scales and measured velocities, the observed values are consistent with a reasonable distribution of Lorentz factors. However ambiguities in defining the variability time scales, in relating the variable component to the moving component, and the apparent differences between the bulk velocity and pattern velocity may preclude any meaningful quantitative conclusions.

Modern VLBA (Kellermann et al; 1998, Zensus et al. 2002; Kovalev, these proceedings, page 65) and VSOP space VLBI (Tingay et al. 2001) observations suggest maximum observed brightness temperatures in the range 10^{11-13}, much as we measured in the first years of VLBI. Although VSOP uses longer baselines, fringe visibilities are more accurately determined on the VLBA so the corresponding size limits for unresolved components are comparable for the two instruments. The Russian *RadioAstron* space VLBI mission which is scheduled for launch in 2006 will give more than an order of magnitude increase in baseline length. This will directly measure brightness temperatures up to 10^{15} K and perhaps allow better comparison with calculated Doppler factors.

The maximum relativistic boosting occurs when the the motion is oriented directly along the line of sight. In this case, $T_b(\text{obs}) \sim 2\gamma T_b(\text{int})$ where γ is the Lorentz factor. The observed range of T_b is consistent with the observed range of apparent velocity and intrinsic values of $T_b \sim 10^{10-11}$ K, or close to both the inverse Compton limit and the equilibrium value. However, perhaps surprisingly, there is no simple one-to-one relation between the observed brightness temperature and measured superluminal motions.

Using the same kind of causality arguments used in the 1960s to "predict" superluminal motion, IDV observations suggest brightness temperatures as much as $\sim 10^{21}$ K they are intrinsic and $T_b \sim 10^{15}$ K if the variations are due to interstellar scintillations as seems likely, (Jauncey et al., these proceedings, page 199). It may be hard to reconcile these numbers with reasonable Doppler factors.

In principle, the intrinsic brightness temperature and Doppler factor can be determined from the ratio of observed inverse Compton X-ray flux density to the radio flux density from synchrotron radiation, (e.g., Unwin et al. 1997). However, there are three problems in implementing this technique. 1) It is difficult to get contemporaneous radio and X-ray light curves. 2) We don't know whether the X-rays are coming from the core or from somewhere along the jet. 3) There may be an unknown additional (non-Compton), amount of X-ray emission for example from the hot ionized torus surrounding the central black hole.

The concept of equipartition in extragalactic radio sources was first introduced by Burbidge (1959) to minimize the total energy content of extended lobe radio sources which typically have ages of 10^{8-9} years. We do not know if it is appropriate to assume that equipartition conditions apply in the short lived compact cores and jets. Indeed, the presence of bulk relativistic motion in the parsec scale jets suggests that the plasma may not be contained by the magnetic field so that there may be an excess of particle energy over magnetic energy. Comparison of Doppler factors deduced from variability time scales and measured proper motions suggests that the intrinsic brightness temperatures may be closer to the equipartition value than to the inverse Compton limit (Cohen et al., these proceedings, page 177), but the difference between the two is small and perhaps not meaningful.

I am grateful to the many colleagues with whom I have had the privilege of working; especially to Ivan Pauliny-Toth who could not be here with us, but who collaborated with me for many years in the work described in this paper; as well as Marshall Cohen, who has been a valued colleague and good friend for nearly all of the 40 years that I have been working in radio astronomy, and who organized this meeting. Special thanks go to Dan Homan and Matt Lister who have tried to educate me in the subtleties of the LaTeX language. The NRAO is is a facility of the NSF which is operated under cooperative agreement by AUI.

References

Bennett, A. S. 1962, MNRAS, 68, 1
Burbidge, G. R. 1959, ApJ, 129, 849
Cohen, M. H. et al. 1971, ApJ, 170, 207
Cohen, M. H. et al. 1977, Nature, 268, 405
Day, G. A., et al. 1966, Austr. Jour. Phys. 19, 35
Dent, W. A. 1965, AJ, 70, 672
Fanti, R., Ficarra, A., Mantovani, F., Padrielli, L., & Weiler, K. 1979, A&AS, 36, 359
Hoyle, F., Burbidge, G. & Sargent, W. 1966, Nature, 209, 751
Hunstead, R. W. 1972, ApJ, 12, L193
Kardashev, N. S. 2000, Astron. Rep., 44, 719
Kellermann, K. I., et al. 1968, ApJ, 153, L209
Kellermann, K. I. & Pauliny-Toth, I. I. K. 1969, ApJ, 155, L71
Marscher, A. P. & Gear, W. K. 1985, ApJ, 298, 114
Melrose, D. B. 2002, PASA, 19, 34
Pauliny-Toth, I. I. K., Wade, C. M., & Heeschen, D. S. 1966, ApJS, 13, 65
Pauliny-Toth, I. I. K. & Kellermann, K. I. 1966, ApJ, 146, 634
Pauliny-Toth, I. I. K. & Kellermann, K. I. 1968, ApJ, 152, L169
Rees, M. J. 1967, MNRAS, 135, 345
Readhead, A. C. S. 1994, ApJ, 426, 51
Shaffer, D. B., Kellermann, K. I., & Cornwell, T. J. 1999, ApJ, 515, 558
Shklovskii, I. S. 1965, AZh, 42, 30
Sholomitskii, G. B. 1965, Soviet Ast., 9, 516
Slysh, V. I. 1992, ApJ, 391, 453
Tingay, S. J. et al. 2001, ApJ, 549, L55
Unwin, S. C., et al. ApJ, 480, 596
van der Laan, H. 1966, Nature, 211, 1131
Vermeulen, R. C. & Cohen, M. H. 1994, ApJ, 430, 467
Vermeulen, R. C., et al. 2003, A&A, 401, 113
Woltjer, L. 1966, ApJ, 146, 597
Whitney, A. et al. 1971, Science, 173, 225
Zensus, J. A., et al. 2002, AJ, 124, 662

Radio Astronomy at the Fringe
ASP Conference Series, Vol. 300, 2003
J.A. Zensus, M.H. Cohen, & E. Ros, eds.

Space VLBI and High Brightness Temperature Observations

H. Hirabayashi
Institute of Space and Astronautical Science, 3–1–1 Yoshinodai, Sagamihara, Kanagawa 229-8510, Japan

Abstract. The maximum brightness temperature that can be probed depends only on the maximum baseline length, and thus one of the advantages of space VLBI is its ability to measure brightness temperatures beyond the inverse Compton limit of 10^{12} K. In this paper, results from *TDRSS*-OVLBI observations and the VSOP mission are reviewed, and the impact of the VSOP-2 mission outlined.

1. Introduction

The VLBI Space Observatory Programme (VSOP) is a Japanese-led project to image radio sources at 1.6 and 5 GHz with a linear resolution three-times greater than with ground arrays at these frequencies (Hirabayashi et al. 1998). From February 2002, VSOP observations are being made predominantly for the VSOP Survey Program, an on-going, mission-led survey of Active Galactic Nuclei (AGN) at 5 GHz, the major goals of which are statistical in nature: to determine the brightness temperature and approximate structure; to provide a source list for use with future VLBI and space VLBI observations; and to compare the VLBI properties with data from other wavelengths.

2. Brightness Temperature

The standard inverse-Compton limit for incoherent radio emission is 10^{12} K in the rest frame of the emitting plasma (Kellermann & Pauliny-Toth 1969). A number of alternative, and generally lower, limits have been postulated, arising from equipartition limits (Readhead 1994) or other physical processes (Sincell & Krolik 1994). A detailed discussion of the various proposed limits to brightness temperature are discussed by Kellermann (2002).

Coincidentally, for ground-based interferometric observations the maximum brightness temperature that can be obtained is $\sim 10^{12}$ K, a limit imposed by the size of Earth. The highest interferometric brightness temperatures reported to date are those from the *TDRSS* experiment (Linfield et al. 1989) and from observations on the longest possible ground baselines (Moellenbrock et al. 1996), which were conducted as a VSOP pre-launch survey. A number of sources were found to have source frame brightness temperatures up to $\sim 7 \times 10^{12}$ K. These values were interpreted as being due to bulk relativistic motion close to the line of sight, with the inferred Doppler factors being not inconsistent with those determined from superluminal motion studies.

By these brightness temperature measurements, the time had ripened for space-VLBI. Captain Parker at Lexington, Massachusetts, just before the start of the war of US Independence in 1775 commanded the Minutemen (Colonists) against the troops of approaching Redcoats (British), "Stand your ground. Don't shoot until you see the whites of their eyes." As the bright cores were observed, the spacecraft *HALCA* was launched in 1997.

VSOP observations have allowed brightness temperatures to be measured on longer baselines and with a more robust, better calibrated system than these previous efforts. One of the first published results from the VSOP mission was a measurement of the brightness temperature of NRAO 530 (PKS 1730−130) (Bower & Backer 1998). More recently, Tingay et al. (2001) have derived the distribution of brightness temperatures for a carefully selected sample of 27 AGN and found a significant fraction of sources exceed the inverse Compton limit.

VSOP observations have also been used to measure the brightness temperature for a galactic OH maser (Slysh et al. 2001). The brightest OH maser spot in the star-forming region OH 34.26+0.15 was found to be only partially resolved, and had a brightness temperature in excess of 6×10^{12} K. Results such as this have implications for our understanding of both maser physics and the interstellar medium (Slysh et al. 2001; Slysh, these proceedings, page 239).

3. The VSOP Survey Sample and Observations

The VSOP Survey program is a large, mission-led survey for the statistical study of AGNs. The VSOP Survey sample includes all cataloged extra-galactic radio sources with:

- a cataloged total flux density at 5 GHz > 0.95 Jy
- a spectral index $\alpha > -0.45$ (where $S \propto \nu^{\alpha}$)
- a galactic latitude $|b| > 10°$.

The list contains 402 sources and comprises ~60% of all sources above 1 Jy (Kühr et al. 1981). A full description of the source selection is given in Hirabayashi et al. (2000). Pre-launch observations with the VLBA at 5 GHz for all sources in the list north of declination −44° (Fomalont et al. 2000) revealed that 108 (revised from 113, see Edwards et al. 2002) of the 402 sources were sufficiently resolved on ground baselines that they were unlikely to be detectable on baselines to *HALCA*. The remaining 294 sources are the sub-set for which VSOP observations are being undertaken.

A typical VSOP Survey observation uses ~3 ground telescopes and *HALCA*, co-observing for up to ~6 hours (i.e., one *HALCA* orbit). Participating ground radio telescopes include Hobart, Hartebeesthoek, Mopra, Shanghai, Kashima, Noto and Usuda. Two 16 MHz bandwidths of LCP data are recorded. From February 2002, all real-time *HALCA* tracking and data recording is done with the Usuda tracking station, with observations being correlated at either the Penticton or Mitaka correlator. Data from the correlator are sent to ISAS for distribution to the various Survey Reduction Team members.

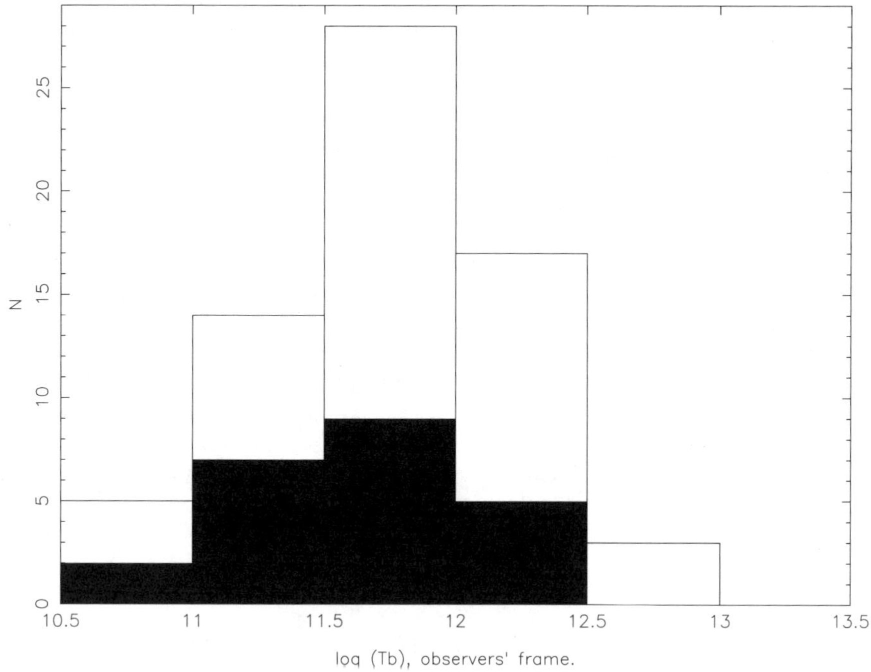

Figure 1. Observers' frame brightness temperature distribution for the first 67 VSOP Survey observations to be reduced. The filled blocks represent measured values, while the unshaded blocks represent sources where only a size upper limit, and hence brightness temperature lower limit, was able to be determined (Hirabayashi et al. 2000).

4. VSOP Survey Results to date

A study of the first 67 sources analyzed revealed that the average visibility amplitude drops more slowly beyond 200–250 Mλ, indicating that a less-resolved component tends to dominate on the space-VLBI scale at 5 GHz and that most of the more extended structure is resolved out (Hirabayashi et al. 2000). A similar measure of the compactness of the typical AGN radio core is that 15% of the sample shows no appreciable decrease in correlated flux density between 200 Mλ and 500 Mλ. The observers' frame brightness temperature distribution is shown in Figure 1. Of the 59 sources for which redshifts were available, about half had source-frame brightness temperatures greater than standard inverse-Compton limit of 10^{12} K. Although the superluminal motion in many of these sources strongly suggests that the high brightness temperatures are associated with Doppler beaming of the components nearly in the line of sight, there is at present no good statistical body of data which can test various competing theories which incorporate the radiative limits of high brightness regions, the distribution and lifetime of superluminal motions, and the connection with short-term variability in many of these sources. As the measurable brightness temperature scales with baseline length, it is only with space VLBI that brightness temperatures significantly in excess of 10^{12} K can be measured and compared with other phenomena in active galaxies and inner jet regions.

The results from the VSOP Survey Program will be enhanced by a comparison with other observational results, such as the arc-second scale radio structure, X-ray flux, morphological classification etc. Efforts have been made to schedule VSOP observations of sources included the VLBA 2cm survey (Kellermann et al. 1998) contemporaneously with one of the VLBA 2cm epochs, as the matched resolutions and spectral information obtained are useful in determining source structure. A first example illustrating the benefits gained from a comparison of a VSOP Survey Program image with a VLBA 2cm Survey image is presented in Hirabayashi et al. (2000). Total flux density and radio source variability programs (e.g. Kovalev et al. 2000; Tingay et al. 2003) provide useful information both for calibration of the VSOP data and also for correlating with other source properties. The study of intra-day variability is one example of an area that VSOP Survey results will have an impact on, as independent investigations (Quirrenbach et al. 1992; Tingay et al. 2001; Lister et al. 2001) have shown a significant link between IDV and the presence of a strong, unresolved VLBI core.

5. VSOP status and future

For the first five years of the mission, the majority of VSOP observing time was given to peer-reviewed proposals solicited from the world-wide astronomical community, designated as General Observing Time (GOT). Five hundred GOT observations, and 200 of the 294 Survey Program observations, were successfully carried out in the first five years of the mission. With the end of the Memorandum of Understanding between ISAS and NASA and corresponding loss of NASA-funded mission elements, the VSOP mission has made the S2- and VSOP-recorder based VSOP Survey Program its main priority.

At the currently sustainable rate (since October 1999, when one of *HALCA*'s four reaction wheels stopped working) of ~2 observations per week, the remaining 60 Survey observations can be made by mid-2003, with occasional GOT and Target of Opportunity observations with the S2/VSOP network telescopes also possible. The Penticton S2 correlator is funded by the Canadian Space Agency until that time, after which all correlating will be carried out at the Mitaka correlator.

A paper describing the results of the 100 survey observations is nearing completion (Scott et al., in preparation). The results from the VSOP Survey will provide a source list which will be useful for planning future VLBI observations, and future space VLBI missions, such as the VSOP-2 mission.

As was questioned already by Hamlet, "Tb or not Tb?", the brightness temperature problem seems to be historical. If we consult the tragedy of Macbeth we can find some hints in the Witches' Apparitions: When Macbeth goes back to speak with the witches again, they call up three apparitions. The first tells Macbeth to beware of Macduff, however the second tells Macbeth that no man of women born can harm him. The last explains to Macbeth that he shall never be vanquished until the Great Birnam Wood marches to Dunsinane Hill. The first seems to tell us to be cautious of ADAF, or CDAF, and the second to suggest that such exotic phenomena are related to SMB. And as we already know, the marching Great Birnam Wood refers to Doppler boosting!

The Russian *RadioAstron* mission (Kardashev 1997) can be regarded as an excellent mission to solve the brightness temperature problem because of its high orbit (4 day to 28 day orbit). It has been recently announced that *RadioAstron* (Spectrum-R) is now the first priority in the Spectrum series and the current launch date is March 2006. Possible problems in this high orbit mission are: source modeling is needed for the nearly 1-dimensional beam; the decrease in the number of detectable sources on the longer baselines; and, because of long orbit period, source structure may be affected by scattering or intrinsic source changes.

6. VSOP-2 and the future

Following the successes of VSOP, a next generation space VLBI mission, VSOP-2, is being planned (see `http://www.vsop.isas.ac.jp/vsop2/` and Hirabayashi et al. 2001). Higher observing frequencies, cooled receivers, increased bandwidths and larger telescope diameters will result in gains in resolution and interferometer sensitivity by factors of ~10 over the VSOP mission. Observing frequencies up to 43 GHz will allow high angular resolution observations of the optically thin emission in many AGN cores. An angular resolution of ~25 microarcseconds at 43 GHz will be achievable, corresponding to ~10 Schwarzschild radii at the distance of M 87. At this stage, the nominal orbit will have an apogee height of ~30,000 km (compared to *HALCA*'s 21,000 km), enhancing both the angular resolution and the sensitivity to high brightness temperatures even further.

What is needed in the future for brightness temperature measurements from an observational point of view? First, measuring T_b values and not lower limits is necessary. In the VSOP Survey, the current situation is that about 50% of brightness temperatures are measured. Here, longer baselines are needed. Secondly, for high T_b, model fitting is used now, but we need to measure the T_b distribution to understand the real morphology and physical parameters. Here, good (u, v) coverage and better sensitivity are needed. Thirdly, we need a time sequence of the T_b distribution, to understand injection/relaxation/ equilibrium timescale or morphology changes including jet direction changes. Then we can start to understand the difference between brightness temperatures inferred from variability (Valtaoja et al., these proceedings, page 169) and those measured with VLBI. These may be coupled with IDV and/or scattering phenomena. So, there remains a lot to do in future by imaging missions. A longer baseline mission allowing quick imaging with good fidelity is necessary, and a sensitive multiple spacecraft mission would be ideal.

7. Conclusions

The first results from the VSOP Survey Program show that a large number of radio sources have significant emission on the longest *HALCA* baselines, corresponding to an angular size < 0.15 mas with a brightness temperature in excess of 5×10^{12} K. A comparison of the VSOP Survey images with other high-resolution ground images and with other astronomical information will clearly aid in understanding the phenomena associated with AGN evolution. Realistic tests of radio source models involving Doppler beaming, superluminal motions, and intra-day variability, can be made with the statistical completeness anticipated with the survey database.

The author would like to present Ken Kellermann, who played an important role in this subject, on the occasion of his 65th birthday with a phrase from Anthony and Cleopatra: "Age cannot wither her." Age could be read as AGN, and her should read not as Cleopatra but as Ken.
The author would like to thank Dr. P. G. Edwards for his kind checking of the tragical English of this paper before its launch.

References

Bower, G.C. & Backer, D.C. 1998, ApJ, 507, L117

Edwards, P.G., et al. 2002, in Proc. IAU Asia-Pacific Regional Meeting, vol. II, ed. S. Ikeuchi, J. Hearnshaw, & T. Hanawa, (Tokyo: Astron. Soc. Japan), 375

Fomalont, E.B., et al. 2000, ApJS, 131, 95

Hirabayashi, H., et al. 1998, Science, 281, 1825 and erratum 282, 1995

Hirabayashi, H., et al. 2000, PASJ, 52, 997

Hirabayashi, H., et al. 2001, in IAU Symposium 205, Galaxies and Their Constituents at the Highest Angular Resolutions, ed. R. T. Schilizzi, S. N. Vogel, F. Paresce, & M. S. Elvis, (San Francisco: ASP), 428

Kardashev, N. S. 1997, Exp. Astron., 7, 329

Kellermann, K. I., & Pauliny-Toth, I. I. K. 1969, ApJ, 155, L71

Kellermann, K. I., et al. 1998 AJ, 115, 1295

Kellermann, K. I. 2002, PASA, 19, 77

Kovalev, Yu. A., Kovalev, Y. Y., & Nizhelsky, N. A. 2000, PASJ, 52, 1027

Kühr, H., et al. 1981, A&AS, 45, 367

Linfield, R. P., et al. 1989, ApJ, 336, 1105

Lister, M. L., Tingay, S. J., & Preston, R. A. 2001, ApJ, 554, 964

Moellenbrock, G. A., et al. 1996, AJ, 111, 2174

Quirrenbach, A., et al. 1992, A&A, 258, 279

Readhead, A. C. S. 1994, ApJ, 426, 51

Sincell, M. W., & Krolik, J. H. 1994, ApJ, 430, 550

Slysh, V. I., et al. 2001, MNRAS, 320, 217

Tingay, S. J., et al. 2001, ApJ, 549, L55

Tingay, S. J., et al. 2003, PASJ, submitted

Radio Astronomy at the Fringe
ASP Conference Series, Vol. 300, 2003
J.A. Zensus, M.H. Cohen, & E. Ros, eds.

Interstellar Scintillation and Radio Intra-Day Variability

D. L. Jauncey
Australia Telescope National Facility

H. E. Bignall
Department of Physics and Mathematical Physics, University of Adelaide

J. E. J. Lovell, L. Kedziora-Chudczer, A. K. Tzioumis
Australia Telescope National Facility

J.-P. Macquart
Kapteyn Astronomical Institute

B. J. Rickett
University of California San Diego

But leave the Wise to wrangle, and with me
The Quarrel of the Universe let be:
And, in some corner of the Hubbub coucht,
Make Game of that which makes as much of Thee.

For in and out, above, about, below,
'Tis nothing but a Magic Shadow-show,
Play'd in a Box whose Candle is the Sun,
Round which we Phantom Figures came and go.

– Rubaiyat of Omar Khayyam, Fitzgerald translation,
Quatrains 45 and 46 of the first edition 1859.

Abstract. Considerable evidence has now accumulated strongly favoring interstellar scintillation as the mechanism responsible for the rapid intra-day variability seen at cm wavelengths in many flat-spectrum extragalactic radio sources. The discovery of three rapid intra-day variability sources, PKS 0405−385, PKS 1267−326 and J1819+3845, has been largely responsible for this new paradigm. We present here recent results for the two southern sources PKS 0405−385 and PKS 1257−326, together with observations and first results from the MASIV (Micro-Arcsecond Scintillation-Induced Variability) survey that we are undertaking at 5 GHz with the VLA.

1. Introduction

Much of the early radio observations of variability, intra-day variability (IDV) and interstellar scintillation (ISS) were made here in Green Bank. Ken Kellermann, with the 140-foot telescope, first demonstrated the value of accurate flux density measurements. Through their monitoring observations of radio source long-term variability, Ken and Ivan Pauliny-Toth showed the value and insight to be achieved from such measurements (Kellermann & Pauliny-Toth 1968).

Some years later, following a suggestion by Lovelace & Backer (1972), the Green Bank interferometer was used to search, unsuccessfully, for interstellar scintillation in 12 flat-spectrum radio sources (Condon & Backer 1975). In the early 1980s short-term variability was discovered in a sample of flat-spectrum radio sources (Heeschen 1982; 1984) from measurements at 3.3 GHz with the 300-foot radio telescope. Heeschen's observations of 226 flat- and steep-spectrum sources revealed the presence of such flickering on time scales of 2 to 20 days amongst some of the flat-spectrum sources As a transit instrument, the 300 foot telescope could sample these rapid flux density variations only once per day. Thus, Heeschen, with Witzel and colleagues continued this research with the Effelsberg 100 m radio telescope, where the flickering phenomenon was studied in detail, and was given the name intra-day variability (Witzel et al. 1986). Over the next decade, the Bonn group investigated this phenomenon in considerable detail with the 100 m Effelsberg radio telescope (Quirrenbach 1989) finding IDV in up to $\sim 20\%$ of flat-spectrum sources (Quirrenbach et al. 1992).

Considerable debate as to the mechanism responsible for this IDV at radio wavelengths followed these discoveries. If it was intrinsic to the sources, then brightness temperatures of up to 10^{18} K were implied. However, if it was caused by modulation of the radio signal along the path to the observer, namely interstellar scintillation (Rickett 1990), then much lower brightness temperatures followed. Interstellar scintillation in the turbulent interstellar medium (ISM) of the Galaxy appeared to be the most likely extrinsic mechanism (Heeschen & Rickett, 1987). Now, two decades after its discovery, overwhelming evidence has accumulated to show that ISS is the principal mechanism responsible.

2. The Fast Variable Sources

Three rapidly variable sources have revolutionized our understanding of IDV: PKS 0405−385 (Kedziora-Chudczer et al. 1997), PKS 1257−326 (Bignall et al. 2003), and J1819+3845 (Dennett-Thorpe & de Bruyn 2000).

PKS 0405−385 The first of these, PKS 0405−385, was discovered during the ATCA IDV Survey (Kedziora-Chudczer 1998; Kedziora-Chudczer et al. 2001a). PKS 0405−385 is a 17.5 magnitude, $z = 1.285$, inverted-spectrum radio quasar that varies at 5 and 8 GHz by up to 50% on a time-scale of less than an hour (Kedziora-Chudczer et al. 1997).

Until the early 1990s, large single dish radio telescopes had been used predominantly for accurate flux density measurements, but the ATCA IDV Survey was undertaken with the Australia Telescope Compact Array (Frater et al. 1992), and marked a move towards using array telescopes for IDV investigations.

The IDV in PKS 0405−385 found in 1996 was quasi-periodic and so rapid that many "cycles" of variability were sampled in a single 12-hour observing session. Figure 1 of Kedziora-Chudczer et al. (1997) reveals beautifully the quasi-periodic nature of the variations, and demonstrates the correlation between the variations at 4.8 and 8.6 GHz, which breaks down at 2.4 and 1.4 GHz. If these variations were intrinsic to PKS 0405−385, then a brightness temperature in excess of 10^{21} K would be implied. To explain this in terms of a beamed synchrotron source requires a bulk motion with a Lorentz factor of at least 1000, well in excess of those seen in VLBI surveys (e.g., Kellermann et al. 2000; Kellermann, these proceedings, page 185).

As an alternative, interstellar scintillation proved capable of explaining the observations; the differing behavior at frequencies above and below 5 GHz was explained in terms of the separation of "strong" and "weak" scattering in the turbulent ISM of our Galaxy. At 4.8 and 8.6 GHz, the source is undergoing weak scintillation, where the phase change through the scattering screen is less than one radian, while at frequencies below ~ 3 GHz, it is undergoing strong, refractive scintillation (Walker 1998). The observed variation of modulation index with frequency was found to be in agreement with a simple one-parameter model where the source contained 15% of its total flux density in a compact, scintillating component (Kedziora-Chudczer et al. 1997).

While ISS provided an excellent description of these observations, it was not able to exclude the idea that the variations were some how intrinsic to the source. In the meantime, weekly monitoring of PKS 0405−385 proved the IDV to be episodic, and it ceased after a few months. The variability seen in PKS 0405−385 is sufficiently rapid that it is feasible to search for a time delay between the variability patterns at two widely-spaced radio telescopes, like the ATCA and the VLA. The precision with which the variability pattern can be located in time is the characteristic time divided by the signal-to-noise ratio of the measurements. For PKS 0405−385, this implies a precision of 20 to 30 seconds. For the 10,000 km telescope separation, and an ISM velocity of ~ 50 km s^{-1}, a time delay of ~ 3 minutes or more might be expected, depending on the geometry. Such measurements are impossible for the slower IDV sources, with $T_{\rm char} \sim 1$ day, and emphasize the importance of these rapid variables and of the use of array telescopes for quasi-continuous flux density measurements.

A further IDV outburst in PKS 0405−385 (Kedziora-Chudczer et al. 2001b) and the first time delay measurement followed in December 1998 at 5 and 8 GHz using the ATCA and VLA. At the southerly declination of −39 degrees, the geometry is less than ideal; nevertheless a significant time delay was found at both frequencies (Jauncey et al. 2000), further strengthening the case for ISS, at least in this source. Unfortunately, the IDV in PKS 0405−385 again stopped before the observations could be repeated; despite weekly monitoring, no further outburst has been detected since.

A recent analysis of the 1996 ATCA 5 and 8 GHz discovery data for PKS 0405−385 has been undertaken (Rickett et al. 2002; Kedziora-Chudczer et al., these proceedings, page 211), showing the ISS origin of the dramatic variability in this source, a nearby screen distance of ~ 20 pc, an angular size of ~ 30 microarcseconds, and a rest-frame brightness temperature of 2×10^{13} K. Additionally, a comparison of the polarized and the unpolarized scintillation data reveals a triple structure in the linearly polarized component.

PKS 1257−326 IDV was discovered serendipitously in this source from a re-examination of archival data taken in 1995 (Bignall 2003). Like J1819+3845, PKS 1257−326 is a long-lived IDV source, and has proven valuable in unravelling the link between IDV and ISS. PKS 1257−326 is an 18.7 magnitude, flat-

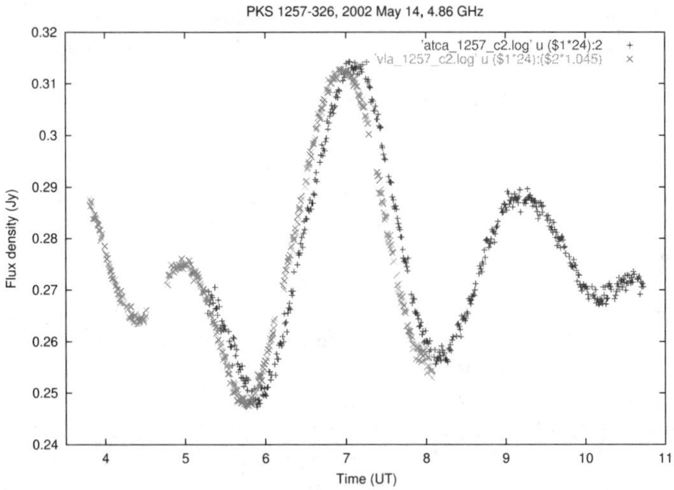

Figure 1. The variability patterns from the ATCA and the VLA at 4.86 GHz from May 14, 2002, for PKS 1257−326.

spectrum, radio and X-ray emitting quasar at $z = 1.256$ (Perlman et al. 1998). The ATCA archival data consisted of two "cuts" made ~6 hours apart, that differed by 20% in flux density. IDV was confirmed in June 2000, and it was then included in our ATCA IDV source flux density monitoring program (Bignall et al. 2003; Bignall 2003).

Two consecutive observations were scheduled between the ATCA and the VLA on May 13 and 14, 2002 at 4.9 and 8.5 GHz, to determine if there was a detectable pattern time delay across this 10,000 km baseline. This source is further north than PKS 0405−385, and the geometry is more favorable with almost 3 hours of common time. The VLA was divided into two sub-arrays, one at each frequency, and Figure 1 shows the data directly from the two telescopes; the setting scan, given by $\times$, is from the VLA and the rising scan, $+$, from the ATCA. There is a time delay present, with the VLA pattern leading the ATCA pattern by $\sim$ 7 minutes. Also, the patterns are similar on both days and at both frequencies. This leaves no doubt that ISS is the principal cause of the IDV. Moreover, the resultant ISM velocity is $\sim$ 25 km s^{-1}, close to the Earth's speed in its orbit around the Sun.

For a source to scintillate in the ISM, the angular size must be comparable to or smaller than the angular size, θ, of the first Fresnel zone. Following Walker (1998), $\theta = \frac{8}{\sqrt{D\nu}}$, where θ is in microarcseconds, D, the screen distance, is in kpc, and ν is in GHz. For a screen distance of 0.1 kpc and a frequency of 10 GHz, the size of a scintillating source is $\sim$ 5 microarcseconds. Such a source size exceeds the sub-nano-arcsecond expected if the IDV were intrinsic to these sources, and reduces the implied brightness temperatures by many orders of magnitude.

It is remarkable that accurate flux density measurements of the rapid variability of these sources yields information on their microarcsecond structure. It is as though the baselines of the ATCA, the VLA, and the WSRT were extended beyond their present limits all the way to the Moon!

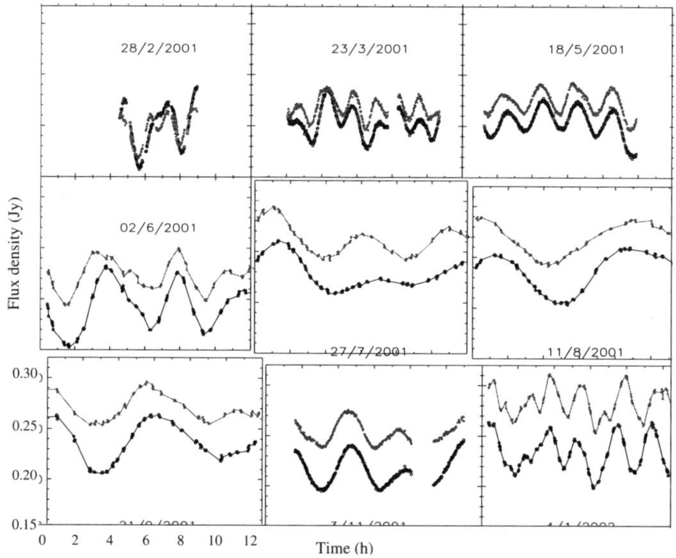

Figure 2. Light curves over the course of a year for PKS 1257−326. All plots are on the same scale, shown on the bottom left plot. 4.8 GHz data are represented by black open circles, and the 8.6 GHz data by filled circles. *(Taken from Bignall et al. 2003, ApJ, 585, 653.)*

At certain times during the year the Earth moves with the ISM irregularities, and the scintillation pattern changes only slowly. Six months later, the Earth is moving opposite to the ISM, the relative speed is high and the scintillation pattern changes more rapidly. Accordingly, in February 2001 we commenced a year-long flux density monitoring program with the ATCA at 4.8 and 8.6 GHz, to measure directly this changing annual cycle in the scintillation pattern. Observing sessions were scheduled approximately every six weeks. Figure 2 presents nine of the light curves at both frequencies made at approximately six week intervals over the course of the year.

The change in timescale of the variations at both frequencies is immediately apparent. From February through May, the flux density varies rapidly with less than an hour between excursions. In June they begin a slowdown that lasted through September. November sees them speed up again, while by 2002 January they have returned to much the same rate as 2001 February (Bignall et al. 2003). The timescales estimated from the earlier discovery observations of 2000 June and September, as well as more recent observations of 2002 February and April, are consistent with this annual cycle in the timescale of variations. The annual cycle is due to the changing relative velocities of the Earth and the ISM, and completely confirms ISS as the dominant source of the variations.

To quantify these changes, Figure 3 shows the characteristic timescale plotted against day of year at 4.8 and 8.6 GHz. The characteristic time, $t_{0.5}$, is defined as the half-width at half maximum of the autocorrelation function (Rickett et al. 2002). Plotted below, is the expected scintillation speed versus DOY for a scattering medium moving with the local standard of rest (LSR). The ISM

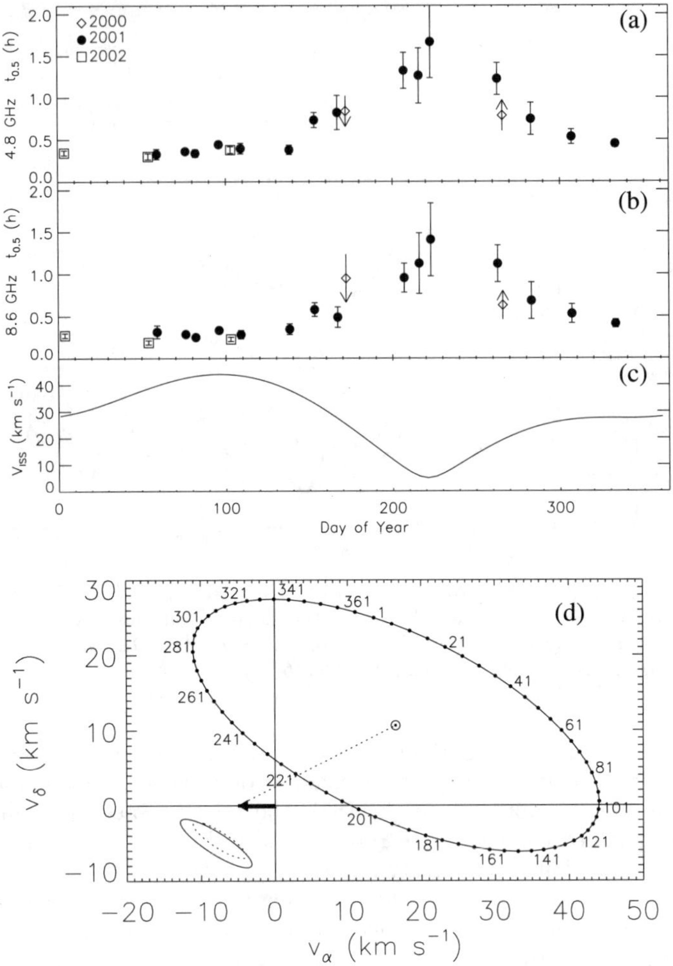

Figure 3. Characteristic timescale $t_{0.5}$, versus day of year at (a) 4.8 and (b) 8.6 GHz. (c) Expected scintillation speed $v_{\rm ISS}$ versus day of year for a scattering medium moving with the LSR. (d) Corresponding scintillation velocity $v_{\rm ISS}$ projected onto the plane transverse to the source line of sight, in components of right ascension and declination. The arrow shows the best fit offset velocity of -5 km s^{-1}. The small ellipse shows a contour of the best fit scintillation pattern. *(Taken from Bignall et al. 2003, ApJ, 585, 653.)*

is moving close to, but not exactly at the velocity of the LSR. Also plotted is the scintillation velocity projected onto the plane transverse to the source line of sight, in components of right ascension and declination velocity. The arrow shows the best fit offset velocity of -5 km s^{-1} in right ascension.

Such an annual cycle had already been found in two other IDV sources, J1819+3845 (Dennett-Thorpe & de Bruyn 2001) and B0917+624 (Rickett et al. 2001; Jauncey & Macquart 2001), and establishes ISS as the principal mechanism responsible for the IDV. The strong correlation in Figure 2 between the variations at the two frequencies is a pattern commonly seen in other IDV sources, and is indicative that at 5 and 8 GHz the scintillations are in the regime of weak scattering (Walker 1998). The presence of such an annual cycle in PKS 1257−326 gives an estimate of the linear size of the first Fresnel zone as the ISM velocity times the characteristic timescale. In the present case this is $\sim 5 \times 10^4$ km. This implies a screen distance of 10–15 pc, a source size of 30–37 microarcseconds and a brightness temperature of 4×10^{12} to 2×10^{13} K (Bignall et al. 2003).

Another finding is the longevity of the scintillation in some of these sources. B0917+624, for example, has been scintillating for close on 15 years (Jauncey & Macquart, 2001), although it recently ceased (Fuhrmann et al. 2002; M. J. Klein, 2002, private communication). Archival ATCA data for PKS 1257−326 suggests that it has been scintillating since at least 1995, while J1819+3845 has been scintillating for at least 4 years (Dennett-Thorpe & de Bruyn 2000, 2002). Despite their microarcsecond angular sizes and light month physical sizes, these scintillating components appear to remain ultra-compact for many years or even decades.

The presence of an annual cycle can, in principle, be established regardless of the time scale of the source. It occurs because the velocity of the ISM, relative to an Earth-based observer, changes over the course of a year. This property can in turn be used to determine the source structure, in effect to "image" the source, with microarcsecond resolution, as has been achieved for PKS 0405−385 (Rickett et al. 2002; Kedziora-Chudczer et al., these proceedings, page 211), and for J1819+3845 (Dennett-Thorpe & de Bruyn, 2001). Macquart & Jauncey (2002) present a formal outline of the technique, dubbed Earth Orbital Synthesis because of the similarities with conventional Earth rotation synthesis. Earth Orbital Synthesis is a powerful technique derived directly from flux density monitoring. Observations of scintillating sources provide a method of achieving micro-arcsecond resolution at cm wavelengths with modest telescopes, resolution unachievable with current VLBI techniques and comparable with that proposed for future space-based optical and X-ray interferometers.

The case for interstellar scintillation as the principal cause of the intra-day variability seen at cm wavelengths in many extragalactic sources, is now so strong that we suggest that this phenomenon be referred to as the physical phenomenon of interstellar scintillation, rather than the observational phenomenon of intra-day variability. Such a change has value in focusing on the physics involved. While the principal cause of IDV is scintillation, not all scintillation is necessarily IDV. For example, in the case of strong, refractive scattering, the characteristic timescale increases with decreasing frequency as $\nu^{-11/5}$, leading to time scales which can easily be an order of magnitude longer than intra-day. Note too, that any extragalactic source that did vary intrinsically on a time scale of days, would be so small that it would, of necessity, scintillate in the ISM.

The screen distance of 10−15 pc found for PKS 1257−326, implies that the scattering occurs in a very localized region of turbulence, as has been suggested for the two other fast scintillators J1819+3845 (Dennett-Thorpe & de Bruyn, 2000) and PKS 0405−385 (Rickett et al. 2002). From an IDV perspective, their properties appear extreme. From a scintillation perspective, however, are they really so extreme? The characteristic timescale is the scintillation length scale divided by the ISS velocity in the observer's frame. In order to observe an annual cycle, it follows that the ISM velocity must be close to the Earth's orbital speed of 30 $\mathrm{km\,s^{-1}}$. As noted above, for weak scattering, the scintillation length scale is proportional to $D^{1/2}$, where D is the screen distance. It follows that the fast scintillators, those with small t_{ISS}, will be those behind nearby scattering screens, as is the case for PKS 0405−385, PKS 1257−326 and J1819+3845. From a scintillation perspective, these fast variables are extreme only that they happen to be behind nearby scattering material.

This has an implication for the ISM, since the small number of fast scintillators found to date suggests that there may be few regions of the Galaxy where such nearby screens are to be found. Moreover, the presence of a nearby screen relaxes the angular size constraint on the angular size necessary for a source to scintillate. Thus it may be that there are more sources with angular sizes of order 20−30 microarcseconds which can scintillate behind a screen at < 30 pc, than there are sources of size 5−10 microarcseconds to scintillate behind screens at, say, 500 pc. Alternatively, the discovery of more such rapid scintillators can reveal the presence of more nearby scattering material in the Galaxy. As is outlined below, we are currently undertaking such a survey of the northern sky using the VLA to look for scintillation-induced variability.

3. The MASIV (Micro Arcsecond Scintillation-Induced Variability) VLA Survey

The realization that ISS is the physical process suggests that the results of such a large-scale scintillation survey may address numerous unanswered questions. To date, IDV surveys have been relatively small in size and restricted to about 100 of the strongest flat-spectrum sources (Heeschen 1984; Quirrenbach et al. 1989; Kedziora-Chudczer et al. 2001a). These surveys have shown that some $\sim 20\%$ of the strong flat-spectrum sources scintillate, but their small size leads to numbers of scintillators that are too small for reliable statistical investigations.

In January 2002 we commenced a large variability survey of the northern sky with the VLA at 5 GHz (Lovell et al. 2002; 2003). The aim of this MASIV survey is to construct a sample of 100–150 scintillating sources with which to examine both the microarcsecond structure and parent populations of these sources, and to probe the turbulent interstellar medium responsible for the scintillation. The existing surveys have shown that, away from the Galactic plane, 5 GHz is close to the transition frequency between weak and strong scintillations, and here the amplitude of the scintillations are usually at their greatest (Walker 1998).

In order to understand the sources themselves, it will be necessary to also understand the effects of the ISM, since the two are linked via the scintillation process. Assuming that the extragalactic sources are distributed uniformly, then the observed distribution of 100 or more scintillators will reveal their distribution on the sky, and hence the distribution of scattering material over the northern Galaxy. In particular, the presence of new fast scintillators will reveal the presence of nearby scattering material. And once the annual cycle distribution has been determined, the overall distribution of the velocity of the ISM will follow.

For the sources themselves, the survey results should lead directly to an understanding of the scintillation dependence on quantities such as flux density, spectral index, optical identification, redshift, luminosity, VLBI structure and evolution, and so on. Moreover, considerable information on the brightness temperature distribution may be expected as the properties of the scattering process become better understood. All of this information will become available at microarcsecond resolution, several orders of magnitude higher than is presently achievable with VLBI.

ISS thus gives us access to an interstellar radio telescope with Earth-Moon dimensions. And this can be achieved, albeit under restricted conditions, through precision flux density measurements with moderate sized radio telescopes.

To make sure that the survey is as unbiased as possible, the sources were observed over four well-spaced epochs over the year, ensuring that both fast and slow scintillation cycles were sampled. Additionally, four well spaced epochs allowed us to better find those sources like PKS 0405−385 whose scintillation outbursts are episodic. Each session extended over three days so that the slower scintillators would not be missed. The 2-hour minimum observing interval gives sufficient coverage each day so as to reliably recognize any variability.

Any new fast scintillators will reveal the presence of nearby regions of interstellar turbulence in the Galaxy. Both of the long-lived fast variables are amongst the weakest known scintillators, so that the survey also included a selection of both strong, > 0.5 Jy, and weak, > 0.1 Jy, sources. Moreover, as the VLA was in a variety of configurations during the extended observations, we selected only point sources at all VLA resolutions. This was done by selecting flat-spectrum, unresolved sources from the JVAS (Patnaik et al. 1992; Browne et al. 1998; Wilkinson et al. 1998) and CLASS (Myers et al. 1995) catalogs.

The spectral indices were determined from the quoted 8.5 GHz VLA and 1.4 GHz NVSS catalog measurements (Condon et al. 1998); we chose a spectral index lower limit of -0.3. Finally, to cover such a large number of sources, we divided the VLA into 5 sub-arrays, with each sub-array observing a set of sources in a limited declination range either side of the zenith, to avoid long slews through transit.

3.1. Results

We were allocated four sessions and the observations started 2002 January 19 to 22 during the reconfiguration from D to A array. Each sub-array was scheduled so that every source was observed for one minute every $\sim$2 hours while the source was above 15 degrees elevation. Here we present some preliminary results from this first session.

Flux density errors are made up of two components, a percentage component, $p\%$, which is proportional to flux density, and another, s Jy, due to noise and confusion. Figure 4 shows the observed root-mean-square (rms) scatter divided by the mean flux density, plotted against the mean flux density, for each of the 710 sources in the first survey session. It is clear from the strong sources that p is very close to 1% and appears to result primarily from small pointing errors, as not all telescopes which had been moved had their full pointing solutions determined. From the weak sources, s is 1.5 mJy, as expected from the receiver noise, confusion being negligible at this frequency. As a first step, we selected those sources which satisfied RMS $\geq \sqrt{(2s)^2 + \left(\bar{S}\,\frac{2p}{100}\right)^2}$ Jy. This corresponds to a simple χ^2 test to identify those sources whose scatter is well in excess of the overall population errors, and serves as a good starting point to identify the variables. In this way we identified 99 sources as potential scintillators, and these were individually examined in detail.

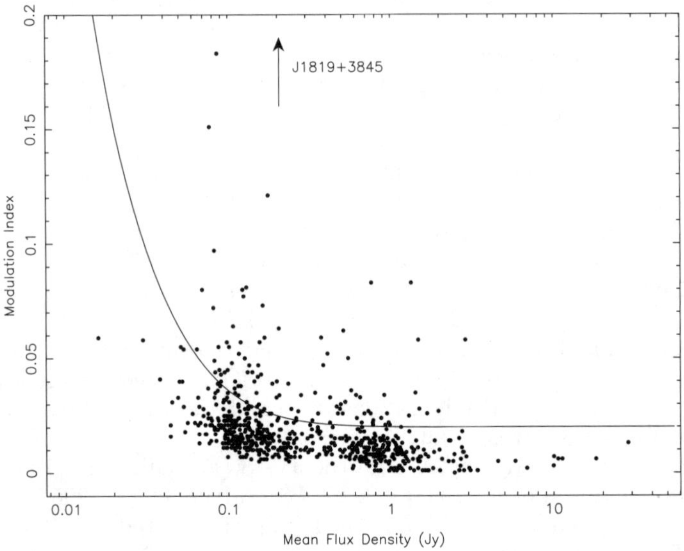

Figure 4. Results from the first epoch of the MASIV Survey: variability index (rms/$\bar{S}$) as a function of mean flux density. Sources above the solid line are classified as variable.

It is noticeable in Figure 4 that there appears to be more variation amongst the weaker sources. We divided the list into strong, $S > 0.3$ Jy, and weak, $0.06 < S < 0.3$ Jy, samples at 5 GHz, and selected only those sources with greater than 4% variability. A χ^2 test shows that the two populations differ at the 99.5% significance level. Moreover, not only are there more variables amongst the weaker sources, but a further test demonstrates that they are significantly more variable than the strong sources. This is perhaps not unexpected if the sources were brightness temperature limited. However, it may also result from there being a higher fraction of "naked cores" amongst the weaker sources. Future VLBA observations are planned to address this question.

Figure 5 shows a selection of 6 of the variables observed over the three days. It is noticeable that, aside from J1819+3845, we found only two other fast variables with time scales significantly less than a day. This reflects the small fraction of the sky covered by nearby scattering screens. It is remarkable that, of all of the 710 sources, none varies to the same degree as J1819+3845, a highly inverted spectrum source that was found serendipitously to exhibit IDV that was more rapid than PKS 0405−385 (Dennett-Thorpe & de Bruyn 2000), and whose scattering takes place in a screen a mere 20–30 pc away.

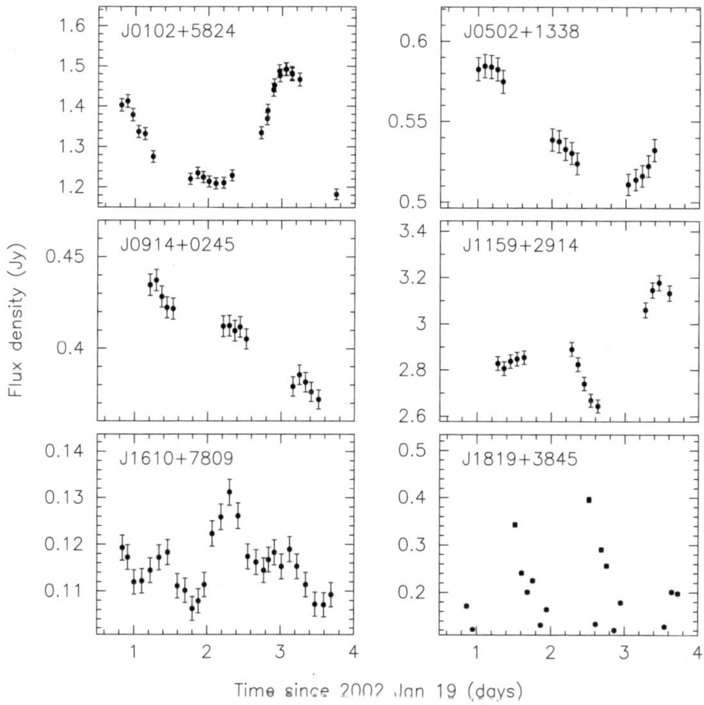

Figure 5. Light curves for a selection of 6 sources found in the first epoch of the VLA 5 GHz MASIV Survey. The sources range in intensity from ~100 mJy to ~3 Jy, and exhibit timescales from less than 2 hours to longer than 3 days.

4. Conclusions

There is now persuasive evidence that ISS is the principal mechanism responsible for the IDV seen in many flat- spectrum radio sources at cm wavelengths. The measured pattern time delays seen in the three fast variables, PKS 0405−385, PKS 1257−326, and J1819+3845 (Dennett-Thorpe & de Bruyn 2002), and the presence of an annual cycle in the variability characteristics, leave little doubt as to the origin of this variability. As the physical phenomenon responsible for IDV at cm wavelengths is now clearly identified, we echo Bignall et al. (2003) suggestion that the phenomenon be hereafter referred to simply as interstellar scintillation.

The Compact Array is part of the Australia Telescope, which is funded by the Commonwealth of Australia for operation as a National Facility managed by CSIRO. The NRAO is a facility of the US National Science Foundation and is operated by Associated Universities Inc., under a cooperative agreement. We thank Ken Sowinski, Miller Goss, Mark Claussen, and Jim Ulvestad, and the NRAO staff at Socorro for their assistance with the VLA observations. HEB acknowledges the support of a Faculty of Science Scholarship from the University of Adelaide. BJR thanks the US National Science Foundation for support under grant AST 9988398.

References

Bignall, H. E., et al. 2003, ApJ, 585, 653

Bignall, H. E. 2003, PhD Thesis, University of Adelaide, Australia

Browne, I. W. A., et al. 1998, MNRAS, 293, 257

Condon, J. J., & Backer, D. C. 1975, ApJ, 197, 31

Condon, J. J., et al. 1998, AJ, 115, 1693

Dennett-Thorpe, J., & de Bruyn, A. G. 2000, ApJ, 529, L65

Dennett-Thorpe, J., & de Bruyn, A. G. 2001, Ap&SS, 278, 101

Dennett-Thorpe, J., & de Bruyn, A. G. 2002, Nature, 415, 57

Frater, R. H., Brooks, J. W., & Whiteoak, J. B. 1992, Journal of Electrical and Electronics Engineering, Australia, 12, 103

Fuhrmann, L., et al. 2002, PASA, 19, 64

Heeschen, D. S. 1982, in Proc. IAU 97, Extragalactic Radio Sources, ed. D. S. Heeschen & C. M. Wade, (Dordrecht: Kluwer), 327

Heeschen, D. S. 1984, AJ, 89, 1111

Heeschen, D. S., & Rickett, B. J. 1987, ApJ, 93, 589

Jauncey, D. L., et al. 2000, in Astrophysical Phenomena Revealed by Space VLBI, ed. H. Hirabayashi, P. G. Edwards, & D. W. Murphy, (Tokyo: ISAS), 147

Jauncey, D. L., & Macquart, J.-P. 2001, A&A, 370, L9

Kedziora-Chudczer, L. 1998, PhD thesis, Sydney University, Australia

Kedziora-Chudczer, L., et al. 1997, ApJ, 490, L9

Kedziora-Chudczer, L., et al. 2001a, MNRAS, 325, 1411

Kedziora-Chudczer, L., et al. 2001b, in ASP Conf. Ser. Vol. 250, Particles & Fields in Radio Galaxies, ed. R. A. Laing & K. M. Blundell, (San Francisco: ASP), 128

Kellermann, K. I., & Pauliny-Toth, I. I. K. 1968, ARA&A, 6, 417

Kellermann, K. I., & Pauliny-Toth, I. I. K. 1969, ApJ, 155, L71

Kellermann, K. I., et al. 2000, in Astrophysical Phenomena Revealed by Space VLBI, ed. H. Hirabayashi, P. G. Edwards, & D. W. Murphy, (Tokyo: ISAS), 159

Lovelace, R. V. E., & Backer, D. C. 1972, Astrophysics Letters, 11, 135

Lovell, J. E. J., et al. 2002, in ASP Conf. Ser., Active Galactic Nuclei: from Central Engine to Host Galaxy, ed. S. Collin, F. Combes, & I. Shlosman, (San Francisco: ASP), 43

Lovell, J. E. J., et al. 2003, AJ, submitted

Macquart, J.-P., et al. 2000, ApJ, 538, 623

Macquart, J.-P., & Jauncey, D. L. 2002, 572, 786

Myers, S. T., et al. 1995, ApJ, 447, L5

Patnaik, A. R., et al. 1992, MNRAS, 254, 655

Perlman, E. S., et al. 1998, AJ, 115, 1253

Quirrenbach, A., et al. 1989, Nature, 337, 442

Quirrenbach, A., et al. 1992, A&A, 258, 279

Rickett, B. J. 1990, ARA&A, 28, 561

Rickett, B. J., et al. 2001, ApJ., 550, L11

Rickett, B. J., Kedziora-Chudczer, L., & Jauncey, D. L. 2002, ApJ., 581, 103

Taylor, J. H., & Cordes, J. M. 1993, ApJ., 411, 674

Walker, M. A., 1998, MNRAS, 294, 307

Wilkinson, P. N., et al. 1998, MNRAS, 300, 790

Witzel, A., et al. 1986, Mitteilungen der Astronomischen Gesellschaft, 65, 239

Radio Astronomy at the Fringe
ASP Conference Series, Vol. 300, 2003
J.A. Zensus, M.H. Cohen, & E. Ros, eds.

Seeing the Cores of Active Galaxies in the Sharpest Detail

L. Kedziora-Chudczer
Institute of Astronomy, School of Physics A28, The University of Sydney, NSW, 2006, Australia

B. J. Rickett
Department of Electrical and Computer Engineering, University of California at San Diego, La Jolla, CA 92093

H. E. Bignall
Department of Physics & Mathematical Physics, University of Adelaide, SA 5005, Australia

D. L. Jauncey, J. E. J. Lovell, A. K. Tzioumis
Australia Telescope National Facility, P.O. Box 76, Epping NSW 2121, Australia

J.-P. Macquart
Kapteyn Astronomical Institute, University of Groningen, Postbus 800 9700 AV, Groningen, The Netherlands

Abstract. Only the smallest and hottest radio sources "twinkle", or scintillate, when observed through the ionized interstellar medium of our Galaxy. We describe how observations of interstellar scintillation (ISS) can be used to learn about the inner regions of AGN, BL Lacs and quasars. We demonstrate the use of scintillation to model the microarcsecond scale components of several sources in both total and polarized flux density. At present such resolution cannot be obtained with any other direct imaging instruments, and is almost 100 times better than can be achieved with Space VLBI.

1. Introduction

Active Galactic Nuclei (AGN) are thought to consist of a super-massive black hole surrounded by an accretion disk of infalling matter which, in turn feeds a radio jet which carries away charged particles at relativistic speeds. Despite several decades of extensive studies, the issues of jet composition, acceleration and collimation all remain unclear. Magnetic fields must play an important part in jet formation, but their exact role is still uncertain.

Observations at the very highest resolutions are needed to address these questions because the powerful jets associated with bright AGN are most common at cosmological distances. The highest resolutions afforded by telescopes presently available, such as VSOP, hardly match this need, as they do not resolve structure on linear scales below $\sim 1\,\mathrm{pc}$ in most AGN. However a model of more detailed structure can be uncovered by the use of information from the intensity fluctuations in such sources.

The last few years have seen the accumulation of undeniable evidence (see e.g., Jauncey et al., these proceedings, page 199) to show that interstellar scintillation (ISS) is the principal mechanism responsible for the intra-day variability (IDV) seen at radio frequencies above 1 GHz in many compact, extragalactic sources (Heeschen 1984; Quirrenbach 1989; Kedziora-Chudczer et al. 2001; Lovell et al. 2003). This evidence follows from the measurement of time delays in the variability patterns seen in widely spaced telescopes for three sources (Jauncey et al. 2000; Dennett-Thorpe & de Bruyn 2002; Bignall et al. 2003), plus the discovery of an annual cycle in the variations in the scintillation pattern over the course of the year (Dennett-Thorpe & de Bruyn 2000; Rickett et al. 2001; Jauncey & Macquart 2001; Bignall et al. 2003).

In order to exhibit interstellar scintillation (ISS) the source angular size must be of the order of the angular size of the first Fresnel zone (e.g., Narayan 1992). This depends on the distance to the scattering material, which is in the range of tens to hundreds of pc, and implies microarcsecond source sizes.

2. Variability of AGN - Encoded Information about the Source Structure.

Radio variability with a time scale of months to years was discovered in 1965 (Dent 1965), and was assumed to be intrinsic to the sources. However, causality arguments, that the physical source size had to be of the same order as the light travel time, implied brightness temperatures in excess of the inverse Compton limit (Kellermann & Pauliny-Toth 1969). Rees (1966) had suggested that this could be resolved if the jet material were moving relativistically. This was later confirmed through VLBI imaging of the first quasar 3C 273, which revealed super-luminal expansion in this source (Pearson et al. 1981).

The problem of excessive brightness temperatures reappeared after the discovery of IDV at cm wavelengths in a number of sources (Heeschen 1984, Witzel et al. 1986). The most extreme case of large amplitude intra-hourly variations discovered in the strong quasar PKS 0405−385 (Kedziora-Chudczer et al. 1997), seriously challenged the idea that such rapid variability can originate within the source. Intrinsic variability implied a brightness temperature of 10^{21} K, nine orders of magnitude higher than the inverse Compton limit for synchrotron emission. To avoid this problem we postulated that the IDV in this source was not intrinsic, but rather was caused by interstellar scintillation. This was later vindicated by the measurement of a time-delay in the variability pattern of this source between the ATCA and the VLA (Jauncey et al. 2001).

In the following sections we discuss the ways of using observations of ISS to learn about the structure of extragalactic sources. However it should be noted that the same observations can help to understand the nature of the interstellar medium itself (Rickett, Kedziora-Chudczer, & Jauncey 2002).

3. Using Interstellar Scintillation to measure AGN Size and Brightness

The idea of measuring source size using scintillation variability was first proposed by Hewish, Scott & Wills (1964) in the context of interplanetary scintillation (IPS). IPS is strongest at frequencies less than 1 GHz, and its intensity depends on the source angular distance from the Sun. Cohen et al. (1967) derived size upper limits from $0\rlap{.}''5$ to $0\rlap{.}''04$, for those extragalactic sources in their survey that showed IPS. Moreover, they found excellent agreement between diameters derived from scintillation observations and those obtained from interferometer observations. Over the last three decades, however, VLBI has displaced the use of IPS to determine source sizes.

Scheuer (1968) and Salpeter (1969) showed that at frequencies above 1 GHz where IPS is almost negligible, the effects of ISS become important. Rickett (1969) was first to discuss the effects of ISS on the minimum observable size of radio sources as a function of frequency.

The phenomena associated with scintillation depend on the number density and distribution of scattering electrons in interstellar medium. Narrow-band, diffractive (interference) effects and, separately, as broad-band, refractive lens-like phenomena are observed in the "strong" scattering regime. Above the transition frequency, ν_0, which depends on the scattering properties of the interstellar medium, in the "weak" scattering regime, broad-band variations arise due to mild focusing and defocusing over the Fresnel scale ($r_F = \sqrt{cD/(2\nu\pi)}$). D denotes the distance of the scattering material assumed to be in a form of a "thin disk". For a source not larger than r_F, the transition frequency is where the ISS is fastest and strongest, and is found to be 3 to 5 GHz at most Galactic latitudes, although it can be much higher close to the Galactic plane.

Multi-frequency monitoring observations on both sides of ν_0 can provide an upper limit on the observed angular size of the radio source at the frequency ν. The r_F scale can be derived from the timescale of variability and the relative transverse velocity between the ISM and the observer. The angular size of the source is then found by dividing r_F by the distance to the scattering medium. For example we find this to be 30 μarcsec for the scattering screen at the derived distance of no more than 30 pc for quasar PKS 0405−385 (Rickett, Kedziora-Chudczer, & Jauncey 2002).

Diffractive scintillation observed in "strong" scattering provides the most stringent limit on the source size ($\leq 10^{-9}$ arcsec). For pulsars, which have the smallest angular sizes of all known radio sources, both diffractive and refractive ISS are observed (Cordes & Rickett 1998). However there is, as yet, no conclusive evidence of diffractive scintillation in compact, extragalactic radio sources, suggesting that the bulk of their radiation comes from much larger angular sizes (Kedziora-Chudczer et al. 2001).

The source size requirements for refractive scintillation in "strong" scattering regime are much less restrictive. Low-frequency variability with a timescale of the order of weeks has been successfully interpreted as refractive ISS of milliarcsecond-sized radio sources (Hunstead 1972; Rickett 1986). The angular dimensions derived from scintillation are up to three orders of magnitude larger than source sizes implied assuming intrinsic variability. In consequence the apparent brightness temperatures of scintillating radio sources are much less extreme, although still higher than the inverse Compton limit, which suggests that they come from the population of the brightest, most energetic sources (Jauncey et al., these proceedings, page 199).

4. Clues from the Annual Modulation of Variability Timescale

More detailed information about the structure of scintillating sources can be obtained from the annual modulation in variability timescales (Rickett et al. 2001; Jauncey & Macquart 2001; Dennett-Thorpe & de Bruyn 2002; Bignall et al. 2003). This arises because the velocity of the scattering medium is comparable to Earth's, and changes relative to Earth as it orbits the Sun. Changes in the direction of the apparent scintillation velocity during the course of a year allow one to obtain detailed information about the structure of the scintillation sources. The observed form of the annual cycle can then be used to model both the source structure, on microarcsecond scales, and the structure and velocity of the ISM. This Earth Orbit Synthesis (EOS) (Macquart & Jauncey 2002) has been used very successfully for the rapidly scintillating quasars J1819+3845 (Dennett-Thorpe & de Bruyn 2002) and PKS 1257−326 (Bignall et al. 2003).

However the anisotropy in the turbulence pattern of the ISM can mimic the anisotropic source structure in the shape of annual cycle. To avoid this problem one could derive information about the anisotropy of the ISM from measurements of the temporal power-spectrum during separate monitoring epochs, as was done successfully for PKS 0405−385 (Rickett, Kedziora-Chudczer, & Jauncey 2002).

5. Delay in the Variability Pattern at Different Frequencies

Additional information about the structure of sources can be derived from the cross-correlation of light curves at different frequencies. For the scintillating quasar PKS 1257−326, in addition to the annual cycle in the variability timescale, we also found an annual cycle in the time delay between variability patterns at our two observing frequencies, 8.6 and 4.8 GHz. This is displayed in Figure 1, which is adapted from Bignall et al. (2003). We have interpreted this annual cycle as being produced in a jet which has an offset of 12 μas between the regions where the jet becomes optically thick at the two frequencies due to opacity effects in the source. If this is correct, then it follows that the base of the jet, namely the point that becomes optically thick at infinite frequency, lies no more than $\sim$ 10 μarcsec from the scintillating component at 8.6 GHz. However, this offset may possibly be due to refractive effects in the ISM behind the scattering medium, and we are endeavoring to establish the correct interpretation.

6. Interstellar Scintillation of the Polarized Flux Density

Simultaneous measurements of the Stokes parameters I, Q, U and V, are proving to be a powerful probe of the detailed structure of scintillating sources. For example, Figure 2 shows our measurements of both the total and circularly polarized flux density in the BL Lac object PKS 1519−273. Here it can be seen

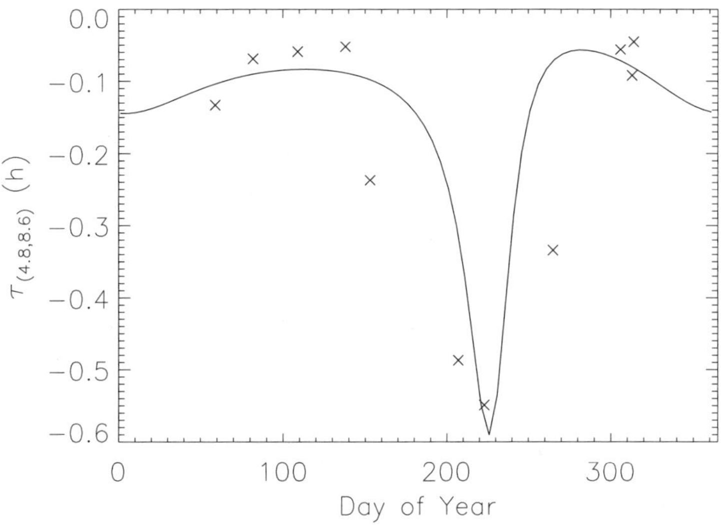

Figure 1. The model and observations of the annual cycle in the time delay between variability pattern at 4.8 and 8.6 GHz for PKS 1257−326 *(Taken from Bignall et al. 2003, ApJ, 585, 653.)*

that the variations in $|V|$ and I are closely correlated. There is no measurable time offset between the two, which demonstrates that they are spatially coincident, and the high cross-correlation shows that they have the same shape and size on microarcsecond scales, as seen through the scintillation telescope. Taking this a step further, it follows that the circular polarization in the scintillating component is remarkably high, 4% at 4.8 GHz (Macquart et al., 2000).

Thus the simultaneous multi-Stokes observations resemble "phase reference" interferometer observations, with the direct comparison between I and V revealing the relative size and location of the scintillating circularly polarized component with respect to the stronger, scintillating unpolarized component. This is also similar to the cross-correlation we made above between the scintillation patterns at 4.8 and 8.6 GHz in PKS 1257−326 (Bignall et al. 2003). The principal advantage of these "phase reference" observations is that they achieve microarcsecond resolution compared to VLBI's milliarcsecond resolution.

In many cases, however, the relationship is more complex than that seen above between the variations in I and V in PKS 1519−273. In fact, the scintillation seen in the linearly polarized flux density is often more rapid than that in the unpolarized scintillating component (Wagner & Witzel 1995). The natural explanation of this is that the linearly polarized source consists of a few components, which are smaller than the unpolarized component. We have developed this idea in considerable detail for the quasar PKS 0405−385, where the ISS in the linear polarization is much faster than the ISS in the total flux density, as shown in Figure 3 from Rickett, Kedziora-Chudczer, & Jauncey (2002). Using a non-linear least-squares optimization scheme, we found a three-component

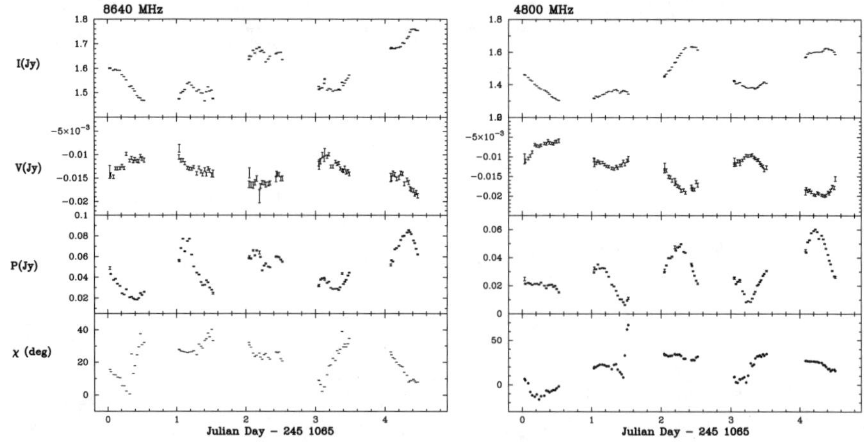

Figure 2. Time series for I, V, $P = \sqrt{Q^2 + U^2}$ and the position angle of P for PKS 1519−273 observed with the ATCA over 5 days at 8.6 and 4.6 GHz.

source model shown in Figure 4, that best fits the observed auto- and cross-correlations of the I, Q, and U Stokes parameters in PKS 0405−385 at 8.6 GHz.

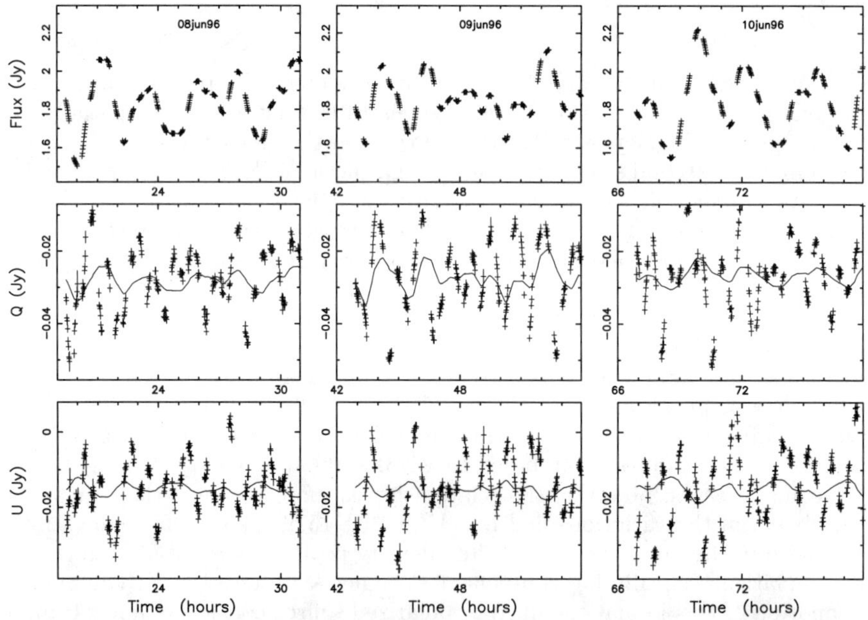

Figure 3. Time series of I, Q, U Stokes for PKS 0405−385 at 8.6 GHz observed with the ATCA between June 8–10, 1996. The solid lines represent the expected variability due to a linearly polarized one-component source with uniform degree and angle of polarization. *(Taken from Rickett, Kedziora-Chudczer, & Jauncey, 2002, ApJ, 581, 103.)*

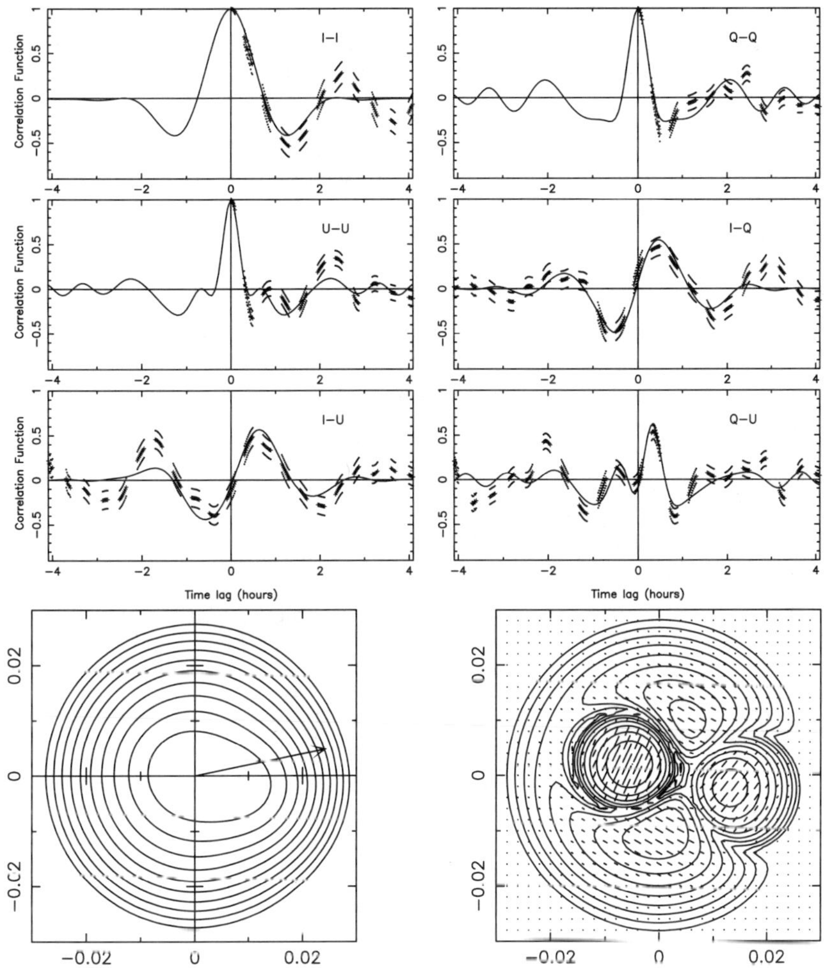

Figure 4. In the upper six panels auto- and cross correlations at 8.6 GHz are over-plotted with a best fit model with 3 scintillating source components. In the left bottom panel the brightness distribution for the three component source model is shown. The peak brightness temperature is about 2×10^{13}K. The right bottom panel shows the polarized components with maximum $\sim$ 70% of the total brightness. The angular scale is given in mas. *(Taken from Rickett, Kedziora-Chudczer, & Jauncey 2002, ApJ, 581, 103.)*

Although our model is not unique due to the considerable number of fitted parameters, it clearly demonstrates that the linearly polarized scintillating source is considerably more complex and more compact than the unpolarized source. Moreover, the success of this fitting-process further reinforces the veracity of the scintillation origin of the IDV seen at GHz frequencies in this source.

Earth Orbital Synthesis techniques clearly can provide powerful probes of the microarcsecond structure of these sources, and we are only just beginning to explore their wider potential. More sophisticated observations and applications of correlation analysis and of the annual cycles of scintillating sources will be very effective ways to determine the structure of the sources and magnetic fields, and will greatly advance our understanding of the formation processes and the development of AGN jets.

The Compact Array is part of the Australia Telescope, which is funded by the Commonwealth of Australia for operation as a National Facility managed by CSIRO. L. K.-C. would like to thank the organizers of this workshop for support received during her visit to Green Bank.

References

Bignall, H. E., et al. 2003, ApJ, 585, 653

Cohen, M. H., Gundermann, E. J., & Harris, D. E. 1967, ApJ, 150, 767

Cordes, J. M., & Rickett, B. J. 1998, ApJ, 507, 846

Dennett-Thorpe, J., & de Bruyn, A. G. 2000, ApJ, 529, L65

Dennett-Thorpe, J., & de Bruyn, A. G. 2002, Nature, 415, 57

Dent, W. A. 1965, Science, 148, 1458

Heeschen, D. S. 1984, AJ, 89, 1111

Hewish, A., Scott, P. F. & Wills, D. 1964, Nature, 203, 1214

Hunstead, R. W. 1972, ApL, 12, 193

Jauncey, D. L., et al. 2000, in Astrophysical Phenomena revealed by Space VLBI, ed. H. Hirabayashi, P. G. Edwards, & D.W. Murphy, (Tokyo: ISAS), 147

Jauncey, D. L., & Macquart, J.-P. 2001, A&A, 370, L9

Kedziora-Chudczer, L., et al. 1997, ApJ, 490, L9

Kedziora-Chudczer, L., et al. 2001, MNRAS, 325, 1411

Kedziora-Chudczer, L., Macquart, J.-P. & Jauncey, D. L. 2001, in IAU Symp. 205, Galaxies and their Constituents at the Highest Angular Resolution, ed. R. T. Schilizzi, S. N. Vogel, F. Paresce, & M. Elvis, (San Francisco: ASP), 90

Kellermann, K. I., & Pauliny-Toth, I. I. K. ApJ, 1969, 157, 1

Lovell, J. E. J., et al. 2003, ApJ, submitted

Macquart, J.-P., Kedziora-Chudczer, L., Rayner, D. P., & Jauncey, D. L. 2000, ApJ, 538, 623

Macquart, J.-P. & Jauncey, D. L. 2002, ApJ, 572, 786

Narayan R. 1992, Phil. Trans. Roy. Soc., 341, 151

Quirrenbach, A., et al. 1989, Nature, 337, 442

Pearson et al. 1981, Nature, 290, 365

Rees, M. J. 1966, Nature, 211, 468

Rickett, B. J. 1969, Nature, 221, 158

Rickett, B. J. 1986, ApJ, 307, 564

Rickett, B. J., et al. , 2001, ApJ, 550, L11

Rickett, B. J., Kedziora-Chudczer, L., & Jauncey, D. L. 2002, ApJ, 581, 103

Salpeter, E. E. 1969, Nature, 221, 31

Scheuer, P. A. G. 1968, Nature, 218, 920

Wagner S. J., & Witzel, A. 1995, ARA&A, 33, 163

Witzel, A., Heeschen, D. S., Schalinski, C., & Krichbaum, T. 1986, Mitteilungen der Astronomischen Gesellschaft, 65, 239

Radio Astronomy at the Fringe
ASP Conference Series, Vol. 300, 2003
J.A. Zensus, M.H. Cohen, & E. Ros, eds.

The Energetics and Apparent Superluminal Motions of Extragalactic Radio Sources

G. BURBIDGE
Center for Astrophysics and Space Sciences 0424, University of California, San Diego, CA 92093-0424, USA

Abstract. A critical discussion is given of the ideas that have been developed to explain the very large energies which are emitted from active galaxies and QSOs. It is likely that the flux of energy observed to be emitted from the nuclei of galaxies is generated at a much lower efficiency than the canonical 10% often assumed in the conversion of gravitational energy to whatever flux is seen. In section 4 it is shown that there is good observational evidence suggesting that many of the compact radio sources which show large apparently superluminal speeds are likely to be associated with low redshift spiral galaxies. If this is correct, the observed velocities, while they are only mildly relativistic, are comparable to those detected in low redshift radio galaxies like Centaurus A and M 87 and in "microquasars" in our own Galaxy.

1. Introduction

I was delighted when I was invited to attend and speak at this meeting honoring Ken Kellermann. He is a good friend, and I have known him since the early days at Caltech. He has done many notable things and has remained at the top of his game for more than thirty years. When I was asked to talk about the energetics of compact radio sources, I was forced to face up to the fact that while I have always maintained a close interest in radio source energetics I have done nothing constructive or original in this area for many years. Why is this? It would be easy to answer this question by claiming that I have been working in other (related) areas, on cosmology and the properties of QSOs and other exotic extragalactic phenomena, and that this has taken up all of my time. In part that is true, but it's not the real reason. It's also honest to admit that while I am a pure theorist at bottom, I have found myself more and more driven to study observational phenomena.

Like the great theorist Viktor Ambartsumian, I have found that observational phenomena which are very striking have tended more and more to dominate my professional life, and I believe, as did Ambartsumian that if the simplest interpretations of images (of galaxies or QSOs), leads to the possibility that very new ideas are required, then we should take these ideas seriously, unless our sense of what is good physics is seriously violated. This approach goes contrary to what is going on today in extragalactic astronomy, when theories based on a few very simple ideas are often maintained in spite of the observations.

Luckily for me, I met up with the late Fred Hoyle early in my career, and gradually realized that he took a similar approach. With those remarks as a background, I turn to radio source problems, and closely related to them the problems of the very small sources, and so-called superluminal motions.

ejected from galaxies in the Virgo cluster, and lie at distances no greater than 20 Mpc. If this is the case the values of $\beta_{\rm app}$ for them will also be less than unity.

In summary I have shown that there is strong evidence that a number of apparently superluminal QSOs and BL Lac objects are closely associated with spiral galaxies with low redshifts, so that the velocities of expansion of the jets are only mildly relativistic. It appears reasonable to extrapolate this argument to the general class of so-called superluminal radio sources, where it is clearly the large redshifts which have led directly to the belief in the existence of large superluminal motions.

It is important to stress that the mildly relativistic speeds that are then found (last column) in Tables 1 and 2 are very comparable to the speeds detected in the so-called micro-quasars in our own galaxy such as GRS 1915+105 and XTE J1550−564 (Mirabel & Rodríguez 1999; Corbel et al. 2002) and objects like SS 433.

However, there are still a few really anomalous cases which cannot easily be explained in this way. Perhaps coherent radiation mechanisms play a role in those sources.

5. The Real Energy Budget of Non-Thermal Sources

Radio Galaxies I believe that the very large powerful radio galaxies are still the most difficult objects to explain using the conventional models. Even if the existence of a massive black hole in the center is postulated, and the usual models are used, the efficiency of the transformation of gravitational energy into relativistic particles and magnetic flux must be very very small.

Let us assume, following the earlier discussion, that the initial process of accretion has an efficiency of 10% from gravitational energy to relativistic particles. Thus $f_1 = 0.1$.

To explain an extended radio source associated with a radio galaxy, we follow the conventional Rees-Blandford model and assume that the energy is then transported via a relativistic beam of particles assumed to be stable, through a distance $\simeq$ 100 kpc or more into a radio emitting lobe. Let us suppose that the efficiency factor associated with its transmission is f_2. This energy is then dumped into the low density plasma in the lobes and generates shocks in which electrons (and protons) are re-accelerated. Suppose that this final efficiency is $f_3 = 0.1$. If the total energy in relativistic particles and magnetic flux is 10^{61} ergs, the accreting central black hole rest mass M is given by

$$M = 10^{61}(\mathrm{M}_{\odot}c^2)^{-1}(f_1 f_2 f_3)^{-1} \tag{15}$$

If $f_1 = f_2 = f_3 = 0.1, M = 10^{10}\,\mathrm{M}_{\odot}$. Thus using these very reasonable efficiency factors only a very <u>small</u> fraction of the accretion power has gone into the radio flux observed. Where has all of the rest of the power gone? Of course it must have gone into the equivalent in the cosmos of the dissipation in the magnets in a man-made accelerator. This means that it has mostly gone into the galaxy and the intergalactic gas.

It has been clear for some time that a large number of the extended powerful radio sources arise in the central galaxies in clusters which also contain extended

2. Energetics

Soon after it became clear that the extended extragalactic radio sources are emitting by the incoherent synchrotron process I made some of the first calculations of the energetics (Burbidge 1956, Burbidge 1958, 1959). The very large energies required for powerful sources like Cygnus A clearly posed some problems. I presented these results at the Solvay Conference in 1964 (Burbidge & Burbidge 1964). This entailed convincing the physicists on the Solvay Commission that these astrophysical results were to be believed. The President of the commission at that time was Robert Oppenheimer, and others I remember well were Emilio Segré, Werner Heisenberg and Christian Møller. Since the synchrotron theory is a comparatively straightforward piece of physics I had no great difficulty in persuading them that we had the theory right. But when I started talking about 10^{59}–10^{61} ergs as minimum total energies in magnetic flux and relativistic particles, I was questioned very severely. I was asked if I knew how efficiently relativistic (GeV) particles could be generated on Earth (in accelerators). I found out in the ensuing discussion and in later correspondence with Emilio Segré that the efficiency is only $\sim 1\%$ for linear accelerators and 10^{-1}–10^{-2} % for synchrotrons. As thermodynamics requires, the bulk of the energy is dissipated in heat (in the magnets). The strong implication is that nature can do no better. I have never forgotten this argument. Also when nature makes a machine, when we are considering the efficiency of the process, we must take into account the energy dissipated in making that machine, an argument that Dick Feynman often emphasized when I was working on the jet in M 87 (Burbidge 1956).

Oppenheimer's last question to me was "Are there any ways of reducing these energies"? My answer to him was that the only way that I could see was to bring the sources closer, and this was for radio galaxies!

In the 1960s I proposed that chain reactions of supernovae in the nuclei of galaxies might give rise to the energy observed (Burbidge 1961). My close collaborators Fred Hoyle and Willy Fowler were very skeptical about this scheme, and out of many discussions came the first proposals by Hoyle and Fowler that the energy must be gravitational in origin associated with large collapsing superstars (Hoyle & Fowler 1963a; 1963b).

There were two basic difficulties with this proposal. To start with no one could come up with a realistic scenario about the way in which such stars with masses 10^6–$10^9\,M_\odot$ could be formed. Secondly, we realized that to get significant amounts of energy out, the stars must have contracted close to their Schwarzschild radii, but at this stage it is extremely difficult to extract the energy.

Despite these difficulties, discovery of the QSOs in the early 1960s led to the general belief that gravitational energy released by some processes involving massive black holes in the centers of galaxies must be responsible for them, and for activity in galaxies and radio sources in general (cf. Robinson, Schild & Schucking 1964).

In the late 1960s the first X-ray sources were identified as coming from very close binary systems in our own galaxy. Here it soon became clear that the X-rays arise in accretion disks about highly evolved stars, either a neutron star or a black hole, where the accreting gas has flowed from the extended atmosphere of the other member of the binary system, often a highly luminous star which can be optically identified.

While the scales associated with the X-ray luminosity of binary star systems in galaxies, and energetic events in the nuclei of galaxies, differ by many orders of magnitude, as do the types of photons and ejecta, my own view is that the fact that it could reasonably be established that binary X-ray sources were accreting objects came at about the same time, led many theorists to the view that a similar accretion process involving accretion on to a massive black hole must be the answer for radio sources, QSOs and active galaxies. And despite the uncertainties, this point of view has been generally accepted. Key papers which are often quoted are those of Lynden-Bell (1969) about a black hole at the galactic center, Blandford & Rees (1974) on a model for extended radio sources, and a review by Rees (1984). But starting in the 1960s Hoyle and I continued to worry about the efficiency of the process which could turn gravitational energy into the forms of the energy that we observe. We first began to consider alternatives in 1964 (see p. 40 in Hoyle et al. 1964, and p. 26 in Hoyle & Fowler 1963b).

There are three scenarios that can be envisioned in the conventional scheme. The first is simply energy released from matter falling directly into the black hole (Bondi 1952). Studies of this (cf. Shapiro & Teukolsky 1983) using the Schwarzschild solution show that for particles with finite rest mass, the maximum energy which can be released is $0.057mc^2$.

If the central mass (black hole) is rotating it is possible in principle to extract a great deal more energy. This is due to the general relativistic effect known as the frame dragging or Lense-Thirring Effect. It was Penrose (1969) who first pointed out that in principle energetic particles can be extracted because negative energy orbits can exist within the ergosphere. In the limit the maximum energy release is increased from $0.057mc^2$ to $0.42mc^2$. However the maximum can never be attained (cf. Shapiro & Teukolsky 1983; Krolik 1999; De Young 2001) since the limiting case is never realized.

Within the general relativistic framework a more realistic approach is to consider magnetic hydrodynamical processes involving a rotating (Kerr) black hole (cf. Blandford & Znajek 1977). This is similar to mechanisms proposed by Goldreich & Julian (1969) for pulsars. The mechanism requires the presence of an exterior magnetic field, charged particles, and a selectively distant region far from the rotating black hole where the energy is dissipated. A fairly rigorous study has been made by Thorne, Price, & McDonald (1986).

This "homopolar generator" is capable of accelerating charged particles to very high energies, but a source of such particles is required inside the ergosphere, and then a pathway out. Blandford and Zjanek argued that the charged particles arise because the electric field is so strong near the event horizon that the vacuum is unstable to pair production. The path to the outside is taken to be the same closed path that is used to extricate the total emf available using the homopolar generator model along the surface just outside which the integral was done to find the emf.

The region of dissipation then has to be the place where the particles responsible for the observed source are accelerated.

This model leads to an estimate of the power output given by

$$P = 10^{20} M^2 B^2 \left(\frac{J}{J_{\rm max}}\right)^2 \text{erg s}^{-1}\,. \quad (1)$$

Here M is the mass of the black hole in solar units, B is the magnetic field strength and $J_{\rm max}$ is the maximum possible angular rotation.

$$\left(\frac{J}{J_{\rm max}}\right) = \frac{\omega_{\rm bh} r_{\rm h}}{c}\,, \quad (2)$$

where $r_{\rm h}$ is the event horizon for a rotating black hole and is given by

$$r_{\rm h} = \frac{2G}{c^2}\left(M + (M^2 - a^2)^{1/2}\right)\,, \quad (3)$$

and $a = J/M$.

If $M = 10^8\,{\rm M}_\odot$, and $B = 10^4\,{\rm G}$

$$P = 10^{44}\left(\frac{J}{J_{\rm max}}\right) \text{erg s}^{-1}\,. \quad (4)$$

Thus if $J \simeq J_{\rm max}$ the power is comparable to the power output from the most luminous radio sources, the extended sources associated with galaxies at high redshifts. For the much smaller compact radio sources associated with the nuclei of active galaxies, and those sources at comparatively low redshifts, the luminosities lie in the range 10^{41}–$10^{43}\,{\rm erg\,s^{-1}}$. However, for other objects, with large redshifts, assuming that they are cosmological in origin, e.g., the QSOs, the luminosities lie in the range 10^{45}–$10^{47}\,{\rm erg\,s^{-1}}$. Radio sources only comprise a small fraction of the QSOs. For most QSOs the luminosities are largely in the form of optical radiation or X-rays.

Since these compact sources often show flux variations, polarization, etc., by analogy with the radio sources it was generally argued from the earliest days that the bulk of the radiation is non-thermal in origin, and thus that very large fluxes of very high energy electrons must be present with magnetic fields in the compact sources in the range 10^{-1}–10^{-3} gauss. The variability led to the conclusion that the size of the typical region in which the energy is generated is no larger than about 10^{15} cm. and this is approximately the value of $r_{\rm h}$, the event horizon, for a black hole with $M \simeq 10^8\,{\rm M}_\odot$.

However, all that theory has shown so far is that a significant fraction of the rest mass energy can be extracted from a rotating black hole, and most likely emerges as high-energy particles. The real problem lies in working out how is this energy propagated outward, and what fraction of it do we observe?

The most popular idea is that a massive black hole lies at the center of a dense cloud of gas and stars (a galaxy in other words) and that as matter tends to fall into the black hole an accretion disk is formed and as individual particles fall into the hole, some fraction of the energy is dissipated in the outer parts of the disk.

If a steady state can be reached, then the luminosity of the disk (assuming that all of the energy is converted into radiation) is given by

$$L_{\rm disk} = \frac{G M_{\rm bh} \dot{M}}{2 r_\star}\,, \quad (5)$$

where $\dot{M}$ is the rate at which matter is falling in and $r_\star$ is the inner radius of the disk. The total amount of potential energy is $GM/r_\star$, and the accretion luminosity

$$L_{\rm acc} = \frac{GM_{\rm bh}\dot{M}}{r_\star} \,. \tag{6}$$

Thus a maximum of 50% of the available energy in the disk is available to transmit to the outside. The other half is rotational energy that is lost as the matter falls in. If $r_\star = 10\, r_{\rm h}$, and $\dot{M} = dM/dt = 1\,{\rm M}_\odot\ {\rm yr}^{-1}$, $L_{\rm disk} = 2 \times 10^{45}$ erg sec^{-1}. This estimate has led to the general belief that the energy of an accretion disk is responsible for what we see in active galaxies and radio sources. The view appears to be that if you need to explain a higher luminosity you simply argue that the accretion rate is higher. Of course the argument only applies for these accretion disks.

Since it is assumed that the heating in this disk is due to viscous processes we are dealing with a purely thermal situation, and an estimate of the temperature of the thermal radiation is straightforward since the surface of the disk is optically thick and $L = \sigma T^4$. It turns out that the temperature at the inner radius is

$$T_\star = \left[\frac{3GM\dot{M}}{8\pi\sigma\, r_\star^3}\right]^{1/4} , \tag{7}$$

which gives a value of about 10^6 degrees for the parameters used above.

3. The Observations to be Explained

1. Radio, Optical and X-ray Continuum Emission There is a wealth of observational data that began with the discovery of the extended radio sources, which were shown to be radiating by non-thermal processes, with luminosities in the range 10^{40-45} erg sec^{-1}.

It has now become clear that thermal and non-thermal processes also give rise to photons with wavelengths over a wide range of the electromagnetic spectrum from infrared wavelengths to γ-rays, and very large fluxes are emitted in the radio, X-ray and optical energy bands. Apart from the extended radio sources, most of these sources are extremely compact, and from direct measurement of radio frequencies, and from the measurement of flux variations at all wavelengths, it has become clear that most of these very large fluxes arise in very small volumes, with sizes ranging from $\sim 10^{15}$–10^{18} cm. The forms of the spectra, the variations, and other indicators - particularly the polarization, show that the bulk of the energy is of non-thermal origin. For the extended sources the energy must often be generated, transported, and released over much larger volumes, where radiation is also largely non-thermal in origin. The extended radio galaxies have typical lobe size $\sim 100 - 200$ kpc, and in a small fraction of cases 10^6 pc. They are able to generate and/or re-accelerate large fluxes of Gev particles very far from the central source.

2. Ejection of Gas at High Speeds The most detailed studies showing that shells of gas are being ejected from the compact centers of active galaxies and QSOs have come from UV and X-ray spectroscopy of low redshift active galaxies and QSOs (cf. Crenshaw, Kraemer, & George 2003).

The outflow is manifested by blue-shifted absorption lines observed at optical, UV and X-ray wavelengths. Outflowing clouds of ionized gas have been detected in a majority of moderate luminosity Seyfert galaxies. About 10% of the QSOs show broad absorption lines, (the BAL QSOs), which are blue-shifted with respect to the emission spectra, and which give outflow velocities $\leq 0.1c$. It is often assumed that this outflow only takes place over limited regions of the surface, and that the fact that only 10% of QSOs show the BAL characteristic this is an orientation effect.

The mass involved in ejected material is hard to estimate. But, for example, for the specific case of NGC 3516 (cf. Crenshaw et al.), it is estimated that the mass loss rate lies in the range 3×10^{24}–$6 \times 10^{25}\,\mathrm{g\,s^{-1}}$ where the outflow velocity $\sim 500\,\mathrm{km\,s^{-1}}$. For the much higher velocity of ejecta in BAL QSOs with velocities in the range $10{,}000\,\mathrm{km\,s^{-1}}$–$30,000\,\mathrm{km\,s^{-1}}$, the kinetic energies are much larger.

It is generally argued that the gas is being driven outward directly by radiative acceleration from the central source (Arav & Li 1994) or indirectly in the form of a thermal wind arising in the assumed thermal accretion disk (Begelman et al. 1983).

In any case this energetics of the gas must derive its energy from the energy source at the center. In addition to this the broad emission line spectra, found both in active galaxies and QSOs, must also get kinetic energy from the central source. While the mass involved is quite small $\leq 100\,\mathrm{M_\odot}$ the velocities are typically $\sim 3 \times 10^3\,\mathrm{km\,s^{-1}}$. Thus this kinetic energy is of the order 10^{41}–10^{42} ergs. The characteristic time for renewal of these energies is $\sim$ crossing time, $\sim 10 - 1000$ yr.

3. Jets and Relativistic Effects Over the past thirty years very small (parsec) scale radio jets have been found to be very common in various kinds of active galaxies and QSOs. These include the radio galaxies with extended radio sources, QSOs, and BL Lac objects. Here I use the term BL Lac object in its classical definition, a compact source which is rapidly variable at radio frequencies and shows very few, if any lines in its optical spectrum. Such sources may be variable at optical wavelengths and/or X-rays. Such BL Lac objects are mostly likely compact (elliptical) galaxies containing powerful non-thermal energy sources of the same kind that power the nuclei of active galaxies and QSOs.

The methods of VLB1 have shown that in the radio regime tiny, but very powerful radio emitting jets are seen in both lobe dominated and core dominated sources.

The most remarkable result of these studies was the discovery of apparent superluminal motions in the early 1970s (cf. Pearson & Zensus 1987). The observers measured changes in angular size of jets as a function of time. By using the redshifts of the sources to determine their distances the change in angular size could be connected to a change in linear sizes, and thus apparently motions very much greater than the velocity of light were observed. Following

Rees (1966) this result was immediately interpreted as arising from a shell or plasmoids moving at relativistic speed away from a stationary center. In this case the velocity measured is $v_{\rm app} = \beta_{\rm app}c$. This is related to the true speed $v = \beta c$ and the angle to the line of sight θ by

$$\beta_{\rm app} = \frac{\beta \sin\theta}{1-\beta\cos\theta}\,; \tag{8}$$

$\theta = 0$ corresponds to motion directly along the line of sight, and $\theta = \pi$ to motion directly away from the observer.

The apparent transverse speed is γv where

$$\gamma = \left(1 - \frac{v^2}{c^2}\right)^{-1/2} . \tag{9}$$

If β is a fixed parameter, $\beta_{\rm app}$ has a maximum value of $\gamma\beta = (\gamma^2-1)^{1/2}$ when $\cos\theta = \beta$. $\beta_{\rm app}$ can exceed unity and then give rise to superluminal motions when $\beta > 1/\sqrt{2}$, or $\gamma > \sqrt{2}$. If a sample of sources are emitting plasmoids with the same speed β in many different directions, the fraction with apparent speed greater than $\beta_{\rm app}$ is $(\cos\theta_2 - \cos\theta_1)/2$ where θ_2 and θ_1 are the solutions of equation (8) for the given β and $\beta_{\rm app}$. This fraction is given by

$$p\,(>\beta_{\rm app}) = \left[1 - \beta_{\rm app}^2\left(\gamma^2-1\right)\right]^{1/2}\left(1+\beta_{\rm app}^2\right)^{-1} . \tag{10}$$

Thus for a uniform model this means that the fraction of sources with a given value of $\beta_{\rm app}$ decreases rapidly with increasing $\beta_{\rm app}$,

$$p\,(>\beta_{\rm app}) < (1+\beta_{\rm app})^{-1} . \tag{11}$$

Since the value $\beta_{\rm app}$ derived from the observations depends on the distance of the source, the value of the Hubble constant is required. Currently the most likely value is $H_\circ = 60\,{\rm km\,s^{-1}\,Mpc^{-1}}$. Using this, typical values of $\beta_{\rm app}$ for a large number of sources now lie in the range 2–20.

These values in turn require that the bulk velocities of the plasmoids are extremely large, since for $\gamma = 10$, $v = 0.9949c$ and for $\gamma = 20$, $v = 0.9987c$. If the plasmoid is moving relativistically towards us at an angle θ with a speed βc and is emitting an isotropic flux density $S_0(\nu)$, the observer will measure a flux density $S(\nu)$ where

$$S(v) = S_o(\nu/\delta)\delta^3\,, \tag{12}$$

where δ is the ratio of observed to emitted frequency

$$\delta = [\gamma\,(1-\beta\cos\theta)]^{-1} . \tag{13}$$

This Doppler boosting has a very strong effect for highly relativistic plasmoids moving at small angles towards us.

A number of theoretical models which try to explain the observations using these superluminal parameters have been made by Martin Rees, Alan Marscher, A. Königl, M. Begelman, R. Blandford and many others. The models all involve a relativistic jet with an axis at a small angle to the line of sight. Within this jet are blobs (the plasmoids) or relativistic shocks and it is the radiation from them that we see.

4. A New Approach for the Compact Jet Sources

I have outlined the energy requirements for the sources in general, and the requirements on the geometry for the very compact sources. The difficulties that I have always had with the conventional view can be stated very simply:

1. I believe that the efficiency with which energy can be put into the modes in which we see it is very much less than the $\sim 10\%$ which has been continuously assumed to be reasonable by the community at large.

2. I have never been convinced that the extremely high velocities of plasmoids or shocks which are required to explain the superluminal sources can be realized in practice.

The Problems Associated with Highly Superluminal Speeds It is extremely difficult to see how plasmoids with $v \geq 0.95c$ could be sustained or propagate for periods of years or more in the central region of a galaxy or QSO. The matter making up the plasmoid, even if it contains only relativistic particles contained in magnetic structures (which must be anchored) is bound to interact with the radiation field it is moving through and the matter which is present in the nucleus.

The physics behind these problems was briefly discussed by Jones & Burbidge (1973) and by Phinney (1987). Phinney pointed out that the limit will be set by the radiation pressure on the electrons which are responsible for the jet that we see. The intense radiation from the central source will kill off any jet with a γ value greater than some critical value by converting the kinetic energy into a directed beam of scattered photons. Of course the riddle is that the weaker the intensity of the central radiation field is, the higher is the value of γ that can be attained. And yet all of the ideas require that the acceleration starts very close to the center. How does nature keep the plasmoids out of the way of the central radiation field? It is possible, but looks very unlikely.

Interactions with matter are even more difficult to deal with. The region of acceleration and propagation of the superluminal jet contains a considerable amount of ambient gas which gives rise to the emission-line spectrum of the QSO or galaxy nucleus. Even in BL Lac objects where the emission lines are weak or non-existent, it is quite unreasonable to argue that the gas density is extremely small. It might be argued that the jets arise inside the broad emission line regions which for cosmologically distant QSOs have gas masses $\sim 10^4$–10^5 $M_\odot$ contained in a radius $\sim 100\,\mathrm{pc}$ ($\rho \sim 10^{-25}\,\mathrm{g\,cm^{-3}}$), but it is quite artificial to argue that the density of gas closer in is many orders of magnitude lower. But an extremely low density is required, otherwise the plasmoid would be very rapidly decelerated.

Thus I believe that the increasingly large superluminal speeds calculated using these models are not real.

From the earliest days I have thought the answer must lie in the fact that the distances of the sources which apparently show large superluminal motions must be wrong. Not surprisingly this leads me back to the old argument which some of us have advanced since the 1970s that the redshifts of QSOs and related objects are of non-cosmological origin. The evidence for this radical view has always been that there are far too many QSOs with large redshifts closely associated with comparatively bright and often active galaxies with low redshifts (cf. Burbidge et al. 1971, Burbidge 2001; Arp 1967; 1987) than is expected by accident. By now the observational evidence has become very strong, but the community still does not want to accept it. It means, of course, that the bright QSOs are not very far away and their redshifts are not measures of their distances. With this background I decided that I would take a sample containing many superluminal sources and investigate how many of them are associated with low redshift galaxies.

Associations Between Vermeulen-Cohen Sources & Low Redshift Galaxies For the sources in which rapid changes in angular size have been measured I have used the compilation of Vermeulen & Cohen (1994). In their Table 1 they give a list of 66 sources, classified as 47 QSOs, 13 BL Lac objects and 6 radio galaxies. All of the radio galaxies have $z < 0.1$. To them I have added one more BL object 3C 66B (0219+422) (Marchenko et al. 1998). From this table we find that 45 QSOs and BL Lac objects, and one radio galaxy, 3C 120, have $\beta_{\rm app} > 1$.

I have compared the positions of all of these sources with the positions of the galaxies listed in the revised Shapley-Ames catalog compiled by de Vancouleurs et al. (1991) which contains 23024 galaxies and is complete down to $m_b = 15$. I have also compared the positions of these objects with the positions of all QSOs and AGN with measured redshifts compiled by Véron-Cetty & Véron (2001).

I have first looked for apparent associations between Vermeulen - Cohen sources and galaxies where the separations $\Delta\theta$ are $\sim 10'$ or less.

Table 1. Apparently Superluminal QSOs close to Galaxies

Name		z_q	Galaxy	cz_g (km s^{-1})	m_g	$\Delta\theta$ ($'$)	μ (mas yr^{-1})	$\beta_{\rm app}$
PKS 0106+01		2.107	UGC 711	1980	14.4	3.7	0.20	0.10
0140+123[1]	= 3C 245	1.028	CGCG 66−1	7900	15.0	11.4	0.11	0.21
0219+422	= 3C 66B	0.444	UGC 1837	6300	14.8	6.4	0.75	0.53
3C 66B)			(3C 66A)					
0836+71	= 4C 71.07	2.218	UGC 4522	?	14.8	6.9	0.23	0.2*
0851+202	= 0J 287	0.306	IC 2423	9100	14.5	6.8	0.20	0.47
1308+326[2]		0.996	UGC 8250	?	15.5	8.4	0.13	0.15*
1458+718	= 3C 309.1	0.904	NGC 5832	6000	12.9	6.3	0.25	0.36
1618+177	= 3C 334	0.555	Mrk 877	34200	15.39	12.2	0.10	0.87
1641+399[3]	= 3C 345	0.594	NGC 6212	9000	13	4.7	0.47	0.33
1749+701		0.77	NGC 6503	1000	11	5.5	0.26	0.07
2230+114	= CTA 102	1.037	NGC 7305	7800	15	5.5	0 ± 0.5 (variable)	≤ 0.95

* Assuming the absolute magnitude of galaxy $M = -19$.

1. Four more QSOs nearby.
2. Two other QSOs at distances of 4.′8 and 9.′4.
3. Many QSOs around NGC 6212.

The results are shown in Table 1. Eleven pairs involving 3 BL Lac objects and 8 QSOs have been found. This is a very large number. Is it possible that these are accidental configurations? For random distributions on the sky of both galaxies and radio sources the number of accidental configurations (n) separated by an angular distance $\Delta\theta$ (in arc minutes) is given by

$$n = 8.64 \times 10^{-4}\Gamma\ (\Delta\theta)^2 N\,, \tag{14}$$

where Γ is the surface density of galaxies (per square degree) and N is the number of superluminal sources in the list. If we put $\Gamma = 0.57$, and $\Delta\theta = 10'$, $N = 45$, we find that $n = 2.2$, which is to be compared with the 11 pairs which are found. Thus this result is highly significant. This a conservative conclusion for the following reasons:

1. All of the galaxies found close to sources in Table 1 are spirals, or galaxies of later type. In many other studies of QSOs associated with galaxies they have always been late-type spirals, sometimes with active nuclei. It seems likely therefore that the ellipticals in the de Vaucouleurs catalog should be excluded from the analysis. Since they make up a fraction $\geq 30\%$ of the galaxies in the catalog, the real value that should be used for Γ is less than 0.57.

2. For a uniform distribution of sources over the sky the areal numbers are proportional to cos δ, so that there will be fewer sources at high declinations than are at low declinations. But one third of the sources listed by Vermeulen and Cohen have $\delta > 60°$ (this is clearly due to observational selection). Thus the selection of sources in which superluminal effects have been detected tends to minimize the number of associations.

Comparison of the positions of the sources with the Veron catalogue of QSOs shows that three of the associations shown in Table 1 have several other QSOs very close-by. The most remarkable case is that of the Seyfert galaxy NGC 6212 and 3C 345. I have discussed this in detail elsewhere (Burbidge 2003). As well as 3C 345 and a second QSO with exactly the same redshift equally close to NGC 6212, there are 40 more QSOs all lying within 1 degree of NGC 6212.

In the case of 1308+326, there are two more QSOs within 10′ of the galaxy, and for 3C 245 there are four more QSOs nearby. These associations are also highly significant.

Thus we conclude that for all of the sources listed in Table 1 it is entirely reasonable to suppose that they have been ejected from the nearby (parent) galaxies and thus are at the distances of those galaxies.

Deriving the distances now from the redshifts of the galaxies, and putting $H_\circ = 60\,\mathrm{km\,s^{-1}\,Mpc^{-1}}$, we convert the proper motions μ to speeds $= \beta_{\rm app}c$. In the last column of Table 1 we see that $\beta_{\rm app} \leq 1$ for all of these sources, i.e. while the speeds are still high, large values of γ have gone away.

What about all of the other sources? There are a number of sources where galaxies are comparatively close-by, but further away than $\sim 10'$, but the pairings in these cases are not statistically significant. There are also some sources which have several close companion QSOs. We also have several of the well known low redshift radio galaxies in the sample. These are shown in Table 2. Not surprisingly they all have $\beta_{\rm app} \leq 1$.

Table 2. The Properties of Genuine Radio Galaxies

Name		z	μ (mas yr^{-1})	β_{app}
0055+301	NGC 315	0.067	< 0.7	< 0.88
0316+413 (3C 84)	NGC 1275	0.017	0.54	0.69
1228+127	M 87	0.0032	1.1	0.33[1]
1322−427 (Cen A)	NGC 5128	0.0008	4.0	0.18[2]
1637+826	NGC 6251	0.023	0.05	0.09

1. Distance of M 87 assumed to be 20 Mpc
2. Distance of NGC 5128 assumed to be 3 Mpc

The remaining sources which have comparatively large values of γ, are nearly all QSOs or BL Lac objects, with large redshifts, and it is clear that it is the large distances derived from these redshifts which lead to the apparent superluminal motions. Thus I believe that, like the sources listed in Table 1, these objects also have been ejected from comparatively nearby galaxies, and that the ejection speeds are only mildly relativistic. It is reasonable to conclude that since ejection they have moved considerable distances from their parent galaxies. Thus their parent galaxies remain unidentified.

We are still left with a few sources with apparent superluminal motions which cannot be explained away in this fashion. They include two classical BL Lac objects, Mrk 421 and BL Lac, and two peculiar radio galaxies, 3C 111 and 3C 120. All of these sources have small redshifts but still have large values of γ.

As far as the BL Lac objects are concerned I believe that the redshifts are cosmological in origin, and thus that the distances are correctly calculated. Thus these objects' values of $\gamma \simeq 3$–5 are real. Perhaps in such cases the beaming of the non-thermal continuum <u>is</u> arranged in such a way that very high velocity plasmoids can be accelerated and survive. Also the near absence of excited gas in these objects may help the acceleration mechanism. But 3C 120 and 3C 111 cannot be explained in this way. They are both classified as peculiar radio galaxies.

3C 120 is the most difficult object to understand. It has many of the emission line properties of a QSO, but it has a very small redshift $z = 0.033$, and some evidence for a stellar component. 3C 111 has $z = 0.0485$. To reduce the apparent speeds in these objects to values such that $\gamma < 1$, means that their distances must be at least ten times less than the values indicated from their redshifts. They remain a puzzle.

Finally we have to mention 3C 273 and 3C 279. Both of these are bright QSOs with apparently very large superluminal speeds with values of $\gamma \simeq 15$. Since both of them lie in directions within the central field of the Virgo cluster and are very bright, I believe that the most likely answer for them lies in the suggestion made long ago by Arp that they are indeed sources that have been

X-ray sources with $L_{\rm X} \gtrsim 10^{43}\,{\rm erg\,s^{-1}}$ arising from hot gas, for historical reasons called "cooling flows". They must also ultimately get their energy from the central machine (cf. Binney & Tabor 1995). Recently Churazov et al. (2001) have discussed this energy input in terms of the mechanical power generated by Bondi-type accretion onto a black hole.

They show that in the case of M 87 most of the power is carried away by an outflow of gas. For M 87 the mass of the central black hole is estimated to be about $3 \times 10^9\,{\rm M_\odot}$ (Macchetto et al. 1997). Churazov et al. concluded that for Bondi accretion a large fraction of the energy is carried by an outflow and a few percent of the rest mass has to be channeled into a jet. The amount depends on the density of gas in the center. The X-ray luminosity of M 87 is $3 \times 10^{43}\,{\rm erg\,s^{-1}}$, a value much larger than the luminosity of the synchrotron jet or the more extended radio source. M 87 is not a very powerful radio galaxy. Thus in this prototype object it is clear that even with a large central black hole, the overall efficiency is very low.

Compact Non-Thermal Sources How well do these models involving accretion onto black holes work for the compact sources?

In nearby Seyfert galaxies where real galaxies can be seen, and BL Lac objects where the luminosities in very small volumes are $\lesssim 10^{42}\,{\rm erg\,s^{-1}}$ and the jets etc. have high but not apparently superluminal speeds, realistic black hole masses $\lesssim 10^8\,{\rm M_\odot}$ (based on galaxies which are not active, but in which black holes were present) may very well be able to explain the phenomena.

But for the QSOs in general, where optical luminosities in the range 10^{46}–$10^{48}\,{\rm erg\,s^{-1}}$ are routinely assumed, there are real problems.

1. First it has to be assumed that all of these objects lie at the center of galaxies which contain massive black holes. This is very far from being firmly established.

   Often the "fuzz" around the QSO images has an emission line spectrum associated with hot gas. Only in one case, 3C 48, has an early type absorption line spectrum denoting a stellar content been found. In many investigations it has been claimed that the fuzz has the luminosity distribution typical of a classical elliptical galaxy. Of course the central QSO luminosity has to be subtracted to determine the light distribution, and even those who would like to believe, consider that this is a bootstrap process. Direct imaging using HST of bright QSOs has shown that many appear to lie in, or are interacting with, galaxies, but some are bare and the redshifts of the galaxies are in general not known. thus the observational case for assuming that QSOs in general lie in galaxies (with the same redshifts) remains very weak.

2. Estimates of black hole masses in Seyfert galaxies and very low redshift QSOs have been made by using the recently developed method of reverberation mapping. Here the time-lag between flux variations of the continuum and the strength of the Hβ emission line is measured giving the radius of the Hβ emitting region. It is then assumed that the widths of the emission lines give a measure of the velocity dispersion and that the region is stationary so that the mass can be obtained using the virial theorem (Wandel,

Peterson, & Malkan 1999). Despite the uncertainties (cf. Krolik 2001) it is claimed that the results can be trusted (McClure & Jarvis 2002). By using the Mg II λλ 2798, 2800 doublet MacClure & Jarvis have extended this study to much higher redshift QSOs. They find that for QSOs with redshifts in the range $0.20 < z < 0.66$ the black hole masses lie in the range 10^8–$10^{10}\,M_\odot$. Of course since the luminosities are determined from the redshifts, the larger the redshifts, the larger will be the value of the masses of the black hole claimed to be present. But it is on arguments like this, and the belief that accretion onto a black hole (in a galaxy) with an efficiency $\sim 10\%$ is responsible for what we see, that the observers are making their extravagant claims (cf Carnegie Conference on Black Holes and Active Galaxies, Pasadena, October 2002).

3. Continuum Energy from Compact Sources – It is often supposed that the continuum energy is thermal radiation from a thin accretion disk. However the general forms of the spectra do not support this interpretation. In general the spectra are proportional to ν^{-1} in the infrared, flattening towards $\nu^{-\frac{1}{2}}$ in the optical. But this is roughly orthogonal to the spectrum of an accretion disk. Much has been made of the possibility that the "blue bump" in the UV spectra of some QSOs may be due to a disk but this is only one feature.

In addition, it is necessary to invoke a large non-thermal component in the infrared, optical and UV together with an X-ray component in some objects. Sometimes it is claimed that there is a coronal component generated above and below the disk which can do this. But all of these ideas require a further set of unspecified acceleration processes which must take energy at low efficiency from the thermal accretion disk. For example the magnitude of any coronal emission must be much less than the thermal radiation of a disk or a photosphere: thermodynamics again. All of these problems remain unsolved and seem to be impossible to deal with unless ever larger accretion rates are invoked.

6. Conclusions

I have given a review of what I believe are some of the real problems associated with our understanding of what gives rise to energetic events in galaxies and QSOs. The major new conclusion that I have reached is that it is that large superluminal motions are not real and like many other QSOs, these sources have been ejected from low redshift spiral galaxies. The observational evidence suggest that values of $\beta_{\rm app} \leq 1$ are common, and that in genuine BL Lac objects a real absence of gas allows jets to propagate with values of γ as large as 3–5.

As far as the large, powerful radio sources are concerned their energy requirements have been underestimated. It is probable that the release of energy does require the existence of massive black holes in the parent galaxies but the release may involve creation processes very close to the black hole (cf. Hoyle, Burbidge, & Narlikar 2000). Such processes may give rise to larger energy sources. However it is important to study the total energy budget and not just that fraction that is easily detectable. Also no one understands the origin of the massive black holes.

Comparatively low redshift Seyfert galaxies are also generating their energy from black holes in the galaxy centers. At a level of 10^{39}–$10^{42}\,\mathrm{erg\,s^{-1}}$ this can be understood as due to classical accretion processes, ejection of gas shells, etc. However the efficiency even in this case is not well determined.

As far as the very compact sources with high redshifts are concerned, the observations more and more point to the view that they are being ejected from active galaxies and their redshifts are not due to the expansion of the universe. There is a large body of observational evidence supporting this view. Thus while the redshifts are large, most of the redshift is intrinsic and the luminosities are comparatively low, $\lesssim 10^{41}\,\mathrm{erg\,s^{-1}}$. Consequently a great deal of the work that is being done on assumption that these objects all lie in distant galaxies will only turn out to lead to more and more contradictions.

References

Arav, N., & Li, Z.-Y. 1994, ApJ, 427, 700

Arp, H. 1987, Quasars, Redshifts and Controversies, (Berkeley: Interstellar Media)

Begelman, M. C., McKee, C. F., & Shields, G. A. 1983, ApJ, 271, 70

Binney, J., & Tabor, G. 1995, MNRAS, 276, 663

Blandford, R. D., & Rees, M. J. 1974, MNRAS, 169, 395

Blandford, R. D., & Znajek, R. L. 1977, MNRAS, 179, 433

Bondi, H. 1952, MNRAS, 112, 195

Burbidge, G. 1956, ApJ, 124, 416

Burbidge, G. 1958, ApJ, 127, 48

Burbidge, G. 1959, ApJ, 129, 849

Burbidge, G. 1961, Nature, 190, 1053

Burbidge, G., & Burbidge, E. M. 1964, in The Structure and Evolution of Galaxies, ed. I. Prigogine, Proc. 13th Solvay Conf. on Physics, 137

Burbidge, E. M., et al. 1971, ApJ, 170, 233

Burbidge, G. 2001, PASP, 113, 899

Burbidge, G. 2003, ApJL, submitted

Churazov, E., et al. 2002, MNRAS, 332, 729

Corbel, S., et al. 2002, Science, 298, 196.

Crenshaw, D. M., Kraemer, S. B., & George, I. M. 2003, ARA&A, 41, in press

de Vaucouleurs, G., et al. 1991, Third Reference Catalogue of Bright Galaxies, (Springer: New York)

de Young, D. S. 2002, The Physics of Extragalactic Radio Sources, (Chicago: University of Chicago Press)

Goldreich, P., & Julian, W. H. 1969, ApJ, 157, 869

Hoyle, F., & Fowler, W. A. 1963a, MNRAS, 125, 169

Hoyle, F., & Fowler, W. A. 1963b, Nature, 197, 533

Hoyle, F., et al. 1964, ApJ, 139, 909

Hoyle, F., Burbidge, G., & Narlikar, J. V. 2000, A Different Approach to Cosmology: from a Static Universe through the Big Bang Towards Reality, (Cambridge: Cambridge Univ. Press)

Jones, T. W., & Burbidge, G. R. 1973, ApJ, 186, 791

Krolik, J. H. 1999, Active Galactic Nuclei: From the Central Black Hole to the Galactic Environment, (Princeton: Princeton Univ. Press)

Krolik, J. H. 2001, ApJ, 551, 72

Lynden-Bell, D. 1969, Nature, 223, 690

Macchetto, F., et al. 1997, ApJ, 489, 579

Marchenko, S. V., et al. 1998, in ASP Conf. Ser. Vol. 144, Radio Emisssion from Galactic and Extragalactic Compact Sources, ed. J. A. Zensus, G. B. Taylor & J. M. Wrobel, (San Francisco: ASP), 67

Mirabel, I. F., & Rodríguez, L. F. 1999, ARA&A, 37, 409

McLure, R. J., & Jarvis, M. J. 2002, MNRAS, 337, 109

Penrose, R. 1969, Nuovo Cimento, 1, 252

Phinney, S. 1987, in Superluminal Radio Sources, ed. J. A. Zensus & T. J. Pearson, (Cambridge: Cambridge Univ. Press), 301

Rees, M. J. 1984, ARA&A, 22, 471

Robinson, I., Schild, A., & Schucking, E. (ed.) 1964, Proc. First Texas Conference on Relativistic Astrophysics (Chicago: University of Chicago Press)

Shapiro, S. L., & Teukolsky, S. A. 1983, Black Holes, White Dwarfs, and Neutron Stars, (New York: John Wiley)

Thorne, K. S., Price, R. H., & MacDonald, D. A. 1986, Black-Holes - The Membrane Paradigm, (New Haven: Yale Univ. Press).

Vermeulen, R. C., & Cohen, M. H. 1994, ApJ, 430, 467

Véron-Cetty, M.-P., & Véron, P. 2001, A&A, 374, 92

Wandel, A., Peterson, B. M., & Malkan, M. A. 1999, ApJ, 526, 579

Zensus, J. A., & Pearson, T. J. (ed.) 1987, "Superluminal Radio Sources", (Cambridge: CUP)

Radio Astronomy at the Fringe
ASP Conference Series, Vol. 300, 2003
J.A. Zensus, M.H. Cohen, & E. Ros, eds.

The KelPau Limit and Sources of Cosmic Rays

N. S. Kardashev
Astro Space Center of P.N. Lebedev Physical Institute, Profsoyuznaya 84/32, 117810 Moscow, Russia

Abstract. The limiting value of intensity for synchrotron emission determined by its reabsorption and the self Compton scattering catastrophe (KelPau limit) is of fundamental significance for models of radio sources. This limit for relativistic electrons (10^{12} K) has been confirmed by many observations. Practically, we observe this value in many nuclear regions of AGNs. But also for many extragalactic radio sources, whose nuclei are unresolved by Earth VLBI, we expect a much higher brightness temperature. Some new results of the KelPau model, particularly the strong synchrotron emission of relativistic protons and the higher intensity determined by proton KelPau limit, will be important. Such emission is expected in the nearest to super massive black hole regions, corresponding to the cosmic rays acceleration regions, and this is the prospect for future Space VLBI explorations.

Introduction The 1960s are related to many outstanding scientific discoveries in astronomy, physics, and the first explorations by rockets and satellites. Just in this years were discovered quasars, space masers and pulsars—objects with extremely small angular sizes and probably very high intensity of radio emission (brightness temperature). An indirect indication of this was the detection of fast radio flux variability and strict measurements of an upper limit of angular size for many extragalactic radio sources, by using interferometers with Very Long Bases, which were realized for the first time also in these same years. The searching enthusiasm has been warmed up by the analogy between observing distant sources and the giant intensity of radio stations on the Earth—in the same years for the first time the possibility for detection of radio signals from extraterrestrial civilizations has been strongly debated.

The history of the exploration of super compact radio sources and their brightness temperatures is excellently described in a recent paper (Kellermann 2002). Here I wish to fix attention on the great importance of the paper by Kellermann & Pauliny-Toth (1969), which demonstrated that the brightness temperature $T_{\rm b}$ of synchrotron emission by relativistic electrons is limited by the self-Compton scattering of the same emission and its self absorption. In this model a variable magnetic field amplitude of synchrotron emission is near the same value as the constant magnetic field. The limit is $T_{\rm b} = 10^{12}$ K and this value, as has been demonstrated by many observations over more than 30 years, very nicely corresponds to the brightest components of many extragalactic radio sources. I wish to demonstrate here (more details in Kardashev 2000; Kardashev 2001) some prospects for future explorations, connected with a principal limitation on $T_{\rm b}$ in the Kellermann/Pauliny-Toth model.

T_b for synchrotron emission For isotropic synchrotron emission in the frame of the emitting volume (and an isotropic or anisotropic distribution of particle velocities) and at observations in a spectral band where the optical depth of reabsorption is near unity, the KelPau limit is

$$T_b = a(\gamma)(mc^2/k)(m^2c^4e^{-3}H_\perp^{-1})^{1/7}, \quad (1)$$

where $a(\gamma)$ is 0.85, 0.28, and 0.19 accordingly for the spectrum relativistic particles $N(E) \propto E^{-\gamma}$, and γ is 1, 2, and 3 (Kardashev 2000). At the monoenergetic particles spectrum $a = 0.725$. For relativistic electrons $T_b = 1.07 \cdot 10^{12}a(\gamma)H_\perp^{-1/7}$ K. $H_\perp$ is the magnetic field tension component orthogonal to the particle velocity motion. Readhead (1994), Begelman et al. (1994) and Guijosa & Daly (1996) showed that for many synchrotron sources an equipartition model is more convenient; i.e., the particle and magnetic field energy densities are equal. T_b found this way has the limit $T_b = 10^{11}$ K. This value is also probably confirmed by observations, but it corresponds to the more external parts of the sources. But for the central regions near the super massive black hole of AGNs we expect a very big value ($10^4 - 10^{10}$ G) of the regular magnetic field, which is much higher in energy density than the accelerating particles in the same place. The KelPau model seems more convenient for such conditions.

The highest lower limits of T_b from direct ground based and Space VLBI measurements are 10^{13} K. Numerous data on IDV radio sources point to much higher values T_b than 10^{12} K, if the IDV are due to intrinsic variability. As has been demonstrated in the last years, the part of this variation that changes with season and frequency of observations is completely explained by interstellar scintillations and orbital motion of the Earth (Shapirovskaya 1982; Heeschen & Rickett 1987; Macquart & Jauncey 2002; Rickett et al. 2002; Dennett-Thorpe & de Bruyn 2002). This phenomenon indicats very small angular size (smaller than 10 μas) and very high T_b. But for some sources correlations between radio and optical variations (Wagner & Witzel 1995; Peng et al. 2000) were found. This variability is probably intrinsic, and indicates angular sizes smaller 1 μas and extremely high T_b (up to 10^{16} K and higher).

Korchak & Terletsky (1952) proposed that galactic radio waves can be explained by synchrotron emission of relativistic protons. The same was proposed for quasars (Jukes 1967) and for pulsars (Pacini & Rees 1970). In the last paper it was noted that the T_b for synchrotron emission scales with particle mass as $m^{3/2}$. From eq. (1) and, if the KelPau model for protons is acceptable, it follows that $T_b = 1.68 \cdot 10^{16}a(\gamma)H_\perp^{-1/7}$ K.

T_b for curvature radiation From eq. (1), it follows that the smaller the magnetic field projection $H_\perp$ is, the higher T_b is. The value of $H_\perp$ for central regions of AGNs can be very high, but the particle acceleration, the big value of synchrotron losses and conservation of adiabatic invariant will constrain the outward motion strongly along magnetic field lines. At this circumstance, for a monoenergetic particle spectrum

$$T_b = a^{7/8}(mc^2/k)(0.75mc^2R_ce^{-2})^{1/8}, \quad (2)$$

where R_c is the curvature radius of magnetic field lines. This relation gives at $R_c = 4 \cdot 10^{15}$ cm (black hole of few billions Solar masses) $1.4 \cdot 10^{13}$ K for electrons and $6.7 \cdot 10^{16}$ K for protons.

Permanent losses and injection of relativistic particles Slysh (1992) proposed a model of a very high intensity synchrotron source in which Compton scattering losses of particles is compensated by permanent injection of new relativistic particles. For such a model and a monoenergetic particle spectrum

$$T_b = 0.75 m^2 c^5 (4\pi e^2 k)^{-1} \nu_c^{-3/2} (2t/3)^{-1/2}\,, \tag{3}$$

where ν_c is a critical frequency of synchrotron emission and t is the time for Compton losses or emission volume crossing. It gives $3 \cdot 10^{13}$ K for electrons and $1 \cdot 10^{20}$ K at $\nu_c = 30$ GHz and $t = 1$ day.

Coherent emission The limiting value of T_b can increase also if fluctuations of relativistic particle density exist on scales smaller than the emitting wavelength. If N is the number of particles in one fluctuation, then for eq. (1) by replacing charge e with eN and mass m with mN, we find for the whole source, as an ensemble of such fluctuations, T_b increasing proportional to $N^{6/7}$. For relation (2) we find T_b proportional to $N^{7/8}$, and (3) does not change. This type of emission can be responsible for radio emission of pulsars, and also, possibly, for most central regions of AGNs.

Doppler boosting If the emitting region has systematic motion, then T_b is increased proportionally to $\delta^{n-\alpha}$ and the emission spectrum is shifted to higher frequencies proportional to δ. Here the Doppler factor $\delta = \Gamma^{-1}(1-\beta\cos\theta)^{-1}$, $\Gamma = (1-\beta^2)^{-1/2}$, $\beta = v/c$, $n = 3$ for ballistic motion and $n = 2$ for jet, spectral index a corresponds to a spectrum proportional of ν^{α}, and θ is the angle between velocity of systematic motion and direction to the observer (Rees 1966; Woltjer 1966; Blandford & Rees 1974).

Prospects and conclusion The KelPau model predicts the existence in the central regions of some specific type of sources with a very high T_b and some new processes described above. Future exploration of these objects probably will reveal the structures near super massive black holes (Kardashev 1995). At least for some objects, we expect that the conditions for radio wave propagation will allow us to observe these regions. It seems that IDV sources are the best candidates for such a program, and probably it is not by chance that the shortest time variability of these sources (in radio, optical and X-ray bands) is comparable with the expected light gravitational radius R_g. The strict VLBI sizes of the Sgr A*, M 87 nucleus and some others near extragalactic objects are near a few R_g.

It is necessary to search for the expected correlation of brightness temperature and super luminal motion (or expansion) and this searching is critical for the determination of Doppler boosting amplification. If this factor can be excluded, the proton synchrotron model can be checked. The value, structure and variability of magnetic field will be determined by future polarization exploration of the same sources in all spectral bands.

The most interesting future results connected with this program are expected at explorations with higher angular resolution. The question of first VLBI observations on intercontinental bases were discussed in 1963 by Sir Bernard Lovell, Josef Shklovsky, Leonid Matveyenko, Gennadi Sholomitskii and me. Lovell said that such an interferometer can be constructed, but where is the guarantee of the existence of radio sources with such small angular sizes. The answer was found in the paper by Kellermann & Pauliny-Toth (1969) and confirmed by the following years of VLBI observations.

For observations by a simple two antenna interferometer of a compact source with circular Gaussian intensity distribution, the maximum brightness temperature $T_{\max} = \pi B_{\perp}^2 F_{\nu}(2k \ln(\Upsilon^{-1}))^{-1}$, where $B_{\perp}$ is the base line projection, Υ is the fringe visibility function, and F_{ν} is the total flux of the source. For $B_{\perp} = 10^4$ km, $\Upsilon = 0.9$, and $B_{\nu} = 1$ Jy we find $T_{\max} = 10^{12}$ K. Observations on the Earth are strongly restricted in the detection and exploration of sources with much higher $T_{\rm b}$, and this is independent of the wavelength of observations. It's seems in conclusion that future explorations of the most central regions of AGNs with different modifications of the KelPau model are strongly dependent on the realization of base lines much larger than an Earth diameter. In particular, we hope for the *RadioAstron* mission (`http://www.asc.rssi.ru/radioastron/index.asp`) at maximum $B_{\perp} = 350,000$ km) to increase the threshold of investigations up to 10^{15} K. This is enough for the detection and exploration of cosmic ray acceleration regions.

The author would like to thank Ken Kellermann for several years' very productive cooperation with USSR and Russian scientists.

References

Begelman, M. C., Rees, M. J., & Sikora, M. 1994, ApJ, 429, L57
Blandford, R. D., & Rees, M. J. 1974, MNRAS, 169, 395
Dennett-Thorpe, J., & de Bruyn, A. G. 2002, Nature, 415, 57
Guijosa, A., & Daly, R. A. 1996, ApJ, 461, 600
Heeschen, D. S., & Rickett, B. J. 1987, AJ, 93, 589
Jukes, J. D. 1967, Nature, 216, 461
Kardashev, N. S. 1995, MNRAS, 276,515
Kardashev, N. S. 2000, AZh, 77, 813 (translation in Astron. Rep., 44, 719)
Kardashev, N. S. 2001, MNRAS, 326, 1122
Kellermann, K. I., & Pauliny-Toth, I. I. K. 1969, ApJ, 155, L71
Kellermann, K. I. 2002, Publ. Astron. Soc. Aus., 19, 77
Korchak, A. A., & Terletsky, Y. P. 1952, ZhETF, 22, 507
Macquart, J.-P., & Jauncey, D. L. 2002, ApJ, 572, 786
Pacini, F., & Rees, M. J. 1970, Nature, 226, 622
Peng, B., et al. 2000, A&A, 353, 937
Readhead, A. 1994, ApJ, 426, 51
Rees, M. J. 1966, Nature, 211, 468
Rickett, B. J., Kedziora-Chudczer, L., & Jauncey, D. L. 2002, ApJ, 581, 103
Shapirovskaya, N. Y. 1982, AZh, 59, 246 (translation in Sov. Astron., 26, 151)
Slysh, V. I., 1992, ApJ, 391, 453
Wagner, S. J., & Witzel, A. 1995, ARA&A, 33, 163

Radio Astronomy at the Fringe
ASP Conference Series, Vol. 300, 2003
J.A. Zensus, M.H. Cohen, & E. Ros, eds.

Self-Stark Limit on Brightness Temperature in Cosmic Masers

V. I. Slysh
Arecibo Observatory, HC3 Box 53995, Arecibo PR 00612
Astro Space Center, Profsoyuznaya st. 84/32, 117997 Moscow, Russia

Abstract. The radiation broadening of narrow maser lines by its own emission which can also be described as resonance Self-Stark effect puts an upper limit on the brightness temperature of cosmic masers. It is shown that available measurements of the brightness temperature of OH and water masers are consistent with the observed narrow line widths only if the maser emission is highly directive.

1. Introduction

Methanol, OH and water masers are generally seen toward star-forming regions and exhibit narrow, strong emission line features coming from bright small spots. The brightness temperature T_b of the spots is very high, from 10^9 K for methanol masers, to 6×10^{12} K for OH masers, and 10^{15} K for water masers. In many cases the above values are only lower limits since maser spots are barely resolved with VLBI. Variability time scale often indicates, in an indirect way, a smaller linear size and still higher brightness temperature. There must be physical limits on the brightness temperature as there are for thermal and synchrotron emission in continuum sources. The brightness temperature of the thermal free-free emission from ionized gas depends on the kinetic temperature T_e of electrons and optical depth. It is the highest for optically thick sources and is equal to the kinetic temperature. In H II regions the kinetic temperature of electrons is limited to T_e=10,000 K by the energy losses from electrons through the line emission of heavy elements. Similarly, in synchrotron sources the brightness temperature is at maximum in optically thick sources, and is equal to the kinetic temperature of relativistic electrons. The kinetic energy of relativistic electrons is limited to 10^{11}–10^{12} K by the energy loss through inverse Compton emission on synchrotron photons (Kellermann & Pauliny-Toth 1969). The same arguments can be applied to molecular line emission: in the optically thick sources the brightness temperature in the line is limited by the gas kinetic temperature. The molecular maser sources make an exception because the absorption coefficient and optical depth are both negative due to the population inversion: the brightness temperature of the maser line emission may increase indefinitely with the optical depth. However other mechanisms can limit the brightness temperature of the maser lines, such as the radiation line broadening, or the resonance Stark-effect caused by the maser radiation field (we call it Self-Stark limit by analogy with Self-Compton limit).

2. Radiation, or resonance Stark broadening of maser lines

The natural band width of a molecular emission line in the absence of external field is $\Delta\nu = A/2\pi$, where $A = 64\pi^4\nu_0^3\mu^2/3hc^3$ is spontaneous emission rate at the frequency of the line ν_0, μ is the dipole-moment matrix element for the transition, and h is Planck constant. This is a very small band width compared to the observed line widths. For OH line at 1665 MHz, $A = 7.1 \times 10^{-11}\,\mathrm{s}^{-1}$, and for water line at 22.235 GHz, $A = 2 \times 10^{-9}\,\mathrm{s}^{-1}$, and the natural band widths are 10^{-11} Hz and 3×10^{-10} Hz, respectively. In strong masers like W3 (OH) or W 49 every OH or H_2O molecule is exposed to the intense radiation field, produced by all other molecules in the maser emission region. That causes frequent up and down transitions between the energy levels involved, due to induced emission and absorption. The time which a molecules spends on the energy level is greatly reduced, by a factor $2kT_\mathrm{b}/h\nu_0$. The linewidth is increased by the same factor which can be very large when the brightness temperature is high. For OH maser with brightness temperature 6×10^{12} K the linewidth is 1,700 Hz ($0.3\,\mathrm{km\,s^{-1}}$), and for water maser with the brightness temperature 10^{15} K it is 0.6 MHz ($8\,\mathrm{km\,s^{-1}}$). This linewidth is comparable to or even greater than observed line widths in masers. Einstein induced emission and absorption coefficients can be used only in the weak field case. For the strong field which is the case in water masers a correct description of the interaction between a molecule and the narrow band radiation field is the resonance Stark effect approach. In the weak field the resonance Stark effect is quadratic with respect to the electric field strength E, and causes the line broadening. In the strong field the resonance Stark effect becomes linear, and causes line splitting into a main line at the transition frequency ν_0 and two satellite lines at frequencies $\nu_0 \pm \mu E/h$ for non-degenerate energy levels such as are levels of the OH line transition at 1665 MHz. The energy levels of the H_2O transition at 22.235 GHz are degenerate, and are split in several satellite components (Slysh 1973). AC electric field strength E is related to the brightness temperature T_b of the maser radiation of band width $\delta\nu$ as $E = 8\pi(kT_\mathrm{b})^{0.5}\nu_0(\delta\nu)^{0.5}c^{-1.5}$. This relation holds for the isotropic distribution of the radiation intensity. If the radiation is concentrated in the solid angle Ω, the electric field strength must be multiplied by $(\Omega/4\pi)^{0.5}$, with the corresponding reduction of the line splitting. For the line broadening the reduction factor is $\Omega/4\pi$.

3. Brightness temperature of H_2O and OH masers

The strongest spectral feature in H_2O maser W3 OH is completely unresolved on all VLBA baselines (Fig. 1), and its brightness temperature exceeds 4×10^{15} K. Another strong H_2O maser, an outburst in Orion-KL, was measured with the space radio telescope *HALCA* to have brightness temperature $T_\mathrm{b}=10^{16}$ K (Kobayashi et al 2000). OH maser OH 34.26+0.15 was unresolved on baselines from ground to space radio telescope *HALCA*, and the brightness temperature exceeds 4×10^{12} K (Fig. 2).

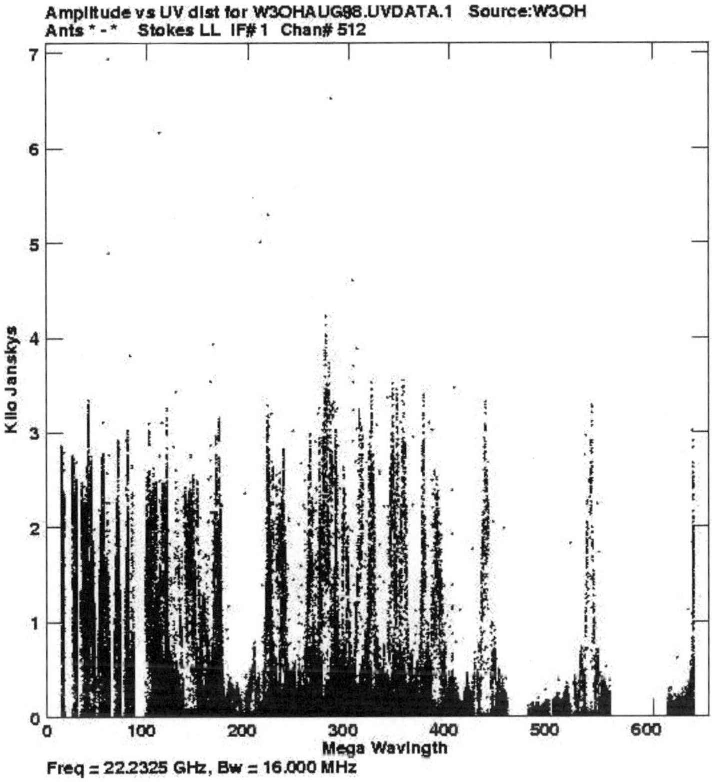

Figure 1. Correlated flux vs baseline for the strongest H_2O maser spectral feature in W3OH obtained from VLBA archive data

4. Directivity of the maser emission

No line splitting due to the resonance Stark effect was observed in H_2O or OH masers. Also, no significant radiation line broadening is seen in the line profiles. This means that the electric field strength in the masers is weak enough in spite of the very high brightness temperature, because of the small solid angle in which emission is directed. Assuming that the the radiation broadening contributes to the total linewidth less than 1/5 of the linewidth, one can put an upper limit on the solid angle. One has for OH maser OH 34.26+0.15 an upper limit for the emission solid angle $\Omega/4\pi$=0.02. For H_2O masers in W3 OH one has $\Omega/4\pi$=0.005, and for the outburst H_2O maser in Orion-KL $\Omega/4\pi$=0.002 as the upper limits.

5. Conclusion

The brightness temperature of cosmic masers is limited by the radiation broadening in the intense radiation field at the frequency of the maser emission. Up

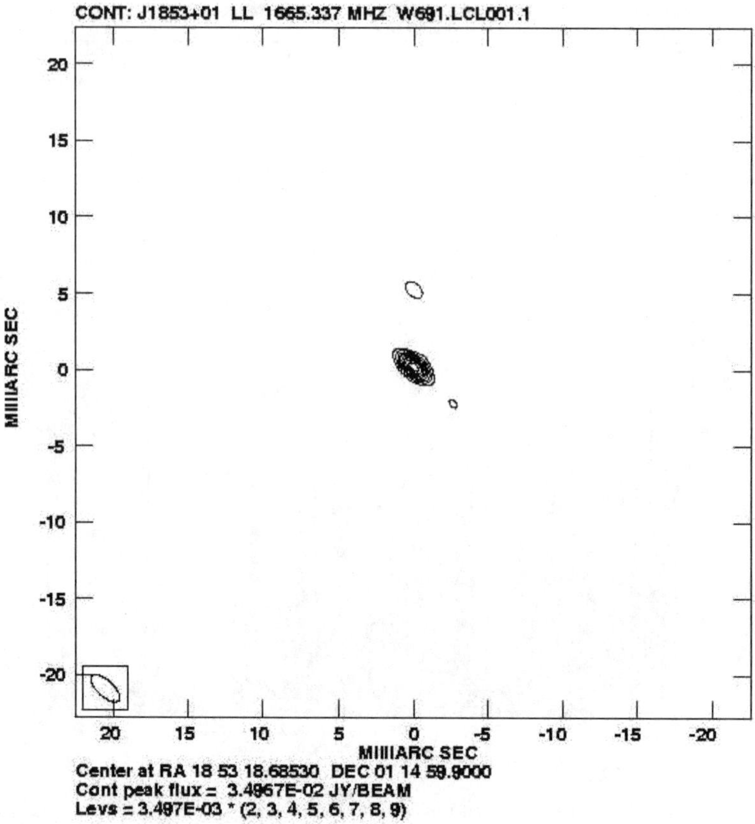

Figure 2. A map of the most compact OH-maser emission spectral feature in OH 34.26+0.15. Participating telescopes were phased VLA, 70 m Tidbinbilla and 8 m space telescope *HALCA*. Synthesized beam was 2.4×1.1 mas

to now no radiation broadening has been observed in OH and H_2O masers, which can be due to the very small solid angle in which the emission is directed. The theory of the resonance radiation broadening allows us to put stringent upper limits on the emission solid angle. This may be important for the further development of cosmic maser models.

References

Kellermann, K. I.& Pauliny-Toth, I. I. K. 1969, ApJ, 155, L71

Kobayashi, H., Shimoikura, T., Omodaka, T. & Diamond, P.J. 2000 in Astrophysical Phenomena Revealed by Space VLBI, ed. H. Hirabayashi, P. G. Edwards & D. W. Murphy, 109

Slysh, V. I. 1973, Astrophysical Letters, 14, 213

Radio Astronomy at the Fringe
ASP Conference Series, Vol. 300, 2003
J.A. Zensus, M.H. Cohen, & E. Ros, eds.

The Space Distribution and Evolution of Radio Sources

J. J. CONDON
National Radio Astronomy Observatory, 520 Edgemont Road, Charlottesville, VA 22903 USA

Abstract. Ken Kellermann has played an important role in measuring and explaining the space distribution and evolution of extragalactic radio sources. A clear picture has emerged from initial chaos during his career. Once at the fringe of astronomy because it was focused on AGNs in distant galaxies, extragalactic radio astronomy is becoming an integral part of multi-wavelength studies of galaxy and even stellar evolution.

1. Introduction

The "golden age" of a new scientific field begins with bits of apparently contradictory data partially explained by competing theories, grows through an exciting period of discovery, and concludes with the much clearer understanding that characterizes a mature field. Thus the 1920s and 1930s were the golden age of quantum mechanics. It is a fortunate scientist whose career spans such a golden age, as Ken's has for exploring the space distribution and cosmological evolution of radio sources.

To set the stage, we go back to Ken's 1971 Warner Prize lecture on "Radio Galaxies, Quasars, and Cosmology" at the AAS meeting in San Juan (Kellermann 1972), when he could summarize in one sentence what was known about the space distribution of radio sources:

"So, while we radio astronomers may be able to see further out in the universe than anyone, unfortunately, we don't know *where* we are looking, and we don't know *what* we are looking at."

The apparently contradictory bits of data included:

(1) The steep slope of the strong-source counts suggested cosmological evolution of luminous radio galaxies, but the non-Euclidean part was confined to source densities $N \leq 10\,\mathrm{sr}^{-1}$ in the statistically independent differential counts (Jauncey 1967). Anisotropy in source distributions on angular scales both large (differences between the numbers of strong sources in the northern and southern hemispheres) and small (differences between the deeper 5C1 and 5C2 surveys) suggested a very inhomogeneous universe. We might live in "local hole" and be misled by the relative scarcity of very strong sources. Ken concluded:

"Clearly it is difficult to reconcile the presence of these apparent non-uniformities in the angular distribution of radio sources on scales ranging from a few square degrees to 2π sr, with the source counts being cosmologically significant."

(2) The first extensive surveys at 2.7 GHz (Wall et al. 1971) and 5 GHz (Kellermann, Davis, & Pauliny-Toth 1971) found many flat-spectrum sources

whose $V/V_{\rm m}$ distribution was consistent with a non-evolving local population, unlike steep-spectrum radio galaxies whose large $\langle V/V_{\rm m}\rangle > 0.5$ indicated strong cosmological evolution. Might these flat-spectrum quasars be nearby objects with non-cosmological redshifts? In addition, independent cosmological tests using radio sources alone gave negative results:

"It is indeed disappointing, and perhaps cause for concern, that so far it has not been possible to detect the expected effect of the red shift on the observed intensity, angular size, or spectra of the extragalactic radio sources."

(3) The local radio luminosity function had huge statistical uncertainties because few extragalactic sources are "local." Consequently, early evolutionary models were constrained more by the source counts and luminosity distributions of strong but distant sources than by the local luminosity function. The biggest uncertainties were in the here and now, not the long ago and far away. The universe of extragalactic radio astronomers is a nearly hollow shell at redshift $z \sim 1$ which has little overlap with the optical astronomers' universe of bright, nearby galaxies. The symposium title "Radio Astronomy at the Fringe" refers to interferometry but has a sociological meaning as well. Ken hoped to bypass the difficulties of making optical identifications and interpreting redshifts, and he speculated about

"what can be done by using the radio data alone without making direct use of individual optical identifications."

What use can be made of the radio data alone? How have these contradictions have been resolved? Is extragalactic radio astronomy still at the fringe of astronomy?

2. Using the Radio Data Alone

A good example of using the radio data alone is exploiting the unique isotropy of radio sources on the sky to demonstrate (1) that we don't live in a local hole, (2) radio sources evolve, and (3) most radio sources really are at cosmological distances. The uniquely smooth sky distribution of radio sources (Fig. 1) in the NVSS (Condon et al. 1998) is quite different from the clumpy distributions of the brightest galaxies selected at optical or infrared wavelengths. If we were living in a local hole large enough to affect the counts of strong radio sources, similar holes elsewhere in the universe would have been conspicuous in Figure 1. If radio sources didn't evolve significantly, the stronger radio sources would be about as close as optically bright galaxies, and clustering with correlation lengths $L \sim 10\,{\rm Mpc}$ would have dominated the appearance of Figure 1. Likewise, if a moderately large fraction of radio sources had non-cosmological redshifts and were associated with nearby galaxies, radio sources would have about the same clumpy distribution on the sky as optical galaxies. Only if most radio sources really are cosmologically distant can they be distributed so isotropically that sensitive statistical tests are needed to reveal clustering with normal correlation lengths (Blake & Wall 2002a). Additional evidence for the cosmological nature of extragalactic radio sources comes from the small dipolar anisotropy in the sky density of radio sources caused by the Earth's motion. This effect was predicted by Ellis & Baldwin (1984) and recently measured in the NVSS source counts by Blake & Wall (2002b).

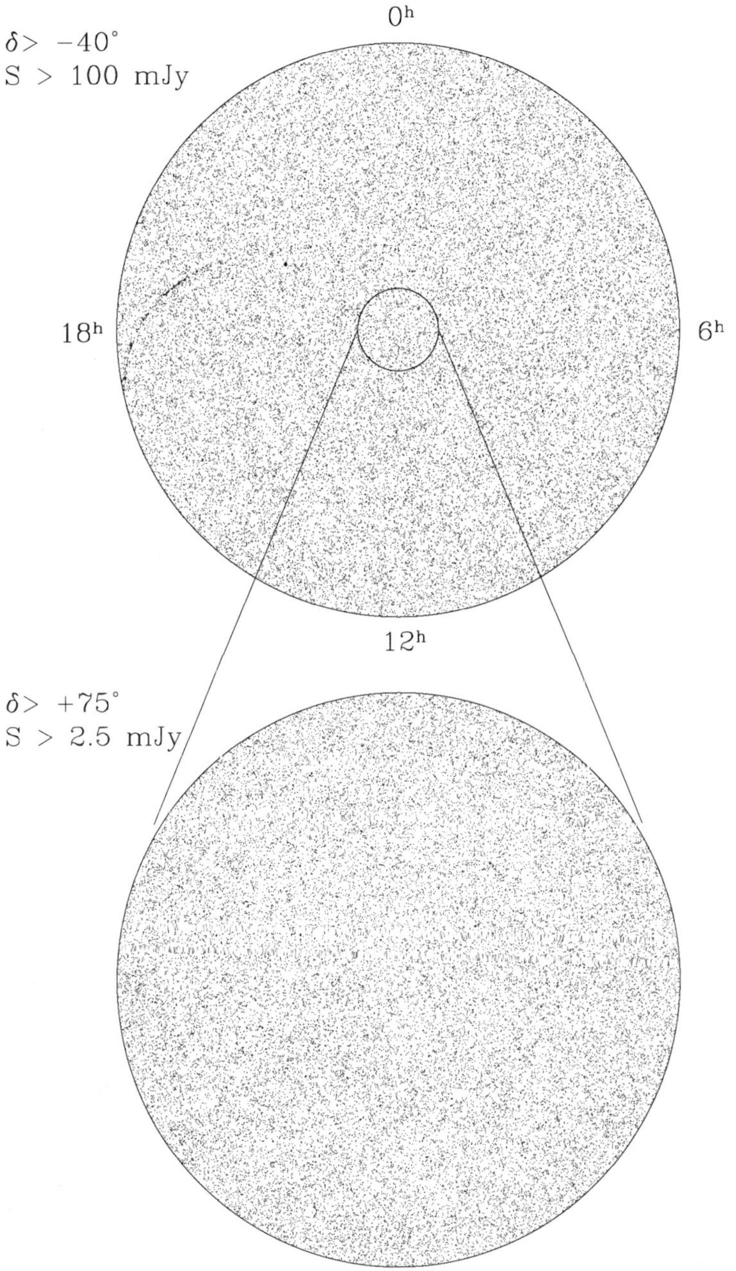

Figure 1. Equal-area plots show that radio sources with $\delta > -40°$ and stronger than 100 mJy (upper circle) and sources with $\delta > +75°$ and stronger than 2.5 mJy at 1.4 GHz (lower circle) have nearly isotropic distributions on the sky.

3. The Local Luminosity Function

The evolution of a source population is specified by its time-dependent luminosity function. The present, or local, radio luminosity function is difficult to measure because $< 1\%$ of radio sources are reasonably local ($D \lesssim 100\,\mathrm{Mpc}$) at *any* flux level (Fig. 2). This statistical problem can be solved only by samples containing very large numbers of radio sources. Recently, NVSS sources have been cross-correlated with UGC galaxies to yield accurate local luminosity functions for star-forming galaxies and AGNs.

4. Star-forming Galaxies

Deep surveys in mid-80's finally broke through the 1 mJy barrier (e.g., Fomalont et al. 1984) which marks a division in source populations. Many of the microJy radio sources are produced by an evolving population of star-forming galaxies that, at least locally, obey the remarkably tight far-infrared (FIR) to radio flux correlation. Consequently, there is a nearly complete overlap of these radio and FIR populations (Fig. 3). Extragalactic radio astronomy isn't just about AGNs any more; it also provides an extinction-free measure of star-formation rates in the universe.

5. No Longer At The Fringe?

Extragalactic radio astronomy is no longer "at the fringe" for two more reasons: (1) radio astronomers have succeeded in detecting and identifying large populations of reasonably nearby radio sources and (2) astronomers observing in other wavebands are catching up with radio astronomers in their ability to detect complete samples of galaxies at cosmological distances (e.g., Richards et al. 1998). I believe that most future radio surveys and studies of source populations will be made as parts of multi-wavelength collaborations. A good example is the SIRTF First-Look Survey (FLS, see `http://sirtf.caltech.edu/SSC/fls/`) designed to explore the infrared sky two orders of magnitude deeper than IRAS reached. Optical and radio surveys are an essential part of the FLS. If local IRAS sources with the highest FIR/optical flux ratios (Arp 220, for example) were moved to cosmological distances where they would be just above the FLS sensitivity limit, they would still be brighter than $R \approx 25.5\,\mathrm{mag}$. Optical images of this depth have been obtained with the CCD Mosaic camera on the NOAO 4 m telescope. Accurate ($\sigma_{\alpha,\delta} < 1''$) positions are needed to make complete and reliable associations of infrared or radio sources with such faint galaxies. However, most of the detected FIR sources will be just above the 5σ detection limit, so their root-mean-square (rms) positional uncertainties will be about 10% of the SIRTF beamwidth, $\sigma_{\alpha,\delta} \approx 2''$ at $\lambda = 70\,\mu\mathrm{m}$ and $\sigma_{\alpha,\delta} \approx 5''$ at $\lambda = 160\,\mu\mathrm{m}$.

A deep radio image covering the FLS area is crucial for two reasons: (1) Nearly all far-infrared (FIR) sources obey the tight FIR/radio flux correlation and most radio sources fainter than 1 mJy at 1.4
,GHz also appear to be star burst galaxies. We therefore expect that the deep FIR and radio source populations are nearly identical. Only radio observations can provide pre-launch images and catalogs that simultaneously select most of

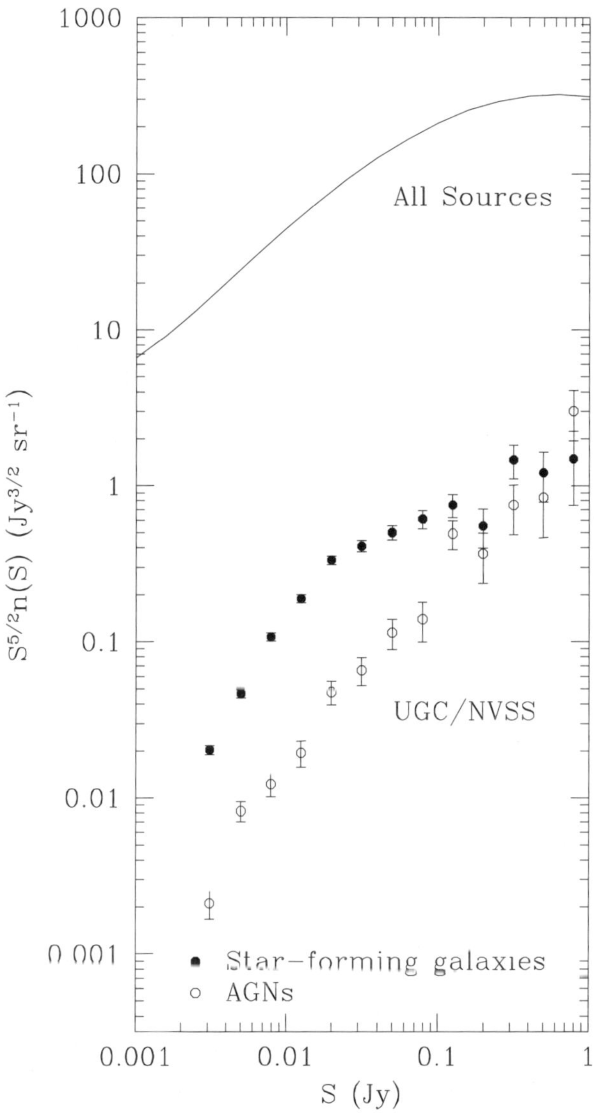

Figure 2. Extragalactic radio astronomy was "at the fringe" for so long because nearly all radio sources are at cosmological distances, unlike the brightest optical galaxies which tend to be nearby. Less than 1% of NVSS radio sources at any flux level can be identified with UGC galaxies brighter than $m_p = 14.5$ (Condon, Cotton, & Broderick 2002). Abscissa: 1.4 GHz flux density (Jy). Ordinate: Differential source counts normalized to the static Euclidean slope ($Jy^{3/2}$ sr^{-1}). Parameter: Energy source (stars and stellar remnants or active galactic nuclei) responsible for the radio emission.

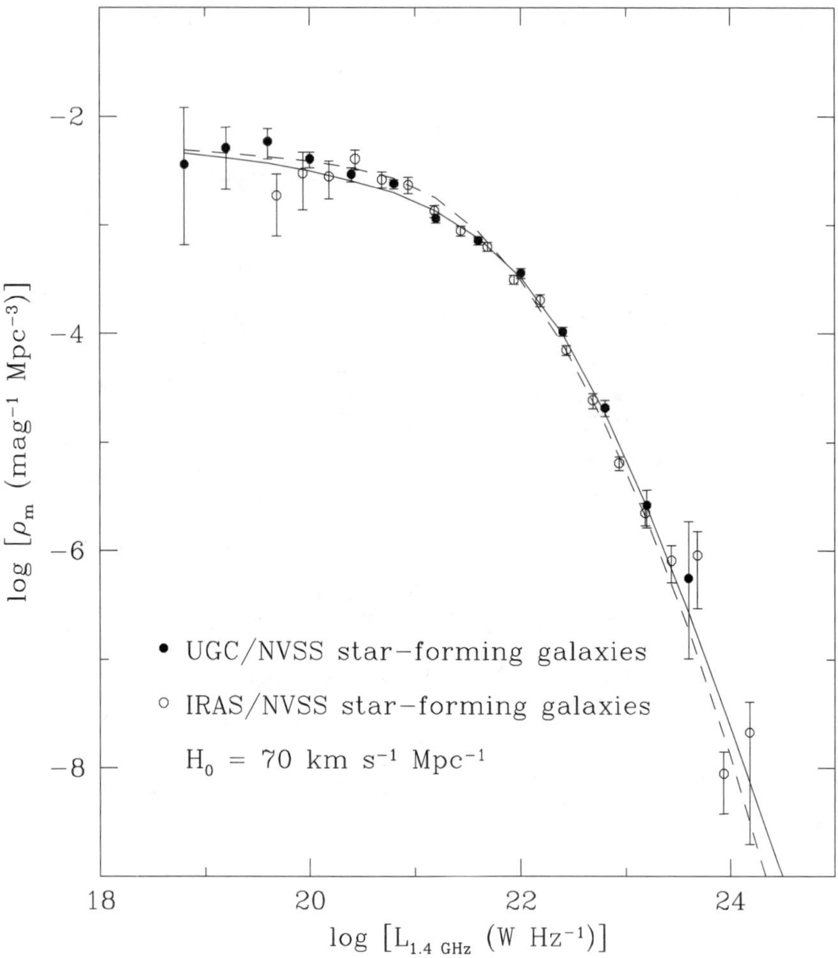

Figure 3. The local radio luminosity function of star-forming galaxies (data points and fitted solid curve) is nearly identical to the FIR luminosity function converted to 1.4 GHz by the mean FIR/radio flux ratio (dashed curve) owing to the tight FIR/radio correlation. Extragalactic radio and far-infrared astronomers now have a lot in common.

the several thousand galaxies that will appear in the FIR images and reject the millions of galaxies brighter than $R \approx 25.5$ that will not. (2) The radio image can have the higher resolution and positional accuracy required for making reliable optical identifications on the basis of position coincidence alone. The radio data can be used to select and identify those optical galaxies most likely to be FLS sources for further study before launch. Later, the FIR/radio flux ratios can also be used to distinguish dust-shrouded AGN from star burst galaxies. The second law of thermodynamics notwithstanding, the VLA image covering the

FLS (Fig. 4) shows that it is possible to extract useful work from extragalactic radio astronomers.

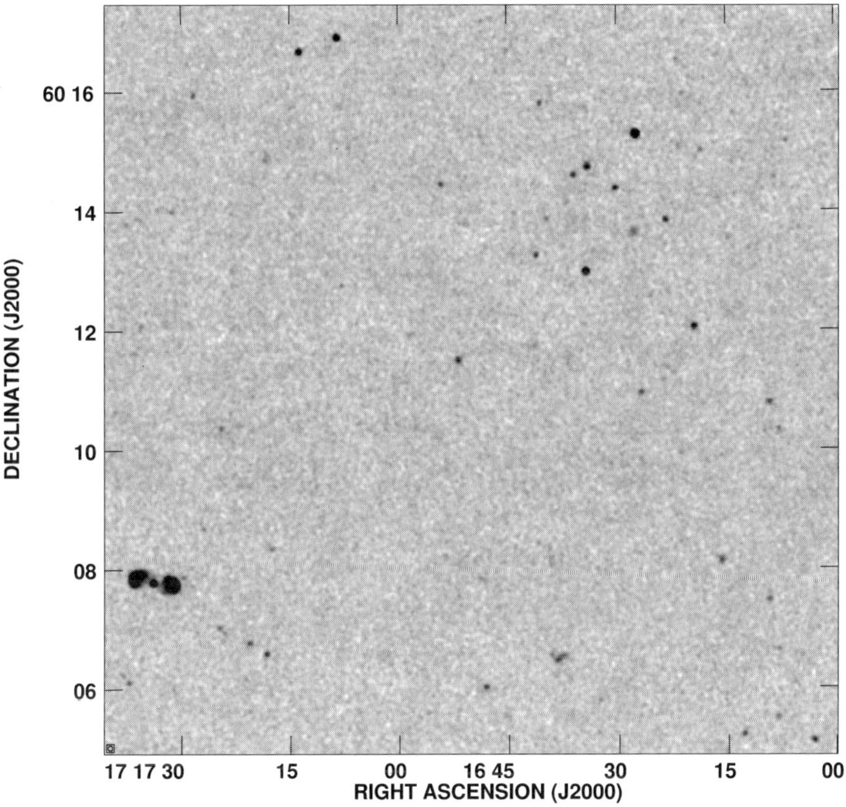

Figure 4. This subimage shows $\approx 1\%$ of the FLS area at 1.4 GHz with $\theta = 5''.0$ resolution (see the small inset in the lower left corner) and nearly uniform $\sigma_n \approx 23\,\mu$Jy noise. The full image and source catalog are available on the web at `http://www.cv.nrao.edu/sirtf_fls/`.

In the next few years the major space observatories SIRTF (via the SWIRE legacy project), GALEX (an ultraviolet explorer), and *XMM-Newton* (X-ray) will make large-scale sky surveys capable of detecting millions of galaxies, tens of thousands of quasars, and hundreds of clusters of galaxies at cosmological distances in the infrared, ultraviolet, and X-ray bands. They will concentrate on the $\sim 100\,\mathrm{deg}^2$ of sky in several "cosmic windows" having the lowest Galactic extinction and cirrus emission to in order to study the history of star formation, super-massive black holes, and massive galaxy clusters. Complementary optical and radio surveys will provide essential source identifications, redshifts, morphologies, and continuum spectra. A VLA "Cosmic Windows Survey" has been proposed to cover $\Omega \sim 52\,\mathrm{deg}^2$ with $\theta = 5''$ FWHM resolution and detect $\sim 2.5 \times 10^4$ sources having $S \geq 200\,\mu$Jy, and similar radio surveys are sure to follow in efforts to pin down the space distribution and evolution of radio sources.

References

Blake, C., & Wall, J. 2002a, MNRAS, 337, 993
Blake, C., & Wall, J. 2002b, Nature, 416, 150
Condon, J. J., Cotton, W. D., & Broderick, J. J. 2002, AJ, 124, 675
Condon, J. J., et al. 1998, AJ, 115, 1693 (NVSS)
Ellis, G. F. R., & Baldwin, J. E. 1984, MNRAS, 206, 377
Fomalont, E. B., et al. & Weistrop, D. 1984, Science, 225, 23
Jauncey, D. L. 1967, Nature, 216, 877
Kellermann, K. I. 1972, AJ, 77, 531
Kellermann, K. I., Davis, M. M., & Pauliny-Toth, I. I. K. 1971, ApJ, 170, L1
Richards, E. A., et al. 1998, AJ, 116, 1039
Wall, J. V., Shimmins, A. J., & Merkelijn, J. K. 1971, AuJPAS, 19, 1

Radio Astronomy at the Fringe
ASP Conference Series, Vol. 300, 2003
J.A. Zensus, M.H. Cohen, & E. Ros, eds.

Cosmology from the NRAO VLA Sky Survey

J. V. Wall, C. A. Blake
Astrophysics: Oxford, Denys Wilkinson Building, Keble Road, Oxford, OX1 3RH, UK

Abstract. The clustering of radio sources in the NRAO VLA Sky Survey has been investigated on all scales with three techniques: angular two-point correlation, counts-in-cells, and the angular power spectrum. Instrumental effects, surface-density gradients and over-resolution, affect these measures and required careful consideration before cosmological results could be deduced. These results include a measurement of the correlation length $r_o \sim 8h^{-1}$ Mpc for radio galaxies; an estimation of the initial mass perturbation spectrum from the angular power spectrum; and detection of the cosmological velocity dipole in the surface density of NVSS sources.

When Ken Kellermann was much younger and gainfully employed surveying the sky at cm wavelengths, we had several discussions, verbally and in print, about large-scale anisotropies at the level of $2.5 - 3\sigma$ in the sky distribution of radio sources. These anisotropies gave us much unease: they were apparent between different hemispheres, different telescopes, different frequencies and different surveyors. Smoke hung in the air from old and acrimonious debates on instrumental effects, source counts, confusion, and non-cosmological redshifts. Lost in the smoke was a suggestion by Ivan Pauliny-Toth and Ken (1972): "This raises again the old question as to what extent the observed source counts are due to a local anomaly, as for example the effect of the local super-cluster..."; and it was an old question—Hanbury Brown (1962) had discussed the possibility in detail, even concluding "that there should be significant differences between the NGH and SGH". Shaver & Pierre (1989) showed conclusively how the supergalactic plane did indeed account for the apparent anisotropies. At the same time deep pencil-beam surveys were detecting distant super-clusters (e.g., Windhorst et al. 1990). Armed with the indication that radio surveys could 'see' both local and distant clustering (Benn & Wall 1995), it seemed worth while to investigate the cosmology available from the latest generation of deep sky surveys (Cress et al. 1996; Loan, Wall, & Lahav 1997; Baleisis et al. 1998; Magliocchetti et al. 1998, 1999). Here we summarize recent results for the NVSS survey.

1. NVSS angular clustering measurements

The NVSS (Condon et al. 1998) is a stunning achievement of the VLA and the NRAO team, the benchmark radio survey in terms of sky coverage (total down to dec $-40°$), number of sources (1.8×10^6), and depth (complete to ~ 3.5 mJy at 1.4 GHz). It is superbly complemented by the FIRST (Becker et al. 1995) survey, also with the VLA and also at 1.4 GHz, which currently covers less than half as much sky, but at resolution of 5 arcsec, nearly an order of magnitude smaller than the 45 arcsec resolution of NVSS. Detailed comparison of the two surveys enables the effects of resolution and surface brightness on source statistics to be understood (Blake & Wall 2002c). The FIRST survey has undergone considerable clustering analysis (Cress et al. 1996; Cress & Kamionkowski 1998; Magliocchetti et al. 1998, 1999), via the angular correlation function ($w(\theta)$) and counts-in-cells (c-in-c). We have now explored clustering in the NVSS with three classical clustering measures, (1) the two-point angular correlation function, (2) counts-in-cells, and (3) the angular power spectrum.

Before embarking on the clustering analyses, several issues, instrumental and astronomical, required consideration. Firstly, we masked regions within 5° of the Galactic plane in order to 'remove' Galactic sources. Secondly, we masked regions about 22 very bright extragalactic sources whose extended structures (and side-lobe responses) are represented in the NVSS by 'clusters' of point-source responses. Thirdly we discovered that the original NVSS catalog showed severe discontinuities in surface density at declinations where array configurations were changed between D and C, to provide improved north-south resolution. These effects were traced by Jim Condon and Bill Cotton to difficulties in reconstructing radio images from the snapshot VLA observations with sparse (u, v)-coverage. A new version of the catalog software which they issued as a result of this analysis reduced these discontinuities very substantially.

(1). The **two-point angular correlation function** $w(\theta)$ measures the fractional increase in the probability δP of finding objects in two solid-angle elements separated by θ: $\delta P = \varsigma^2[1 + w(\theta)]d\Omega_1 d\Omega_2$ where ς is the object surface density. It is far from a complete description of clustering; it has drawbacks in error estimation and insensitivity on the larger scales—but it is simple, intuitive, and easy to measure, particularly for irregularly-shaped areas. The simplest estimator of $w(\theta)$ is $w_0 = DD(\theta)/RR(\theta) - 1$, with DD the number of observed separations in the range θ to $\theta + d\theta$, and RR the corresponding number for a Monte-Carlo-generated set of random galaxy positions. Better estimators are available and in fact after considerable experimentation we settled on the Landy-Szalay (1993) estimator. The results are shown in Figure 1.

The dominant feature of the measured $w(\theta)$ is the steep slope (~ -3.0) at small separations; moreover the amplitudes of power laws fitted in this region show a clear dependence on flux-density level (Fig. 1). Suppose that the survey has resolved some large sources into two components, cataloged independently. This must produce an 'intrinsic' contribution to the small-angle $w(\theta)$ – and it can be shown (Blake & Wall 2002b) that this contribution $\Delta w(\theta) \propto \theta^{-\beta-1}/\varsigma$, where the size distribution of large double sources is a power-law of slope $-\beta$ and the surface density of sources is ς. The predicted flux-density dependence is exactly that shown by the measured $w(\theta)$ at angles $< 0\overset{\circ}{.}1$; moreover the angular size distribution derived on this assumption predicts a number of doubles in close agreement with the findings by Lara et al. (2001). We concluded that this portion of the measured $w(\theta)$ is dominated entirely by 'intrinsic' correlation, a few per cent of the NVSS sources resolved as giant doubles serving to overwhelm the true clustering signal at small separations.

In contrast, the power-law at separations beyond $0\overset{\circ}{.}1$ shows the canonical slope of ~ -0.8 found for optically-selected galaxies and QSOs, together with a reassuring lack of amplitude-dependence on flux-density level (Fig. 2).

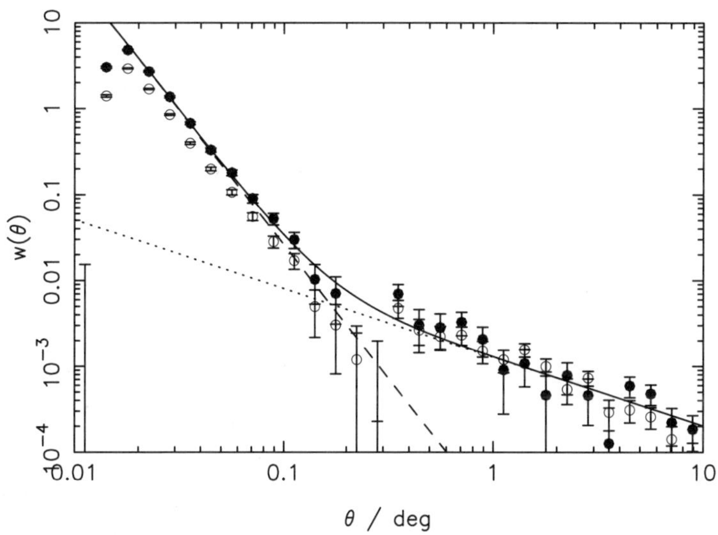

Figure 1. The NVSS angular correlation function, measured at $S_{1.4\mathrm{GHz}} = 20\,\mathrm{mJy}$ (dots) and $10\,\mathrm{mJy}$ (circles). The best-fit sum of two power laws for the 20 mJy data is shown as the solid line; the dashed small-angle power law is due to large multiple-component sources while the dotted power law is due to galaxy clustering.

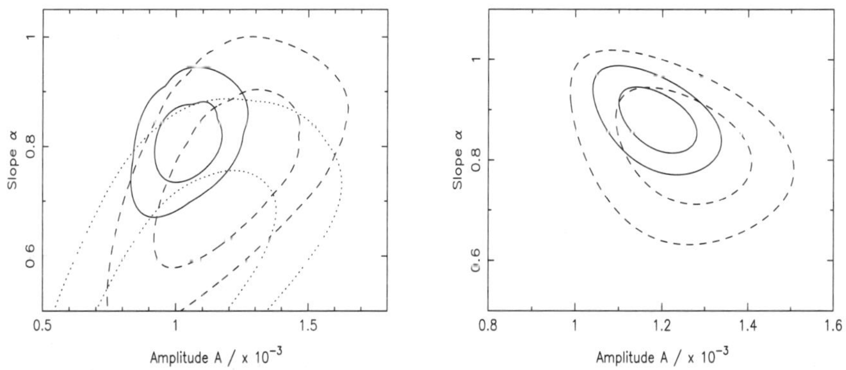

Figure 2. Constraints on the clustering parameters for $w(\theta) = A\theta^{-\alpha}$, θ in degrees. Left: from fitting the large-angle power law to $w(\theta)$ (Fig. 1); right: from converting the measured $w(\theta)$ to an equivalent counts-in-cells variance for comparison with c-in-c measurements. The contours represent 1σ and 2σ contours, solid for $S_{1.4\mathrm{GHz}} \geq 10\,\mathrm{mJy}$, dashed for 20 mJy, dotted for 30 mJy.

(2). **Counts-in-cells (c-in-c)** is a second simple way to quantify galaxy clustering, done by dividing the sky into cells of fixed area and shape and counting cell contents for a range of cell sizes. It works for irregular areas because partial cells may be considered. Angular clustering is completely characterized by the resultant distributions which thus contain more information than $w(\theta)$; but extracting the information from frequency distributions that are inevitably close to Poissonian is not easy. Both surface density gradients and multiple-component sources affect the c-in-c analyses as for $w(\theta)$, and the effects may be similarly quantified (Blake & Wall 2002c). A relation exists between the c-in-c variances and $w(\theta)$, and this relation can be derived for a double-power-law $w(\theta)$ of the type shown in Figure 1. As a separate estimate of clustering parameters, we obtained the best-fit values for amplitude and slope of the correlation law from the c-in-c variance results, by fixing the steep portion of the slope at the best-fit values found directly from the measured $w(\theta)$. The results are shown in Figure 2. The methods yield consistent values for the power-law parameters amplitude A and slope $-\alpha$; and this with residual instrumental effects affecting c-in-c and $w(\theta)$ in different manners.

(3). Measurement of the NVSS **angular power spectrum** constituted a third classical method to investigate clustering. This analysis represents the source surface-density field over the sky as the sum of sinusoidal angular density fluctuations of different wavelengths. The base functions are the spherical harmonic functions, analogous to sine and cosine in standard Fourier analysis; the quantities c_ℓ express the amplitudes of the ℓth multi-pole, whose wavelength is $\theta \sim 180°/\ell$. It is well known that the c_ℓ spectrum is directly related to $w(\theta)$ through integrals involving Legendre polynomials. However, $w(\theta)$ can be determined accurately at small angles only, whereas c_ℓ has highest signal at small ℓ corresponding to large angular scales. The two statistics are thus highly complementary. There are two statistical advantages with c_ℓ: firstly, the multi-poles have generally uncorrelated errors whereas adjacent bins in $w(\theta)$ are inevitably highly correlated; secondly, a natural relation exists between the c_ℓ spectrum and the spatial power spectrum, a very convenient method of describing structure in the Universe.

The estimation of c_ℓ over a survey with incomplete sky coverage together with previously-described instrumental effects requires several steps, detailed by Blake (2002). The results are shown in Figure 3. Highly significant signal is present, and the NVSS angular power spectrum decreases with ℓ approximately as a power law. It is clear from Figure 3 (left panel) that the predictions from transforming the $w(\theta)$ best-fit power law match this spectrum well.

2. Cosmology from the NVSS

The three angular clustering measures produce results which under appropriate transformations are statistically identical. The instrumental effects, in particular over-resolution and surface-density gradients, produce different responses for each of these measures; the consistency in the measures is a necessary (but perhaps not sufficient) condition to show that instrumental effects are understood. It is now not unreasonable to use the measures to derive cosmological results. We present three.

(1). **The clustering length** r_0. Angular clustering properties may be de-projected to infer spatial clustering. The process requires knowledge of—or a prescription for—(a) the redshift distribution of the sample, (b) the evolution of spatial clustering with redshift, (c) the evolution of Universal geometry, i.e., specification of parameters such as Ω_m and Ω_λ, and (d) the bias and evolution of bias for radio galaxy 'light' with respect to mass over-densities. The

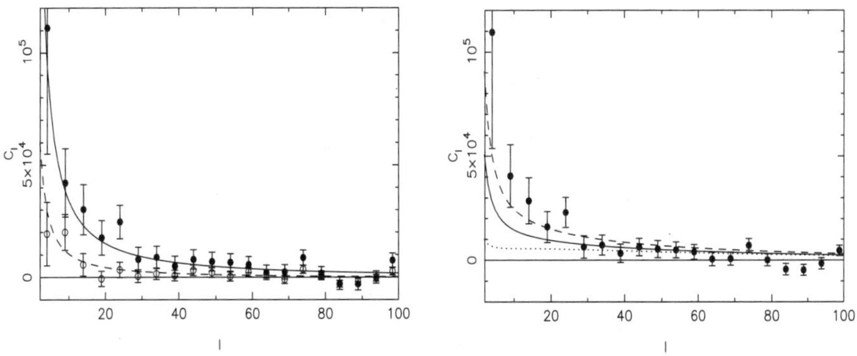

Figure 3. Left: measurement of the NVSS c_ℓ spectrum, at thresholds of 10 mJy (dots) and 20 mJy (circles). The solid and dashed lines show the spectra calculated from $w(\theta) = 1 \times 10^{-3}\theta^{-0.8}$. Right: the measured c_ℓ spectrum compared with model predictions using three different redshift distributions and assuming a spatial power spectrum with scale-invariant COBE normalization, constant bias factor of 1.9, and cosmology $\Omega_m = 0.3, \Omega_\lambda = 0.7, h = H_\circ/100 = 0.7$.

redshift distribution is a key requirement, essential in the use of the cosmological Limber's equation (Loan, Wall, & Lahav 1997) to de-project $w(\theta)$ into the common parameterization of the redshift-dependent spatial correlation function $\xi(r, z) = (\frac{r}{r_o})^{-\gamma}(1+z)^{(3-\gamma-\epsilon)}$. In this formulation, $\gamma = \alpha + 1$, ϵ describes clustering evolution, and the *clustering length* r_o describes the strength of clustering. Using new redshift data from the 2dF survey together with reasonable assumptions for other parameters, we find $6 < r_o/h^{-1} < 12$ Mpc. The range is due to different assumptions for the redshift distribution and for the evolution of bias; actual statistical error on each estimate of r_o is $\sim 0.4\,\mathrm{Mpc}\,h^{-1}$. The range for r_o encompasses values found directly for radio galaxies (Peacock 1999), exceeds the clustering length of local optically-selected galaxies (e.g., Davis & Peebles 1983), and is less than that of galaxy clusters (Dalton et al. 1992). It is consistent with the general nature of NVSS galaxy identifications as optically-luminous ellipticals inhabiting moderately-rich environments.

(2). **The spatial matter power spectrum**. An angular matter power spectrum may be predicted given the initial power spectrum and reasonable assumptions about the transfer function; some results are shown in Figure 3 (right panel). The sensitivity to the form of the redshift distribution is now extreme, to the extent that it is tempting to use this calculation to constrain it. However, there is a far more interesting experiment. If the redshift distribution were known, then it is possible to calculate a region of constraint in the slope - amplitude plane of the initial large-scale matter power spectrum. For any adopted redshift distribution, such regions show strong correlation between amplitude and slope. Moreover, the region of constraint is entirely independent of that found from power-spectrum measurements on the CMB – which also shows a strong correlation between amplitude and slope. *These regions intersect*; the correlations lie in significantly different directions (Blake & Ferreira, in preparation), unsurprising given that the dependence of c_ℓ on ℓ is from different physical phenomena. An accurate determination of the redshift distribution, perfectly possible with the modern generation of large optical telescopes, will thus yield a measure of the initial power spectrum, crucial in examining the validity of inflation models.

(3). **The cosmological velocity dipole**. It is generally accepted that the dipole measured in the temperature of the CMB (Smoot, Gorenstein & Muller 1977) is a velocity dipole due to the earth's motion. If so, a parallel dipole imprint must be present in the surface density of faint galaxies; such a detection would provide independent confirmation of the CMB anisotropy as a velocity dipole, as well as other fundamental tenets of modern cosmology such as homogeneity. Instrumental difficulties and inhomogeneous obscuration preclude detection of the dipole in surveys of optically-selected galaxies, but wide-angle radio surveys represent an ideal search tool (Ellis & Baldwin 1984). The expected peak-to-trough amplitude in surface density is given by $A = 2(v/c)[2+x(1+\alpha)]$, where v is our velocity with respect to the isotropic frame, and radio sources are assumed to have a power-law spectrum of slope $-\alpha$ and an integral source-count of slope $-x$. Previous attempts to detect the dipole in other radio surveys have failed, because of instrumental effects or too few sources (e.g., Baleisis et al. 1998). We believe that we have succeeded in doing so with the NVSS (Blake & Wall 2002a).

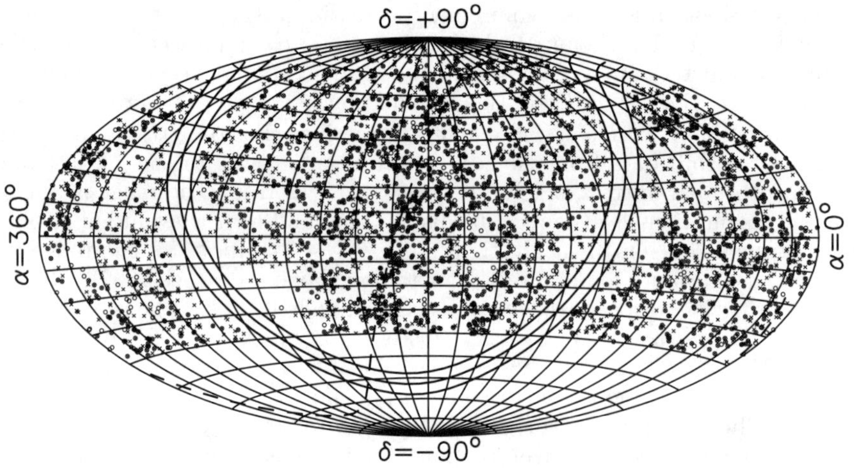

Figure 4. NVSS sources with $S_{1.4\mathrm{GHz}} > 10\,\mathrm{mJy}$ matched with PSCz galaxies (crosses) and RCBG3 galaxies (circles). There is a clear concentration of these sources towards the Super-galactic plane (dashed line).

We confined the dipole search to NVSS flux densities above 15 mJy, the threshold at which surface-density gradients become apparent. The particular challenge in detecting the velocity dipole is that there are two dipoles: the cosmic velocity dipole, and the clustering dipole evident in local galaxies selected either optically or in the IR (the 'IRAS' galaxies). *This latter dipole causes the former*; the local clumps produce the motion of our Galaxy in the direction of the CMB dipole. Any vestige of this clustering dipole in the NVSS must be removed before a claim may be made for a velocity dipole. We did this by comparing

NVSS positions with catalogs of local galaxies (Fig. 4); and we demonstrated that these local catalogs went deep enough to see through the clustering dipole so that we could 'remove' the local sources accordingly.

The formalism of the angular spectrum analysis was then adapted to carry out a minimum-χ^2 model-fitting process with dipole vector parameters as variables . The best fit (Fig. 5) was significantly better than an isotropic universe, rejected at the 2.5σ level; the direction of this fit is within 1σ of the CMB direction and the amplitude within 1.5σ of that predicted. Moreover the measured coefficients $\ell > 1$ up to $\ell = 3$ should show some small signal because incomplete sky coverage means that these higher harmonics are influenced by a dipole distribution; we found that the fit for the 15 coefficients involved is significantly better than that for an isotropic distribution. Some results of the search are

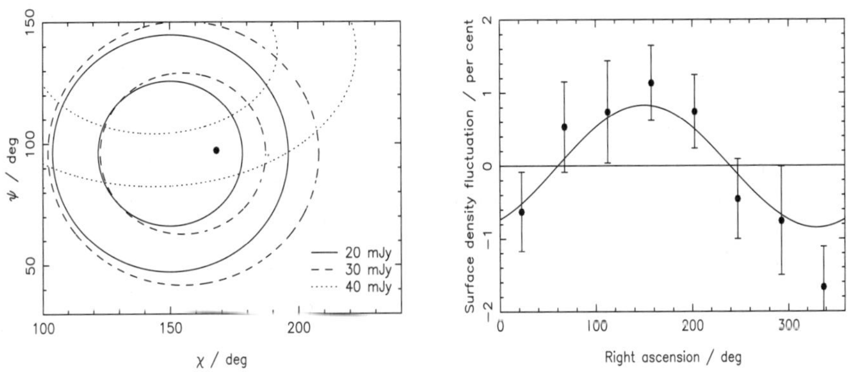

Figure 5. Left: Direction of best-fit NVSS dipole vector for different flux-density thresholds; 1σ and 2σ contours are shown, and the dot gives the direction of the CMB dipole. Right: The dipole imprint in the distribution of NVSS sources. A dipole pointing close to dec 0° should produce an approximately sinusoidal distribution in surface density with ra. Here the 20 mJy NVSS sample is binned in ra bands, with error bars following $\sqrt{N}$. The curve is the prediction of the best-fitting dipole model.

shown in Figure 5 and these strongly suggest that the velocity dipole has been detected.

We conclude that we can now pursue searches for (a) clustering properties of individual AGN populations in deep wide radio surveys; (b) the dipole at much higher levels of significance, and (c) the epoch dependence of clustering. And (d) – with well-observed redshift compilations, we can contemplate measuring the initial power spectrum from which all clustering developed.

We can also conclude that the distant arguments with Ken Kellermann led to real cosmology from studies of wide-area extragalactic source surveys. But arguments with Ken are invariably fruitful. He is almost always right, infuriatingly so, his instinct the hallmark of a great scientist. It just so happens that Ken is a great colleague as well.

References

Baleisis, A., Lahav, O., Loan, A. J., & Wall, J. V. 1998, MNRAS, 297, 545

Becker, R. H., White, R. L., & Helfand, D. J. 1995, ApJ, 450, 559

Benn, C. R., & Wall, J. V. 1995, MNRAS, 272, 678

Blake, C. 2002, Ph.D. thesis, (Oxford, UK: Univ. of Oxford)

Blake, C., & Wall, J. 2002a, Nature, 416, 150

Blake, C., & Wall, J. 2002b, MNRAS, 329, L37

Blake, C., & Wall, J. 2002c, MNRAS, 337, 993

Brown, R. H. 1962, MNRAS, 124, 35

Condon, J. J., et al. 1998, AJ, 115, 1693

Cress, C. M., Helfand, D. J., Becker, R. H., Gregg, M. D., & White, R. L. 1996, ApJ, 473, 7

Cress, C. M. & Kamionkowski, M. 1998, MNRAS, 297, 486

Dalton, G. B., Efstathiou, G., Maddox, S. J., & Sutherland, W. J. 1992, ApJ, 390, L1

Davis, M., & Peebles, P. J. E. 1983, ApJ, 267, 465

Ellis, G. F. R., & Baldwin, J. E. 1984, MNRAS, 206, 377

Landy, S. D., & Szalay, A. S. 1993, ApJ, 412, 64

Lara, L., et al. 2001, A&A, 370, 409

Loan, A. J., Wall, J. V., & Lahav, O. 1997, MNRAS, 286, 994

Magliocchetti, M., Maddox, S. J., Lahav, O., & Wall, J. V. 1998, MNRAS, 300, 257

Magliocchetti, M., Maddox, S. J., Lahav, O., & Wall, J. V. 1999, MNRAS, 306, 943

Pauliny-Toth, I. I. K. & Kellermann, K. I. 1972, AJ, 77, 797

Peacock, J. A. 1999, in The Most Distant Radio Galaxies, ed. H. J. A. Röttgering, P. N. Best, & M. D. Lehnert, (Amsterdam: Royal Netherlands Academy of Arts and Sciences), 377

Shaver, P. A., & Pierre, M. 1989, A&A, 220, 35

Smoot, G. F., Gorenstein, M. V., & Muller, R. A. 1977, Phys. Rev. Lett, 39, 898

Windhorst, R. A., Mathis, D., & Neuschaefer, L. 1990, in ASP Conf. Ser. Vol. 10, Evolution of the Universe of Galaxies, ed. R. G. Kron, (San Francisco: ASP), 389

Radio Astronomy at the Fringe
ASP Conference Series, Vol. 300, 2003
J.A. Zensus, M.H. Cohen, & E. Ros, eds.

The Radio and Optical Morphologies of Micro-Jansky Radio Sources

E. B. Fomalont, K. I. Kellermann
National Radio Astronomy Observatory, 520 Edgemont Rd., Charlottesville, VA 22903, USA

L. L. Cowie, A. J. Berger, P. Capak
Institute for Astronomy, 2680 Woodlawn Dr., Honolulu, HI 96822, USA

R. B. Partridge
Haverford College, Haverford, PA 19041, USA

R. A. Windhorst
Arizona State University, Tempe, AZ 85287, USA

Abstract. We have imaged a 17′ region in the SSA13 field with the VLA at 1.4 GHz with a resolution of 1.″8, and obtained sensitive Subaru images at R-band (0.9 μm) and Z-band (1.3 μm) with a seeing of 1.″1. There are 548 radio sources detected with an image flux density $> 27.5\,\mu$Jy, and over 95% are identified on the Z-band image, to a limiting magnitude of 25.8 mag. The average radio angular size is 1.″35 and very few sources are larger than 5″. The differential count of radio sources has a slope of -2.3. The radio spectral index steepens considerably for the weaker sources with flux density $< 100\,\mu$Jy at 1.4 GHz. Many radio and optical morphologies have similar orientations, suggesting that the emission is associated with the star formation process.

1. Introduction

Surveys of faint sources made from radio to X-ray frequencies (egs. Hasinger et al. 1998; Lilly et al 1996; Hughes et al. 1998; Richards et al. 1998) suggest that star formation is an important component in the energetics of galaxies in the early universe. At the micro-Jansky levels, radio sources are predominantly identified with objects which are between 19 and 25 mag at redshift from 0.3 to 2, and they are often red in color and thus bright at mm-wavelengths, suggesting the presence of warm dust (e.g., Condon 1992). Although star formation activity in the early universe is perhaps most readily observed at sub-millimeter wavelengths, sensitive radio observations detect many of these objects, and the accurate radio positions give unambiguous identification of the objects.

Recent deep integrations using the VLA with $< 2''$ angular resolution at 1.4 GHz (Richards et al. 1998; Smail et al. 2002) resolve many of the radio sources; hence, comparison of the radio and optical morphologies will provide a more detailed understanding of the evolution and dynamics of the star formation process in these distant galaxies. This paper summarizes these results in Fomalont et al. (2003, hereinafter Paper I).

2. The Radio Images

The SSA13 field was observed with the VLA at 1.4 GHz for a total of 104 hours. On eight days between 1999 June 22 and 1999 August 30, the A-configuration (maximum baseline of 35 km) as used. One day of B-configuration (maximum baseline of 7 km) observations was added on 2000 January 14 in order to increase the image sensitivity to sources larger than about 5″ in angular size. The field center of the radio observations was RA=$13^h23^m17\overset{s}{.}4$, DEC=42°38′05″.0. The observations were made at two frequencies, 1.355 GHz and 1.455 GHz, each with dual circular polarization with a 44 MHz bandwidth. The 8.4 GHz image was observed in 1996 and is described in Fomalont et al. (2002).

Details of the observations, calibration, and imaging of the SSA13 field are given in Paper I. This most sensitive image has a resolution of 1″.8 and an root-mean-square (rms) noise of 5.0 μJy. Although the image extends over 30′ from the field center, only the inner 17′ radius, comparable with the optical field of view, was used for the catalog. At this distance from the field center, the sensitivity is 24% of that of the field center.

An initial list of about 1000 radio sources was compiled in the 17′-region for all sources with peak flux density greater than 20 μJy on the 1″.8-image, or above 65 μJy on the 6″-image, both somewhat below the completeness flux density sensitivity of about 5.2 times the rms noise. The flux density, position and angular size of each source were determined for each source, and the parameters were corrected for the effects of the primary beam attenuation of the 25-m VLA antennas, the radial beam smearing caused by finite bandwidth, and the small azimuthal smearing caused by the finite integration time.

3. The Optical Images and Identifications

Optical observations made with the the Subaru Telescope and the Suprime-Cam prime focus camera (Miyazaki et al. 1999). Data were collected on 2001 February 2, April 22, and April 23. All nights except April 22 were photometric. During the February run the Suprime-Cam was still in a commissioning phase. The fully-reduced images have a usable field of view of 34′ × 27′ in R and 34′ × 34′ in Z. Seeing in the final images is 0″.74 in R band and 0″.79 in Z band, although the resolution was degraded to 1″.1 for the radio comparison. The 3-sigma limiting magnitude is 26.1 in R and 24.9 in Z band calculated from blank 3″ apertures on the image at least 15″ from a detected object. Additional details are given in Paper I.

With the initial list of 1000 radio source positions, we searched for identifications on the Subaru optical images before compiling the final SSA13 catalog of radio sources. Because of the high density of radio sources, many of which are extended, identification with a optical counterpart is necessary to determine the nature of the radio/optical emission. After all of the identifications were made, a total of 914 individual radio sources were cataloged, of which 548 sources were above the completeness limit at 1.4 GHz of 27.5 μJy. Only these

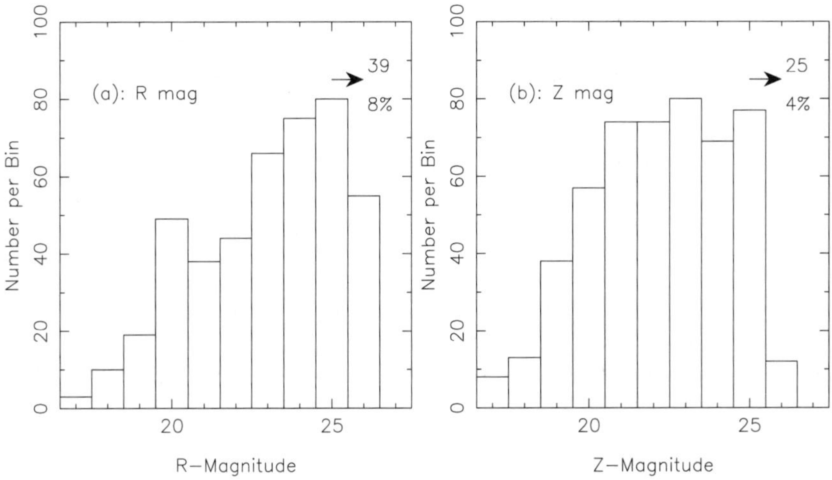

Figure 1. **The Magnitude Distribution of Identifications in the SSA13 field:** (Left) The R-Band distribution and (right) the Z-Band distribution.

sources should be considered for any systematic study of the nature of these micro-Jansky sources. The catalog is given in Paper I.

4. Radio/Optical Properties

The distributions of the R-Band and Z-Band magnitude for the identified radio sources in the complete sample are shown in Figures 1a and 1b, respectively. Between 16 mag to 21 mag, there is a marked increase in the number of galaxies at both bands, and this radio source density is about 3% of the optical density of galaxies in this magnitude range. Fainter than 21 mag, the Z-Band magnitude distribution remains relativity constant per magnitude interval, until 25.5 mag, when the density of fainter galaxies drops significantly. On the other hand, the R-Band galaxy distribution slightly increases in density until 25.5 mag, with a similar drop-off.

The percentage of unidentified sources in the complete sample is 8% (39 of 490 sources) fainter than 26.3 mag at R-Band, and 4% (25 of 565 sources) fainter than 25.7 mag at Z-Band. At I-Band the empty field rate fainter than 25.5 mag is about 20% (Richards et al. 1999). This increase in the identification rate at longer wavelengths clearly shows the redness of the optical counterparts of the micro-Jansky radio population. About 30% of the identifications have (R-Z)> 1.1 mag and are considered as Extremely Red Objects (ERO).

The number distribution of the largest angular size of the radio sources in the complete sample of 548 sources is given in Figure 2a. Only 5 sources, are larger than $5''$, and four of these are extended AGN (see below). Sixty-two sources have upper limits to their size and all of them are placed in the $0''.5$ to $1''.0$ bin. The median angular size is $1''.35$. The curve shows a simple model which fits the angular size distribution, and the distribution can be fit to a truncated Gaussian with mean 1.″0 and a sigma of 0.″5.

Figure 2b shows the comparison of the source count of radio sources in the SSA13 field, and that of other fields at 1.4 GHz. The integral source count from the SSA13 field alone is $N = 1.2(S/40)^{-1.3}$ where N is the number of sources per arcmin-squared with a flux density greater than S micro-Jy. At the flux density limit of

27.5 μJy, the average source separation is 43″. At a flux density of 1 μJy, the average source separation would be 6″. The average slope indicates that the density of sources, even as low as 40 μJy, is only 0.2 less than the Euclidean slope of the uniformly filled universe. The source count must decrease at a flux density < 0.1 μJy, otherwise the integrated flux density of these faint sources would contribute a significant fraction of the cosmic background emission of 2.7 K. There is considerable scatter in the density of sources between 100 μJy and 1000 μJy, and may be caused by clustering of distant galaxies at the 10′ to 60′ angular scale-size.

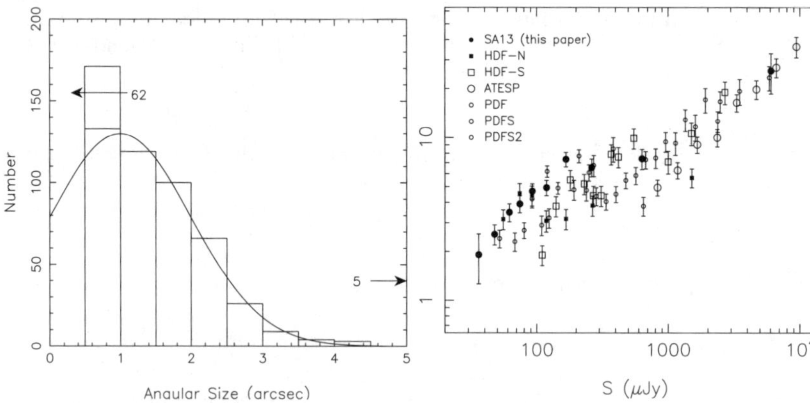

Figure 2. **(a) Left: Angular Size Distribution of Radio Sources at 1.4 GHz:** The histogram shows the number distribution of the angular size of radio sources in the SSA13 complete sample. **(b) Right: The Normalized Differential Source Count at 1.4 GHz:** The count for several different surveys are indicated by the plot symbol.

The 1.4 GHz-8.4 GHz spectral index distribution is shown in Figure 3a and 3b for sources stronger than, and weaker than 100 μJy, respectively. There is a marked contrast in the spectral index distribution, even though the inverse-mean flux density difference differs by only a factor of 5. For the bright sources, the spectral index distribution is relatively flat and is in agreement from radio surveys at somewhat higher flux densities (Donnelly et al. 1987). Below 100 μJy the majority of sources have a spectral index steeper than −0.70.

5. The Radio/Optical Morphologies

Although there is a variety of radio and optical morphologies, their general morphological properties have been gathered into a relatively small number of

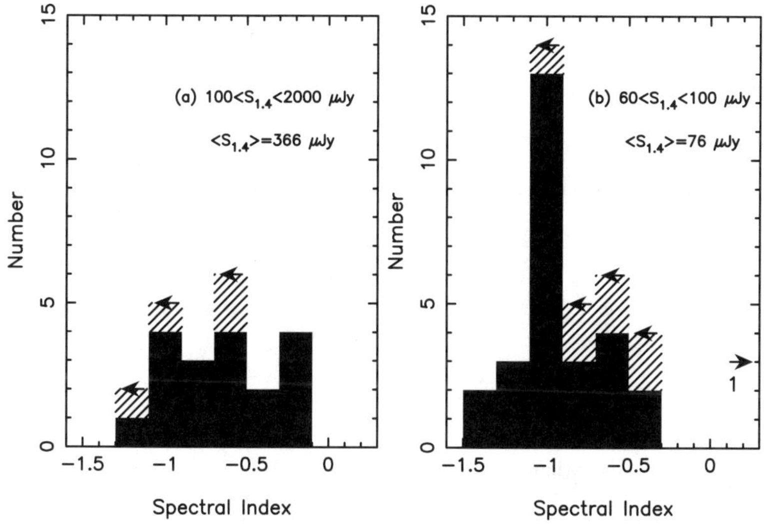

Figure 3. **Source Spectral Index Distribution between 1.4 to 8.4 GHz in SSA13 Field: (left)** The distribution for sources stronger than 100 μJy, **(right)** The distribution for sources weaker than 100 μJy. The cross-hatched regions with the arrow indicated upper limits. The spectral index scale ($S \propto \nu^{\alpha}$) is $\alpha = -1.6$ on the left and $\alpha = +0.3$ on the right.

classes which are shown in Table 1. The class and subclass categories are listed in column 1. The number of sources in each class and subclass for the entire catalog (N_t) and for the complete sample (N_s) are given in the next two columns. The last column describes each class and subclass.

Seven extended AGN were found in the SSA13 field, and these display several types of morphology. The radio luminosities range between 3 and 30 $\times 10^{21}$ w Hz^{-1}, the linear size range between 3 and 80 kpc, and the radio angular sizes range between 8″ and 80″. All of these extended AGN are brighter than 400 μJy, nearly 20 times larger than the completeness limit of this survey, and suggests that about 10% of the sources at 1000 μJy are extended AGN, but $< 1\%$ of the sources are extended AGN below 400 μJy. Observations of the FIRST survey (White et al. 1997) show that about 30% of the sources near 3000 μJy were larger than 5″ in angular size, with many as extended AGN.

About half of the radio sources are identified with a relatively isolated galaxy; that is, one with no neighbors of comparable brightness within about 10″ separation. The radio emission is generally confined to the central part of the galaxy (Class= Sc, Se, Sep; 232 out of 284), although the radio size is often $> 1''$ (Se), and some radio and optical orientations are nearly parallel (Sep). About 20% of the single galaxy identifications have radio components well-displaced from the galactic nucleus (Sd or Sx), sometimes near the edge of the detectable galaxy emission. Often, there is optical emission enhancement associated with the radio component.

Over 20% of the radio sources are associated with a binary system—two galaxies which often overlap. Usually, one of the galaxies contains the bulk of

Table 1. Radio/Optical Morphological Classifications

Class	Number N_t	N_s	Description
ALL	915	548	
AGN	7	4	EXTENDED AGN. (Core with radio lobes)
S	465	284	SINGLE GALAXY
Sc	99	72	Radio source located near nucleus and $< 1''$ in size
Se	151	102	Radio source located near nucleus and $> 1''$ in size
Sep	86	58	Same as Se with radio and optical orientations nearly parallel
Sd	78	30	Radio source somewhat displaced from nucleus, but within galaxy
Sx	51	22	Radio source well displaced from nucleus, but id is secure
B	206	123	BINARY SYSTEM
Bc	38	28	Radio source located near nucleus of one galaxy and $< 1''$ in size
Be	45	31	Radio source located near nucleus of one galaxy and $> 1''$ in size
Bep	45	34	Radio orientation nearly parallel with binary system or galaxy orientation
Bd	8	4	Radio source displaced from nucleus of one galaxy
B2	70	26	Both galaxies contain radio emission (each galaxy included in count)
C	71	37	COMPLEX OPTICAL FIELD
F	78	62	FAINT OPTICAL OBJECT (≈ 24.5 to $26.0\,\mathrm{mag}$)
U	45	25	UNIDENTIFIED
N	43	13	NOT WITHIN OPTICAL IMAGE REGION

the radio emission, but in 35 of the 171 binary systems in the entire catalog, both galaxies have comparable radio emission. Radio sources which are labeled as complex generally contain more than two galaxies, and their emission properties are often similar to those in the binary system.

A trip through the radio/optical morphology 'zoo' is given in Figure 4. The top line shows four examples of a radio source located near the center of an isolated galaxy. The radio sizes are about one to two arc seconds. For source 396, the orientation of the radio source is similar to that of the optical galaxy. The second line shows radio sources associated with single galaxies, but the radio/optical comparison is more complicated. Source 41 is well offset from the galaxy and perhaps associated with a filamentary optical feature to the west. Source 91 is associated with an asymmetry in the galaxy. Source 157 is offset from the center of a 17 mag galaxy (saturated on the image), and source 503 is associated with a faint feature near the edge of a bright galaxy. It could be a background galaxy.

On the third line, some examples of radio emission associated with binary galaxy systems are shown. Source 159 suggests that a faint, elliptical extended galaxy has been disrupted by a bright galaxy and the radio emission in the faint galaxy is extended in the same direction. Source 284 has radio emission near the centers of both binary galaxies. On the bottom line, optical and radio morphologies are shown. These comparisons suggest galaxies are associated with star burst activity.

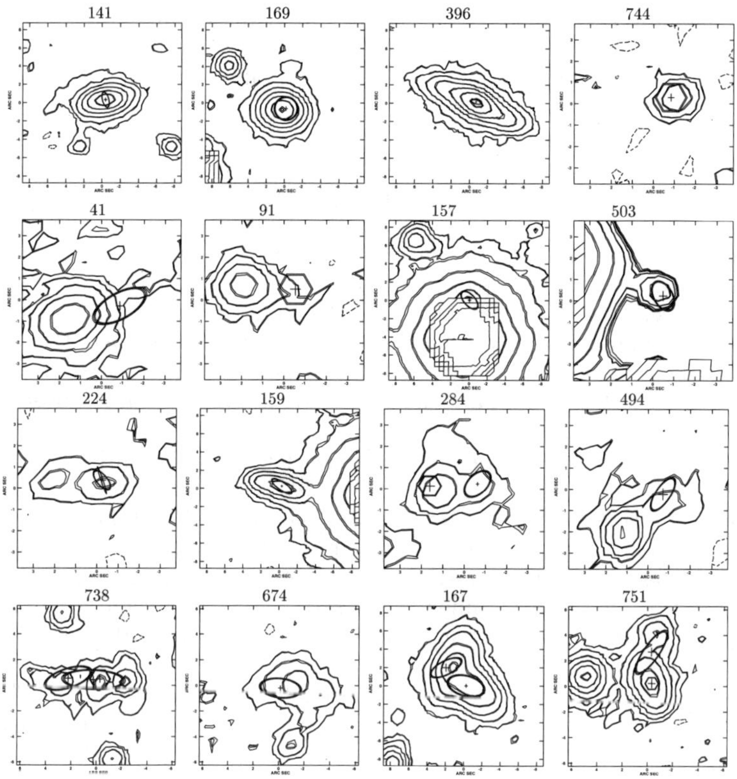

Figure 4. **Selected Radio/Optical Comparisons:** The Z-band optical emission is shown by the contour levels with a minimum contour level at 25.5 mag and a separation of 1 mag per beam (1.″1 seeing). The field of view of each image is $10'' \times 10''$. The approximate parameters of the radio emission are shown by the superposed symbols. The ellipse shows the full-width half-power size of a radio component. If the component is unresolved, the hexagon symbol shows the upper limit size of the radio component. Some fields contain more than one radio component.

6. Discussion

Most of the micro-Jansky radio sources are identified at Z-band, and many of the optical counterparts are red. About 30% of the SSA13 sample have an (R-band - Z-band) flux density > 1.1 mag and fall under the classification of extremely red objects (ERO). This percentage is similar to other investigations (Small et al. 2002). The radio spectral index between 1.4 and 8.4 GHz becomes significantly steeper below about 100 μJy. This 'reddening' of the radio spectrum suggests that the fainter objects may be more dominated by ERO's which tend to have a steeper radio spectrum than the more normal starburst population or AGN.

The source count at 1.4 GHz below 30 μJy shows no sign of converging (running out of radio sources), and the sky at the one micro-Jansky level will be crowded. However, even with a modest amount of source overlap, observations of radio resolution of $< 0.5''$ can distinguish individual sources. However, below about 0.1 μJy, the source density must drop-off considerably in order to avoid a contribution of more than 0.01 K to the integrated sky brightness, based on accurately-know spectrum of the CMB (Paper I).

This paper has shown that the typical micro-Jansky radio source is extended about $1\overset{''}{.}35$ and often shows a similarity in its structure with that of the optical counterpart. Since about 75% of the radio sources are larger than $1''$, it is likely that the bulk of the radio emission is associated with the starburst phenomenon. The radio emission comes from relativistic electrons generated from supernovae of massive stars, while the optical emission is associated with the present star formation rate. Hence, the similarities and differences of the radio and optical emission contain information about the evolution and time-scale of the starburst activity. Since starburst activity may be strongly associated with mergers and galaxy interactions, the comparison of radio structure and binary orientation is also important.

About 25% of the galaxies are associated with radio emission of angular size $< 1''$ near their nucleus, and are, thus, AGN candidates. However, unless their radio spectral index is flat or their angular size $< 0\overset{''}{.}04$ (Garrett 2002), the compact components may be dense starburst regions. We believe that less than 10% of the micro-Jy radio sources are associated with significant AGN activity. The proportion of radio-extended AGN (core plus significant lobe emission) is even rarer and decreases precipitously below 3000 μJy and none have been detected below 400 μJy.

The wealth of detail in the comparison between the radio emission and the optical emission is illustrated by examples in Figure 4. It is clear that the radio emission from starburst galaxies is complex and probably contains several components of emission, as shown in M 82 and Arp 220. With the higher resolution and sensitivity that proposed new radio arrays will reach, accurate modeling of the evolution and dynamics of starburst and galaxy mergers should be possible using the radio and optical images.

References

Condon, J. J. 1992, ARA&A, 30, 575

Donnelly, R. H., Partridge, R. B., & Windhorst, R. A. 1987, ApJ, 321, 94

Fomalont, E. B., et al. 2003, ApJ, submitted (Paper I)

Fomalont, E. B., Kellermann, K. I., Partridge, R. B., Windhorst, R. A. & Richards, E. A. 2002, AJ, 123, 2402

Garrett, M. A. 2002, A&A, 384, L19

Hasinger, G., et al. 1998, A&A, 329, 482

Hughes, D. H., et al. 1998, Nature, 394, 241

Lilly, S. J., LeFevre, O., Hammer, F. & Crampton, D. 1996, ApJ, 460, L1

Miyazaki, S., et al. 1998, in Proc. SPIE 3355: Optical Astronomical Instrumentation, ed. S. D'Odorico, (Bellingham, WA: SPIE), 363

Richards, E. A., et al. 1999, ApJ, 526, L73

Richards, E. A., Kellermann, K. I., Fomalont, E. B., Windhorst, R. A., & Partridge, R. A. 1998, AJ, 116, 1039

Smail, I., et al. 2002, ApJ 581, 844

White, R. L, Becker, R. H., Helfand, D. J., & Gregg, M. D. 1997, ApJ 475, 497

Radio Astronomy at the Fringe
ASP Conference Series, Vol. 300, 2003
J.A. Zensus, M.H. Cohen, & E. Ros, eds.

High-Redshift Radio-Loud QSOs: Unification and the Space Density Turnover

C. A. Jackson
Research School of Astronomy & Astrophysics, Mount Stromlo Observatory, The Australian National University, Canberra, ACT, Australia

J. V. Wall
Department of Physics, University of Oxford, Nuclear and Astrophysics Laboratory, Keble Road, Oxford OX1 3RH, UK

P. A. Shaver
European Southern Observatory, Karl-Schwarzschild-Str. 2, 85748 Garching bei München, Germany

K. I. Kellermann
National Radio Astronomy Observatory, 520 Edgemont Rd., Charlottesville, VA 22903-2475, USA

I. M. Hook
Department of Physics, University of Oxford, Nuclear and Astrophysics Laboratory, Keble Road, Oxford OX1 3RH, UK

Abstract. We discuss results from a large completed project which takes advantage of the characteristics of compact, high-power radio-loud QSOs identified by Ken Kellermann early in his career. We set out to (i) search for high-redshift radio-loud QSOs and (ii) study the evolution of the flat-spectrum QSO population. Following a 6 yr campaign we have obtained accurate radio positions and optical data for a large sample of Parkes flat-spectrum radio sources. The resultant sample enables us to trace the space density turnover of flat-spectrum QSOs. We conclude that there is a space density turnover in the QSO population and that optically-selected QSO samples are little affected by dust obscuration.

1. High-redshift QSOs and the RLF

Since the first identification of the spectral features of the radio-loud QSO 3C 273, bright QSOs have been targeted in the quest to find the highest-redshift object in the Universe. Historically, the blue-ribbon for QSO redshifts was long held by radio-loud sources. This bias was due to a large fraction of radio-bright ($S_{178\,\mathrm{MHz}} > 10$ Jy) sources lying at high redshift, as evidenced by the redshift distribution for the Cambridge 3C sample and known to be due to a strong evolution in the powerful radio source population. (Longair 1966). More recently optical color selection techniques and narrow-band Lyα searches have proved highly successful in detecting the highest-redshift objects: The current record-

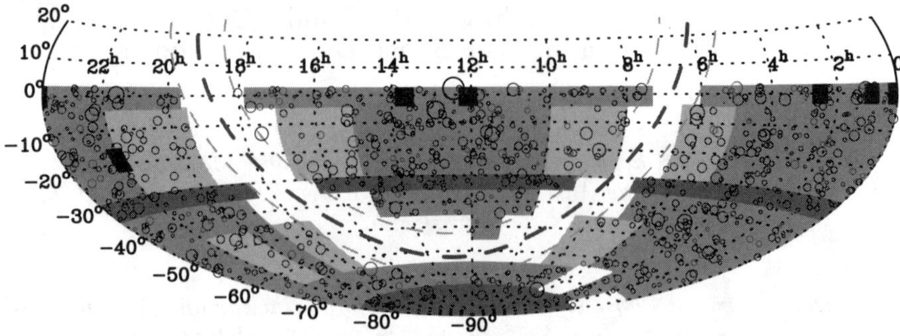

Figure 1. Sky distribution of the Parkes quarter-Jansky flat-spectrum sources within the Parkes survey regions. Radio sources identified with galaxies and stellar objects are shown as circles with the size of the circle being proportional to the 2.7 GHz flux density. Parkes 2.7 GHz survey limits are shown in shades of grey, from dark to light, equating to $S_{2.7\,\mathrm{GHz}}$ of 0.10, 0.20, 0.25 and 0.60 Jy. Sources in the Magellanic Cloud regions and near the Galactic equator ($|\,b\,| < 10°$) are excluded from the sample.

holders are a galaxy identified by Hu et al. (2002) at z=6.56 and a QSO in the SDS at z=6.43 (Fan et al. 2003).

Aligned with the detection of high-redshift QSOs is the question of whether there is a diminution in the space density of powerful QSOs at high redshift ($z > 2$). For both radio and optically-selected QSOs the form of the luminosity function (LF) and its redshift dependence to $z \sim 1$ are well established. In both, powerful evolution is required to explain the changing LF so that by $z \sim 1$ there is an increase of order of $10^2 - 10^3$ in the space density of the most luminous sources. But what happens at higher redshifts is not entirely clear-cut: there were no sizable radio-selected samples. Optically-selected QSO samples show there is a cutoff, (Schmidt et al. 1991; Kennefick et al. 1995; Warren et al. 1994) which, when combined with the detection of $z > 6$ QSOs from the Sloan Digital Sky Survey (Fan et al. 2003), yields strong evidence of a drop in space density beyond $z \sim 3$. However, concerns have been raised that the diminution in optically-selected QSOs is wholly due to dust obscuration (Heisler & Ostriker 1988; Fall & Pei 1993) – a matter which can only be resolved by a large sample of QSOs selected in an unobscured waveband such as radio.

We set out to obtain a spectroscopically-identified sample of radio-selected QSOs, large enough to directly map their space density behavior and to resolve whether the optically-selected QSO LF results are affected by dust.

Throughout this paper we use $h = 0.5$ and Ω=1.0.

2. The Parkes flat-spectrum QSO sample

In selecting our radio sample we have to account for the radio luminosity function (RLF) being both broad and steep: Across a wide range of radio powers there is a mix of both radio galaxies and QSOs. The current unification paradigm (Urry

& Padovani 1995, and references therein) links radio galaxies and radio-loud QSOs as manifestations of the same underlying parent object due to orientation and beaming effects. Thus radio galaxies should exhibit the same space-density behavior as QSOs, albeit with the caveat that we would have to 'de-beam' the QSO radio powers to be able make the comparison. As most powerful radio galaxies have extended radio structures and relatively faint optical hosts we focus on a QSO sample to ensure we can obtain near-complete optical identifications.

To select a sample which is dominated by QSOs we use the 'flat-spectrum' criterion where compact radio-loud QSOs characteristically have flat, undulating or inverted spectra at cm-wavelengths (Kellermann & Pauliny-Toth 1969; 1971). The considerable advantage with a sample of compact radio sources is the relative ease of determining their arcsecond radio position and hence the secure optical identification of their host galaxy. The contribution from flat-spectrum sources to the total source count at 5 GHz has been determined by Kellermann & Pauliny-Toth (1981), showing that the flat-spectrum count peaks near $S_{5\,\mathrm{GHz}} \sim 1$ Jy and then converges rapidly below this value. These authors recognized that compact radio sources are intrinsically powerful QSOs in the redshift range $0.5 \leq z \leq 3$, and that by $S_{5\,\mathrm{GHz}} \sim 1$ Jy the majority of powerful radio QSOs are detected—i.e., we 'see through' the whole of the QSO RLF—so that going deeper in flux density would not yield many more QSOs.

We therefore chose a flux density-limited sample ($\sim$0.25 Jy) which preferentially selects high-power QSOs ($P_{2.7\,\mathrm{GHz}} > 10^{26}$ W Hz^{-1} sr^{-1}). At these flux densities radio sources are sparse so we require a survey which covers a large sky area to ensure we obtain a statistically useful sample.

We selected our sample from the machine-readable version of the Parkes radio source catalog (PKSCAT90: Wright & Otrupcek 1990). This catalog contains radio and optical data for 8264 radio sources south of +27° and was compiled from a series of major radio surveys at 2.7 GHz performed by John Bolton and colleagues between 1969 and 1979 (Bolton, Savage, & Wright 1979, and references therein). The PKSCAT90 also contains data from 5 GHz surveys which were undertaken simultaneously or within a few months of the 2.7 GHz surveys. The time-lag between the 2.7 and 5 GHz surveys is important as QSOs are often variable so that simultaneous flux density measurements are the only guarantee of obtaining a meaningful radio spectra (Kellermann & Pauliny-Toth 1981). We selected all flat-spectrum sources with $\alpha_{2.7\,\mathrm{GHz}}^{5\,\mathrm{GHz}} > -0.4$, where $S \propto \nu^{\alpha}$, lying between $+2\overset{\circ}{.}5$ and $-80°$ declination at all right ascensions, excluding Magellanic Cloud regions and galactic latitudes below 10°. This gave us a sample of 878 sources and their sky distribution is shown in Figure 1. Of these 645 have optical identifications in PKSCAT90, with 535 (82%) being QSOs.

The optical imaging and spectroscopy for the sample are discussed in a pair of papers (Jackson et al. 2002; Hook et al. 2003) and the data are available on-line via CDS and ADS.

Our imaging and spectroscopy included a full review of all source identifications and spectra held in the PKSCAT90 catalog. Our final dataset comprises identifications for 677 stellar (QSOs and BL Lacs) sources. From this large set of sources we can use sub-samples to map the space density turnover (section 5).

3. Tests of the unified scheme & evolution

We compare the predictions of models described by Wall & Jackson (1997) and Jackson & Wall (1999) to the results from our complete flat-spectrum sample. The models incorporate parameterized beaming and evolution for two underlying populations of radio sources —FRI/BL Lacs and FRII/QSO/BL Lacs. The evolution parameters are fitted at 151 MHz where the contribution from beamed objects is negligible, and the Doppler beaming parameters are then fit at 5 GHz. In this manner these models are independent of the PKS 2.7 GHz-survey data.

Firstly we compare a simple prediction from the dual-population unified scheme of Wall & Jackson (1997) and Jackson & Wall (1999) to a set of flux-density limited samples from the Parkes Survey. We find that the contribution from flat-spectrum sources and its rate of the decline towards lower flux densities is in close agreement with that predicted by the model (Fig. 2). The shape and magnitude of the model flat-spectrum fraction is a direct product of the strong evolution of high-excitation FRII radio galaxies and the magnitude of the Doppler beaming parameters —namely the jet opening angle and gamma factor.

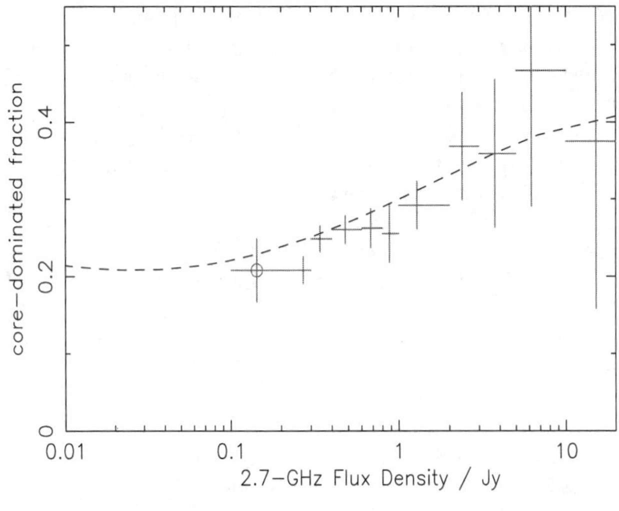

Figure 2. Core-dominated fractions at 2.7 GHz. The dashed line is the model prediction (Jackson & Wall 1999) with data points derived from the PKSCAT90 sample (Jackson et al. 2002; Wall & Jackson 1997) and the Parkes selected regions (Wall & Peacock 1985).

We also compare the observed redshift distribution, $N(z)$, of all stellar (QSOs and BL Lacs) sources with $S_{2.7\,\mathrm{GHz}} \geq 0.25$ Jy in the PKS flat-spectrum to that predicted by the model. We find (Fig. 3) that $N(z)$ is reasonably good with a correct normalization. There are fewer observed high-redshift ($z > 2$) objects than predicted. This is due in part to the lack of steepening in the underlying radio spectra adopted in the predictive model.

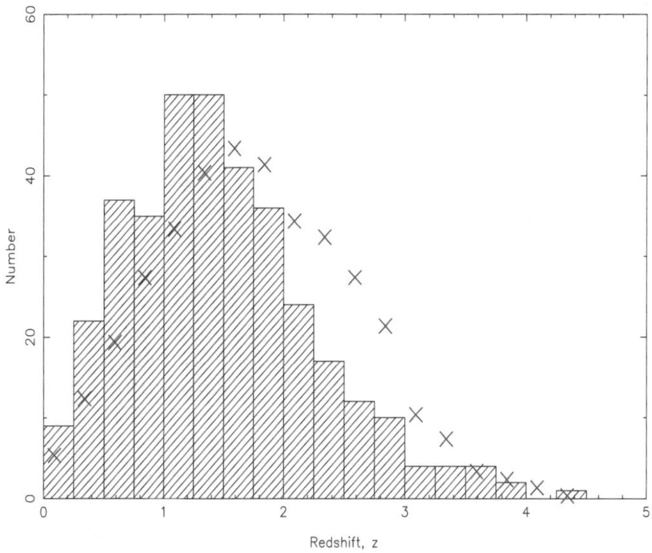

Figure 3. The redshift distribution, $N(z)$, for flat-spectrum sources, $S_{2.7\,\mathrm{GHz}} \geq 0.25$ Jy and $\alpha_{2.7\,\mathrm{GHz}}^{5\,\mathrm{GHz}} \geq -0.4$ over a survey region of 3.987 sr. The observed redshift distribution for all QSOs and BL Lacs is shown hatched. The crosses show the prediction from the dual-population model of Jackson & Wall (1999).

4. High-redshift QSOs—a blue ribbon!

The original goal of the project was to detect high-redshift QSOs free of dust obscuration. Optical counterparts for essentially all sources were obtained and no QSO at $z > 5$ was found. The highest-redshift source in the sample was PKS B1251−407 at z=4.46 (Shaver, Wall & Kellermann 1996) which claimed the blue-ribbon for radio-loud QSOs. More recently higher-redshift radio-loud QSOs have been found using a combination of optical color and flat-spectrum radio characteristics. The most distant radio-loud QSO found by this method is PMN J1451−1512 at z=4.763 (Hook et al. 2002).

5. Is there a space density turnover ?

Our initial analysis of the flat-spectrum QSOs (Shaver et al. 1996) considered the powerful QSOs in a complete declination-limited sub-sample, with $\log_{10} P_{2.7\,\mathrm{GHz}} \geq 27.0$ W Hz^{-1} sr^{-1} and lying above −45°. There are 25 powerful QSOs lying at $1 \leq z \leq 4$. If the space density remained constant to high redshift we would have expected to find 15 lying between $5 \leq z \leq 7$. As no QSO above z=4.46 was found in the whole Parkes flat-spectrum sample the difference is significant at the 99.96%-level.

Whilst this result is unequivocal in that there is a space density turnover for the very powerful sources, there were two niggling concerns: (i) the curved nature of the radio spectra of many QSOs would cause some of them to naturally drop out of our sample at high redshifts (Jarvis & Rawlings 2000) and (ii) that this analysis was limited to a small number of very luminous QSOs, taking no account of our large sample which extends to much lower radio powers. There is the danger being that we are tracing the space density behavior of a singular class of powerful sources.

There is good evidence that some QSOs have flat radio spectra which remain at least approximately flat spectrum to tens or hundreds of GHz. However, there are a fair number which are only just flat-spectrum, i.e., $\alpha < -0.3$ in our sample. We carefully considered both concerns and have devised a methodology which accounts for the individual spectra of the QSOs and includes as many sources as possible in the analysis. We name this the 'single source survey' ('sss', Wall et al. in preparation). each individual source is considered 'complete' if it lies above the Parkes survey region limit in that area. Moreover, this approach does not rely on a model of the luminosity function or evolution in determining if there is a space density turnover.

The 'sss' approach selects all QSOs lying above $-40°$ declination with flux densities above the PKS survey limit. As this limit varies across the various PKS survey regions from $S_{2.7\,\mathrm{GHz}} = 0.1$ to 0.6 Jy it is vitally important to assign this correctly for each source. Because identification-completeness was achieved, there is no optical limit applicable. We find 239 QSOs which satisfy these criteria. We calculate the 'observable volume' of each source by comparing it to our sample selection criteria using flux density and spectral index limits: firstly for the volume from $1 \leq z \leq 3$, then for the volume between $3 \leq z \leq 8$. If we assume that the space density from $1 \leq z \leq 3$ remains constraint to $z = 8$ then the ratio of the two volumes gives the expected contribution to the high redshift count for that source.

Our initial 'sss' analysis uses only the PKS 2.7 GHz and 5 GHz survey data (case 0 in Table 1), predicting 53.5 sources at $3 \leq z \leq 8$. We then improve our analysis by tracing the spectral shape of each source using data from two lower-frequency surveys —Texas: 365 MHz (Douglas et al. 1996) and NVSS: 1.4 GHz (Condon et al. 1998). This modifies the 'observable volume' for a number of sources up to their observed redshift. However, the modifications are both in the positive and negative senses so that the overall result is that the total high-redshift prediction is only marginally different than before (case 1 in Table 1).

To account for the high-frequency spectral shapes of the sources we can incorporate the PKS 8.4 GHz survey data (Wright et al. 1990) This modifies the 'observable volume' at high redshift for 116 sources. This drastically reduces the predicted number of sources at high redshift and the total high-redshift prediction is now only 30.1 (case 2 in Table 1).

As already noted, there is a fundamental problem in using non-simultaneous flux density measurements for compact flat-spectrum QSOs due to the high incidence of source variability. In this case the 8.9 GHz survey was performed in 1989, some 17 yr after the PKS 2.7 GHz and 5 GHz surveys. The 17 yr gap introduces a severe bias as the sources were in an up-state when detected by the original Parkes surveys. In contrast, the 8.9 GHz survey sampled the sources in a random state. Spectra estimated from the 8.9 GHz flux densities will be too steep, leading to serious under-prediction of high-redshift sources.

To investigate the magnitude of the flux density variations we incorporated 8.9 GHz flux density data from Shimmins & Wall's survey of 1972 (Shimmins &

Wall 1973) which are almost contemporary to the PKS 2.7 GHz survey. There are marked differences between the flux density measurements for individual sources at 8.4 GHz and 8.9 GHz and this increases the predicted high-redshift count (case 3 in Table 1). If we drop the 8.4 GHz data completely then the prediction increases again (case 4 in Table 1), and interestingly is close to those for cases 0 and 1.

Table 1. Parkes flat-spectrum QSOs: Single source analysis

	Case	Predicted # of QSOs $3 \leq z \leq 8$
0.	Using just PKS 2.7 & 5 GHz survey data	53.3
1.	Using low-frequency survey data (up to 5-GHz)	54.0
2.	Including PKS 8.4 GHz survey data	30.1
3.	Using PKS 8.9 GHz + 8.4 GHz data	38.6
4.	Using PKS 8.9 GHz data, no 8.4 GHz data	52.9

We conclude that the predicted number of QSOs at $3 \leq z \leq 8$ is somewhere between 38.6 and 54.0. We ignore the result for case 2 using just the 8.4 GHz data due to the variability of individual sources. In this redshift range we observed 16 QSOs and infer another 1.8 from the small (unbiased) set of candidate QSOs with no spectroscopy. The difference between the observed and predicted QSO number is significant at $> 3\sigma$.

6. Discussion

We achieved the two main goals of our project, finding the highest-redshift radio-loud QSO at the time (Shaver, Wall, & Kellermann 1996) and also establishing that there is a space density turnover in the flat-spectrum QSO population (Shaver et al. 1996; Wall et al. in preparation). We have determined the reality of a redshift cut-off in the radio-loud QSO LF which is unbiased by dust-obscuration. Furthermore we suggest that the similarity between the radio-selected and optically-selected QSO redshift cut-offs implies that optically-selected samples are not seriously biased by dust. This is consistent with the results from the CORALS survey (Ellison et al. 2001; 2002) where damped Lyα systems were used to compare the difference between the neutral gas density traced by radio-selected and optically-selected samples. No significant difference is detectable, again giving the conclusion that there is no missing population of high-column-density absorbers obscured by dust. A more detailed discussion of these results will be given by Wall et al. (in preparation).

References

Bolton, J. G., Savage, A., Wright, A. E., 1979, AuJPA, 46, 1B
Condon, J. J., et al. 1998, AJ, 115, 1693

Douglas, J. N., et al. 1996, AJ, 111, 1945
Ellison, S. L., et al. 2001, A&A, 379, 393
Ellison, S. L., et al. 2002, A&A, 383, 91
Fan, X., et al. 2003, AJ, 125, 1649
Hook, I. M., et al. 2002, A&A, 391, 509
Hook, I. M., et al. 2003, A&A, 399, 469
Hu, E. M., et al. 2002, ApJ, 568, L75 (erratum, ApJ, 576, L99)
Jackson, C. A., & Wall, J. V. 1999, MNRAS, 304, 169
Jackson, C. A., et al. 2002, A&A, 386, 97
Jarvis, M., & Rawlings, S., 2000, MNRAS, 319, 121
Kellermann, K. I., & Pauliny-Toth, I. I. K., 1981, ARA&A, 19, 373–410
Longair, M. S., MNRAS, 1966, 133, 421
Shaver, P. A., Wall, J. V., & Kellermann, K. I., 1996 MNRAS, 278, L11
Shaver, P. A., et al. 1996, Nature, 384, 439
Shimmins, A. J., & Wall, J. V., 1973, Austr J Phys, 26, 93
Urry, C. M., & Padovani, P., 1995, PASP, 107, 803
Wall, J. V., & Jackson, C. A., 1997, MNRAS, 290, L17
Wall, J. V., & Peacock, J. A., 1985, MNRAS, 216, 173
Wright, A. E., & Otrupcek, R. E., 1990, Parkes Catalogue, (Sydney: Australia Telescope National Facility, CSIRO)
Wright, A. E., et al. 1990, PASA., 8, 261

Radio Astronomy at the Fringe
ASP Conference Series, Vol. 300, 2003
J.A. Zensus, M.H. Cohen, & E. Ros, eds.

Observations at the Edge

P. A. Shaver
European Southern Observatory, Karl-Schwarzschild-Str. 2, 85748 Garching bei München, Germany

Abstract. There have been dramatic new developments in the search for the reionisation epoch over the last few years. Both QSOs and galaxies have now been found up to $z \sim 6.5$, and there is evidence that the end of the reionisation epoch may have occurred at about that redshift. In that case the search now enters the reionisation epoch itself. Prospects for detecting the first luminous objects and the neutral medium beyond the reionisation epoch are discussed. Radio observations may play a very important role in this endeavour.

1. Introduction

Following recombination at $z \sim 1000$, the Universe went into a protracted dark and neutral phase, and only emerged from this state when the first luminous objects reionised the Universe. This "reionisation epoch" was essentially a fundamental phase transition for the Universe, the most significant development since recombination. It was preceded by the period known as the "Dark Ages", and followed by the familiar Universe we know today. Thus, as we observe outwards, back in time and increasing in redshift, we approach an "edge", before which the Universe was dark.

Analytical and numerical estimates suggest that reionisation occurred sometime in the period $6 < z < 15$ (e.g., Gnedin & Ostriker 1997; Haiman & Loeb 1998; Valageas & Silk 1999; Gnedin 2000). Gnedin (2000) has described three stages in the reionisation process: (i) pre-overlap ($z \sim 15 \rightarrow 7$), which occurred in about the (then) Hubble time, (ii) overlap ($z \sim 7$), when the H II regions forming around the ionising sources overlapped each other, and which occurred very suddenly (in about 10% of the Hubble time), and (iii) post-overlap ($z \sim$ 7–>). The reionisation redshift z_i is defined to be that of the rapid overlap phase.

Major developments have taken place over the last few years in this field of research. The evolution of populations of objects since reionisation has been increasingly well defined, the trailing edge of the reionisation epoch may now have been discovered, and a growing list of possibilities for observations into the reionisation epoch and beyond is stimulating new efforts. These developments are briefly reviewed in this paper.

2. Observing the Near Side

2.1. High Redshift QSOs

While the Sloan Digital Sky Survey (SDSS) has continued to find quasars at record redshifts, it has also confirmed that their space density drops dramatically with redshift. The highest redshift QSO now known has a redshift of $z = 6.43$ (Fan et al. 2003), but the space density of QSOs at that redshift is 20 times less than at $z = 3$. The "QSO epoch" is now very well defined.

The high redshift fall-off in space density is also evident in both radio and X-ray selected samples, showing that it is real and not merely due to dust obscuration. The final analysis of the Parkes quarter-Jansky flat-spectrum sample of radio sources (Wall et al. 2003) provides confirmation for the radio population (see Jackson et al. these proceedings, page 267).

New results from *CHANDRA* observations give a consistent story for X-ray selected AGN (Hasinger et al. 2002; Barger et al. 2003). Barger et al. find that there is no $z > 6$ X-ray AGN in their sample, compared with the large numbers predicted earlier, and conclude that reionisation could not have been caused by AGN of even modest luminosity.

As the intergalactic medium was already highly ionised at $z = 5$, it seems clear that the AGN population—whether selected in the optical, radio, or X-ray—peaked at a redshift well below that of reionisation.

2.2. High Redshift Galaxies and Star Formation

Large numbers of galaxies are now known at high redshifts $z \sim 3 - 6$. However, their evolution with redshift is still not obvious. The cosmic star formation rate deduced from observed galaxies increased strongly from $z = 0$ to $z \sim 1 - 2$, but whether it then flattened off at earlier times or actually decreased towards higher redshifts is not yet clear. Uncertain corrections for the effects of dust have complicated the issue.

However, the same dust that causes obscuration in the ultraviolet and optical radiates brightly at far-infrared and millimetre wavelengths. Strong high-redshift sources of millimetre and sub-millimetre radiation have been found using the SCUBA and MAMBO bolometer arrays on the JCMT and IRAM 30 m telescopes. These appear to be galaxies with huge rates of star formation, and yet are barely visible in the optical/UV. They indicate a large global star formation rate at $z \sim 2 - 3$, but again the evolution at higher redshift is unclear.

Thus, although many high-redshift galaxies are now known from searches at various wavelengths, they have not yet revealed the evolutionary pattern at the highest redshifts.

2.3. Detection of the Reionisation Epoch?

It has been obvious for some years that the "forest" of H I absorption lines in QSO spectra – the "Lyman-alpha forest" – becomes denser towards high redshift. A broadband measure of this effect is given by the drop in the continuum flux on the blue side of the Lyman-alpha emission line relative to that extrapolated from the red. This ratio was clearly approaching unity at $z \sim 5 - 6$, indicating more and more neutral gas along the intervening line of sight.

In 2001, with the discovery of the then highest-redshift QSO, a Sloan QSO at $z = 6.28$ (Fan et al. 2001), analysis of this and other spectra indicated that the rate at which this ratio was approaching unity appeared to be increasing sharply. While the observations could only give a lower limit on the optical depth of about 20 (Becker et al. 2001) (compared with the 10^5 optical depth for a completely neutral IGM), it was widely concluded that the trailing edge of the reionisation epoch had probably been detected.

Since then, three more $z > 6$ Sloan QSOs have been found (Fan et al. 2003), the most distant being at $z = 6.43$. These are of great importance in confirming the universality of the sudden increase in opacity of the IGM. The "redshift of reionisation", defined as that when the ionised regions around different luminous sources completely merged together, has been estimated to be $z = 6.2$ (Gnedin 2001).

Meanwhile, there is increasing evidence that the He II reionisation epoch may already have been found, at $z \sim 3$ (Reimers et al. 1998; Theuns et al. 2002; and references therein). This appears to be consistent with the peak in the QSO space density at a similar redshift, and with the hydrogen reionisation redshift mentioned above.

3. Observing the Far Side

3.1. Discrete Sources

If indeed the near side of the reionisation epoch has already been detected, then the next frontier is the search for the ionising sources embedded within the reionisation epoch itself, and the neutral intergalactic medium beyond.

For some years it was thought that the ionising sources within the reionisation epoch would have spectra that were essentially continuum step functions – no emission whatever blue-ward of Lyman-alpha at the source redshift, and a continuum to the red (e.g., Haiman & Loeb 1998). This would happen because the strong damping wings of the intergalactic H I absorption would completely obliterate any Lyman alpha emission from the source itself.

A curious development in this context, then, was the discovery a year ago of a galaxy at $z = 6.56$ with a prominent Lyman-alpha emission line (Hu et al. 2002). However, it was already becoming realized that any luminous QSO within the reionisation epoch would ionise such a large region around it that the Lyman alpha photons from the source itself would be able to escape absorption by the immediate neutral medium (Madau & Rees (2001). The same in fact would be true for galaxies (Haiman 2002), as they would also create large ionised bubbles around themselves, although over longer timescales.

Many possibilities have been discussed for the detection of discrete sources in the reionisation epoch. Present large-area searches (radio, optical) may find rare objects at high redshifts. In spite of current limits, there could still be many low-luminosity AGN at high redshift, detectable with future instruments (NGST, SKA). ALMA may detect the heated dust associated with star formation around these objects. Present dedicated searches (e.g., deep narrow-band Lyα) may reach to galaxies at higher redshifts. Again, ALMA may detect dust emission from such early star-forming galaxies, and NGST may detect as many as $10^3 - - 10^4$ mini-galaxies at $z > 5$ per field of view (Haiman & Loeb, 1999).

Clusters of metal-free massive stars at $z \sim 10$ may be detectable with NGST (Bromm et al. 2000), although there could be possible confusion with lower-z globular clusters (Carlberg, 2002). With $T_{\rm eff} \sim 10^5$ K they would produce strong H and He ISM recombination lines.

Supernovae and gamma-ray bursts are also strong possibilities for detection in this epoch of massive star formation. NGST may detect 20 supernovae $\rm yr^{-1}$ at $z > 5$ per $5' \times 5'$ FOV (Rees 1998). And, if they are associated with massive stars, GRB afterglows may be detectable to high redshifts (Lamb & Reichart 2000; Ciardi & Loeb 2000; Bromm & Loeb 2002).

Emission and absorption lines may be detectable from from halos around discrete sources. H_2 line emission and absorption from cooling shells may be detectable in the mid-infrared (Ciardi & Ferrara 2001) and sub-millimetre (Shibai et al. 2001). Hα and Hβ will be produced by the ionised sources (Oh 1999). He II recombination lines may also be detectable from sources of hard radiation (Oh, Haiman, & Rees 2001). Spectral features of infalling gas around early super-massive black holes may produce distinctive Lya double spectral feature (Barkana & Loeb 2003).

Continuum emission may be detectable from ionised halos. Oh (1999) has estimated that free-free emission from $\sim 10^4$ sources at $z > 5$ may be detectable above 70 nJy in the field of the Square Kilometre Array. Diffuse ($\sim 15''$) Lyα halos produced by resonant scattering around luminous sources before reionisation may be detectable (Loeb & Rybicki, 1999). And stimulated radio recombination line emission may be produced by early H II regions of high emission measure (Spaans & Norman, 1997). $n \sim 100$ states might produce lines at 50–200 MHz detectable with SKA.

Finally, a strong possibility much explored is that H I emission and absorption shells may be generated around the first bright quasars, and detectable with SKA (Madau et al. 1997; Tozzi et al. 2000).

3.2. Large-scale Features

Rather than looking for discrete sources and their immediate environs, a very different possibility is to try to detect the neutral intergalactic medium itself at and before reionisation.

A global H I signal in the form of a broadband spectral step must in principle be present over the whole sky. At $z = 6.2$, the current estimate for the "overlap redshift" at which reionisation is essentially complete, the H I line is redshifted to 197 MHz. If reionisation began at $z \sim 10$, the IGM would have been completely neutral before this redshift, at which the H I line would be redshifted to an observed frequency of 129 MHz. Therefore, the transition from ionised to neutral would correspond to a transition from negligible H I emission above 197 MHz to continuous H I emission below 129 MHz – a "step" in the broadband spectrum. The step may actually cover a much smaller frequency range, depending on the details and synchronisation of the reionisation process over the sky, in which case it may conceivably be detectable, even in the presence of the contaminating foreground (galactic and extragalactic) emissions (Shaver et al. 1999). A global spectral feature such as this is particularly interesting because, as an all-sky signal, it would in principle be detectable even with small telescopes. This global feature could also appear in absorption (against the microwave background) over at least part of the transition frequency range (Tozzi et al. 2000); this absorption feature could be quite sharp. A global spectral feature in the near-infrared may also be present due to Lyα at reionisation, when the rate of recombinations peaked (Baltz, Gnedin, & Silk 1999; Shaver et al. 1999).

Fluctuations in redshifted 21 cm emission from the neutral medium before reionisation may be detectable with the LOFAR and the SKA, and give the matter density field (Madau et al. 1997; Tozzi et al. 2000). Foreground source confusion may, however, be a significant problem in this endeavour (di Matteo et al. 2001).

"Forests" of emission and absorption lines may be detectable if there are sufficiently luminous objects in the early stages of reionisation to serve as background sources. A widespread forest of H I emission features due to mini-halos may be detectable with LOFAR or the SKA (Iliev et al. 2002). A widespread forest of H I absorption features may also be present in the spectra of early radio sources due to mini-halos, proto-galactic disks, and the irregular intergalactic medium (Carilli et al. 2002; Furlanetto & Loeb 2002). There could also be forests of metal absorption lines from polluted over-dense regions (Oh, 2002).

Finally, the reionisation epoch will leave its imprint on the Cosmic Microwave Background. Thomson scattering from the free electrons will result in damping, Doppler and polarisation effects (e.g., Haiman & Knox 1999; Benson et al. 2000). Any extended period of "smoldering reionisation" may best be detected from the morphology of the secondary CMB anisotropies (Gnedin & Shandarin 2002).

4. Summary

These are exciting times in the search for the earliest luminous sources, and for the universal phase transition known as the "reionisation epoch". Current observations now extend out to $z \sim 6.5$, and it appears that the trailing edge of the reionisation epoch may have already been found. In that case the search now enters the reionisation epoch itself and beyond. Many observational possibilities have been suggested, and radio observations may play an important role.

References

Baltz, E. A., Gnedin N. Y., & Silk J. 1999, ApJ 493, L1

Barger, A. J., et al. 2003, ApJ, 584, L61

Barkana, R. & Loeb, A. 2003, Nature, 421, 341

Becker, R. H., et al. 2001, AJ, 122, 2850

Benson, A. J., Nusser, A., Sugiyama, N., & Lacey, C. G. 2001, MNRAS, 320, 153

Bromm, V., Kudritzki, R. P., Loeb, A. 2001, ApJ 552, 464

Bromm, V., & Loeb, A. 2002, ApJ, 575, 111

Carilli, C., Gnedin, N. Y., & Owen, F. 2002, ApJ, 577, 22

Carlberg, R. G. 2002, ApJ, 573, 60

Ciardi, B., & Loeb, A. 2000, ApJ 540, 687

Ciardi, B., & Ferrara A. 2001, MNRAS, 324, 648

Di Matteo, T., Perna, R., Abel, T., & Rees, M. J. 2002, ApJ, 564, 576
Fan, X., et al. 2001, AJ, 122, 2833
Fan, X., et al. 2002, AJ, 123, 1247
Fan, X., et al. 2003, AJ, 125, 1649
Furlanetto, S. R., & Loeb, A. 2002, ApJ 579, 1
Gnedin, N. Y. 2000, ApJ, 535, 530
Gnedin, N. Y. 2001, MNRAS, submitted (astro-ph/0110290)
Gnedin, N. Y., & Ostriker, J. P. 1997, ApJ, 486, 581
Gnedin, N. Y., & Shandarin, S. F. 2002, MNRAS, 337, 1435
Haiman, Z. 2002, ApJ, 576, L1
Haiman, Z., & Knox, L. 1999, in ASP Conf. Ser. Vol. 181, Microwave Foregrounds, ed. A de Oliviera-Costa & M. Tegmark, (San Francisco: ASP), 227
Haiman, Z., & Loeb, A. 1998, ApJ, 503, 505
Haiman, Z., Madau, P., & Loeb, A. 1999, ApJ 514, 535
Hasinger, G., et al. 2002, Messenger, 108, 11
Hu, E. M., et al. 2002, ApJ, 568, L75
Iliev, I. T., Shapiro, P. R., Ferrara, A., & Martel, H. 2002, ApJ, 572, L123
Lamb, D. Q., & Reichart, D. E. 2000, in AIP Conf. Ser. Vol. 526, Gamma-ray Bursts, ed. R. M. Kippen, R. S. Mallozzi, & G. J. Fishman, (Melville, NY: American Institute of Physics), 658
Loeb, A., & Rybicki, G. B. 1999, ApJ, 524, 527
Madau, P., Meiksin, A., & Rees, M. J. 1997, ApJ, 475, 429
Madau, P., & Rees, M. J. 2000, ApJ, 542, L69
Oh, S. P. 2002, MNRAS, 336, 1021
Oh, S. P. 1999, ApJ, 527, 16
Oh, S. P., Haiman, Z., & Rees, M. J. 2001, ApJ, 553, 73
Rees, M. J. 1998, in The Next Generation Space Telescope: Science Drivers and Technological Challenges, 34th Liege Astr. Coll., ESA SP-429, 5
Reimers, D., et al. 1997, A&A 327, 890
Shaver, P. A., Windhorst, R. A., Madau, P., & de Bruyn, A. G. 1999, A&A, 345, 390
Shibai, H., Takeuchi, T., Rengarajan, T. N., & Hirashita, H. 2001, PASJ, 53, 589
Spaans, M., & Norman, C. A. 1997, ApJ, 488, 27
Theuns, T., et al. 2002, ApJ, 567, L103
Tozzi, P., Madau, P., Meiksin, A., & Rees, M. J. 2000, ApJ, 528, 597
Valageas P., & Silk, J. 1999, A&A, 347, 1

Radio Astronomy at the Fringe
ASP Conference Series, Vol. 300, 2003
J.A. Zensus, M.H. Cohen, & E. Ros, eds.

Radio Galaxies as Probes of Distant Proto-clusters

G. Miley
Sterrewacht, Postbus 9513, NL-2300 RA, Leiden, The Netherlands

Abstract. I shall review some highlights of the Leiden Distant Radio Galaxy Programme. In particular, I shall describe recent results of a large program with the VLT, which demonstrate that luminous distant radio galaxies are excellent signposts for finding galaxy over-densities in the early Universe.

1. Happy Birthday Ken

I am delighted to be here and have the opportunity of paying homage to Ken Kellermann. It was as a graduate student in the mid sixties that we first met, when I was asked to show him around Jodrell Bank Since then he has been a respected colleague and a dear friend. Not only is he one of the greatest living radio astronomers, but he is also an extremely nice person. Ken, I wish you a belated happy birthday and many more stimulating radio astronomical years.

2. The Leiden Distant Radio Galaxy Program

About 15 years ago, at Leiden, we started a program to find distant radio galaxies, study their properties and use them as probes of the early Universe. This has developed into a set of mutually dependant collaborations and provided the food for several theses, including those of Ken Chambers, Huub Röttgering, Rob van Ojik, Roeland Rengelink, Laura Pentericci and Carlos De Breuck. Some of our other collaborators are Wil van Breugel, Chris Carilli, Philip Best, Tim Heckman, Holland Ford and the ACS Instrument Definition Team.

Detailed optical and infrared studies of high-redshift radio galaxies (HzRGs) revealed that they were an extremely important class of objects for studying the formation and evolution of galaxies. Some of the results that we were involved in obtaining include the following:

• HzRGs trace the bright envelope in the infrared Hubble diagram, implying that they are amongst the **most massive** galaxies in the early Universe (e.g., De Breuck et al. 2000).

• HzRGs appear extremely clumpy on *HST* images, strikingly similar to simulations from hierarchical models of **forming** massive galaxies, e.g., brightest cluster galaxies (e.g., Pentericci et al. 1999)

• HzRGs are usually surrounded by **giant Ly α emitting gas halos**, whose sizes ($\sim$ 100 kpc) are comparable to those of cD galaxy envelopes (e.g., van Ojik et al. 1997).

• HzRGs are associated with radio sources having extremely large depolarisations, indicating that they are **embedded in a dense hot gas** (Carilli et al. 1997).

These properties, taken together, provided strong indirect evidence that luminous distant radio sources pinpoint forming massive galaxies at the centres of proto-clusters. We therefore instigated a program to search for direct evidence of galaxy over-densities in around distant radio galaxies. This programme has been the subject of an ESO VLT Large Program and was allocated 55 orbits of guaranteed GTO time on the Advanced Camera for Surveys.

3. Discovery of Proto-clusters with $2 < z < 4.1$: A Large VLT Project

The VLT projects are presently food for the Ph.D theses of Jaron Kurk and Bram Venemans. The discovery technique (narrow-band Ly α imaging of a $7'$ by $7'$ field around the radio galaxy at the appropriate redshift) has been extremely successful in finding proto-clusters. All 5 $z > 2$ targets studied with sufficient depth have > 20 spectroscopically–confirmed Ly α and/or Hα companions, with galaxy over-densities of ~ 10, structure sizes of $\sim 3\,$Mpc and velocity dispersions of $\sim 300 - 1000\,\mathrm{km\,s^{-1}}$, centred at the velocity of the radio galaxies.

The masses of the structures, estimated from the measured galaxy over-density, the volume occupied by this over-density, the mean density of the Universe and the bias parameter are $\gtrsim 1 - 2 \times 10^{15}\,\mathrm{M_\odot}$, comparable to the mass of the Coma cluster. This provides further evidence that our galaxy structures are indeed the ancestors of present-day rich clusters.

4. Lyman Break Galaxies in the $z = 4.1$ Proto-cluster: ACS GTO Programme

Our most spectacular field is 1338−19 at $z \sim 4.1$, presently the most distant structure of galaxies known (Venemans et al. 2002) with 21 spectroscopically confirmed Lyman α emitters. This object has long been a target of the ACS GTO Team, led by Holland Ford and Garth Illingworth and was observed during the first few months after this new camera was installed on the *HST*. The main members of the team who have worked on planning and reducing these data, besides myself are Roderik Overzier, a Leiden graduate student and "Zlatko" Tsvetanov.

Our observations were multicolour ones, using 3 filters: g (4-orbits), r (5-orbits), i (4-orbits). All 12 confirmed VLT Ly α proto-cluster members located in the $3'$ by $3'$ ACS field were detected as "g–dropout" objects and spatially resolved. Further, the ACS observations show that the number of objects with g-band breaks ($\mathrm{g} - \mathrm{r} > 1.5$) is both anomalously high (factor of $4 - 6$) compared with that in similar–exposure ACS fields and non-uniformly distributed, with a higher density close to the radio galaxy. This implies the presence of an additional ~ 50 Lyman break proto-cluster candidates and is additional evidence that 1338−26 is a proto-cluster at $z \sim 4.1$.

5. Conclusions

We have discovered substantial galaxy over-densities around all the distant radio galaxy targets that we have been observed to sufficient depth with the VLT. From a preliminary analysis of the data, we draw the following conclusions:

- HzRGs are generally associated with the most massive forming galaxies at the centre of proto-clusters and can be used to pinpoint such structures.
- Because the radio lifetimes (few $\times 10^7$ yr) are small by the standard of cosmic evolution, there must be orders of magnitude more "dead" radio sources than living ones. When the proto-clusters have evolved, the luminous radio sources are likely to be extinguished, explaining why $z < 1$ clusters do not generally harbour active luminous radio sources.
- The luminosity functions and lifetimes of luminous distant radio sources are consistent with such objects being associated with every pronounced velocity spike in the space density of Lyman break galaxies.

Our distant proto-clusters are unique laboratories for studying the most extreme over-dense regions in the early Universe and crucial locations for tracing the formation and evolution of clusters and galaxies. We are in the process of analysing more than 100 confirmed Lyman alpha-emitting galaxies that lie within our proto-clusters and in addition we are obtaining observations with other facilities, including the Advanced Camera for Surveys on the *HST* to find and study the various other populations of galaxies expected within the proto-clusters. A detailed study of the morphologies and spectral energy distributions of proto-cluster members between $z \sim 4$ and $z \sim 1$ can be used as a unique method for tracing the history of galaxy assembly and star formation.

References

Carilli, C. L., Röttgering, H. J. A., van Ojik, R., Miley, G. K., & van Breugel, W. J. M. 1997, ApJS, 109, 1

De Breuck, C., van Breugel, W., Röttgering, H. J. A., & Miley, G. 2000, A&A, 143, 303

Pentericci, L., et al. 1999, A&A, 341, 329

van Ojik, R., Röttgering, H. J. A., Miley, G. K., & Hunstead, R. W. 1997, A&A, 317, 358

Venemans, B. P., et al. 2002, ApJ, 569, L11

Radio Astronomy at the Fringe
ASP Conference Series, Vol. 300, 2003
J.A. Zensus, M.H. Cohen, & E. Ros, eds.

Milliarcsecond Radio Structures in AGN as Cosmological Probes

L. I. Gurvits
Joint Institute for VLBI in Europe, P.O. Box 2, 7990 AA Dwingeloo, The Netherlands

Abstract. An "angular size-redshift" relation for radio structures in extragalactic radio sources is reviewed with reference to its potential as a cosmological test. The test is providing a useful consistency check on other methods of measuring cosmological parameters, but can be based on data obtained from non-cosmological observational projects. A future dramatic increase in the number of VLBI-imaged extragalactic sources with known redshifts will likely allow the relation to remain in the arsenal of cosmological tests.

1. Introduction

It was exactly ten years ago, at the Sub-Arcsecond Radio Astronomy conference in Manchester, when Ken Kellermann and the author realized that their presentations, albeit conceived independently and differing in some technical details, were rather similar: they were both about attempts to use the "angular size – redshift" ("$\theta - z$") relation for milliarcsecond-scale radio structures in AGN of different types as a means of estimating the cosmological deceleration parameter, $q_\circ$ (Kellermann 1993a; Gurvits 1993).

The idea of an "angular size – distance" test for cosmology was probably first discussed by Tolman (1930) with reference to "nebulæ", which by that time were known to populate the Universe up to the redshift of $z = 0.04$ as standard objects. The essence of such a test is very simple and attractive. Suppose an object of a known size is observed from different distances. Its apparent angular size will depend on the metric of space-time. For the Euclidean metric, the apparent angular size will decrease as r^{-1}, where r is the distance between the object and observer. However, this is not the case in other metrics, and by measuring the behaviour of "standard objects" in the Universe, one can make a conclusion on the metric "implemented" in reality. The metric, in turn, depends on the cosmological model which governs the Universe.

Despite its apparent simplicity, this idea was not suitable for use with optical images of galaxies ("nebulæ"), the only extragalactic objects detectable in Tolman's time and over the next several decades: too many masking factors prevented cosmology from being visible in the observed "angular size – distance" dependencies for galaxies. Furthermore, it is not trivial to define the size of a galaxy, especially in view of the K-correction complications.

In the 1950s, after the first aperture synthesis radio telescopes began producing images of extragalactic radio sources, Hoyle (1959) proposed a similar test for powerful double-lobe radio galaxies (later classified as FRII radio sources). He suggested considering these sources as standard "rods", characterised by the distance between their two radio lobes. Notably, Hoyle's suggestion was made at a time when the number of radio images of extragalactic radio sources hardly exceeded a dozen. Quasars were discovered five years later.

It took another quarter of a century before Hoyle's idea materialised in the studies of hundreds of extragalactic radio sources with arcsecond-scale angular resolution. These studies were based on radio images made with the Cambridge, Caltech, and Green Bank interferometers, and later Westerbork and the VLA. They provided a wealth of information on the sources themselves but remained mostly inconclusive as cosmological data sets. As was shown repeatedly by many authors (e.g., Kapahi 1987; Barthel & Miley 1988; Nilsson et al. 1993 and references therein) the $\theta - z$ relation for arcsecond (kiloparsec) scale radio structures is an amalgam of intrinsic object evolution, projection and selection effects, which masks cosmological effects and results in a decrease of apparent angular sizes much like the simple Euclidean case. I leave out of this brief review topics related to the kpc-scale $\theta - z$ studies. Note that in recent works by Buchalter et al. (1998), Guerra, Daly & Wan (2000), and Podariu et al. (2003), the search for cosmological imprints in the kpc-scale $\theta - z$ dependence for QSOs and radio galaxies has continued with more optimistic outcomes. However, it was precisely the quasi-Euclidean appearance of early kpc-scale $\theta - z$ studies that triggered an interest in similar tests at a thousand times smaller angular scale.

The first demonstration of $\theta - z$ tests at the milliarcsecond (mas) scale were based on VLBI images at 5 GHz of 79 extragalactic radio sources with known redshift (Kellermann 1993a, 1993b) and non-imaging model-fitting of VLBI visibility amplitudes of 337 sources obtained by Preston et al. (1985) at 2.3 GHz (Gurvits 1993, 1994). The main outcome of these first demonstrations was a striking difference between kiloparsec- and parsec-scale $\theta - z$ dependences: the former decreased monotonically to the highest available redshift, while the latter clearly flattened out at redshift $z \geq 1$. This kind of flattening, or reaching a minimum, is predicted for true standard rods by many cosmological models. Thus, it was very tempting to consider the mas-scale $\theta - z$ dependence as a true cosmological test. The first attempts of such tests led to the conclusion that on the largest observable scale, the mas-scale $\theta - z$ data were consistent with the case of an Einstein-de-Sitter Universe with $q_\circ = 0.5$ (Kellermann 1993), and actually favoured somewhat smaller values ($q_\circ = 0.16 \pm 0.71$, Gurvits 1994). Wilkinson et al. (1998) came to a similar conclusion based on the $\theta - z$ test for sources from the Caltech–Jodrell Bank VLBI surveys.

It is a rather widely spread misconception that the first mas-scale $\theta - z$ attempts were aimed at achieving conclusive estimates of cosmological parameters. In fact they were only an indicator of the potential of the method and underlined the difference between the arcsecond- and milliarcsecond-scale $\theta - z$ relations.

Interestingly, the three mas-scale $\theta - z$ data sets mentioned above used different definitions of a source size. Kellermann (1993) measured the maximum distance between the brightest component and the most distant one with the brightness at least 2% of the peak. Gurvits (1994) operated with single-component Gaussian model fitting of visibility amplitudes. Wilkinson et al. (1998) used the same definition as Kellermann (1993) lowering the brightness level from 2% to 1% of the peak.Yet all three data sets have shown similar flattening of the $\theta - z$ relation at $z \gtrsim 1$.

The publication of the first three mas-scale $\theta - z$ studies (Kellermann 1993; Gurvits 1994; Wilkinson et al. 1998) resulted in quite a lively discussion in the literature and several attempts to quantify parameters of various cosmological models (Krauss & Schramm 1993; Stelmah 1994; Dabrowski et al. 1995; Kayser 1995; Stepanas & Saha 1995; Jackson & Dodgson 1996; Banerjee & Narlikar 1999). In particular, Dabrowski et al. (1995) argued that at least 600 properly selected sources were needed to meaningfully constraint $q_\circ$ in the models with $\Lambda \equiv 0$.

Wilkinson et al. (1998) have expressed a less optimistic view on the usefulness of $\theta - z$ tests for cosmology and provided a list of difficulties which face the approach. Of them, the most serious is whether or not the milliarcsecond radio structures in AGN are standard rods.

2. Are parsec-scale radio structures standard?

Simply put, "No, they are not". However, variations from standard are believed to be minimal. This belief is based on the following two essential propositions:

(i) The evolutionary effects, which affect strongly the kpc-scale radio structures of AGN are not important for the short-lived structures deeply embedded in galactic nuclei. This hypothesis is based on the fact that the characteristic lifetime of pc scale radio features is judged to be $10^2 - 10^4$ years, much shorter than the age of the Universe, or the evolution time-scale of host galaxies and their associated kpc-scale radio sources.

(ii) Radio emission on the milliarcsecond scale originates in the nuclear area, which is dominated by the gravitational potential and should be characterised by a limited number of parameters, such as the mass of a central black hole, the accretion rate, the angular momentum, and magnetic field strength. A quantitative model of radio emission from the compact nuclear region would be very helpful either in proving the second proposition or providing a tool for parameterising the dependence of the linear size of a source on its physical attributes. Some promising attempts at such parameterisation have been published recently (e.g., Falcke 1995, 1998 and references therein; Beskin 1997).

As of today, neither of the two propositions above has been proved "beyond a reasonable doubt". Meanwhile, a simple recipe for treating the problem of varying linear sizes of "standard objects" is to allow this size to be one of the fitted parameters of the $\theta - z$ relation. This approach was used by Gurvits (1994) in the form of the following phenomenological representation of the linear size of a source, l:

$$l = l_\circ \left(\frac{L}{L_\circ} \right)^\beta (1+z)^n, \qquad (1)$$

where $l_\circ$ and $L_\circ$ are the normalising linear size and luminosity, respectively, and L is the source luminosity. This representation accounts for a possible dependence of the linear size on luminosity (via parameter β) and redshift (and therefore frequency) dependence (via parameter n). The latter also includes a potential (weak?) cosmological evolution of linear size. As shown by Gurvits

(1994), this parameterisation makes possible simultaneous estimation of cosmological parameters and the values of l, β

Recently, Chen & Ratra (2003) have also demonstrated how to treat the linear size l as a "nuisance" parameter for the cosmologically relevant part of the $\theta - z$ test thus eliminating the uncertainty related to the unknown size of the standard rod.

3. "Maximum use" ad-hoc data set and advanced $\theta - z$ tests

Over the last three decades, the collective efforts of approximately 500 observers have resulted in about $\sim 5,000$ VLBI images of up to $\sim 2,000$ extragalactic radio sources. Of these, some ~ 350 QSOs and other types of AGN have known redshifts and their VLBI images at the most popular VLBI frequency of 5 GHz, have a dynamic range of at least 100. These latter sources compose a useful ad-hoc data set for $\theta - z$ tests. Gurvits, Kellermann, & Frey (1999) compiled from the literature and their own observations a $\theta - z$ sample which contained essentially all of the then available data. Due to its genesis, I call this sample a "maximum use" ad-hoc $\theta - z$ data set. The redshift range of the sources in this sample is from 0.01 to 4.72.

Figure 1 shows the binned $\theta - z$ data for this "maximum use" ad-hoc data set. It confirms that the dependence flattens out at $z \gtrsim 1$ and in this sense is consistent with previous $\theta - z$ tests. In order to quantify the properties of this data set we have applied the same approach as described in Gurvits (1994). It is based on the phenomenological model of the source linear size (equation (1)) and leads to the following relation between measurable angular size θ, and the unknown physical parameters of the source and cosmological model:

$$\theta = lh\frac{H_\circ}{c}\left(\frac{L_c}{L_\circ}\right)^\beta (1+x)^n D^{-1}(z), \tag{2}$$

where $D(z)$ is the angular size distance, which depends on the cosmological model, and the Hubble constant is $H_\circ = 100h\,\mathrm{km\,s^{-1}\,Mpc^{-1}}$. For a homogeneous, isotropic Universe ($q_\circ > 0$) with the cosmological constant $\Lambda \equiv 0$, $D(z)$ is given by the usual expression

$$D(z) = \frac{q_\circ z + (q_\circ - 1)\left(\sqrt{1+2q_\circ z} - 1\right)}{q_\circ^2(1+z)^2} \quad . \tag{3}$$

For quantitative characterisation, the data were fitted with a four-parameter regression model (Gurvits 1994), the variable parameters being lh, β, n and $q_\circ$. In order to minimise possible spectral- and frequency-dependent effects, only 145 sources, with two-point spectral indices between 1.4 and 5 GHz in the range $-0.38 \leq \alpha_{1.4}^5 \leq 0.18$ were used in the regression. The best fit was obtained for $q_\circ \leq 0.5$ and $\beta + n \gtrsim -0.15$. The latter constraint covers a significant fraction of the "$\beta - n$" plane. Thus, this quantitative characterisation provides meaningful support for the models with $q_\circ \leq 0.5$. I note that this result is merely an amplified indication of the potential of the method, rather than a conclusive cosmological result. To date, several attempts to exploit the "maximum use" data set have been undertaken.

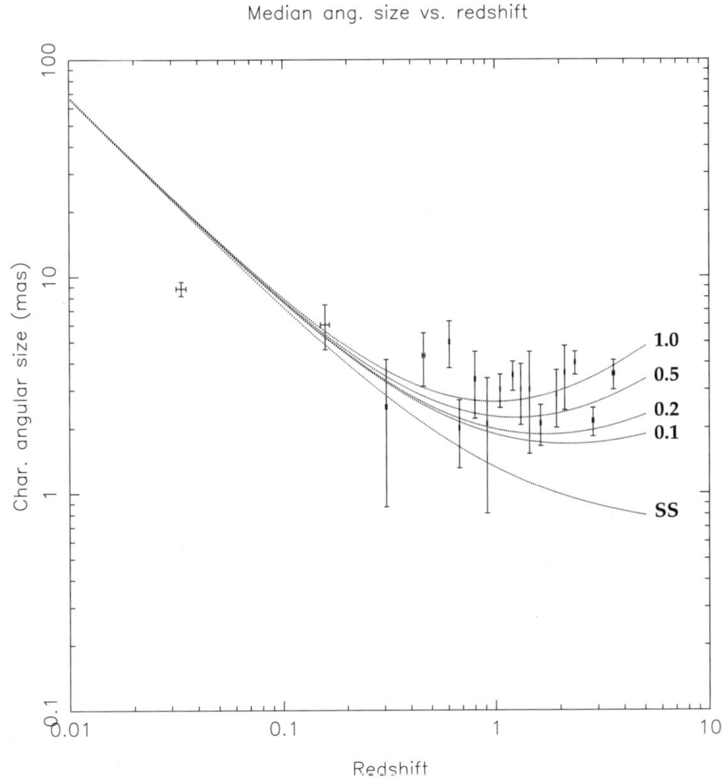

Figure 1. Median angular size versus redshift of 330 sources *(taken from Gurvits, Kellermann, & Frey 1999, A&A, 342, 378)*. The full length of error bars here and in other figures corresponds to 1σ. The solid lines correspond to an arbitrary linear size parameter $l = 9.6h^{-1}$ pc in the Steady-State model (SS), and models of a homogeneous, isotropic Universe with $\Lambda = 0$ and values of $q_\circ = 1.0, 0.5, 0.2, 0.1$ (as marked on the plot). Data are binned into 18 bins nearly equally populated (18–19 sources per bin)

Vishwakarma (2001) used the "maximum use" data set for considering models with $\Lambda \neq 0$ and concluded that for realistic values of the matter density parameter, $\Omega_{\rm m}$, the best-fitting models are those with $\Lambda \propto H^2$, $\Lambda \propto \rho$ or flat models with $\Lambda = \text{const}$.

The same data set was analysed by Lima & Alcaniz (2002) as a test-bed for the models containing the "dark energy" ("quintessence") which is characterised by the equation of state $p_{\rm x} = \omega \rho_{\rm x}$, where $\omega \geq -1$. Under this formalism, the parameter ω becomes a cosmological model parameter alongside the various density parameters Ω. The test was based on the earlier work by the same authors on the behaviour of angular sizes in the "quintessence" cosmology (Lima & Alcaniz 2000). Their result favoured two sets of best fit parameters, $\Omega_\circ \leq 0.62$, $\omega \leq -0.2$ and $\Omega_\circ \leq 0.17$, $\omega \leq -0.65$. These estimates depend strongly on a fixed value of the source linear size, lh, which in this context is a certain liability.

Yet another cosmological formalism, the so called "Cardassian expansion" was confronted with the $\theta - z$ "maximum use" data set by Zhu & Fujimoto (2002). They found (not surprisingly) that the best fit parameters of that model depend heavily on the value of the mean projected linear size of the sources, lh. However, they claim that the data favour the ΛCDM model and make it unnecessary to invoke the Cardassian expansion approach.

As mentioned earlier, Chen & Ratra (2003) described a method which makes the linear size parameter lh a "nuisance" rather than a "liability". Their method was demonstrated by fitting the "maximum use" data set into two low-density CDM models: the first was parameterised by two density parameters, $\Omega_{\rm o}$ and Ω_Λ, the other was spatially flat with a dark energy scalar field, which could be formalised as a time-variable Λ-term. Their conclusion was that the $\theta - z$ test, applied to the "maximum use" data set was consistent with ΛCDM models but less constraining than the test based on the redshift-magnitude Type Ia supernova data.

Another way of effectively eliminating the value of the linear size of the standard object from the fitting procedure of $\theta - z$ data was proposed recently by Wiik & Valtaoja (2001). They proposed considering shocks in AGN jets as standard objects, and estimating their linear sizes from flux density monitoring. The apparent angular sizes of the same shocks could be measured directly by VLBI. The method should work best with mm-wavelength or Space VLBI because the angular sizes of shocks are of the order of 0.1 mas—too small for global cm-wavelength VLBI. The method was demonstrated on a data set of 14 shocks for which all the necessary data were available (the flux density monitoring, VLBI images and redshifts), providing a good fit for Λ-dominated models with $\Omega_{\rm m} + \Omega_\Lambda \simeq \Omega_\Lambda \simeq 1$.

A brief review of various approaches to utilising and enhancing the mas-scale $\theta - z$ dependence seems convincing proof of the usefulness of further studies. What should one expect from these studies in the near future?

4. Future outlook

Due to the statistical nature of the $\theta - z$ test and as correctly pointed out in several recent publications (e.g., Dabrowski, Lasenby, & Saunders 1995; Wiik & Valtaoja 2001), the size of the sample plays an important role: there is a need to increase the number of sources to perhaps several thousand for traditional $\theta - z$ tests, or several hundred for $\theta - z$ tests supplemented by flux density monitoring data.

Figure 2 shows luminosity versus redshift for all 330 sources of the "maximum use" sample. Despite the ad-hoc nature of the sample, it is de-facto flux density limited, with a limit of $\sim$ 300 mJy. The sample is by no means complete, and could be supplemented by another hundred or so sources of flux density $\gtrsim$ 200 mJy. However, the total number of sources in the Universe with this flux density is limited, and the only way to increase the size of the sample dramatically is to include sources at $z \gtrsim 1$ with luminosities in the range $10^{23} - 10^{26}$ W Hz^{-1} (see Fig. 2), which so far has been a "terra incognita" of the "$L - z$" plane. This correspond to flux densities of several tens of mJy and lower. Until recently, this range of flux densities was beyond the reach of VLBI

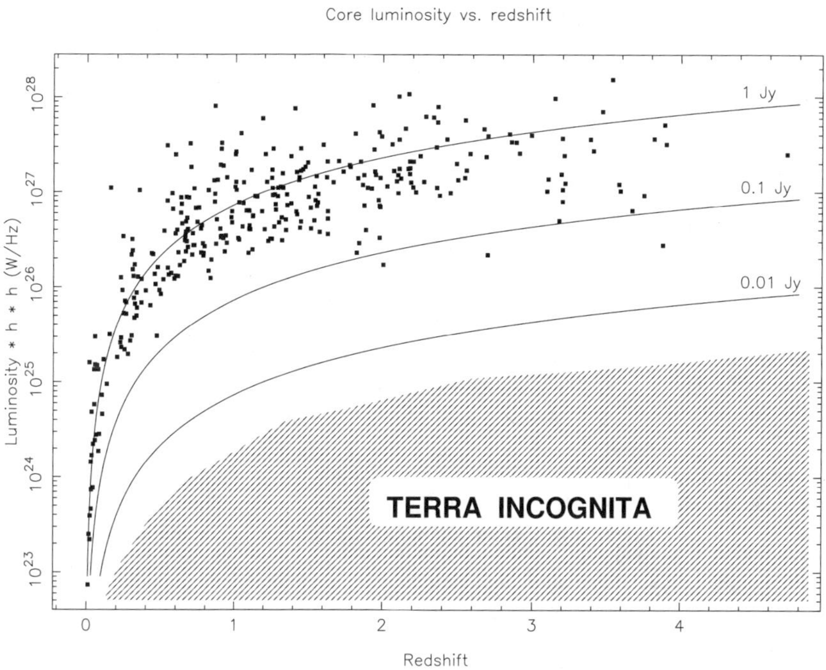

Figure 2. Luminosity of compact cores versus redshift for the "maximum use" sample of 330 quasars and radio galaxies for which VLBI images at 5 GHz are available with a dynamic range of at least 100 ($q_\circ = 0.5$, $H_\circ = 100h\,\mathrm{km\,s^{-1}\,Mpc^{-1}}$). This ad-hoc sample is effectively flux-density-limited at ~ 300 mJy at 5 GHz. Until recently, the shadowed area was not accessible to VLBI studies due to sensitivity limitations. Adopted from Gurvits, Kellermann, & Frey (1999).

imaging. However, recent improvements due in particular to the use of high data rate systems (such as MkIV and MkV, with a data rate per VLBI station measured in Gbit/s) and the technique of phase-referencing make it possible to obtain images of mJy-level sources at the necessary quality. Coupled with advanced optical surveys such as the Sloan Digital Sky Survey (SDSS), which is expected to provide redshifts for 10^5 quasars, one might expect that the sample size for further $\theta - z$ studies could reach $\sim 10^4$ sources. A pilot attempt to obtain VLBI images of mJy-level quasars identified by SDSS has recently been conducted on a sample of 47 quasars (Mosoni et al. 2003, in preparation).

Last but not least: $\theta - z$ could be accumulated from observations conducted for other than cosmological test purposes. In this sense, their cosmological application is "cheap". But their rôle as at least a consistency check for other methods of observational cosmology is far from being exhausted.

I am grateful to Ken Kellermann and Sandor Frey for innumerable stimulating discussions and long-standing collaboration on the subject of this paper, and Ian Avruch for very helpful comments on this paper. I appreciate the hospitality of the National Radio Astronomy Observatory and support from the Netherlands Organization for Scientific

Research (NWO) for a short-term travel subsidy in October 2002 when this contribution was prepared.

References

Banerjee, S. K., & Narlikar, J. V. 1999, MNRAS, 307, 73

Barthel, P. D. & Miley, G. K. 1988, Nature, 333, 310

Beskin, V. S. 1997, Physics-Uspekhi, 40, 659

Buchalter, A., Helfand, D. J., Becker, R. H., & White, R. L. 1998, ApJ, 94, 503

Chen, G., & Ratra, B. 2003, ApJ, 582, 586

Dabrowski, Y., Lasenby, A., & Saunders, R. 1995, MNRAS, 277, 753

Falcke, H., & Biermann, P. L. 1995, A&A, 293, 665

Falcke, H., & Biermann, P. L. 1998, A&A, 342, 49

Guerra, E. J., Daly, R. A., & Wan, L. 2000, ApJ, 544, 659

Gurvits, L. I. 1993, in Sub-Arcsecond Radio Astronomy, ed. R. J. Davies, & R. S. Booth, (Cambridge: CUP), 380

Gurvits, L. I. 1994, ApJ, 425, 442

Gurvits, L. I., Kellermann, K. I., & Frey, S. 1999, A&A, 342, 378

Hoyle, F. 1959, in Proceedings of the Paris Symposium on Radio Astronomy, IAU Symp. No. 9 and URSI Symp. No. 1, ed. R. N. Bracewell, (Stanford: Stanford Univ. Press), p. 529

Jackson, J. C., & Dodgson, M. 1996, MNRAS, 278, 603

Kapahi, V. 1987, in Observational Cosmology, ed. A. Hewitt, G. Burbidge & L. Z. Fang, (Dordrecht: Reidel), 251

Kayser, R. 1995, A&A, 294, L21

Kellermann, K. I. 1993a, in Sub-Arcsecond Radio Astronomy, ed. R. J. Davies and R. S. Booth, (Cambridge: CUP), 386

Kellermann, K. I. 1993b, Nature, 361, 134

Krauss, L. M., & Schramm, D. N. 1993, ApJ, 405, L43

Lima, J. A. S., & Alcaniz, J. S. 2000, A&A, 357, 393

Lima, J. A. S., & Alcaniz, J. S. 2002, ApJ, 566, 15

Nilsson, K., Valtonen, M. I., Kotilainen, I., & Jaakkola, T. 1993, ApJ, 413, 453

Podariu, S., Daly, R. A., Mory, M. P., & Ratra, B. 2003. ApJ, 584, 577

Preston, R. A., et al. 1985, AJ, 90, 1599

Stelmah, J. 1994, ApJ, 428, 61

Stepanas, P. G., & Saha, P. 1995, MNRAS, 272, L13

Tolman, R. C. 1930, Proc. Nat. Acad. Sci., 16, 511

Vishwakarma, R. G. 2001, Class. Quant. Grav., 18, 1159

Wiik, K. & Valtaoja, E. 2001, A&A, 366, 1061

Wilkinson, P. N., et al. 1998, in Observational Cosmology with the New Radio Surveys, ed. M. N. Bremer, N. Jackson, & I. Pérez-Fournon, (Dordrecht: Kluwer), 221

Zhu, Z.-H., & Fujimoto, M.-K. 2002, ApJ, 581, 1

Radio Astronomy at the Fringe
ASP Conference Series, Vol. 300, 2003
J.A. Zensus, M.H. Cohen, & E. Ros, eds.

Radio Galaxies and the Magnetization of the IGM

Gopal-Krishna
National Centre for Radio Astrophysics/TIFR, Post Bag 3, Pune University Campus, Pune 411 007, India

P. J. Wiita
Department of Physics & Astronomy, MSC 8R0314, Georgia State University, Atlanta GA, 30303-3088, USA

Abstract. Observed radio galaxies had a much higher comoving density during the 'quasar era', at $z \sim 2-3$, but these sources are only detectable for small fractions of their active lifetimes at such high z due to expansion losses and increased inverse Compton losses against the cosmic microwave background. Using recent models for the evolution of the size and luminosity of powerful double radio sources, as well as ΛCDM simulations of the cosmic web of baryonic material, we argue that during the quasar era a high volume fraction of this web was occupied by the lobes of double radio sources. They could have seeded the IGM with an average magnetic field approaching 10^{-8}G. Further, these advancing overpressured lobes could compress the denser interstellar gas clouds of the galaxies engulfed by them and thus trigger starbursts. This can probably account for much of the intense star formation activity witnessed beyond $z \sim 1.5$. Also, the sweeping up of the ISM of the gas-rich galaxies by the rapidly advancing radio lobes may well be responsible for the widespread metal pollution of the IGM and proto-galaxies at high redshifts.

1. Introduction

The comoving space density of powerful radio galaxies (RGs) has declined between 100 and 1000 times from redshifts of 2–3 to the present epoch (e.g., Willott et al. 2001). The star formation rate also peaked in roughly the same epoch, and has declined substantially since then (e.g., Archibald et al. 2001; Chary & Elbaz 2001). Severe adiabatic and inverse Compton losses in radio lobes mean that most old and large radio galaxies are very difficult to detect in surveys and only the youngest can be seen at high redshifts (e.g., Blundell et al. 1999, hereafter BRW). Simulations of the growth of structures in cosmological models have indicated that most of the matter that will form galaxies by the current epoch was in the form of filaments that filled only a small portion of the universe at those redshifts (e.g., Cen & Ostriker 1999). Together, these facts lead us to conclude that the formation of many of those galaxies may have been triggered or accelerated by overpressured radio lobes, which probably filled a substantial portion of the cosmic filaments then. Preliminary discussions of these ideas are in Gopal-Krishna & Wiita (2001, hereafter GKW) and Gopal-Krishna, Wiita & Osterman (2003); more extensive calculations are underway (Osterman et al., in preparation). These extremely extended radio lobes are very likely to be responsible for seeding much of the intergalactic medium (IGM) with magnetic fields (GKW; Kronberg et al. 2001; Furlanetto & Loeb 2001). In addition, they may make significant contributions to the spreading of metals widely through the IGM and into protogalaxies.

2. Radio luminosity function – RLF

In general, RLF studies are plagued by uncertainties resulting from incomplete knowledge of the redshifts of the radio sources; however, results based upon the 3CRR, 6CE and 7CRS surveys of different flux limits have the advantage of having 96% of their redshifts known (Willott et al. 2001). In addition, their selection at low frequencies minimizes the bias due to relativistic beaming.

Powerful (FR II type) radio sources are nearly 3 dex above the local RLF by $z \sim 2$, and their RLF varies little out to the beginning of the quasar era at $z \sim 3$, while it appears to decline at higher z (Willott et al. 2001). Furthermore, the RLF for those redshifts is nearly flat for over a decade in radio power above $P_{151} \geq 10^{25.5}$ W Hz^{-1} sr^{-1}, which is where the FR II sources are most numerous.

Combining these results with the correction factor discussed in the following section, we find that at $z = 2.5$ the actual proper density of powerful radio sources lasting for an interval T is

$$\rho \simeq 4 \times 10^{-5}(1+z)^3 T_5 \text{ Mpc}^{-3} (\Delta \log P_{151})^{-1}, \tag{1}$$

where $T_5 \equiv T/(5 \times 10^8\text{y})$. We then integrate over the roughly 1.25 dex of the peak of the RLF. Finally, we take into account the fact that several generations of RGs will be born and die within the ~ 2 Gyr duration of the quasar era. This leads us to the total proper density, Φ, of intrinsically powerful radio sources: $\Phi = 7.7 \times 10^{-3}$ Mpc^{-3}, which is independent of the assumed value of T, and nearly independent of cosmological model parameters (see GKW for details).

3. Restricted visibility of radio galaxies

Various recent models of RG evolution (Kaiser et al. 1997; BRW; Manolakou & Kirk 2002) agree that radio flux declines dramatically with increasing source size (adiabatic losses) and with z (inverse Compton losses off the more intense cosmic background radiation). Theoretical distributions of RG powers, sizes, redshifts and spectral indices can be nicely matched by models that require most RGs to remain active for $T \simeq 5 \times 10^8$y (see, also, Wang, Yamada, & Taniguchi 2003) and to have a distribution of jet powers (Q_0) that goes as $\sim Q_0^{-2.6}$ (BRW). X-ray observations indicate that the density of the matter through which the jets propagate declines with distance roughly as $n(r) = n_0(r/a_0)^{-\beta}$, with $n_0 = 1.0 \times 10^{-2}\text{cm}^{-3}$, $a_0 = 10$ kpc, and $\beta = 1.5$. This leads to the total linear size of the RG being given by

$$D(t) = 3.6 a_0 \left(\frac{t^3 Q_0}{a_0^5 m_p n_0} \right)^{1/(5-\beta)}. \tag{2}$$

We find that in most RGs, particularly those at $z > 2$, the central engines remain active for much longer times than those galaxies are detected in flux limited surveys, and therefore they should grow to very large linear sizes (typically $D(T) > 1\,\mathrm{Mpc}$), although detecting them would require extremely sensitive radio surveys with redshifts. Using the models of BRW we find that the visibility time, $\tau \propto Q_0^{1/2}$, and to properly estimate the actual number of RGs from those detectable in flux-limited surveys, one must multiply by a correction factor (T/τ) of roughly 50 for powerful RGs during the quasar era (GKW). We are currently investigating how this factor depends on the parameters defining the BRW and Manolakou & Kirk (2002) radio source evolution models.

The likelihood of 'missing' faded radio lobes resulting from past activity is underscored by the sophistication needed for detecting even the giant outer lobes of the nearby radio galaxy M 87 (Owen et al. 2000). 'Compact double' radio sources, several of which are found to have faint, diffuse radio lobes (e.g., Schoenmakers et al. 1999; Baum et al. 1990; Stanghellini et al. 1990), provide another example of this difficulty. Associating these faint structures with their true cores, or even with each other, would have been extremely difficult had these sources been imaged only a few hundred years ago when their present central radio components were not yet born. Another indication of the weakening of radio lobes towards higher redshifts is the apparent decrease in the radio-loud fraction of QSOs with redshift, as recently inferred by Impey & Petry (2001).

4. Radio lobes pervade the relevant universe

Recent numerical models of the evolution of ΛCDM universes indicate that at $z \sim 0$, roughly 70% of baryons are in a cosmic web of filaments of warm-hot gas and embedded galaxies and clusters that together occupy only about 10% of the volume of the universe (e.g., Cen & Ostriker 1999; Davé et al. 2001). But at $z \sim 2.5$ the growing network of filaments comprised only about 20% of the baryonic mass, and a quite small fraction, $\eta \simeq 0.03$, of the total volume.

It is certainly to be expected that the massive galaxies that harbor supermassive black holes large enough to form RGs at early times would have typically formed in the densest portions of those filaments, usually at their intersections. The radio lobes ejected from them would mostly remain within the filaments, and since it is in this relatively small, 'relevant universe', that new galaxies formed out of denser gas clumps, we really only need to be concerned with what fraction of this relevant universe the lobes permeated. We find that the mean volume of a radio source is $\langle V(T)\rangle \simeq 2.1 T_5^{18/7}\,\mathrm{Mpc}^3$, and thus, the volume fraction of the relevant universe which radio lobes born during the quasar era cumulatively swept through is (GKW)

$$\zeta = \Phi\ \langle V(5\times 10^8 \mathrm{yr})\rangle\ (0.03/\eta)(5/R_T)^2 \simeq 0.5, \qquad (3)$$

where $R_T \sim 5$ is the typical length to width ratio of an RG; we have conservatively ignored the volume filled by the lobes of the weaker FR I source population, which also evolves rapidly (Snellen & Best 2001). The energy density injected by the lobes into the filaments is $u \simeq 2\times 10^{-16}\,\mathrm{J\,m^{-3}}$ for those same canonical parameters. Because $\langle V(T)\rangle$ is a sensitive function of T, if the typical RG lifetime is $< 10^8$ yr, then $\zeta < 0.01$ and $u < 9\times 10^{-18}\,\mathrm{J\,m^{-3}}$ (GKW).

5. Extensive star formation at high$-z$ is triggered by radio lobes

The discovery of the alignment effect between extended optical emission lines and radio lobe directions (e.g., McCarthy et al. 1987; Chambers et al. 1987) quickly led to calculations (e.g., Begelman & Cioffi 1989; Rees 1989) that indicated that star formation could be triggered by these expanding overpressured lobes. This extended optical emission is produced to some degree by direct ionization by the AGN and perhaps by shock heating, but there is often a significant component best explained as due to radiation from young stars. Recent hydrodynamical simulations including cooling (Mellema et al. 2002) confirm that star formation is likely to occur through cloud fragmentation, cooling and compression. *HST* observations of high-z RGs and associated optical emission (e.g., Best et al. 1996; Bicknell et al. 2000) support this scenario.

To inquire if these lobes are capable of triggering extensive star formation on larger scales we (GKW) used the models of Falle (1991) and BRW to find that: $p_{\rm lobe} \propto t^{(-4-\beta)/(5-\beta)}$, but $D \propto t^{3/(5-\beta)}$, so $p_{\rm lobe} \propto D^{(-4-\beta)/3}$. The decline in the external pressure is slower, $p_{\rm ext} \propto D^{-\beta}$, so $p_{\rm lobe}/p_{\rm ext} \propto D^{(-4+2\beta)/3}$. For $\beta = 3/2$, $p_{\rm lobe}/p_{\rm ext} \propto D^{-1/3}$. The values of Q_0, ρ_0 and a_0 appropriate for FR II sources produce overpressures at $D = 50$ kpc amounting to factors of 10^2–10^4, corresponding to Mach numbers of 10–100 for the bow shock.

Thus, these powerful RGs create lobes which typically remain rapidly expanding, with overpressures of factors > 10 (and Mach numbers above 3) out to distances of well over 1 Mpc. Supersonic expansion into a two-phase circumgalactic medium will compress many of the cooler gas clumps, rapidly reducing the Jeans mass by factors of 10–100 and thereby triggering starbursts (Rees 1989; Mellema et al. 2002). The clouds will then remain in substantially overpressured lobes of low density, which can continue to produce extensive starbursts.

Related situations have been considered by Evrard (1991) for the infall of clouds into the hot ICM of clusters of galaxies, and by Jog & Das (1992; 1996) for the infall of a captured galaxy into the ISM of the captor galaxy. The latter authors examined what happens to molecular clouds in a captured galaxy as they enter into the central portions of a larger galaxy where higher pressure gas is awaiting them. Under many reasonable circumstances these clouds undergo radiative shock compression; when the growth time for gravitational instabilities in the shocked outer shell of the cloud becomes smaller than the shock's crossing time (as it usually will), the fragmenting shell should produce many stars in a rather efficient fashion (Jog & Das 1992; 1996).

It is thus fair to argue that many clumps of gas sitting in localized dark matter potential wells may yield extensive starbursts, or even entire galaxies, after being enveloped by an expanding radio lobe. Hence, we predict an enhanced 2-point correlation function (and an even more enhanced 3-point correlation function) between radio galaxies and newly formed galaxies during the quasar era. Unfortunately, these correlations will be extremely difficult to measure, since the relevant radio sources will have typically faded below detectability even while their expanding lobes continue to have a major impact on the surrounding medium at substantial distances from the AGN.

6. Radio lobes magnetize the IGM

One exciting implication of this scenario is that RGs can inject a substantial amount of magnetic energy into the IGM at $z \sim 2-3$. Faraday rotation measurements of quasars provide a nominal upper bound on intergalactic magnetic fields of $B_{IGM} < 10^{-9}$G (e.g., Kronberg et al. 1999). However, if the magnetic field is preferentially distributed in the cosmic filaments where the relevant IGM is also concentrated, then fields within those filaments ranging up to $\sim 10^{-6}$G are allowed by these observations (Ryu et al. 1998). Kronberg et al. (1999) have argued that a substantial fraction of the IGM may have been permeated by magnetized outflows from stars in galaxies. The tentative detection of magnetic fields in high-z galaxies (e.g., Oren & Wolfe 1995) poses considerable difficulties for standard dynamo models, since amplification in galactic disks requires many dynamical times (e.g., Furlanetto & Loeb 2001).

The possibility of jets in radio galaxies magnetizing the IGM has been examined previously, but those earlier investigations concluded that either only minute magnetization levels or insignificant volume coverage would be attained (e.g., Daly & Loeb 1990; Rees 1994). In GKW we showed that during the quasar era, the permeation of the IGM by the expanded lobes of radio galaxies could have seeded the IGM with an average magnetic fields of $\approx 10^{-8}$G (recall that fields of $\approx 1\mu$G exist even inside the lobes of megaparsec RGs, e.g., Kronberg et al. 2001; Ishwara-Chandra & Saikia 1999), which matches the IGM field strengths inferred by Ryu et al. (1998) and by Furlanetto & Loeb (2001). The latter authors, advancing essentially orthogonal arguments to ours (GKW), based on the evolution of isotropized magnetized bubbles fed by quasars, have argued that the quasar population is capable of polluting $\sim 10\%$ of the entire space with magnetic fields. From independent arguments, Kronberg et al. (2001) have concluded that the accretion energy released by radio-loud QSOs at $z \sim 2-3$ is adequate to magnetize the IGM to the level of its thermal energy, provided the radio lobes can expand to fill up the IGM.

7. Enriching new galaxies with metals

Several authors have discussed the issue of 'metal transport' from their production sites, namely the ISM of galaxies, to the Mpc-scale IGM with a mean density $< 10^{-4}$ as high (e.g., Gnedin 1998; Shchekinov 2002), and even farther into the voids (Theuns et al. 2002). Lyman-break galaxies at $z > 3$ often have metalicities around 0.1 solar and damped Lyman-α clouds have metalicities $\sim 10^{-2.5}$ solar (e.g., Steidel et al. 1999; Pettini et al. 2001). The difficulty is underscored further with the recent detection of O VI absorption at $z \sim 2-3$ from under-dense regions ($\rho/\langle\rho\rangle < 1$) representing the true IGM (Schaye et al. 2000). Supernova explosions in star-forming galaxies are found to fail by at least an order-of-magnitude to pollute the whole IGM to the metalicity levels observed within the available time (e.g., Gnedin & Ostriker 1997; Ferrara et al. 2000). A possible alternative could be mergers of protogalaxies in the process of hierarchical clustering (Gnedin & Ostriker 1997; Gnedin 1998).

In accordance with our picture of a radio lobe-filled universe at high-z, we propose here a potentially attractive new mechanism for large-scale metal transport: namely, the sweeping of the ISM of star-forming galaxies by the expanding giant radio lobes during the quasar era, or even earlier. The outflowing radio jets will drag along a significant fraction of the gas in their host galaxy out with them, most of it compressed into a shell along the bow shock outside the lobes, as illustrated by numerical simulations of jets leaving a galaxy's ISM (e.g., Hooda & Wiita 1998). This enriched gas can then be spread over distances exceeding 1 Mpc over the course of $\sim 10^8$ years. Eventually, substantial portions of this gas can be mixed into the radio lobe, but these Rayleigh-Taylor and Kelvin-Helmholtz types of instabilities grow relatively slowly under these conditions, and so it seems that most of the enriched gas will comprise part of a shell that will also include swept up ICM as the lobe continues to expand. Note that the morphology of the line-emitting gas surrounding the radio bubbles of M 87 suggests that much of this material has been excluded from the cocoon (Bicknell & Begelman 1996). When this expanding gaseous shell interacts with denser clouds in the ICM or IGM, not only will extensive star formation be triggered, but this star-forming region will have incorporated a fraction of this swept-up enriched gas. In that much of it is likely to have remained in the bow-shock region, the dilution will not be as severe as it would have been if the enriched gas were spread throughout the immense volume of the radio lobes. An advantage of this mechanism is that heating of the denser gas is less of a problem than when the metals are conveyed by supernova driven winds.

In addition to the radio lobes contributing to "metalization" by dragging along some of the enriched gas from their host galaxy, it is worth considering their impact on other young galaxies which they may envelop. If these young galaxies have a multi-phase medium and are not too dissimilar from local galaxies in this regard, then ram pressure stripping (Gunn & Gott 1972) produced by these lobes can be important. Even if the average density of the ICM has fallen as low as $10^{-5}\mathrm{cm}^{-3}$ at $r = 1\,\mathrm{Mpc}$, as expected for $\beta = 1.5$, the density of the shell will typically be compressed to several times this value. With the expansion velocity of the lobes remaining roughly 10^4 km s^{-1}, the ram pressure, $P_{\mathrm{ram}} \simeq \rho v^2 > 3 \times 10^{-11}$ dyn cm^{-2}. This pressure is adequate to remove most of the gas from a typical spiral galaxy (with $v_{\mathrm{rot}} \simeq 200$ km s^{-1} and $R \simeq 10$ kpc; see Abadi et al. 1999; Fujita 2001), and is likely to be even more effective in stripping the diffuse gas from smaller, recently forming galaxies at $z > 2$, despite our lack of knowledge of the details of the properties of that ISM. However, colder, denser clumps of gas within those galaxies will probably not be pushed out (cf. Mellema et al. 2002); rather, the arriving lobe may trigger yet more star formation in those regions.

Finally, we note that individual AGN may go through several generations of nuclear activity that yield radio jets and lobes. The first such episode could trigger extensive star formation, or even new galaxy formation, in relatively nearby clouds. Any subsequent lobes hitting that newly formed galaxy a few hundred Myr later could sweep out most of the enriched gas it had already produced, thereby propagating these newer metals into yet more distant regions, in the fashion discussed in the last paragraph. These metals could, in turn, contribute to the seeding of additional clouds which are triggered to collapse by this second (or subsequent) period of activity.

8. Conclusions

Even though the local universe is very sparsely populated by powerful radio sources, several large factors conspire to make them remarkably important for galaxy formation during the quasar era. First, their comoving density was roughly 1000 times higher at $\sim 2 < z <\sim 3$. Second, only a small fraction (roughly two percent) of the powerful sources present during that period are detected in the surveys used to produce the RLFs, because of severe inverse Compton and adiabatic losses; these unseen, old radio lobes fill very large volumes. Third, the fraction of the volume of the universe occupied by the material during the quasar epoch that would finally condense into clusters of galaxies was only a few percent, so these lobes only had to permeate this 'relevant universe' rather than the entire universe. In that the best models of RG evolution indicate that the lobes are overpressured and supersonically expanding into the relevant universe, the scenario that has many massive starbursts, and even many galaxies, formed in this fashion at high redshifts, is quite plausible.

Another key result is that radio galaxies were likely to have been capable of seeding the IGM with magnetic fields of the appropriate strength. It is very encouraging that two other independent and quite different arguments lead to similar conclusions (Furlanetto & Loeb 2001; Kronberg et al. 2001). Also, the sweeping up of the ISM of the galaxies and star-forming clouds by the expanding lobes of RGs suggests a natural way to spread metals produced in the first stellar generations over large volumes. Detailed investigations of many aspects of this scenario are clearly warranted.

GK thanks Prof. Marshall Cohen for the invitation to this memorable meeting held in Ken's honor, and NRAO for partial travel support. PJW is grateful for support from RPE funds at GSU and for continuing hospitality at the Department of Astrophysical Sciences at Princeton.

References

Abadi, M. G., Moore, B., & Bower, R. G. 1999, MNRAS, 308, 947

Archibald, E. N., et al. 2001, MNRAS, 323, 417

Baum, S. A., O'Dea, C. P., Murphy, D. W., et al. 1990, A&A, 232, 19

Begelman, M. C., & Cioffi, D. F. 1989, ApJ, 345, L21

Best, P. N., Longair, M. S., & Röttgering, H. J. A. 1996, MNRAS, 280, L9

Bicknell, G. V., & Begelman, M. C. 1996, ApJ, 467, 597

Bicknell, G. V., et al. 2000, ApJ, 540, 678

Blundell, K. M., Rawlings, S., & Willott, C. J. 1999, AJ, 117, 677 (BRW)

Cen, R., & Ostriker, J. P. 1999, ApJ, 514, 1

Chambers, K. C., Miley, G. K., & van Breugel, W. 1987, Nature, 329, 604

Chary, R., & Elbaz, D. 2001, ApJ, 556, 562

Daly, R. A., & Loeb, A. 1990, ApJ, 364, 451

Davé, R., et al. 2001, ApJ, 552, 473

Evrard, A. E. 1991, MNRAS, 248, 8P

Falle, S. A. E. G. 1991, MNRAS, 250, 581

Ferrara, A., Pettini, M., & Shchekinov, Yu. 2000, MNRAS, 319, 539

Fugita, Y. 2001, ApJ, 550, 612

Furlanetto, S. R. & Loeb, A. 2001, ApJ, 556, 619

Gnedin, N. Y. 1998, MNRAS, 294, 407
Gnedin, N. Y., & Ostriker, J. P. 1997, ApJ, 486, 581
Gopal-Krishna & Wiita, P. J. 2001, ApJ, 560, L115 (GKW)
Gopal-Krishna, Wiita, P. J., & Osterman, M. A. 2003, in ASP Conf. Ser., Active Galactic Nuclei: from Central Engine to Host Galaxy, ed. S. Collin, F. Combes, & I. Shlosman (San Francisco: ASP), in press
Gunn, J. E., & Gott, J. R. 1972, ApJ, 176, 1
Hooda, J. S., & Wiita, P. J. 1998, ApJ, 493, 81
Impey, C., & Petry, C. 2001, ApJ, 547, 117
Ishwara-Chandra, C. H., & Saikia, D. J. 1999, MNRAS, 309, 100
Jog, C. J., & Das, M. 1992, ApJ, 400, 476
Jog, C. J., & Das, M. 1996, ApJ, 473, 797
Kaiser, C. R., Dennett-Thorpe, J., & Alexander, P. 1997, MNRAS, 292, 723
Kronberg, P. P., Lesch, H., & Hopp, U. 1999, ApJ, 511, 56
Kronberg, P. P., et al. 2001, ApJ, 560, 178
Manolakou, K., & Kirk, J. G. 2002, A&A, 391, 127
McCarthy, P. J., et al. 1987, ApJ, 321, L29
Mellema, G., Kurk, J. D., & Röttgering, H. J. A. 2002, A&A, 395, 13
Oren, A. L., & Wolfe, A. M. 1995, ApJ, 445, 624
Owen, F. N., Eilek, J. A., & Kassim, N. E. 2000, ApJ, 543, 611
Pettini, M., et al. 2001, ApJ, 554, 981
Rees, M. J. 1989, MNRAS, 239, 1P
Rees, M. J. 1994, in Cosmical Magnetism, ed. D. Lynden-Bell (Dordrecht: Kluwer), 155
Ryu, D., Kang, H., & Biermann, P. L. 1998, A&A, 335, 19
Schaye, J., et al. 2000, ApJ, 541, L1
Shchekinov, Yu. A. 2002, Astron. Astrophys. Transact., in press (astro-ph/0205320)
Snellen, I. A. G., & Best, P. N. 2001, MNRAS, 328, 897
Stanghellini, C., Baum, S. A., O'Dea, C. P., et al. 1990, A&A, 233, 379
Steidel, C. C., et al. 1999, ApJ, 519, 1
Schoenmakers, A. P., et al. 1999, A&A, 341, 44
Theuns, T., et al., 2002, ApJ, 578, L5
Willott, C. J., et al. 2001, MNRAS, 322, 536
Wang, Y. P., Yamada, T., & Taniguchi, Y. 2003, ApJ, in press (astro-ph/0212534)

Radio Astronomy at the Fringe
ASP Conference Series, Vol. 300, 2003
J.A. Zensus, M.H. Cohen, & E. Ros, eds.

Astronomy at the Fringe

R. D. Ekers
CSIRO, Australia Telescope National Facility, Sydney, NSW 2121, Australia

Abstract. Ken Kellermann's research areas involving the VLBI observations and physics of compact radio sources have been covered by other speakers as the central theme of this workshop. This paper touches on some of the other topics that have not been included—the history of discoveries in radio astronomy, planetary radio astronomy, cosmology and SETI.

1. Introduction

Ginzburg said 'The optimum strategy for a scientist is to know something about everything and everything about something'. Ken manages to do both very well! Astronomy at the Fringe, the title of my talk, picks up some of Ken's activities on the fringe of the VLBI based astronomy theme for the workshop. VLBI is the 'know everything about something' bit. These other activities illustrate his knowing something about everything. By the way, I didn't copy my title from the conference name, the organisers copied my talk title!

2. John Bolton's Influence

Ken did his Ph.D. at Caltech where John Bolton had been brought from Australia to establish a radio astronomy group. Having John Bolton as a supervisor leaves its impact. He provided an understanding of what research is really about and a healthy respect for the value of understanding your instrument. John instilled this perspective right from the beginning. Ken could make solder joints that held (he was a radio ham) and he pulled the cables at OVRO. I graded the track for the Parkes interferometer. Looking round the room I see many of John Bolton's students: Marcus Price, Barry Clark, Glenn Berge, Dan Harris, George Seilstad, Jasper Wall and also Radhakrishnan who got the same treatment. They will all have their own stories.

3. Serendipity in Radio Astronomy

For the 50th anniversary of Karl Jansky's discovery of cosmic radio noise Ken and I planned a workshop on the major discoveries in radio astronomy. The workshop was sponsored by NRAO and held at Green Bank in May 1983. Ken was the driving force behind the meeting and editor of the Proceedings (Kellermann & Sheets 1983). The following extract from Ken's Preface summarises the outcomes:

At least in the past, radio astronomy has been a technique-oriented science. Starting with Jansky, the major discoveries have been made primarily by skilled scientists who, because they completely understood their equipment, were able to spot and correctly interpret their unexpected results. Although, as many of the speakers at this workshop related, the path toward the correct answer was often circuitous, and there was a strong consensus that major discoveries require the 'right person, in the right place, doing the right thing, at the right time.' Several speakers also noted that it sometimes helped 'not to know too much'.

3.1. Serendipity has dominated discoveries in radio astronomy

Radio telescopes have produced a stream of discoveries including an unusually high fraction which are serendipitous. Here is my current list of major discoveries. The majority are serendipitous (shown in italics).

- *Radio galaxies*
- *Quasars*
- 21 cm H I line
- *Cosmological evolution required to explain steep slope of radio source counts*
- 2/3 spin orbit locking of Mercury
- *3K cosmic microwave background*
- *Super-luminal motions of the jets in radio galaxies*
- *First evidence for the need for dark matter in spiral galaxies from the 21 cm H I rotation curves*
- *Masers and megamasers*
- Mass of the black hole in NGC 4258 using H_2O megamaser emission
- *Pulsars*
- *Gravitational lenses*
- Gravitational relativistic time delay measured by planetary radar
- Gravitational radiation (pulsar timing)
- *First extra-solar planetary system*

Kellermann (1994) includes an excellent discussion of the events surrounding the identification of the first quasars.

4. Radio Observations of the Planets

4.1. Mercury

John Bolton made sure the engineers and technicians always had unfettered access to the Parkes telescope for development and maintenance. This meant that day times were not available for astronomers. However, Ken, an addicted observer, found a way to get round this day time observing restriction—he proposed observations of the planet Mercury which of course could only be made during the day.

The Italian observer Schiaparelli had announced in 1890 that Mercury spins on its axis once every 88 days, keeping one side of the planet synchronously locked towards the Sun, and this had been confirmed repeatedly by many eminent planetary observers e.g., Dollfus (1953) concluded an analysis of his visual observations of the planet with the statement that Mercury's spin period equaled its orbital period 'with a precision greater than one part in ten thousand.' Ken observed Mercury to search for the change in emission with the planet's phase as our view changed from the day to the night side of Mercury. He found no such phase effect and published this surprising result in Nature (Kellermann 1965a). To solve this dilemma Ken invoked the atmosphere proposed by Field (1962) to convect heat from the day to the night side. The alternative explanation of a period not synchronised to the 88 day orbit was not considered. However, this is exactly what was found shortly afterwards by the radar determination of a 59 day rotation period (Pettengill & Dyce 1965). Had Ken questioned the rotation period, this discovery could have gone to him (Welch, private communication). Ken has pointed out that if he had read the original papers (in French!) on Mercury's rotation (Dollfus 1953) he would have realized how weak the evidence was. The moral is 'don't believe what people say', a variation on 'knowing too much'.

4.2. Other Planets

Ken went on to observe the radio emission between 6 and 21 cm from Venus, Mars, Saturn, and Uranus with the CSIRO 210-ft radio telescope at Parkes (Kellermann 1965b, 1966), making the first detection of Uranus and showing they were all consistent with thermal emission despite some claims to the contrary. Ken later teamed up with Ivan Pauliny-Toth and extended the planetary observations to 15 GHz using the Green Bank 140-foot. They bagged another detection, Neptune, and set a limit on emission from Pluto (Kellermann & Pauliny-Toth 1966). The observing technique employed so successfully for these observations used the movement of the planet to separate out signals which were well below the confusion limit of the telescope used.

While Ken's career as an observational planetary astronomer seems to have ended here, he has indirectly influenced NASA's future planetary exploration programs by taking on a Ph.D. student, Barry Geldzahler, who is now NASA's program executive for space operations.

5. Kellermann: Warner Prize 1971

At the time of the 1971 Warner Prize Lecture (Kellermann 1972) radio astronomy was the darling of both the relativistic astrophysicists (AGNs and black holes) and the cosmologists because the radio sources were easily seen at any distance and the energy requirements prodigious. This lead to considerable over interpretation which Ken wanted to put in perspective. Following the 'best not to know too much' dictum he questioned the evidence for evolution of the radio source population which had resulted in the demise of the Steady State Theory. He pointed out that the large excess of faint sources could really be just a small (2σ) deficiency of bright sources. Because of the cosmic variance we will never know if that 2σ deficiency of strong sources in our region of the Universe is significant. He also pointed out that the relation between angular size and redshift was easier to explain without the cosmological factors included and he questioned the interpretation of Doppler boosting. His conclusion (Table 1) that four paradoxes were removed if quasar redshifts weren't cosmological must have brought a smile to the faces of the Burbidges and Arps at the time.

Table 1. Paradoxes: As presented in Kellermann's 1971 Helen B Warner Lecture

Either	Or
The density of radio sources depends on red shift in just such a way to cancel the geometrical effects and the effect of the red shift.	The effect of the red shift on the observed fluxes for most of the observed sources is negligibly small.
The intrinsic spectral index distribution depends on redshift in just such a way to cancel the effect of the redshift.	The effect of the redshift on the spectra for most of the observed sources is negligibly small.
The linear dimensions of radio sources depend on redshift in such a way to cancel the geometrical effects of the red shift.	The geometric effect of the red shift on apparent size is negligibly small
The intrinsic intensity of the two components of 3C 279 differ by just the right amount to cancel the effect of the Doppler shifts.	The distance of 3C 279 is much closer than indicated by the red shift and the motion is actually nonrelativistic.

6. Super Light Expansion

A number of papers at this meeting discuss superluminal motion. This is a phenomenon now almost universally accepted to result from relativistic beaming of the AGN jets. Despite the attraction of this model, and its incorporation into the unified AGN paradigm, I still feel uneasy about it. Even though this Doppler boosting model makes well determined predictions, few of the *predicted* consequences have been observed. Instead, the model has grown increasingly complex with additional degrees of freedom (obscuring torus, distribution of β , mixture of bulk motion and moving shocks ...) and these parameters are adjusted to *explain* the ever improving observations.

Although Ken's original provocative query about decreasing the distance to decrease β (see Section 5)is no longer a serious option, Ken has remained one of the few who question the underlying theory (e.g., Kellermann 1994). I was therefore delighted when Ken sent me a note suggesting I look at how the new observations had changed the distribution of $\beta = v_{\rm app}/c$. Ekers & Liang (1990) had proposed a simple variant on the 'light echo' model in which trigger pulses move out in opposite directions at the speed of light but there is no bulk motion of the radiating particles and no Doppler boosting. This model didn't do too bad a job of fitting the general shape of the distribution of β as it was known in 1988 but it had no free parameters and so required an implausibly high $H_\circ$ (120 km s^{-1} Mpc^{-1}) and some orientation bias from a dust torus to fit the peak in $v/c \approx 3$ (Fig. 1).

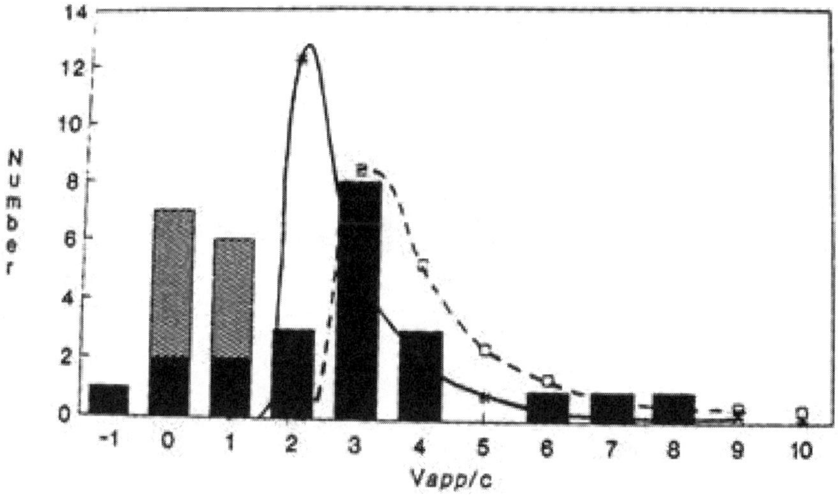

Figure 1. From Ekers & Liang 1990. v_{app}/c data plotted for $H_\circ = 120$. Dashed curve is the effect of a 45° dust torus.

Figure 2 shows the β distribution as presented by Zensus (this workshop). It now peaks at much lower β and the Ekers & Liang model fits with $H_\circ = 65$ and no dust torus. The message to take from this is not whether the simple light echo model is correct but that the observations of superluminal motion do not *require* Doppler boosting as an essential element of the explanation.

7. SETI

7.1. PKS 1934–63 spectrum

While at Parkes in the mid 60s Ken made very detailed observations of the spectrum of the remarkable GHz peaked radio source PKS 1934–63. In addition to his well known analysis of the physical processes which could cause the low frequency turnover in the spectrum, Ken also noted that the spectrum of PKS 1934–63 was remarkably similar to that predicted by Kardashev (1964) for signals transmitted by extraterrestrial 'super civilisations'. To follow up on this Ken, together with Frank Gardener, used the 48 channel HI receiver at Parkes to look for any unusual characteristics near the 21 cm line (Kellermann 1965c). This was the first publication of a SETI observation in a refereed journal! Ken modestly concluded that it was not possible to draw any conclusion about the artificial nature of the source on the basis of this observation. This was Ken's only venture into spectral line astronomy.

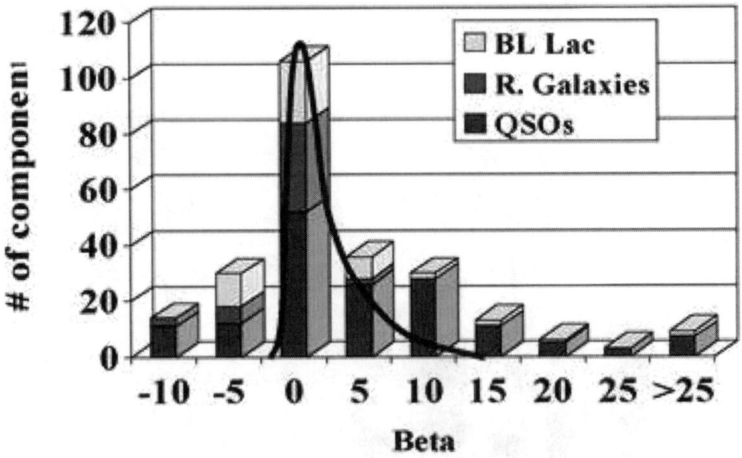

Figure 2. All components with measured speeds using $H_\circ = 65\,\mathrm{km\,s^{-1}\,Mpc^{-1}}$ and $\Omega_\mathrm{m} = 0.3$; data from Zensus et al. (these proceedings, page 27), with light echo model superposed.

7.2. Nature of SETI signals

One of the signs of an extraterrestrial intelligence would be the display of a signal which could not be generated by natural processes. What better than to generate an effect which appeared to violate the speed of light constraint! Ken didn't find the 21 cm spectral notch but that notch would be a bit tricky for a super civilisation that didn't know our redshift —but using $v > c$ should do the trick? Ken's speculations on better ways to explain the superlight expansion phenomenon included possible SETI signals

Ken's interest in SETI has continued throughout his career. He has participated in a string of studies of the strategy and technology needed to detect evidence of extraterrestrial intelligence (Morrison, Billingham, & Wolfe 1977; Drake, Wolfe, & Seeger 1983; Ekers et al. 2002). He also argued for inclusion of SETI in the US decadal plan.

8. Future Facilities

Encouraged by John Bolton's *dictum* that to do better astronomy you need to build better telescopes, Ken has always been enthusiastically involved in the development of new facilities. He has participated in the US decadal reviews for 1990 and 2000, chairing the radio astronomy panel in 1990.

8.1. VLBA

Ken was the champion for the VLBA during the era when NRAO was pushing for the big millimetre dish. He had a vision of VLBI research conducted like research at any other observational facility rather than an arcane area of science only accessible to highly specialised practitioners. He also promoted the vision of a facility transcending national barriers and truly becoming International Science.

8.2. SKA

The SKA project (see `http://www.ska.org` and Taylor & Braun 1998) embodies many of Ken's ideals and he is one of the leading advocates. It provides the tools to explore sources fainter than anything seen before, and to do so with exquisite angular resolution. It pushes the technology barriers and provides innovative new capabilities and the potential for discovery of unexpected phenomena. As Ken said in Kellermann (1994) 'real research starts with building a better machine than the next guy'.

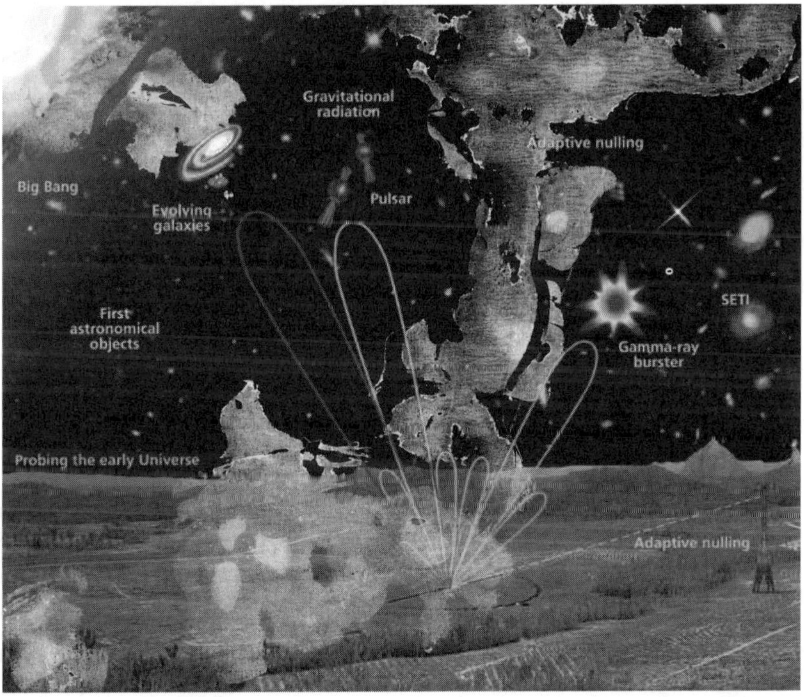

Figure 3. SKA Multi-beaming to simultaneously observe disparate phenomena while nulling unwanted signals.

References

Dollfus, A. 1953, Bull. Soc. Astron. France, 67, 61

Drake, F., Wolfe, J. H., & Seeger, C. L. (ed.) 1983, SETI Science Working Group Report, (Washington, DC: NASA)

Ekers, R. D., Cullers, D. K., Billingham, J., & Scheffer, L. K. (ed.) 2002, SETI 2020: A Roadmap for the Search for Extraterrestrial Intelligence, (Mountain View, CA: SETI Press).

Ekers, R. D., & Liang, H. 1990, in Proc. Workshop on Parsec-Scale Radio Jets, Socorro, NM, October 1989, (Cambridge: CUP), 333

Field, G. B. 1962, AJ, 67, 575

Kardashev, N. S. 1964, Soviet Ast., 8, 217-221.

Kellermann, K. I. 1965a, Nature, 205, 1091

Kellermann, K. I. 1965b, Nature, 206, 1034

Kellermann, K. I. 1965c, Aust. J. of Phys., 19, 195

Kellermann, K. I. 1966, Icarus, 5, 478

Kellermann, K. I. 1972, AJ, 77, 531

Kellermann, K. I. 1994. Aust. J. of Phys., 47, 599

Kellermann, K. I., & Pauliny-Toth, I. I. K. 1966, ApJ, 145, 954

Kellermann, K. I., & Sheets, B. (ed.) 1983, Workshop May 4-6, 1983, (Green Bank, WV: NRAO)

Morrison, P., Billingham, J., & Wolfe, J. (ed.) 1977, The Search for Extraterrestrial Intelligence, (Washington, DC: NASA)

Pettengill, G. H., & Dyce, B. R. 1965, Nature, 206, 1240

Taylor, A. R., & Braun, R. 1998, `http://www.skatelescope.org/ska_science.shtml`

Radio Astronomy at the Fringe
ASP Conference Series, Vol. 300, 2003
J.A. Zensus, M.H. Cohen, & E. Ros, eds.

Bibliography

This bibliography combines all references from these Proceedings. The status of "in press" or "submitted" indicates the publication status confirmed as of March 2003. The listing was checked for internal consistency, but not all items were verified against the original publication (especially those cited only once).

References

Aaron, S. E. 1996. "VLBA Observations of Highly Misaligned Radio Jets", Ph.D. thesis, (Waltham, MA: Brandeis Univ.).

Abadi, M. G., Moore, B., & Bower, R. G. 1999. "Ram pressure stripping of spiral galaxies in clusters", MNRAS, 308, 947–954.

Ablyazov, V. S., Alekseev, V. A., Antonets, M. A., Ariskin, V. I., Vekshin, V. P., Gatelyuk, E. D., Demin, V. V., Efanov, V. A., Kutuza, B. G., Lipatov, B. N., Matveenko, L. I., Mkrtchan, S. M., Moiseev, I. G., Nikonov, V. N., Oganesjan, V. A., Sanamjan, V. A., Sizov, A. S., Sorochenko, R. L., Sosnin, V. P., Troitskiy, V. S., Fateev, B. P., Chikin, A. I., & Schekotov, B. V. 1974. "Observations of H_2O sources on Pushino-Simeiz interferometer", Radiofizika, 17, 10, 1431–1437.

Adgie, R. L., Gent, H., Slee, O. B., Frost, A. D., Palmer, H. P., & Rowson, B. 1965. "New limits to the angular sizes of some quasars", Nature, 208, 275–276.

Agol, E., & Kamionkowski, M. 2002. "X-rays from isolated black holes in the Milky Way", MNRAS, 334, 553–562.

Aitken, D. K., Greaves, J., Chrysostmou, A. Jennes, T., Holland, W., Hough, J. H., Pierce-Price, D., & Richer, J. 2000. "Detection of Polarized Millimeter and Submillimeter Emission from Sagittarius A*", ApJ, 534, L173–L176.

Allen, L. R., Anderson, B., Conway, R. G., Palmer, H. P., Reddish, V. C., & Rowson, B. 1962. "Observations of 384 radiosources at a frequency of 158 Mc/s with a long baseline interferometer", MNRAS, 124, 477–499.

Aller, H. D., Aller, M. F., & Hughes, P. A. 1985. "Polarized radio outbursts in BL Lacertae. I - Polarized emission from a compact jet. II - The flux and polarization of a piston-driven shock", ApJ, 298, 296–315.

Aller, H. D., Aller, M. F., Latimer, G. E., & Hodge, P. E. 1985. "Spectra and linear polarizations of extragalactic variable sources at centimeter wavelengths", ApJS, 59, 513–768.

Aller, H. D., Aller, M. F., & Plotkin, R. 2003. "Circular Polarization Variability in Extragalactic Sources on Time Scales of Months to Decades", in: Circular Polarization of Relativistic Jet Sources, ed. R. P. Fender & J.-P. Macquart, Ap&SS, in press.

Aller, M. F., Aller, H. D., & Hughes, P. A. 2003. "Pearson-Readhead survey sources. II: The longterm centimeter-band total flux and linear polarization properties of a complete radio sample", ApJ, 586, 33-51.

Aller, M. F., Aller, H. D., Hughes, P. A., & Latimer, G. E. 1999. "Cm-wavelength total flux density and linear polarization properties of radio-loud BL Lacertae objects", ApJ, 512, 601–622.

Anderson, B., Donaldson, W., Palmer, H. P., & Rowson, B. 1965. "Observations of the quasi-stellar and other radio sources with an interferometer of resolving power 0.4 sec of arc", Nature, 205, 375–376.

Antonucci, R. 1993. "Unified models for active galactic nuclei and quasars", ARA&A, 31, 473–521.

Arav, N., & Li, Z.-Y. 1994. "The role of radiative acceleration in outflows from broad absorption line QSOs. 1: Comparison with O star winds", ApJ, 427, 700–707.

Archibald, E. N., Dunlop, J. S., Hughes, D. H., Rawlings, S., Eales, S. A., & Ivison, R. J. 2001. "A submillimetre survey of the star formation history of radio galaxies", MNRAS, 323, 417–444.

Arp, H. 1987. "Quasars, Redshifts and Controversies", (Berkeley: Interstellar Media).

Arshakian, T. G., & Longair, M. S. 2000. "An asymmetric relativistic model for classical double radio sources", MNRAS, 311, 846–860.

Asada, K., Inoue, M., Uchida, U., Kameno, S., Fujisawa, K., Iguchi, S., & Mutoh, M. 2002. "A helical magnetic fiels in the jet of 3C 273", PASJ, 54, L39–L43.

Attridge, J. M., Roberts, D. H., & Wardle, J. F. C. 1999. "Radio Jet-Ambient Medium Interactions on Parsec Scales in the Blazar 1055+018", ApJ, 518, L87–L90.

Backer, D. C., & Sramek, R. A. 1982. "Apparent proper motions of the galactic center compact radio source and PSR 1929+10", ApJ, 260, 512–519.

———. 1999. "Proper Motion of the Compact, Nonthermal Radio Source in the Galactic Center, Sagittarius A*", ApJ, 524, 805–815.

Baganoff, F. K., Bautz, M. W., Brandt, W. N., Chartas, G., Feigelson, E. D., Garmire, G. P., Maeda, Y., Morris, M., Ricker, G. R., Townsley, L. K., & Walter, F. 2001. "Rapid X-ray flaring from the direction of the supermassive black hole at the Galactic Centre", Nature, 412, 45–48.

Bailyn, C. D., Jain, R. K., Coppi, P., & Orosz, J. A. 1998. "The Mass Distribution of Stellar Black Holes", ApJ, 499, 367–374.

Baldwin, J. 1982. "Evolutionary tracks of extended radio sources", in IAU Symp. 97, Extragalactic Radio Sources, ed. D. S. Heeschen & C. M. Wade (Dordrecht: Reidel), 21–24.

Baleisis, A., Lahav, O., Loan, A. J., & Wall, J. V. 1998. "Searching for large-scale structure in deep radio surveys", MNRAS, 297, 545–558.

Balick, B., & Brown, R. L. 1974. "Intense sub-arcsecond structure in the galactic center", ApJ, 194, 265–270.

Ballantyne, D. R., Ross, R. R., & Fabian, A. C. 2002. "On the hard X-ray spectra of radio-loud active galaxies", MNRAS, 332, L45–L49.

Baltz, E. A., Gnedin, N. Y., & Silk, J. 1998. "Spectral Features from the Reionization Epoch", ApJ, 493, L1–L4.

Banerjee, S. K., & Narlikar, J. V. 1999. "The quasi-steady-state cosmology: a study of angular size against redshift", MNRAS, 307, 73–78.

Bare, C. C., Clark, B. G., Kellermann, K. I., Cohen, M. H., & Jauncey, D. L. 1967. "Interferometer experiment with independent local oscilator", Science, 157, 14, 189–191.

Barger, A. J., Cowie, L. L., Capak, P., Alexander, D. M., Bauer, F. E., Brandt, W. N., Garmire, G. P., & Hornschemeier, A. E. 2003. "Very High Redshift X-ray Selected AGN in the *CHANDRA* Deep Field-North", ApJ, 584, L61–L64.

Barkana, R., & Loeb, A. 2003. "Spectral Signature of Cosmological Infall of Gas Around the First Quasars", Nature, 421, 341–343.

Barthel, P. D. & Miley, G. K. 1988. "Evolution of radio structure in quasars : a new probe of protogalaxies ?", Nature, 333, 310–325.

Batchelor, R., Jauncey, D. L., Johnston, K. J., Knowles, S. H., Papatsenko, A. Kh., Preston, R., Spenser, J., Timofeev, A. N., Fourikis, N., & Schilizzi, R. W. 1976. "The First Global Radio Telescope", Soviet Ast. Lett., 2, 10, 467–473.

Baum, S. A., O'Dea, C. P., Murphy, D. W., & de Bruyn, A. G. 1990. "0108+388: a compact double source with surprising properties", A&A, 232, 19–26.

Becker, R. H., Fan, X., White, R. L., Strauss, M. A., Narayanan, V. K., Lupton, R. H., Gunn, J. E., Annis, J., Bahcall, N. A., Brinkmann, J., Connolly, A. J., Csabai, I., Czarapata, P. C., Doi, M., Heckman, T. M., Hennessy, G. S., Ivezic, Z., Knapp, G. R., Lamb, D. Q., McKay, T. A., Munn, J. A., Nash, T., Nichol, R., Pier, J. R., Richards, G. T., Schneider, D. P., Stoughton, C., Szalay, A. S., Thakar, A. R., & York, D. G. 2001. "Evidence for Reionization at $z \sim 6$: Detection of a Gunn-Peterson Trough in a $z = 6.28$ Quasar", AJ, 122, 2850–2857.

Becker, R. H., White, R. L., & Helfand, D. J. 1995. "The FIRST Survey: Faint Images of the Radio Sky at Twenty Centimeters", ApJ, 450, 559–577.

Beckert, T., & Falcke, H. 2002. "Circular polarization of radio emission from relativistic jets", A&A, 388, 1106–1119.

Begelman, M. C. 1996. "Baby Cygnus A's", in Cygnus A: Study of a Radio Galaxy, ed. C. Carilli & D. Harris, (Cambridge: CUP), 209–214.

Begelman, M. C., & Cioffi, D. F. 1989. "Overpressured cocoons in extragalactic radio sources", ApJ, 345, L21–L24.

Begelman, M. C., McKee, C. F., & Shields, G. A. 1983. "Compton heated winds and coronae above accretion disks. I Dynamics", ApJ, 271, 70–88.

Begelman, M. C., Rees, M. J., & Sikora, M. 1994. "Energetic and radiative constraints on highly relativistic jets", ApJ, 429, L57–L60.

Belczynski, K., & Bulik, T. 2002. "Formation and Evolution of Microquasar GRS 1915+105", ApJ, 574, L147–L150.

Benn, C. R., & Wall, J. V. 1995. "Structure on the largest scales: constraints from the isotropy of radio source counts", MNRAS, 272, 678–698.

Bennett, A. S. 1962. "The revised 3C catalogue of radio sources", MmRAS, 68, 163–172.

Benson, A. J., Nusser, A., Sugiyama, N., & Lacey, C. G. 2001. "Non-uniform reionization by galaxies and its effect on the cosmic microwave background", MNRAS, 320, 153–176.

Beskin, V. S. 1997. "Axisymmetric stationary flows in compact astrophysical objects", Physics–Uspekhi, 40, 659–688.

Best, P. N., Longair, M. S., & Röttgering, H. J. A. 1996. "Evolution of the aligned structures in $z \sim 1$ radio galaxies", MNRAS, 280, L9–L12.

Bicknell, G. V., & Begelman, M. C. 1996. "Understanding the kiloparsec-scale structure of M 87", ApJ, 467, 597–621.

Bicknell, G., Dopita, M. A., & O'Dea, C. P. 1997. "Unification of Radio and Optical Properties of Gigahertz Peaked Spectrum and Compact Steep-Spectrum Radio Sources", ApJ, 485, 112–124.

Bicknell, G. V., Sutherland, R. S., van Breugel, W. J. M., Dopita, M. A., Dey, A., & Miley, G. K. 2000. "Jet-induced emission-line nebulosity and star formation in the high redshift radio galaxy 4C 41.17", ApJ, 540, 678–686.

Bignall, H. E. 2003. "Radio Variability and Interstellar Scintillation of Blazars", Ph.D. Thesis, (Adelaide, Australia: Univ. of Adelaide).

Bignall, H. E., Jauncey, D. L., Lovell, J. E. J., Tzioumis, A. K., Kedziora-Chudczer, L., Macquart, J-P, Tingay, S. J., Rayner, D. P., & Clay, R. W. 2003. "Rapid Variability and Annual Cycles in the Characteristic Time-scale of the Scintillating Source PKS 1257−326", ApJ, 585, 653–664.

Binney, J., & Tabor, G. 1995. "Evolving Cooling Flows", MNRAS, 276, 663–678.

Biretta, J. A., Junor, W., & Livio, M. 2002. "Evidence for initial jet formation by an accretion disk in the radio galaxy M 87", New Astron. Rev., 46, 239–245.

Biretta, J. A., Sparks, W. B., & Macchetto, F. 1999. "*Hubble Space Telescope* Observations of Superluminal Motion in the M 87 Jet", ApJ, 520, 621–626.

Blake, C. 2002. "Large-scale structure via wide-angle radio surveys", Ph.D. thesis, (Oxford, UK: Univ. of Oxford).

Blake, C., & Wall, J. 2002. "A velocity dipole in the distribution of radio galaxies", Nature, 416, 150–152.

———. 2002. "Measurement of the angular correlation function of radio galaxies from the NRAO VLA Sky Survey", MNRAS, 329, L37–L41.

———. 2002. "Quantifying angular clustering in wide-area radio surveys", MNRAS, 337, 993–1003.

Blandford, R. D., & Rees, M. J. 1974. "A 'twin-exhaust' model for double radio sources", MNRAS, 169, 395–415.

Blandford, R. D., & Znajek, R. L. 1977. "Electromagnetic extraction of energy from Kerr black holes", MNRAS, 179, 433–456.

Blundell, K. M., Mioduszewski, A. J., Muxlow, T. W. B., Podsiadlowski, P., & Rupen, M. P. 2001. "Images of an Equatorial Outflow in SS 433", ApJ, 562, L79–L82.

Blundell, K. M., Rawlings, S., & Willott, C. J. 1999. "The nature and evolution of classical double radio sources from complete samples", AJ, 117, 677–706.

Boehringer, H., Voges, W., Fabian, A. C., Edge, A. C., & Neumann, D. M. 1993. "A *ROSAT* HRI study of the interaction of the X-ray-emitting gas and radio lobes of NGC 1275", MNRAS, 264, L25–L28.

Bolton, J. G., Savage, A., & Wright, A. E. 1979. "The Parkes 2700-MHz Survey - Part Fourteen - Catalogue and New Optical Identifications", Aust. Jou. Phys, Astroph. Sup., 46, 1–17.

Bondi, H. 1952. "On spherically symmetrical accretion", MNRAS, 112, 195–204.

Bower, G. C., & Backer, D. C. 1998. "Space VLBI Observations Show $T_b > 10^{12}$ K in the Quasar NRAO 530", ApJ, 507, L117–L120.

Bower, G. C., Falcke, H., & Backer, D. C. 1999. "Detection of Circular Polarization in the Galactic Center Black Hole Candidate Sagittarius A*", ApJ, 523, L29–L32.

Bower, G. C., Falcke, H., Sault, R. J., & Backer, D. C. 2002. "The Spectrum and Variability of Circular Polarization in Sagittarius A* from 1.4 to 15 GHz", ApJ, 571, 843–855.

Bower, G. C., Wright, M. C. H., Falcke, H., & Backer, D. C. 2003. "Interferometric Detection of Linear Polarization from Sagittarius A* at 230 GHz", ApJ, in press (astro-ph/0302227).

Bradshaw, C. F., Fomalont, E. B., & Geldzahler, B. J. 1999. "High-Resolution Parallax Measurements of Scorpius X-1", ApJ, 512, L121–L124.

Bridle, A. H. 1996. "Observations of Energy Transport", in ASP Conf. Ser. Vol. 100, Energy Transport in Radio Galaxies, ed. P. E. Hardee, A. H. Bridle, & J. A. Zensus, (San Francisco: ASP), 383–394.

Brinkmann, W., Siebert, J., Feigelson, E. D., Kollgaard, R. I., Laurent-Muehleisen, S. A., Reich, W., Fuerst, E., Reich, P., Voges, W., Truemper, J., & McMahon, R. 1997. "Radio-loud active galaxies in the northern *ROSAT* All-Sky Survey. II. Multi-frequency properties of unidentified sources", A&A, 323, 739–748.

Brisken, W. F., Benson, J. M., Goss, W. M., & Thorsett, S. E. 2002. "Very Long Baseline Array Measurement of Nine Pulsar Parallaxes", ApJ, 571, 906–917.

Britzen, S., Vermeulen, R. C., Taylor, G. B., Pearson, T. J., Readhead, A. C. S., Wilkinson, P. N., & Browne, I. W. A. 1999. "The CJF survey: first results on superluminal motion, in ASP Conf. Ser. Vol. 159, BL Lac Phenomenon, ed. L. O. Takalo, A. Sillanpää (San Francisco: ASP), 431–434.

Britzen, S., Vermeulen, R. C., Taylor, G. B., Campbell, R. M., Browne, I. W. A., Wilkinson, P. N., Pearson, T. J., & Readhead, A. C. S. 2001. "CJ–F: The Kinematics of 241 AGN", in IAU Symp. 205, Galaxies and Their Constituents at the Highest Angular Resolutions, ed. R. T. Schilizzi, S. Vogel, F. Paresce, & M. Elvis (San Francisco: ASP), 106–109.

Britzen, S., Witzel, A., Krichbaum, T. P., Muxlow, T., & Matveyenko, L. I. 2001. "Studies of the Fine Structure of the Object 1803+784", Astron. Letters, 27, 1, 1–14.

Broderick, J. J., Vitkevitch, V. V., Jauncey, D. L., Efanov, V. A., Kellermann, K. I., Cohen, M. H., Clark, B. G., Kogan, L. R., Kostenko, V. I., Matveenko, L. I., Moiseev, I. G., Pine, J., Hansen, B. 1970. "Observations of compact radio sources in radio interferometer Green Bank–Crimea", AZh, 47, 4, 784–786.

Bromm, V., Kudritzki, R. P., Loeb, A. 2001. "Generic Spectrum and Ionization Efficiency of a Heavy Initial Mass Function for the First Stars", ApJ, 552, 464–472.

Bromm, V., & Loeb, A. 2002. "The Expected Redshift Distribution of Gamma-Ray Bursts", ApJ, 575, 111–116.

Broten, N. W., Legg, T. H., Locke, J. L., McLeish, C. W., Chisholm, R. M., Gush, H. P., Yen, J. L., & Galt, J. A. 1967. "Observations of quasars using interferometer baselines up to 3,074 km", Nature, 215, 38.

Broten, N. W., Legg, T. H., Locke, J. L., McLeish, C. W., Richards, R. S., Chisholm, R. M., Gush, H. P., Yen, J. L., & Galt, J. A. 1967. "Long baseline interferometry: a new technique", Science, 156, 1592–1593.

Brown, G. W., Carr, T. D., & Block, W. F. 1968. "Long-baseline interferometry of S-bursts from Jupiter", Astrophysical Letters, 1, 89–94.

Brown, R. H. 1962. "Clustering, cosmology & source counts", MNRAS, 124, 35–49.

Brown, R. H., & Twiss, R. Q. 1954. "A new type of interferometer for use in radio astronomy", Phil. Mag., 45, 663–708.

Brown, R. L. 1982. "Precessing jets in Sagittarius A - Gas dynamics in the central parsec of the galaxy", ApJ, 262, 110–119.

Brown, R. L., & Lo, K. Y. 1982. "Variability of the compact radio source at the galactic center", ApJ, 253, 108–104.

Browne, I. W. A., Wilkinson, P. N., Patnaik, A. R., & Wrobel, J. M. 1998. "Interferometer phase calibration sources. II - The region $\delta_{\rm B1950}$ from 0 deg to 20 deg", MNRAS, 293, 257–287.

Brunthaler, A., Falcke, H., Bower, G. C., Aller, M. F., Aller, H. D., Terasranta, H., Lobanov, A. P., Krichbaum, T. P., & Patnaik, A. R. 2000. "III Zw 2, the first superluminal jet in a Seyfert galaxy", A&A, 357, L45–L48.

Buchalter, A., Helfand, D. J., Becker, R. H., & White, R. L. 1998. "Constraining Ω_0 with the angular size – redshift relation of double-lobed quasars in the *FIRST* survey", ApJ, 94, 503–522.

Burbidge, E. M., Burbidge, G. R., Solomon, P. M., & Strittmatter, P. A. 1971. "Apparent Associations Between Bright Galaxies and Quasi-Stellar Objects", ApJ, 170, 233–240.

Burbidge, G. 1956. "On Synchrotron Radiation from Messier 87", ApJ, 124, 416–429.

———. 1958. "Particle Energies and Magnetic Energy in the Crab Nebula", ApJ, 127, 48–53.

———. 1959. "Estimates of the Total Energy in Particles and Magnetic Field in the Non-Thermal Radio Sources", ApJ, 129, 849–851.

———. 1961. "Galactic Explosions as Sources of Radio Emission", Nature, 190, 1053.

———. 2001. "Noncosmological Redshifts", PASP, 113, 899–902.

Burbidge, G., & Burbidge, E. M. 1964. In The Structure and Evolution of Galaxies, ed. I. Prigogine, Proc. 13th Solvay Conf. on Phys., 137.

Burke, B. F., Johnston, K. J., Efanov, V. A., Clark, B. G., Kogan, L. R., Kostenko, V. I., Lo, K. Y., Matveenko, L. I., Moiseev, I. G., Moran, J. M., Knowles, C. H., Papa, D. K., & Papadopoulos, G. D. 1972. "Observations of sources of maser radio emission with angular resolution of 0.″0002", AZh, 49, 3, 465–469.

Burke, B. F., Papa, D. C., Papadopoulos, G. D., Schwartz, P. R., Knowles, S. H., Sullivan, W. T., Meeks, M. L., & Moran, J. M. 1970. "Studies of H_2O sources by means of a very-long interferometer", ApJ, 160, L63–L68.

Carilli, C. L., Gnedin, N. Y., & Owen, F. 2002. "HI 21 Centimeter Absorption beyond the Epoch of Reionization", ApJ, 577, 22–30.

Carilli, C. L., Röttgering, H. J. A., van Ojik, R., Miley, G. K., & van Breugel, W. J. M. 1997. "Radio Continuum Imaging of High-Redshift Radio Galaxies", ApJS, 109, 1–44.

Carlberg, R. G. 2002. "Globular Clusters at High Redshift", ApJ, 573, 60–65.

Carr, T. D., May, J., Olsson, C. N., & Walls, G. F. 1965. "Post-Detection correlation interferometry of Jupiter at 18 Mc/s", IEEE NEREM Record, 7, 222–223.

Castro-Tirado, A. J., Greiner, J., & Paredes, J. M. (ed.) 2001. "Proceedings of the third Microquasar Workshop", Ap&SS, 276.

Cawthorne, T. V., & Cobb, W. K. 1990. "Linear Polarization of Radiation from Oblique and Conical Shocks", ApJ, 350, 536–544.

Celotti, A., Ghisellini, G., & Chiaberge, M. 2001. "Large scale jets in AGN: multiwavelength mapping", MNRAS, 321, L1–L5.

Cen, R., & Ostriker, J. P. 1999. "Where are the baryons?", ApJ, 514, 1–6.

Chambers, K. C., Miley, G. K., & van Breugel, W. 1987. "Alignment of radio and optical orientations in high-redshift radio galaxies", Nature, 329, 604–606.

Chary, R., & Elbaz, D. 2001. "Interpreting the cosmic infrared background: constraints on the evolution of the dust-enshrouded star formation rate", ApJ, 556, 562–581.

Chen, G., & Ratra, B. 2003. "Cosmological constraints from compact radio source angular size versus redshift data", ApJ, 582, 586–589.

Cheng, A. F. 1989. "Radio pulsar spin-down - Tests of pulsar models", ApJ, 337, 803–813.

Churazov, E., Sunyaev, R., Forman, W., & Böhringer, H. 2002. "Cooling flows as a calorimeter of active galactic nucleus mechanical power", MNRAS, 332, 729–734.

Ciardi, B., & Ferrara, A. 2001. "Detecting the first objects in the mid-infrared with the Next Generation Space Telescope", MNRAS, 324, 648–652.

Ciardi, B., & Loeb, A. 2000. "Expected Number and Flux Distribution of Gamma-Ray Burst Afterglows with High Redshifts", ApJ, 540, 687–696.

Ciliegi, P., & Maccacaro, T. 1997. "The X-ray flux and spectral variability of the Einstein Extended Medium Sensitivity Survey AGN", MNRAS, 292, 338–348.

Clark, B. G., Broderick, J. J., Efanov, V. A., Kellermann, K. I., Cohen, M. H., Kogan, L. R., Kostenko, V. I., Matveenko, L. I., Moiseev, I. G., Muhina, M. M., Shteishleiger, V. B., & Jauncey, D. L. 1972. "Observations of compact objects of cosmic radio emission at 3.55 cm wavelength with maximum angular resolution. AZh, 49, 4, 700–703.

Clark, B. G., Kellermann, K. I., Bare, C. C., Cohen, M. H., & Jauncey, D. L. 1968. "Radio interferometry using a base line of 20 million wavelengths", ApJ, 153, L67–L68.

———. 1968. "High-Resolution Observations of Small-Diameter Radio Sources at 18-cm Wavelength", ApJ, 153, 705–714.

Clark, B. G., Kellermann, K. I., Cohen, M. H., Shaffer, D. B., Broderick, J. J., Jauncey, D. L., Matveenko, L. I., Moiseev, I. G. 1973. "Variations in the radio structure of BL Lacertae", ApJ, 182, L57–L60.

Claussen, M. J., Diamond, P. J., Braatz, J. A., Wilson, A. S., & Henkel, C. 1998. "The Water Masers in the Elliptical Galaxy NGC 1052", ApJ, 500, L129–L132.

Coburn, W., Heindl, W. A., Rothschild, R. E., Gruber, D. E., Kreykenbohm, I., Wilms, J., Kretschmar, P., & Staubert, R. 2002. "Magnetic Fields of Accreting X-Ray Pulsars with the Rossi X-Ray Timing Explorer", ApJ, 580, 394–412.

Cohen, M. H. 1965. "Scintillations on the small diameter radio sources", Nature, 208, 277.

———. 1989. "Statistics of Superluminal Sources", in Lecture Notes in Phys. Vol. 334, BL Lac Objects, ed. L. Maraschi, T. Maccacaro, & M.-H. Ulrich (Berlin: Springer), 13–21.

Cohen, M. H., Cannon, W., Purcell, G. H., Shaffer, D. B., Broderick, J. J., Kellermann, K. I., & Jauncey, D. L. 1971. "The Small-Scale Structure of Radio Galaxies and Quasi-Stellar Sources at 3.8 Centimeters", ApJ, 170, 207–217.

Cohen, M. H., Gundermann, E. J., & Harris, D. E. 1967. "New Limits on the Diameters of Radio Sources" ApJ, 150, 767–782.

Cohen, M. H., Linfield, R. P., Moffet, A. T., Seielstad, G. A., Kellermann, K. I., Shaffer, D. B., Pauliny-Toth, I. I. K., Preuß, E., Witzel, A., & Romney, J. D. 1977. "Radio sources with superluminal velocities", Nature, 268, 405–409.

Condon, J. J. 1992. "Radio emission from normal galaxies", ARA&A, 30, 575–611.

Condon, J. J., & Backer, D. C. 1975. "Interstellar scintillation of extragalactic radio sources", ApJ, 197, 31–38.

Condon, J. J., Cotton, W. D., Greisen, E. W., Yin, Q. F., Perley, R. A., Taylor, G. B, & Broderick, J. J. 1998. "The NRAO VLA Sky Survey", AJ, 115, 1693–1716.

Condon, J. J., Cotton, W. D., & Broderick, J. J. 2002. "Radio Sources and Star Formation in the Local Universe", AJ, 124, 675–689.

Conway, R. G., Kellermann, K. I., & Long, R. J. 1963. "The radio frequency spectra of discrete radio sources", MNRAS, 125, 261–284.

Corbel, S., Fender, R. P., Tzioumis, A. K., Nowak, M., McIntyre, V., Durouchoux, P., & Sood, R. 2000. "Coupling of the X-ray and radio emission in the black hole candidate and compact jet source GX 339−4", A&A, 359, 251–268.

Corbel, S., Fender, R. P., Tzioumis, A. K., Tomsick, J. A., Orosz, J. A., Miller, J. M., Wijnands, R., & Kaaret, P. 2002. "Large-Scale, Decelerating, Relativistic X-ray Jets from the Microquasar XTE J1550−564", Science, 298, 196–199.

Cordes, J. M., & Rickett, B. J. 1998. "Diffractive Interstellar Scintillation Timescales and Velocities", ApJ, 507, 846–860.

Cotton, W. D., Counselman, C. C., Geller, R. B., Shapiro, I. I., Wittels, J. J., Hinteregger, H. F., Knight, C. A., Rogers, A. E. E., Whitney, A. R., & Clark, T. A. 1979. "3C 279 - The case for 'superluminal' expansion", ApJ, 229, L115–L117.

Counselman, C. C., Kent, S. M., Knight, C. A., Shapiro, I. I., Clark, T. A., Hinteregger, H. F., Rogers, A. E. E., & Whitney, A. R. 1974. "Solar gravitational deflection of radio waves measured by very-long-baseline interferometry", Phys. Rev. Lett, 33, 1621–1623.

Crenshaw, D. M., Kraemer, S. B., & George, I. M. 2003. "Mass Loss from the Nuclei of Active Galaxies", ARA&A, 41, in press.

Cress, C. M., Helfand, D. J., Becker, R. H., Gregg, M. D., & White, R. L. 1996. "The Angular Two-Point Correlation Function for the FIRST Radio Survey", ApJ, 473, 7–14.

Cress, C. M., & Kamionkowski, M. 1998. "Interpreting the clustering of radio sources", MNRAS, 297, 486–492.

Dabrowski, Y., Lasenby, A., & Saunders, R. 1995. "Testing the angular-size versus redshift relation with compact radio sources", MNRAS, 277, 753–757.

Dallacasa, D., Stanghellini, C., Centonza, M., & Fanti, R. 2000. "High frequency peakers. I. The bright sample", A&A, 363, 887–900.

Dalton, G. B., Efstathiou, G., Maddox, S. J., & Sutherland, W. J. 1992. "Spatial correlations in a redshift survey of APM galaxy clusters", ApJ, 390, L1–L4.

Daly, R. A., & Loeb, A. 1990. "A possible origin of galactic magnetic fields", ApJ, 364, 451–455.

Daly, R. A., & Marscher, A. P. 1988. "The gas dynamics of compact relativistic jets", ApJ, 334, 539–551.

Davé, R., Cen, R., Ostriker, J. P., Bryan, G. L., Hernquist, L., Katz, N., Weinberg, D. H., Norman, M. L., & O'Shea, B. 2001. "Baryons in the warm-hot intergalactic medium", ApJ, 552, 473–483.

Davies, R. D., Walsh, D., & Booth, R. S. 1976. "The radio source at the galactic nucleus", MNRAS, 177, 319–333.

Davis, M., & Peebles, P. J. E. 1983. "A survey of galaxy redshifts. V - The two-point position and velocity correlations", ApJ, 267, 465–482.

De Breuck, C., van Breugel, W., Röttgering, H. J. A., & Miley, G. 2000. "A Catalog of 1.4 GHz Radio Sources from the FIRST Survey", A&A, 143, 303–333.

de Bruyn, A. G. 1990. "Limits on the Expansion of the Radio Source OQ 208 in Mrk 668", in Variability of Active Galaxies, ed. W. J. Duschl, S. J. Wagner & M. Camenzind, (Heidelberg: Springer Verlag), 105–106.

de Vaucouleurs, G., de Vaucouleurs, A., Corwin, H. G. Jr., Buta, R. J., Paturel, G., & Fouque, P. 1991. "Third Reference Catalogue of Bright Galaxies", (Springer: New York).

de Vries, W. H., Barthel, P. D., & O'Dea, C. P. 1997. "Radio spectra of Gigahertz Peaked Spectrum radio sources", A&A, 321, 105–110.

de Young, D. S. 2002. "The Physics of Extragalactic Radio Sources", (Chicago: Univ. of Chicago Press).

Denn, G. R., Mutel, R. L., & Marscher, A. P. 2000. "Very Long Baseline Polarimetry of BL Lacertae", ApJS, 129, 61–92.

Dennett-Thorpe, J., & de Bruyn, A. G. 2000. "The Discovery of a Microarcsecond Quasar: J1819+3845", ApJ, 529, L65–L68.

———. 2001. "Monitoring the Microarcsecond Quasar J1819+3845", Ap&SS, 278, 101–104.

———. 2002. "Interstellar scintillation as the origin of the rapid radio variability of the quasar J1819+3845", Nature, 415, 57–60.

Dent, W. A. 1965. "Quasi-Stellar Sources: Variation in the Radio Emission of 3C 273", Science, 148, 1458–1460.

Dhavan, V., Sarma, N. V. G., Ganesan, R., Matveyenko, L. I., Kogan, L. R., Molodjanu, A. P., & Graham, D. 1992. "10-meter radiotelescope - VLBI network element", Soviet Ast. Lett., 18, 4, 391–400.

Di Matteo, T., Perna, R., Abel, T., & Rees, M. J. 2002. "Radio Foregrounds for the 21 Centimeter Tomography of the Neutral Intergalactic Medium at High Redshifts", ApJ, 564, 576–580.

Diamond, P. J., & Kemball, A. J. 1998. "A stellar movie: VLBA monitoring of SiO masers around the Mira variable TX Cam", BAAS, 30, 1349.

Doeleman, S. S., Shen, Z.-Q., Rogers, A. E. E., Bower, G. C., Wright, M. C. H., Zhao, J. H., Backer, D. C., Crowley, J. W., Freund, R. W., Ho, P. T. P., Lo, K. Y., & Woody, D. P. 2001. "Structure of Sagittarius A* at 86 GHz using VLBI Closure Quantities", AJ, 121, 2610–2617.

Dollfus, A. 1953. "Observation visuelle et photographique des planètes Mercure et Vénus à l'Observatoire du Pic du Midi", Bull. Soc. Astron. France, 67, 61–75.

Donnelly, R. H., Partridge, R. B., & Windhorst, R. A. 1987. "6 centimeter radio source counts and spectral index studies down to 0.1 millijansky", ApJ, 321, 94–112.

Douglas, J. N., Bash, F. N., Bozyan, F. A., Torrence, G. W., & Wolfe, C. 1996. "Texas Survey of radio sources at 365 MHz", AJ, 111, 1945–1963.

Downes, D., & Martin, A. H. M. 1971. "Compact Radio Sources in Galactic Nucleus", Nature, 233, 112–114.

Drake, F., Wolfe, J. H., & Seeger, C. L. 1983. "SETI Science Working Group Report", (Washington, DC: NASA).

Dubner, G. M., Holdaway, M., Goss, W. M., & Mirabel, I. F. 1998. "A High-Resolution Radio Study of the W50-SS 433 System and the Surrounding Medium", AJ, 116, 1842–1855.

Eckart, A., Genzel, R., Ott, T., & Schödel, R. 2002. "Stellar orbits near Sagittarius A*", MNRAS, 331, 917–934.

Edelson, R. A., & Krolik, J. H. 1988. "The discrete correlation function: a new method for analyzing unevenly sampled variability data", ApJ, 333, 646–659.

Edwards, P. G., Giovannini, G., Cotton, W. D., Feretti, L., Fujisawa, K., Hirabayashi, H., Lara, L., & Venturi, T. 2000. "A spectral index map from VSOP observations of Markarian 501", PASJ, 52, 1015–1019.

Edwards, P. G., Hirabayashi, H., Fomalont, E. B., Gurvits, L. I., Horiuchi, S., Lovell, J. E. J., Moellenbrock, G. A., & Scott, W. K. 2002. "The VSOP Survey Program of 402 AGN at 5 GHz", in Proc. IAU Asia-Pacific Regional Meeting, vol. II, ed. S. Ikeuchi, J. Hearnshaw, & T. Hanawa, (Tokyo: Astron. Soc. Japan), 375–376.

Ekers, R. D., Cullers, D. K., Billingham, J., & Scheffer, L. K. (ed.) "SETI 2020: A Roadmap for the Search for Extraterrestrial Intelligence", (Mountain View, CA: SETI Press).

Ekers, R. D., & Liang, H. 1990. "Light echoes revisited", in Proc. Workshop on 'Parsec-Scale Radio Jets', (Cambridge: CUP), 333–339.

Ellis, G. F. R., & Baldwin, J. E. 1984. "On the expected anisotropy of radio source counts", MNRAS, 206, 377–381.

Ellison, S. L., Yan, L., Hook, I. M., Pettini, M., Wall, J. V., & Shaver, P. A. 2001. "The CORALS survey I: New estimates of the number density and gas content of damped Lyman alpha systems free from dust bias", A&A, 379, 393–406.

———. 2002. "The CORALS survey. II. Clues to galaxy clustering around QSOs from $z_{abs} \sim z_{em}$ damped Lyman alpha systems", A&A, 383, 91–97.

Esin, A. A., McClintock, J. E., Drake, J. J., Garcia, M. R., Haswell, C. A., Hynes, R. I., & Muno, M. P. 2001. "Modeling the Low-State Spectrum of the X-Ray Nova XTE J1118+480", ApJ, 555, 483–488.

Evrard, A. E. 1991. "Clues to galaxy formation from rich cluster simulations", MNRAS, 248, 8P–10P.

Falcke, H. 1999. "Radio Variability of Sgr A* at Centimeter Wavelengths", in ASP Conf. Ser. Vol. 186, The Central Parsecs of the Galaxy, ed. H. Falcke, A. Cotera, W. T. Duschl, F. Melia, & M. J. Ricke (San Francisco: ASP), 113–117.

———. 1999. "The jet model for Sgr A*", in ASP Conf. Ser. Vol. 186, The Central Parsecs of the Galaxy, ed. H. Falcke, A. Cotera, W. J. Duschl, F. Melia, & M. J. Rieke (San Francisco: ASP), 148–156.

———. 2001. "The silent majority: jets and radio cores from low-luminosity black holes", Rev. in Modern Astron., 14, 15–51.

Falcke, H., & Biermann, P. L. 1995. "The jet/disk symbiosis. I. Radio to X-ray emission models for quasars", A&A, 293, 665–682.

———. 1999. "The jet/disk symbiosis. III. What the radio cores in GRS 1915+105, NGC 4258, M 81 and Sgr A* tell us about accreting black holes", A&A, 342, 49–56.

Falcke, H., & Markoff, S. 2000. "The jet model for Sgr A*: Radio and X-ray spectrum", A&A, 362, 113–118.

Falcke, H., Melia, F., & Agol, E. 2000. "Viewing the shadow of the black hole at the galactic center.", ApJ, 528, L13–L16.

Falcke, H., Nagar, N. M., Wilson, A. S., & Ulvestad, J. S. 2000. "Radio sources in low-luminosity active galactic nuclei: II. Very long baseline interferometry detections of compact radio cores and jets in a sample of LINERs", ApJ, 542, 197–200.

Falle, S. A. E. G. 1991. "Self-similar jets", MNRAS, 250, 581–596.

Fan, X., Narayanan, V. K., Lupton, R. H., Strauss, M. A., Knapp, G. R., Becker, R. H., White, R. L., Pentericci, L., Leggett, S. K., Haiman, Z., Gunn, J. E., Ivezic, Z., Schneider, D. P., Anderson, S. F., Brinkmann, J., Bahcall, N. A., Connolly, A. J., Csabai, I. D. Mamoru, D., Fukugita, M., Geballe, T, Grebel, E. K., Harbeck, D., Hennessy, G., Lamb, D. Q., Miknaitis, G., Munn, J. A., Nichol, R., Okamura, S., Pier, J. R., Prada, F., Richards, G. T., Szalay, A., & York, D. G. 2001. "A Survey of $z > 5.8$ Quasars in the Sloan Digital Sky Survey. I. Discovery of Three New Quasars and the Spatial Density of Luminous Quasars at $z \sim 6$", AJ, 122, 2833–2849.

Fan, X., Narayanan, V. K., Strauss, M. A., White, R. L.; Becker, R. H., Pentericci, L., & Rix, H.-W. 2002. "Evolution of the Ionizing Background and the Epoch of Reionization from the Spectra of $z \sim 6$ Quasars", AJ, 123, 1247–1257.

Fan, X., Strauss, M. A., Schneider, D. P., Becker, R. H., White, R. L., Haiman, Z., Gregg, M., Pentereccí, L., Grebel, E. K., Narayanan, V. K., Loh, Y.-S. Richards, G. T., Gunn, J. E., Lupton, R. H., Knapp, G. R., Ivezic, E., Brandt, W. N., Collinge, M., Hao, L., Harbeck, D., Parada, F., Schaye, J., Strateva, I., Zakamska, N., Anderson, S., Brinkmann, J., Bahcall, N. A., Lamb, D. Q., Okamura, S., Szalay, A., & York, D. G. 2003. "A Survey of $z > 5.7$ Quasars in the Sloan Digital Sky Survey II: Discovery of Three Additional Quasars at $z > 6$", AJ, 125, 1649–1659.

Fanti, C., & Fanti, R. 2001. "From Compact to Large radio sources: the time evolution", in ASP Conf. Ser. Vol. 258, Issues in Unification of Active Galactic Nuclei, ed. R. Maiolino, A. Marconi, & N. Nagar (San Francisco: ASP), 261–272.

Fanti, C., Fanti, R., Dallacasa, D., Schilizzi, R. T., Spencer, R. E., & Stanghellini, C. 1995. "Are compact steep-spectrum sources young?", A&A, 302, 317–326.

Fanti, C., Fanti, R., Parma, P., Schilizzi, R. T., & van Breugel, W. J. M. 1985. "Compact Steep Spectrum 3CR radio sources - VLBI observations at 18 cm", A&A, 143, 292–306.

Fanti, C., Fanti, R., Schilizzi, R. T., Spencer, R. E., Nan Rendong, Parma, P., van Breugel, W. J. M., & Venturi, T. 1990. "On the nature of compact steep spectrum radio sources", A&A, 231, 333–346.

Fender, R. P. 2001. "Powerful jets from black hole X-ray binaries in low/hard X-ray states", MNRAS, 322, 31–42.

Fender, R. P., & Kuulkers, E. 2001. "On the peak radio and X-ray emission from neutron star and black hole candidate X-ray transients", MNRAS, 324, 923–930.

Fender, R. P., & Pooley, G. G. 1998. "Infrared synchrotron oscillations in GRS 1915+105", MNRAS, 300, 573–576.

Fender, R., Rayner, D., Norris, R., Sault, R. J., & Pooley, G. 2000. "Discovery of Circularly Polarized Radio Emission from SS 433", ApJ, 530, L29–L32.

Ferrara, A., Pettini, M., & Shchekinov, Yu. 2000. "Mixing metals in the early universe", MNRAS, 319, 539–548.

Fey, A. L., & Charlot, P. 1997. "VLBA Observations of Radio Reference Frame Sources. II. Astrometric Suitability Based on Observed Structure", ApJS, 111, 95–142.

———. 2000. "VLBA Observations of Radio Reference Frame Sources. III. Astrometric Suitability of an Additional 225 Sources", ApJS, 128, 17–83.

Fey, A. L., Clegg, A. W., & Fomalont, E. B. 1996. "VLBA Observations of Radio Reference Frame Sources. I", ApJS, 105, 299–330.

Field, G. B. 1962. "Atmosphere of Mercury", AJ, 67, 575–576.

Fomalont, E. B. 2000. "The NRAO Radio-Link Interferometer", in NRAO Symp. 27, in Radio Interferometry: The Saga and the Science, ed. D. Finley, & W. M. Goss (Socorro, NM: NRAO), 41.

Fomalont, E. B., Frey, S., Paragi, Z., Gurvits, L. I., Scott, W. K., Taylor, A. R., Edwards, P. G., & Hirabayashi, H. 2000. "The VSOP 5 GHz Continuum Survey: The Prelaunch VLBA Observations" ApJ, 131, 95–183.

Fomalont, E. B., Geldzahler, B. J., & Bradshaw, C. F. 2001. "Scorpius X-1: The Evolution and Nature of the Twin Compact Radio Lobes", ApJ, 558, 283–301.

Fomalont, E. B., Kellermann, K. I., Cowie, L. L., Berger, A. J., Capak, P., Partridge, R. B., & Windhorst, R. A. 2003. ApJ, submitted.

Fomalont, E. B., Kellermann, K. I., Partridge, R. B., Windhorst, R. A., & Richards, E. A. 2002. "The Microjansky Sky at 8.4 GHz", AJ, 123, 2402–2416.

Fomalont, E. B., Kellermann, K. I., Wall, J. V., & Weistrop, D. 1984. "A Deep 6-Centimeter Radio Source Survey", Science, 225, 23–28.

Frater, R. H., Brooks, J. W., & Whiteoak, J. B. 1992. "The Australia Telescope Overview", Jou. of Electrical and Electronics Engineering, Australia, 12, 103–112.

Fugita, Y. 2001. "Ram-pressure stripping of galaxies in high-redshift clusters and the influence of intracluster medium heating", ApJ, 550, 612–621.

Fuhrmann, L., Krichbaum, T. P., Cimò, G., Beckert, T., Kraus, A., Witzel, A., Qian, S. J., & Rickett, B. J. 2002. "Annual Modulation in the Variability Properties of the IDV Source 0917+624?", PASA, 19, 64–68.

Furlanetto, S. R., & Loeb, A. 2001. "Intergalactic magnetic fields from quasar outflows", ApJ, 556, 619–634.

———. 2002. "The 21 Centimeter Forest: Radio Absorption Spectra as Probes of Minihalos before Reionization", ApJ579, 1–9.

Gabel, J. R., Bruhweiler, D. M., Crenshaw, D. M., Kraemer, S. B., & Miskey, C. L. 2000. "*Hubble Space Telescope* Observations and Photoionization Modeling of the LINER Galaxy NGC 1052", ApJ, 532, 883–894.

Gabuzda, D. C. 1999. "VSOP Observations of the Compact BL Lac Object 1803+784", New Astron. Rev., 43, 695–698.

———. 2002. "New VSOP Polarisation Images of BL Lac Objects", in Proc. of the 6th European VLBI Network Symp., ed. E. Ros, R. W. Porcas, A. P. Lobanov, & J. A. Zensus (Bonn: MPIfR), 83–86.

———. 2003. "VSOP Polarisation Images of Three BL Lac Objects", in ASP Conf. Ser. Vol. 299, High Energy Blazar Astronomy, ed. L. O. Takalo, & E. Valtaoja (San Francisco: ASP), in press.

———. 2003. "VLBI Polarisation Properties of a Complete Sample of Radio-Loud BL Lacs", in Ap&SS, Circular Polarisation from Relativistic Jet Sources, in press.

———. 2003. "The Nature of Compact Parsec-Scale Knots in AGN: Clues from their Polarisation Properties", in New Astron. Rev., in press.

Gabuzda, D. C., Cawthorne, T. V., Roberts, D. H., & Wardle, J. F. C. 1992. "A Survey of the Milliarcsecond Polarization Properties of BL Lacertae Objects", ApJ, 388, 40–54.

Gabuzda, D. C., & Chernetskii, V. A. 2003. "Parsec-scale Faraday Rotation Distribution in the BL Lac Object 1803+784", MNRAS, 339, 669–679.

Gabuzda, D. C., & Gómez, J. L. 2001. "VSOP Polarization Observations of the BL Lacertae Object OJ 287", MNRAS, 320, L49–L54.

Gabuzda, D. C., Gómez, J. L., & Agudo, I. 2001. "Evidence for Parsec-scale Absorption from VSOP Observations of the BL Lacertae Object 0735+178", MNRAS, 328, 719–725.

Gabuzda, D. C., Mullan, C. M., Cawthorne, T. V., Wardle, J. F. C., & Roberts, D. H. 1994. "Evolution of the Milliarcsecond Total Intensity and Polarization Structures of BL Lacertae Objects", ApJ, 435, 140–161.

Gabuzda, D. C., & Pushkarev, A. B. 2002. "Evidence for Helical Magnetic Fields in the Jets of BL Lacertae Objects", in ASP Conf. Ser. Vol. 250, Particles and Fields in Radio Galaxies, ed. R. Laing & K. Blundell (San Francisco: ASP), 180–183.

Gabuzda, D. C., Pushkarev, A. B., & Cawthorne, T. V. 2000. "Analysis of 6 cm VLBI Polarization Observations of a Complete Sample of Northern BL Lacertae Objects", MNRAS, 319, 1109–1124.

Gabuzda, D. C., Pushkarev, A. B., & Garnich, N. N. 2001. "Unusual Radio Properties of the BL Lac Object 0820+225", MNRAS, 327, 1–9.

Gabuzda, D. C., Sitko, M. L., & Smith, P. S. 1996. "Correlations Between the VLBI and Optical Polarization of BL Lacertae Objects", AJ, 112, 1877–1885.

Gallimore, J. F., Baum, S. A., & O'Dea, C. P. 1997. "A direct image of the obscuring disk surrounding an active galactic nucleus", Nature, 388, 852–854.

Galloway, D. K., Chakrabarty, D., Morgan, E. H., & Remillard, R. A. 2002. "Discovery of a High-Latitude Accreting Millisecond Pulsar in an Ultracompact Binary", ApJ, 576, L137–L140.

Garcia, M. R., McClintock, J. E., Narayan, R., Callanan, P., Barret, D., & Murray, S. S. 2001. "New Evidence for Black Hole Event Horizons from *CHANDRA*", ApJ, 553, L47–L50.

Garrett, M. A. 2002. "The FIR/Radio correlation of high redshift galaxies in the region of the HDF-N", A&A, 384, L19–L22.

Geldzahler, B. J., Kellermann, K. I., & Shaffer, D. B. 1979. "High resolution observations of the compact radio sources CL4, Cygnus X-3, and the Galactic Center", AJ, 84, 186–188.

Genzel, R., Downs, D., Moran, J. M., Johnston, K. J., Spencer, J. H., Walker, R. C., Haschick, A., Matveyenko, L. I., Kogan, L. R., Rönnäng, B., Rydbeck, O. E. H., & Moiseev, I. G. 1978. "Structure and Kinematics of H_2O Sources in Clusters of Newly-formed OB Stars", A&A, 66, 13–29.

Genzel, R., Reid, M. J., Moran, J. M., & Downes, D. 1981. "Proper motions and distances of H_2O maser sources I — The outflow in Orion-KL", ApJ, 244, 884–902.

Georganopoulos, M., Aharonian, F. A., & Kirk, J. G. 2002. "External Compton emission from relativistic jets in Galactic black hole candidates and ultraluminous X-ray sources", ApJ, 388, L25–L28.

Georganopoulos, M., & Marscher, A. P. 1998. "Modeling the variability of the BL Lacertae object PKS 2155−304", ApJ, 506, L11–L14.

Ghez, A. M., Morris, M., Becklin, E. E., Tanner, A., & Kremenek, T. 2000. "The accelerations of stars orbiting the Milky Way's central black hole", Nature, 407, 349–351

Ghisellini, G., Maraschi, L., & Treves, A. 1985. "Inhomogeneous synchrotron-self-Compton models and the problem of relativistic beaming of BL Lac objects", A&A, 146, 204–212.

Gierliński, M., & Done, C. 2002. "A comment on the colour-colour diagrams of low-mass X-ray binaries", MNRAS, 331, L47–L50.

Gilmore, G., & Shaw, M. A. 1986. "PKS 1345+125 - A Seyfert nucleus merging with a powerful radio galaxy", Nature, 321, 750–753.

Gnedin, N. Y. 1998. "Metal enrichment of the intergalactic medium", MNRAS, 294, 407–421.

———. 2000. "Effect of Reionization on Structure Formation in the Universe", ApJ, 535, 530–541.

———. 2001. "Redshift of reionization: are we there yet?", submitted to MNRAS, astro-ph/0110290

Gnedin, N. Y., & Ostriker, J. P. 1997. "Reionization of the Universe and the Early Production of Metals", ApJ, 486, 581–598.

Gnedin, N. Y., & Shandarin, S. F. 2002. "Morphology of the secondary cosmic microwave background anisotropies: the key to 'smouldering' reionization", MNRAS, 337, 1435–1440.

Gold, T. 1967. "Radio method for the precise measurement of the rotation period of the earth", Science, 157, 302–304.

Goldreich, P., & Julian, W. H. 1969. "Pulsar Electrodynamics", ApJ, 157, 869–880.

Gómez, J. L., Marscher, A. P., & Alberdi, A. 1999. "86, 43, and 22 GHz VLBI Observations of 3C 120", ApJ, 521, L29–L32.

Gómez, J. L., Marscher, A. P., Alberdi, A., Jorstad, S. G., & Agudo, I. 2001. "Monthly 43 GHz VLBA polarimetric monitoring of 3C 120 over 16 epochs: evidence for trailing shocks in a relativistic jet", ApJ, 561, L161–L164.

Gómez, J. L., Marscher, A. P., Alberdi, A., Jorstad, S. G., & García-Miró, C. 2000. "Flashing superluminal components in the jet of the radio galaxy 3C 120", Science, 289, 2317–2320.

Gopal-Krishna & Wiita, P. J. 2001. "Was the cosmic web of protogalactic material permeated by lobes of radio galaxies during the quasar era?", ApJ, 560, L115–L118.

Gopal-Krishna, Wiita, P. J., & Osterman, M. A. 2003. "Radio galaxies and the star formation history of the universe", in ASP Conf. Ser., Active Galactic Nuclei: from Central Engine to Host Galaxy, ed. S. Collin, F. Combes, & I. Shlosman (San Francisco: ASP), in press.

Goss, W. M., Brown, R. L., & Lo, K. Y. 2003. "The Discovery of Sgr A*", in The Central 300 parsecs of the Milky Way, ed. A. Cotera, H. Falcke, T. R. Geballe, & S. Markoff, Astronomische Nachrichten, 324, in press.

Grandi, P., Sambruna, R. M., Maraschi, L., Matt, G., Urry, C. M., & Mushotzky, R. F. 1997. "*ROSAT*, *ASCA*, and *OSSE* observations of the broad-line radio galaxy 3C 120", ApJ, 487, 636–643.

Green, R. F., Khachikian, E. Ye., & Sanders, D. B. (ed.) 2002. "AGN Surveys", ASP Conf. Proc. Vol. 284, (San Francisco: ASP).

Greenhill, L. J., Kondratko, P. T., Lovell, J. E. J., Kuiper, T. B. H., Moran, J. M., Jauncey, D. L., & Baines, G. P. 2003. "The discovery of H_2O maser emission in seven active galactic nuclei and at high velocities in the Circinus galaxy", ApJ, 582, L11–L14.

Greiner, J., Cuby, J. G., & McCaughrean, M. J. 2001. "An unusually massive stellar black hole in the Galaxy", Nature, 414, 522–525.

Guerra, E. J., Daly, R. A., & Wan, L. 2000. "Global Cosmological Parameters Determined Using Classical Double Radio Galaxies", ApJ, 544, 659–670.

Güijosa, A., & Daly, R. A. 1996. "Equipartition Doppler Factors for a Sample of Active Galactic Nuclei", ApJ, 461, 600–608.

Gunn, J. E., & Gott, J. R. 1972. "On the infall of matter in clusters of galaxies and some effects on their evolution", ApJ, 176, 1–19.

Gurvits, L. I., 1993. "Milliarcsecond Structures of AGN at Different Redshifts", in Sub-Arcsecond Radio Astron., ed. R. J. Davies and R. S. Booth, (Cambridge: CUP), 380–385.

Gurvits, L. I. 1994. "Apparent milliarcsecond sizes of AGN and the geometry of the Universe", ApJ, 425, 442–447.

Gurvits, L. I., Kellermann, K. I., & Frey, S. 1999. "The "angular size – redshift" relation for compact radio structures in quasars and radio galaxies", A&A, 342, 378–388.

Gwinn, C. R., Britton, M. C., Reynolds, J. E., Jauncey, D. L., King, E. A., McCulloch, P. M., Lovell, J. E. J., Flanagan, C. S., & Preston, R. A. 2000. "Size of the Vela Pulsar's emission region at 13 centimeter wavelength", ApJ, 531, 902–916.

Haiman, Z. 2002. "The Detectability of High-Redshift Lyα Emission Lines prior to the Reionization of the Universe", ApJ, 576, L1–L4.

Haiman, Z., & Knox, L. 1999. "Reionization of the Intergalactic Medium and its Effect on the CMB", in ASP Conf. Ser. Vol. 181, Microwave Foregrounds, ed. A de Oliviera-Costa & M. Tegmark, (San Francisco: ASP), 227–250.

Haiman, Z., & Loeb, A. 1998. "Observational Signatures of the First Quasars", ApJ, 503, 505–517.

Haiman, Z., Madau, P., & Loeb, A. 1999. "Constraints from the Hubble Deep Field on High-Redshift Quasar Models", ApJ514, 535–543.

Hardcastle, M. J., Birkinshaw, M., Cameron, R. A., Harris, D. E., Looney, L. W., & Worrall, D. M. 2002. "Magnetic field strengths in the hotspots and lobes of three powerful FRII radio sources", ApJ, 581, 948–973.

Hardee, P. E. 1987. "Spatial stability of relativistic jets - Application to 3C 345", ApJ, 318, 78–92.

Hardee, P. E., Hughes, P. A., Rosen, A., & Gómez, E. A. 2001. "Relativistic jet response to precession and wave-wave interactions", ApJ, 555, 744–757.

Harris, D. E. 2002. "Current Problems for X-ray Emission from Radio Jets", in ASP Conf. Ser. Vol. 250, Particles and Fields in Radio Galaxies, ed. R. A. Laing & K. M. Blundell (San Francisco: ASP) 204–212.

Harris, D. E., Biretta, J. A., Junor, W., Perlman, E. S., Sparks, W. B., & Wilson, A. S. 2003. "Flaring X-ray Emission from HST-1, a Knot in the M 87 Jet", ApJ, 586, L41–L44.

Harris, D. E., & Krawczynski, H. 2002. "X-ray Emission Processes in Radio Jets", ApJ, 565, 244–255.

Hartman, R. C., Bertsch, D. L., Bloom, S. D., Chen, A. W., Deines-Jones, P., Esposito, J. A., Fichtel, C. E., Friedlander, D. P., Hunter, S. D., McDonald, L. M., Sreekumar, P., Thompson, D. J., Jones, B. B., Lin, Y. C., Michelson, P. F., Nolan, P. L., Tompkins, W. F., Kanbach, G., Mayer-Hasselwander, H. A., Mücke, A., Pohl, M., Reimer, O., Kniffen, D. A., Schneid, E. J., von Montigny, C., Mukherjee, R., Dingus. B. L. 1999. "The Third EGRET Catalog of High-Energy Gamma-Ray Sources," ApJS, 123, 79–202.

Hasinger, G., Bergeron, J., Mainieri, V., Rosati, P., Szokoly, G., & Cdfs Team. 2002. "Understanding the sources of the X-ray background: VLT identifications in the *CHANDRA*/*XMM*-Newton Deep Field South", Messenger, 108, 11–16.

Hasinger, G., Burg, R., Giaconni, R., Schmidt, M., Trumper, J., & Zamorani, G. 1998. "The *ROSAT* Deep Survey. I. X-ray sources in the Lockman Field", A&A, 329, 482–494.

Heeschen, D. S. 1982. "Radio Flux Flicker of Extragalactic Sources", in Proc. IAU Symp. 97, Extragalactic Radio Sources, ed. D. S. Heeschen, & C. M. Wade, (Dordrecht: Kluwer), 327–328.

———. 1984. "Flickering of extragalactic radio sources", AJ, 89, 1111–1123.

Heeschen, D. S., & Rickett, B. J. 1987. "The Galactic Latitude Dependence of Centimeter-Wavelength Flicker", ApJ, 93, 589–591.

Henstock, D. R., Browne, I. W. A., Wilkinson, P. N., Taylor, G. B., Vermeulen, R. C., Pearson, T. J., & Readhead, A. C. S. 1995. "The Second Caltech–Jodrell Bank VLBI Survey. II. Observations of 102 of 193 Sources", ApJS, 100, 1–36.

Hewish, A., Scott, P. F., & Wills, D. 1964. "Interplanetary scintillation of small diameter radio sources", Nature, 203, 1214–1217.

Hirabayashi, H., Fomalont, E. B., Horiuchi, S., Lovell, J. E. J., Moellenbrock, G. A., Inoue, M., Burke, B. F., Dewdney, P. E., Gurvits, L. I., Kobayashi, H., Jauncey, D. L., Murata, Y., McCulloch, P., Preston, R. A., Avruch, I. M., Edwards, P. G., Dougherty, S. M., Scott, W. K., Frey, S., Paragi, Z., Kovalev, Y. A., Popov, M., Romney, J. D., Schilizzi, R. T., Shen, Z. Q., Nicolson, G., Quick, J., Costa, M., Dodson, R., Reynolds, J. E., Tzioumis, A. K., Tingay, S. J., Hong, X-Y., Liang, S.-G., Huang, X.-Y., Wei, W.-R., Trigilio, C., Tuccari, G., Nakajima, J., Kawai, E., Umemoto, T., Miyaji, T., Fujisawa, K., Kus, A., Kawaguchi, N. Ghigo, F., Salter, C., Ghosh, T., Kanevsky, B., Slysh, V., Gunn, A., Burgess, P., Carlson, B., Del Rizzo, D., Taylor, R., Cannon, W., Kameno, S., Shibata, K. M., Benson, J., Flatters, C., Hale, A., Lewis, C., Langston, G., Minter, A., Miller, K., Smith, J., Wietfeldt, R., Altunin, V., Meier, D. L., Murphy, D. W., Resch, G., Lister, M. L., Piner, B. G., Jenkins, R., Border, J., & Gimeno, J. 2000. "The VSOP 5 GHz AGN Survey I. Compilation and Observations" PASJ, 52, 997–1014.

Hirabayashi, H., Hirosawa, H., Kobayashi, H., Murata, Y., Edwards, P. G., Fomalont, E. B., Fujisawa, K., Ichikawa, T., Kii, T., Lovell, J. E. J., Moellenbrock, G. A., Okayasu, R., Inoue, M., Kawaguchi, N., Kameno, S., Shibata, K. M., Asaki, Y., Bushimata, T., Enome, S., Horiuchi, S., Miyaji, T., Umemoto, T., Migenes, V., Wajima, K., Nakajima, J., Morimoto, M., Ellis, J., Meier, D. L., Murphy, D. W., Preston, R. A., Smith, J. G., Tingay, S. J., Traub, D. L., Wietfeldt, R. D., Benson, J. M., Claussen, M. J., Flatters, C., Romney, J. D., Ulvestad, J. S., D'Addario, L. R., Langston, G. I., Minter, A. H., Carlson, B. R., Dewdney, P. E., Jauncey, D. L., Reynolds, J. E., Taylor, A. R., McCulloch, P. M., Cannon, W. H., Gurvits, L. I., Mioduszewski, A. J., Schilizzi, R. T., & Booth, R. S. 1998. "Overview and Initial Results of the Very Long Baseline Interferometry Space Observatory Programme", Science, 281, 1825 (erratum 282, 1995)

Hirabayashi, H., Murata, Y., Edwards, P. G., Murphy, D. W., Inoue, M., Kobayashi, H., & Kameno, S. 2001. "Current Status of the VSOP-2 Mission", in IAU Symp. 205, Galaxies and their Constituents at the Highest Angular Resolutions, ed. R. T. Schilizzi, S. N. Vogel, F. Paresce, & M. Elvis, (San Francisco: ASP), 428–429.

Hjellming, R. M., & Johnston, K. J. 1981. "An analysis of the proper motions of SS 433 radio jets", ApJ, 246, L141–L145.

———. 1988. "Radio emission from conical jets associated with X-ray binaries", ApJ, 328, 600–609.

Hjellming, R. M., Rupen, M. P., Hunstead, R. W., Campbell-Wilson, D., Mioduszewski, A. J., Gaensler, B. M., Smith, D. A., Sault, R. J., Fender, R. P., Spencer, R. E., de la Force, C. J., Richards, A. M. S., Garrington, S. T., Trushkin, S. A., Ghigo, F. D., Waltman, E. B., & McCollough, M. 2000. "Light Curves and Radio Structure of the 1999 September Transient Event in V4641 Sagittarii (=XTE J1819−254=SAX J1819.3−2525)", ApJ, 544, 977–992.

Ho, L. C., Filippenko, A. V., & Sargent, W. L. W. 1997. "A search for 'dwarf' Seyfert nuclei. III. Spectroscopic parameters and properties of the host galaxies", ApJS, 112, 315–390.

Ho, L. C., & Peng, C. Y. 2001. "Nuclear luminosities and radio loudness of Seyfert nuclei", ApJ, 555, 650–662.

Homan, D. C. 2002. "Non-Ballistic Motions in Relativistic Jets", BAAS, 34, 1245.

Homan, D. C., Attridge, J. M., & Wardle, J. F. C. 2001. "Parsec-Scale Circular Polarization Observations of 40 Blazars", ApJ, 556, 113–120.

Homan, D. C., Lister, M. L., Kellermann, K. I., Cohen, M. H., Ros, E., Zensus, J. A., Kadler, M., Vermeulen, R. C. 2003. "Jet Collimation in Action: Re-Alignment on the Kiloparsec-Scale 3C 279", ApJ, in press.

Homan, D. C., Ojha, R., Wardle, J. F. C., Roberts, D. H., Aller, M. F., Aller, H. D., & Hughes, P. A. 2001. "Parsec-Scale Blazar Monitoring: Proper Motions", ApJ, 549, 840–861.

———. 2002. "Parsec-Scale Blazar Monitoring: Flux and Polarization Variability", ApJ, 568, 99–119.

Homan, D. C., & Wardle, J. F. C. 1999. "Detection and Measurement of Parsec-Scale Circular Polarization in Four AGNs", AJ, 118, 1942–1962.

———. 2000. "Direct Distance Measurements to Superluminal Radio Sources", ApJ, 535, 575–585.

———. 2003. "Probing the Circular Polarization of Relativistic Jets on VLBI Scales", in: Circular Polarization of Relativistic Jet Sources, ed. R. P. Fender & J.-P. Macquart, Ap&SS, in press (astro-ph/0211183).

Hooda, J. S., & Wiita, P. J. 1998. "Instabilities in three-dimensional simulations of astrophysical jets crossing angled interfaces", ApJ, 493, 81–90.

Hook, I. M., McMahon, R. G., Shaver, P. A., & Snellen, I. A. G. 2002. "Discovery of radio-loud quasars with redshifts above 4 from the PMN sample", A&A, 391, 509–517.

Hook, I. M., Shaver, P. A., Jackson, C. A., Wall, J. V., & Kellermann, K. I. 2003. "The Parkes quarter-Jansky flat-spectrum sample 2. New Optical Spectra and Redshift Measurements", A&A, 399, 469–487.

Hoyle, F. 1959. "The relation of radio astronomy to cosmology", in Proc. of the Paris Symp. on Radio Astron., IAU Symp. No. 9 and URSI Symp. No. 1, ed. R. N. Bracewell, (Stanford: Stanford Univ. Press), 529–532.

Hoyle, F., Burbidge, G., & Narlikar, J. V. 2000. "A Different Approach to Cosmology: from a Static Universe through the Big Bang Towards Reality", (Cambridge: CUP).

Hoyle, F., & Fowler, W. A. 1963. "On the nature of strong radio sources", MNRAS, 125, 169–176.

———. 1963. "Nature of Strong Radio Sources", Nature, 197, 533–535.

Hoyle, F., Fowler, W. A., Burbidge, G. R., & Burbidge, E. M. 1964. "On Relativistic Astrophysics", ApJ, 139, 909–928.

Hu, E. M., Cowie, L. L., McMahon, R. G., Capak, P., Iwamuro, F., Kneib, J.-P., Maihara, T., & Motohara, K. 2002. "A redshift z=6.56 galaxy behind the cluster Abell 370", ApJ, 568, L75–L79 (erratum ApJ, 576, L99).

Hughes, D. H., Serjeant, S., Dunlop, J., Rowan-Robinson, M., Blain, A., Mann, R. G., Ivison, R., Peacock, J., Efstathiou, A., Gear, W., Oliver, S., Lawrence, A., Longair, M., Goldschmidt, P., & Jenness, T. 1998. "High-redshift star formation in the Hubble Deep Field revealed by a submillimetre-wavelength survey", Nature, 394, 241–247.

Hughes, P. A., Aller, H. D., & Aller, M. F. 1985. "Polarized radio outbursts in BL Lacertae. I. Polarized emission from a compact jet. ", ApJ, 298, 296–300.

———. 1985. "Polarized radio outbursts in BL Lacertae. II. The flux and polarization of a piston-driven shock", ApJ, 298, 301–315.

———. 1989. "Synchrotron emission from shocked relativistic jets. II. A model for the centimeter wave band quiescent and burst emission from BL Lacertae", ApJ, 341, 68–79.

———. 1992. "The University of Michigan radio astronomy data base. I. Structure function analysis and the relation between BL Lacertae objects and quasi-stellar objects", ApJ, 396, 469–486.

Hummel, C. A., Schalinski, C. J., Krichbaum, T. P., Rioja, M. J., Quirrenbach, A., Witzel, A., Muxlow, T. W. B., Johnston, K. J., Matveyenko, L. I., & Shevchenko, A. 1992. "The jets of quasar 1928+738: superluminal motion and large-scale structure", A&A, 257, 489–500.

Hunstead, R. W. 1972. "Four variable radio sources at 408 MHz", ApL, 12, 193–200.

Högbom, J. A. 1959. "The Structure and Magnetic Field of the Solar Corona", Ph.D. thesis, (Cambridge: Univ. of Cambridge).

———. 1974. "Aperture Synthesis with a Non-Regular Distribution of Interferometer Baselines", A&AS, 15, 417–426.

Högbom, J. A., & Brouw, W. N. 1974. "The Synthesis Radio Telescope at Westerbork. Principles of Operation, Performance, and Data Reduction", A&A, 33, 289–301.

Iliev, I. T., Shapiro, P. R., Ferrara, A., & Martel, H. 2002. "On the Direct Detectability of the Cosmic Dark Ages: 21 Centimeter Emission from Minihalos", ApJ, 572, L123–L126.

Impey, C., & Petry, C. 2001. "Radio-quiet quasars in the direction of the northern Hubble deep field", ApJ, 547, 117–123.

Ishwara-Chandra, C. H., & Saikia, D. J. 1999. "Giant radio sources", MNRAS, 309, 100–112.

Istomin Ya. N., & Pariev, V. I. 1996. "Stability of a Relativistic Rotating Electron-Positron Jet: Non-Axisymmetric Perturbations", MNRAS, 281, 1–26.

Jackson, J. C., & Dodgson, M. 1996. "On universes with outsides and the angular-size/redshift diagram for milliarcsecond radio-sources", MNRAS, 278, 603–610.

Jackson, C. A., & Wall, J. V. 1999. "Extragalactic radio source evolution under the dual-population unification scheme", MNRAS, 304, 169–174.

Jackson, C. A., Wall, J. V., Shaver, P. A., Kellermann, K. I., & Hook, I. M. 2002. "The Parkes quarter-Jansky flat-spectrum sample. I. Sample selection and source identifications", A&A, 386, 97–113.

Jarvis, M., & Rawlings, S. 2000. "On the redshift cut-off for flat-spectrum radio sources", MNRAS, 319, 121–136.

Jauncey, D. L. 1967. "Re-examination of the Source Counts for the 3C Revised Catalog", Nature, 216, 877–878.

Jauncey, D. L., Kedziora-Chudczer, L. L., Lovell, J. E. J., Nicolson, G. D., Perley, R. A., Reynolds, J. E., Tzioumis, A. K., & Wieringa, M. H. 2000. "The Origin of Intra-Day Variability", Astrophysical Phenomena Revealed by Space VLBI, ed. H. Hirabayashi, P. G. Edwards, & D. W. Murphy, (Tokyo: ISAS), 147–150.

Jauncey, D. L., & Macquart, J.-P. 2001. "Intra-day variability and the interstellar medium towards 0917+624", A&A, 370, L9–L12.

Jeyakumar, S., & Saikia, D. J. 2000. "Collimation of extragalactic radio jets in compact steep-spectrum and larger sources", MNRAS, 311, 397–404.

Jog, C. J., & Das, M. 1992. "Starbursts triggered by central overpressure in interacting galaxies", ApJ, 400, 476–483.

———. 1996. "Starbursts triggered by central overpressure. II. a parameter study", ApJ, 473, 797–805.

Jones, T. W. 1988. "Polarization as a probe of magnetic field and plasma properties of compact radio sources - Simulation of relativistic jets", ApJ, 332, 678–695.

Jones, T. W., & Burbidge, G. R. 1973. "On the Problem of Explaining Decimeter Flux Variations in 3C 454.3 and Other Extragalactic Sources", ApJ, 186, 791–799.

Jones, T. W., Rudnick, L., Fiedler, R. L., Aller, H. D., Aller, M. F., & Hodge, P. E. 1985. "Magnetic field structures in compact radio sources", ApJ, 290, 627–636.

Jorstad, S. G., Marscher, A. P., Mattox, J. R., Aller, M. F., Aller, H. D., Wehrle, A. E., & Bloom, S. D. 2001. "Multi-epoch Very Long Baseline Array observations of EGRET-detected quasars and BL Lacertae objects: connection between superluminal ejections and gamma-ray flares in blazars", ApJ, 556, 738–748.

Jorstad, S. G., Marscher, A. P., Mattox, J. R., Wehrle, A. E., Bloom, S. D, & Yurchenko, A. V. 2001. "Multi-Epoch VLBA Observations of EGRET-Detected Quasars and BL Lac Objects: Superluminal Motion of Gamma-Ray Bright Blazars", ApJS, 134, 181–240.

Jukes, J. D. 1967. "Continuum Radiation from Quasi-Stellar Sources", Nature, 216, 461.

Junor, W., Biretta, J. A., & Livio, M. 1999. "Formation of the radio jet in M 87 at 100 Schwarzschild radii from the central black hole", Nature, 401, 891–892.

Kaiser, C. R., & Alexander, P. 1997. "A self-similar model for extragalactic radio sources", MNRAS, 286, 215–222.

Kaiser, C. R., Dennett-Thorpe, J., & Alexander, P. 1997. "Evolutionary tracks of FR II sources through the $P - D$ diagram", MNRAS, 292, 723–732.

Kayser, R. 1995. "A cosmological test with compact radio sources", A&A, 294, L21–L23.

Kapahi, V. 1987. "The angular size-redshift relation as a cosmological tool", in Observational Cosmology, ed. A. Hewitt, G. Burbidge, & L. Z. Fang, (Dordrecht: Reidel), 251–265.

Kapahi, J. K. 1981. "Westerbork observations of radio sources in the 5 GHz 'S4' survey", A&AS, 43, 381–393.

Kardashev, N. S. 1962. "Nonstationarity of Spectra of Young Sources of Nonthermal Radio Emission", AZh, 39. 393–399.

———. 1964. "Transmission of Information by Extraterrestrial Civilizations", Soviet Ast., 8, 217–221.

———. 1995. "Cosmic supercollider", MNRAS, 276, 515–520.

———. 1997. "*RadioAstron* — a Radio Telescope Much Greater than the Earth", Exp. Astron., 7, 329–343.

———. 2000. "Radio Synchrotron Emission by Protons and Electrons in Pulsars and the Nuclei of Quasars", AZh, 77, 813–818. (translation in Astron. Reports, 44, 719–724.)

———. 2001. "Active galactic nuclei and pulsars as cosmic ray sources", MNRAS, 326, 1122–1126.

Kedziora-Chudczer, L. 1998. "The Study of Intraday Variability of Extragalactic Compact Radio Sources", Ph.D. thesis, (Sydney: Sydney Univ.).

Kedziora-Chudczer, L., Jauncey, D. L., Lovell, J. E. J., Walker, M. A., Macquart, J.-P., Wieringa, M. H., Tzioumis, A. K., Perley, R. A., & Reynolds, J. E. 2001. "Long Term Monitoring of the Extreme Intraday Variable Quasar PKS 0405−385", in ASP Conf. Ser. Vol. 250, Particles & Fields in Radio Galaxies, ed. R. A. Laing & K. M. Blundell, (San Francisco: ASP), 128–132.

Kedziora-Chudczer, L., Jauncey, D. L., Wieringa, M. H., Tzioumis, A. K., & Reynolds, J. E. 2001. "The ATCA intraday variability survey of extragalactic radio sources", MNRAS, 325, 1411–1430.

Kedziora-Chudczer, L., Jauncey, D. L., Wieringa, M. H., Walker, M. A., Nicolson, G. D., Reynolds, J. E., & Tzioumis, A. K. 1997. "PKS 0405−385: the Smallest Radio Quasar?", ApJ, 490, L9–L12.

Kedziora-Chudczer, L., Macquart, J.-P., & Jauncey, D. L. 2001. "Extreme examples of intraday variability - Search for diffractive scintillation in the smallest quasar, PKS 0405−385", in IAU Symp. No. 205, Galaxies and Their Constituents at the Highest Angular Resolutions, ed. R. T. Schilizzi, S. N. Vogel, F. Paresce, & M. S. Elvis, (San Francisco: ASP), 90–91.

Kellermann, K. I. 1964. "The Spectra of Non-Thermal Radio Sources", ApJ, 140, 969–991.

———. 1965. "11-cm Observations of the Temperature of Mercury", Nature, 205, 1091–1092.

———. 1965. "Radio Observations of Mars", Nature, 206, 1034–1035.

———. 1965. "The Spectrum of the Radio Source 1934–63", Aust. Jou. of Phys., 19, 195–207.

———. 1966. "The Radio Source 1934–63", Aust. J. Phys, 19, 195–207.

———. 1966. "The Thermal Radio Emission from Mercury, Venus, Mars, Saturn and Uranus', Icarus, 5, 478–490.

———. 1966. "On the Interpretation of Radio-Source Spectra and the Evolution of Radio Galaxies and Quasi-Stellar Sources", ApJ, 146, 621–633.

———. 1972. "Radio galaxies, Quasars, and Cosmology", AJ, 77, 531–542.

———. 1985. "Superluminal radio sources", Com. Ap., 11, 69–81.

———. 1993. "The angular size – redshift relation for compact radio sources", in Sub-Arcsecond Radio Astronomy, ed. R. J. Davies and R. S. Booth, (Cambridge: CUP), 386–389.

———. 1993. "The cosmological deceleration parameter estimated from the angular-size/redshift relation for compact radio sources", Nature, 361, 134–136.

———. 1994. "Compact Radio Sources, Quasars and Active Galactic Nuclei', Aust. J. of Phys., 47, 599–613.

———. 2002. "Brightness Temperature Constraints to Compact Synchrotron Source Radiation Obtained from IDV and VLBI Observations", PASA, 19, 77–82.

Kellermann, K. I., Clark, B. G., Bare, C. C., Rydbeck, O., Ellder, J., Hansson, B., Kollberg, E., Hoglund, B., Cohen, M. H., & Jauncey, D. L. 1968. "High-Resolution Interferometry of Small Radio Sources Using Intercontinental Base Lines", ApJ, 153, L209–L214.

Kellermann, K. I., Clark, B. G., Jauncey, D. L., Broderick, J. J., Shaffer, D. B., Cohen, M. H., & Niell, A. E. 1973. "Observations of Further Outbursts in the Radio Galaxy 3C 120", ApJ, 183, L51–L55.

Kellermann, K. I., Clark, B. G., Shaffer, D. B., Cohen, M. H., Jauncey, D. L., Broderick, J. J., & Niell, A. E. 1974. "Further Observations of Apparent Changes in the Structure of 3C 273 and 3C 279", ApJ, 189, L19–L22.

Kellermann, K. I., & Cohen, M. H. 1988. "The origin and evolution of the N.R.A.O.-Cornell VLBI System", JRASC, 82, 248–265.

Kellermann, K. I., Davis, M. M., & Pauliny-Toth, I. I. K. 1971. "Counts of Radio Sources at 6 Centimeter Wavelength", ApJ, 170, L1–L5.

Kellermann, K. I., Jauncey, D. L., Cohen, M. H., Shaffer, B. B., Clark, B. G., Broderick, J., Rönnäng, B., Rudbeck, O. E. H., Matveyenko, L., Moiseyev, I., Vitkevitch, V. V., Coper, B. F. C., & Batchelor, R. 1971. "High-Resolution Observations of Compact Radio Sources at 6 and 18 centimeters", ApJ, 169, 1–24

Kellermann, K. I., Lister, M. L., Homan, D. C., Ros, E., Zensus, J. A., Cohen, M. H., Russo, M., & Vermeulen, R. C. 2003. "Superluminal Motion and Relativistic Beaming in Blazar Jets", in ASP Conf. Ser. Vol. 299, High Energy Blazar Astronomy, ed. L. O. Takalo, & E. Valtaoja (San Francisco: ASP), in press (astro-ph/0211398).

Kellermann, K. I., Long, R. J., Allen, L. R., & Moran, M. 1962. "A correlation between spectra of non-thermal radio sources and their brightness temperatures", Nature, 195, 692–693.

Kellermann, K. I., & Moran, J. M. 2001. "The Development of High-Resolution Imaging in Radio Astronomy", ARA&A, 39, 457–509.

Kellermann, K. I., & Pauliny-Toth, I. I. K. 1966. "Observations of the Radio Emission of Uranus, Neptune and Other Planets at 1.9 cm', ApJ, 145, 954–957.

———. 1968. "Variable Radio Sources", ARA&A, 6, 417–448.

———. 1968. "Further Observations of Variable Radio Sources", ApJ, 152, 639–646.

———. 1969. "The Spectra of Opaque Radio Sources", ApJ, 155, L71–L78.

———. 1981. "Compact radio sources", ARA&A, 19, 373–410.

Kellermann, K. I., & Read, R. B. 1965. "A continuum survey for discrete radio sources at 1421 MC/S", Publ. Owens Valley Radio Obs., 1, 1–5.

Kellermann, K. I., Shaffer, D. B., Clark, B. G., & Geldzahler, B. J. 1977. "The small radio source at the galactic center", ApJ, 214, L61–L62.

Kellermann, K. I., & Sheets, B. (ed.) 1983. "Serendipitous Discoveries in Radio Astronomy", Proc. of a Workshop May 4–6, 1983, (Green Bank, WV: NRAO).

Kellermann, K. I., Vermeulen, R. C., Cohen, M. H., & Zensus, J. A. 1997. "Parsec-Scale Imaging of Quasars and AGN", BAAS, 29, 1257.

———. 1998. "Sub-milliarcsecond Imaging of Quasars and AGN", AJ, 115, 1295–1318.

———. 1998. "The Submilliarcsecond Structure of Quasars and AGN", in ASP Conf. Ser. Vol. 144, Radio Emission from Galactic and Extragalactic Compact Sources, ed. J. A. Zensus, G. B. Taylor, & J. M. Wrobel, (San Francisco: ASP) 153–154.

———. 1999. "The Shroud Around the Twin Jets of NGC 1052", BAAS, 31, 856.

———. 2000. "Observations of Relativistic Outflow in AGN and the Brightness Temperature of Synchrotron Sources", in Astrophysical Phenomena Revealed by Space VLBI, ed. H. Hirabayashi, P. G. Edwards, & D. W. Murphy, (Tokyo: ISAS), 159–166.

Kellermann, K. I., Vermeulen, R. C., Zensus, J. A., Cohen, M. H., & West, A. 1999. "Kinematics of quasars and AGN", New Astron. Rev., 43, 757–760

Kelly, B., Hughes, P. A., Aller, H. D., & Aller, M. F. 2003. "The cross-wavelet transform and analysis of quasiperiodic behavior in the Pearson-Readhead VLBI survey sources", ApJ, in press (astro-ph/0301002).

King, A. R., Taam, R. E., & Begelman, M. C. 2000. "The Evolutionary Status of SS 433", ApJ, 530, L25–L28.

Kinney, A. L., Schmitt, H. R., Clarke, C. J., Pringle, J. E., Ulvestad, J. S., & Antonucci, R. R. J. 2000. "Jet directions in Seyfert galaxies", ApJ, 537, 152–177.

Kobayashi, H., Shimoikura, T., Omodaka, T., & Diamond, P. J. 2000. "Monitoring of the Orion-KL water maser outburst", in Astrophysical Phenomena Revealed by Space VLBI, ed. Hirabayashi, P. G. Edwards, & D. W. Murphy (ed.), 109–112.

Kogan, L. R., Linkin, V. M., Matveenko, L. I., Naumov, B. P., Nenarokov, D. F. 1974. "Study of Venus atmosphere circulation by VLBI method", in Soviet-French Symp. on "EOS–Venus" Project, Moscow, 1–10 April, 1974, 7–9.

Kogan, L. R., & Matveenko, L. I. 1976. "VLBI registration and processing system", Radiofizika, 9, 1265–1269.

Komesaroff, M. M., Roberts, J. A., Milne, D. K., Rayner, P. T., & Cooke, D. J. 1984. "Circular and linear polarization variations of compact radio sources", MNRAS, 208, 409–425.

Korchak, A. A., & Terletsky, Y. P. 1952. "Elektromagnitnoe Izluchenie Kosmicheskikh Protonov I Radioizluchenie Galaktiki", Zhurnal Eksperimentalnoi I Teoreticheskoi Fiziki, 22, 507–509.

Kostenko, Val. I., Matveenko, L. I. 1982. "Precision antenna of space radiotelescope", Kosmicheskie Issledovanija, 20, 149–151.

Kovalev, Y. Y., Kovalev, Yu. A., Nizhelsky, N. A., & Bogdantsov, A. B. 2002. "Broad-band Radio Spectra Variability of 550 AGN in 1997–2001", PASA, 19, 83–87.

Kovalev, Y. Y., Nizhelsky, N. A., & Bogdantsov, A. V. 2001. "Strong radio flares occurring in compact relativistic jets of AGNs", in Similarities and Universality in Relativistic Flows, ed. A. Guthmann, K. Manolaku, & M. Georganopoulos, (Berlin: Logos Verlag), 74–81.

Kovalev, Y. Y., Nizhelsky, N. A., Kovalev, Yu. A., Berlin, A. B., Zhekanis, G. V., Mingaliev, M. G., & Bogdantsov, A. V. 1999. "Survey of instantaneous 1–22 GHz spectra of 550 compact extragalactic objects with declinations from -30 to +43 degrees", A&AS, 139, 545–554.

Kovalev, Yu. A., Kovalev, Y. Y., & Nizhelsky, N. A. 2000. "Broad-Band Spectra Study of 213 VSOP 5-GHz Survey Sources", PASJ, 52, 1027–1036.

Kraft, R. P., Forman, W. R., Jones, C., Murray, S. S., Hardcastle, M. J., & Worrall, D. M. 2002. "*CHANDRA* Observations of the X-ray Jet in Centaurus A", ApJ, 569, 54–71.

Kraus, A., Krichbaum, T. P., Wegner, R., Witzel, A., Cimò, G., Quirrenbach, A., Britzen, S., Fuhrmann, L., Lobanov, A. P., Naundorf, C. E., Otterbein, K., Peng, B., Risse, M., Ros,

E., & Zensus, J. A. 2003. "Intraday variability in compact extragalactic radio sources. II. Observations with the Effelsberg 100m radio telescope", A&A, 401, 161–172.

Kraus, A., Quirrenbach, A., Lobanov, A. P., Krichbaum, T. P., Risse, M., Schneider, P., Qian, S. J., Wagner, S. J., Witzel, A., Zensus, J. A., Heidt, J., Bock, H., Aller, M., & Aller, H. 1999. "Unusual radio variability in the BL Lacertae object 0235+164", A&A, 344, 807–816.

Krauss, L. M., & Schramm, D. N. 1993. "Angular diameters as a probe of a cosmological constant and Ω", ApJ, 405, L43–L46.

Krichbaum, T. P., Graham, D. A., Witzel, A., Greve, A., Wink, J. E., Grewing, M., Colomer, F., de Vicente, P., Gómez-Gónzalez, J., Baudry, A., & Zensus, J. A. 1998. "VLBI observations of the galactic center source Sgr A* at 86 GHz and 215 GHz", A&A, 335, L106–L110.

Krolik, J. H. 1999. "Active Galactic Nuclei: From the Central Black Hole to the Galactic Environment", (Princeton: Princeton Univ. Press).

———. 2001. "Systematic Errors in the Estimation of Black Hole Masses by Reverberation Mapping", ApJ, 551, 72–79.

Kronberg, P. P., Dufton, Q. W., Li, H., & Colgate, S. A. 2001. "Magnetic energy of the intergalactic medium from galactic black holes", ApJ, 560, 178–186.

Kronberg, P. P., Lesch, H., & Hopp, U. 1999. "Magnetization of the intergalactic medium by primeval galaxies", ApJ, 511, 56–64.

Kühr, H., & Schmidt, G. D. 1990. "Complete Samples of Radio-Selected BL Lac Objects", AJ, 90, 1–6.

Kühr, H., Witzel, A., Pauliny-Toth, I. I. K., & Nauber, U. 1981. "A catalogue of extragalactic radio sources having flux densities greater than 1 Jy at 5 GHz", A&AS, 45, 367–430.

Lähteenmäki, A., & Valtaoja, E. 1999. "Total Flux Density Variations in Extragalactic Radio Sources. III. Doppler Boosting Factors, Lorentz Factors, and Viewing Angles for Active Galactic Nuclei", ApJ, 521, 493–501.

———. 2003. "Testing of Inverse Compton Models for Active Galactic Nuclei with Gamma-Ray and Radio Observations", ApJ, in press.

Lähteenmäki, A., Valtaoja, E., & Wiik, K. 1999. "Total Flux Density Variations in Extragalactic Radio Sources. II. Determining the Limiting Brightness Temperature for Synchrotron Sources", ApJ, 511, 112–117

Laing, R. A. 1980. "A Model for the Magnetic-Field Structure in Extended Radio Sources", MNRAS, 193, 439–449.

Laing, R. A., Parma, P., de Ruiter, H. R., & Fanti, R. 1999. "Asymmetries in the jets of weak radio galaxies", MNRAS, 306, 513–530.

Laing, R. A., Riley, J. M, & Longair, M. S. 1983. "Bright radio sources at 178 MHz Flux densities, optical identifications and the cosmological evolution of powerful radio galaxies", MNRAS, 204, 151–187.

Lamb, D. Q., & Reichart, D. E. 2000. "Gamma-Ray Bursts as a Probe of the Very High Redshift Universe", in AIP Conf. Ser. Vol. 526, Gamma-ray Bursts, ed. R. M. Kippen, R. S. Mallozzi, & G. J. Fishman, (Melville, NY: American Inst. of Phys.), 658–662.

Lamer, G., Brunner, H., & Staubert, R. 1996. "*ROSAT* observations of BL Lacertae objects", A&A, 311, 384–392.

Landy, S. D., & Szalay, A. S. 1993. "Bias and variance of angular correlation functions", ApJ, 412, 64–71.

Lara, L., Cotton, W. D., Feretti, L., Giovannini, G., Marcaide, J. M., Márquez, I., & Venturi, T. 2001. "A new sample of large angular size radio galaxies. I. The radio data", A&A, 370, 409–425.

Lara, L., Giovannini, G., Cotton, W. D., Feretti, L., Marcaide, J. M., Márquez, I., Taylor, G. B., & Venturi, T. 2002. "Restarting activity in radio galaxies", New Astron. Rev., 46, 89–93.

Lestrade, J.-F., Preston, R. A., Jones, D. L., Phillips, R. B., Rogers, A. E. E., Titus, M. A., Rioja, M. J., & Gabuzda, D. C. 1999. "High-precision VLBI astrometry of radio-emitting stars", A&A, 344, 1014–1026.

Levy, G. S., Linfield, R. P., Edwards, C. D., Ulvestad, J. S., Jordan, J. F., Jr., Dinardo, S. J., Christensen, C. S., Preston, R. A., Skjerve, L. J., Stavert, L. R., Burke, B. F., Whitney, A. R., Cappallo, R. J., Rogers, A. E. E., Blaney, K. B., Maher, M. J., Ottenhoff, C. H., Jauncey, D. L., Peters, W. L., Reynolds, J., Nishimura, T., Hayashi, T., Takano, T., Yamada, T., Hirabayashi, H., Morimoto, M., Inoue, M., Shiomi, T., Kawaguchi, N., Kunimori, H., Tokumaru, M., & Takahashi, F. 1989. "VLBI using a telescope in Earth orbit. I - The observation", ApJ, 336, 1098–1104.

Li, J., & Wilson, G. 1999. "Solutions to the Bernoulli Integral of the Funnel Flow", ApJ, 527, 910–917.

Li, Z., Begelman, M. C., & Chiueh, T. 1992. "The effects of radiation drag on radial, relativistic hydromagnetic winds", ApJ, 384, 567–579.

Lilly, S. J., Le Fèvre, O., Hammer, F., & Crampton, D. 1996. "The Canada-France Redshift Survey: The Luminosity Density and Star Formation History of the Universe to z approximately 1", ApJ, 460, L1–L4.

Lima, J. A. S., & Alcaniz, J. S. 2000. "Angular size in "quintessebce" cosmology", A&A, 357, 393–396.

———. 2002. "Dark energy and the angular size–redshift diagram for milliarcsecond radio sources", ApJ, 566, 15–18.

Lind, K. R., & Blandford, R. D. 1985. "Semidynamical models of radio jets - Relativistic beaming and source counts", ApJ, 295, 358–367.

Linfield, R. P., Levy, G. S., Ulvestad, J. S., Edwards, C. D., Dinardo, S. J., Stavert, L. R., Ottenhoff, C. H., Whitney, A. R., Cappallo, R. J., Rogers, A. E. E., Hirabayashi, H., Morimoto, M., Inoue, M., Jauncey, D. L., & Nishimura, T. 1989. "VLBI using a telescope in Earth orbit. II - Brightness temperatures exceeding the inverse Compton limit", ApJ, 336, 1105–1112.

Lister, M. L., Kellermann, K. I., Homan, D. C., Ros, E., Zensus, J. A., Kadler, M., Cohen, M. H., Russo, M., & Vermeulen, R. C. 2003. "Kinematics of Relativistic Jets in Active Galactic Nuclei", in ASP Conf. Ser., ed. S. Collin, F. Combes, & I. Shlosman, (San Francisco: ASP), in press.

Lister, M. L., Kellermann, K. I., Vermeulen, R. C., Cohen, M. H., Zensus, J. A., & Ros, E. 2003. "4C +12.50: a superluminal precessing jet in the recent merger system IRAS 13451+1232", ApJ, 584, 135–146.

Lister, M. L., Kellermann, K. I., & Pauliny-Toth, I. I. K. 2002. "The Diverse Properties of GPS Sources", in Proc. of the 6th European VLBI Network Symp., ed. E. Ros, R. W. Porcas, A. P. Lobanov, & J. A. Zensus, (Bonn: MPIfR), 135–138.

Lister, M. L., & Marscher, A. P. 1997. "Statistical Effects of Doppler Beaming and Malmquist Bias on Flux-Limited Samples of Compact Radio Sources," ApJ, 476, 572–588.

———. 1999. "Predictions of ECS and SSC Models for Flux- Limited Samples of Gamma-Ray Blazars,", Astroparticle Phys., 11, 65–67.

Lister, M. L., Tingay, S. J., & Preston, R. A. 2001. "The Pearson-Readhead Survey of Compact Extragalactic Radio Sources from Space. II. Analysis of Source Properties", ApJ, 554, 964–980.

Little, L. T., & Hewish, A. 1966. "Interplanetary scintillation and its relation to the angular structure of radio sources", MNRAS, 134, 221–237

———. 1967. "Radio source structure derived from interplanarary scintillation", MNRAS, 138, 393–406

Lo, K. Y. 1974. "Interstellar Microwave Radiation and Early Stellar Evolution", Ph.D. thesis, Mass. Inst. Tech., (Cambridge, MA: MIT).

Lo, K. Y., Backer, D. C., Ekers, R. D., Kellermann, K. I., Reid, M., & Moran, J. M. 1985. "On the size of the galactic centre compact radio source diameter less than 20 AU", Nature, 315, 124–126.

Lo, K. Y., Backer, D. C., Kellermann, K. I., Reid, M., Zhao, J.-H., Goss, W. M., & Moran, J. M. 1993. "High-resolution VLBA imaging of the radio source Sgr A* at the Galactic Centre", Nature, 362, 38–40.

Lo, K. Y., Schilizzi, R. T., Cohen, M. H., & Ross, H. N. 1975. "VLBI observations of the compact radio source in the center of the Galaxy", ApJ, 202, L63–L65.

Lo, K. Y., Shen, Z.-Q., Zhao, J.-H., & Ho, P. T. P. 1998. "Intrinsic Size of Sagittarius A*: 72 Schwarzschild Radii", ApJ, 508, L61–L64.

Loan, A. J., Wall, J. V., & Lahav, O. 1997. "The correlation function of radio sources", MNRAS, 286, 994–1002.

Lobanov, A. P., Krichbaum, T. P., Graham, D. A., Witzel, A., Kraus, A., Zensus, J. A., Britzen, S., Greve, A., & Grewing, M. 2000. "86 GHz VLBI survey of compact radio sources", A&A, 364, 391–408.

Lobanov, A. P., & Zensus, J. A. 2001. "A cosmic double helix in the archetypical quasar 3C 273", Science, 294, 128–131.

Loeb, A., & Rybicki, G. B. 1999. "Scattered Lyα Radiation around Sources before Cosmological Reionization", ApJ, 524, 527–535.

Longair, M. S. 1966. "On the interpretation of radio source counts", MNRAS, 133, 421–436.

Lovelace, R. V. E., & Backer, D. C. 1972. "Extragalactic Radio Sources and Angular Resolution at 25×10^{-6} arc sec", Astroph. Lett., 11, 135–138.

Lovell, A. C. B. 1973. "Out of the Zenith: Jodrell Bank 1957–1970", (Oxford: Oxford Univ. Press; Harper & Row).

Lovell, J. E. J., Jauncey, D. L., Bignall, H. E., Kedziora-Chudczer, L., Macquart, J. P., Rickett, B. J., & Tzioumis, A. K. 2002. "5 GHz VLA IDV Survey of the Northern Sky", in ASP Conf. Ser. Vol., Active Galactic Nuclei: from Central Engine to Host Galaxy, ed. S. Collin, F. Combes, & I. Shlosman, (San Francisco: ASP), 43.

———. 2003. "First Results from MASIV: The Micro-Arcsecond Scintillation-Induced Variability Survey", ApJ, submitted.

Lynden-Bell, D. 1969. "Galactic Nuclei as Collapsed Old Quasars", Nature, 223, 690–694.

Lynden-Bell, D., & Rees, M. J. 1971. "On quasars, dust and the galactic centre", MNRAS, 152, 461–475.

Ma, C., Arias, E. F., Eubanks, T. M., Fey, A. L. Gontier, A.-M., Jacobs, C. S., Sovers, O. J., Archinal, B. A., & Charlot, P. 1998. "The international celestial reference frame as realized by very long baseline interferometry", AJ, 116, 516–546.

Ma, C., Clark, T. A., Ryan, J. W., Herring, T. A., Shapiro, I. I., Corey, B. E., Hinteregger, H. F., Rogers, A. E. E., Whitney, A. R., Knight, C. A., Lundqvist, G. L., Shaffer, D. B., Vandenberg, N. R., Pigg, J. C., Schupler, B. R., & Rönnäng, B. O. 1986. "Radio-source positions from VLBI", AJ, 92, 1020–1029.

Macchetto, F., Marconi, A., Axon, D. J., Capetti, A., Sparks, W., & Crane, P. 1997. "The Supermassive Black Hole of M 87 and the Kinematics of Its Associated Gaseous Disk", ApJ, 489, 579–600.

Macquart, J.-P., & Jauncey, D. L. 2002. "Microarcsecond Radio Imaging using Earth-Orbit Synthesis", ApJ, 572, 786–795.

Macquart, J.-P., Kedziora-Chudczer, L., Rayner, D. P., & Jauncey, D. L. 2000. "Strong, Variable Circular Polarization in PKS 1519−273", ApJ, 538, 623–627.

Madau, P., Meiksin, A., & Rees, M. J. 1997. "21 Centimeter Tomography of the Intergalactic Medium at High Redshift", ApJ, 475, 429–444.

Madau, P., & Rees, M. J. 2000. "The Earliest Luminous Sources and the Damping Wing of the Gunn-Peterson Trough", ApJ, 542, L69–L73.

Magliocchetti, M., Maddox, S. J., Lahav, O., & Wall, J. V. 1998. "Variance and skewness in the FIRST survey", MNRAS, 300, 257–268.

———. 1999. "Constraints on the clustering, biasing and redshift distribution of radio sources", MNRAS, 306, 943–953.

Mahadevan, R. 1997. "Scaling laws for advection-dominated flows: applications to low-luminosity galactic nuclei", ApJ, 477, 585–601.

Maltby, P., & Moffet, A. T. 1965. "Time Dependence of the Radio Emission from CTA 21 and CTA 102", ApJ, 142, 409–412.

Manolakou, K., & Kirk, J. G. 2002. "Modelling the spectral evolution of classical double radio sources", A&A, 391, 127–140.

Marchenko, S. G., Marscher, A. P., Mattox, J. R., Wehrle, J. R., & Xu, W. 1998. "Multi-Epoch VLBA Observations of γ-Ray Bright Blazars", in ASP Conf. Series 144, Radio Emisssion from Galactic and Extragalactic Compact Sources, ed. J. A. Zensus, G. B. Taylor & J. M. Wrobel, (San Francisco: ASP), 67–68.

Marcus, S. L., Chao, Y., Dickey, J. O., & Gegout, P. 1998. "Detection and modeling of nontidal oceanic effects on the Earth's rotation rate", Science, 281, 1656–1659.

Marr, J. M., Taylor, G. B., & Crawford, F. 2001. "Nonuniform Free-Free Absorption in the GPS Radio Galaxy 0108+388", ApJ, 550, 160–171.

Marscher, A. P. 1980. "Relativistic jets and the continuum emission in QSOs", ApJ, 235, 386–391.

———. 2001. "Time delays of blazar flares observed at different wavebands", in ASP Conf. Ser. Vol. 224, Probing the Physics of Active Galactic Nuclei by Multiwavelength Monitoring, ed. B. M. Peterson, R. S. Polidan, & R. E. Pogge (San Francisco: ASP), 23–34.

Marscher, A. P., & Gear, W. K. 1985. "Models for high-frequency radio outbursts in extragalactic sources, with application to the early 1983 millimeter-to-infrared flare of 3C 273", ApJ, 298, 114–127.

Marscher, A. P., Jorstad, S. G., Gómez, J.-L., Aller, M. F., Teräsranta, H., Lister, M. L., & Stirling, A. M. 2002. "Observational evidence for the accretion-disk origin for a radio jet in an active galaxy", Nature, 417, 625–627.

Marscher, A. P., Jorstad, S. G., Gómez, J.-L., Mattox, J. R., Wehrle, A. E. 2002. "High-Frequency VLBA Total and Polarized Intensity Images of Gamma-Ray Blazars", ApJ, 577, 85–97.

Marscher, A. P., Jorstad, S. G., McHardy, I. M., Aller, M. F., Balonek, T. J., Villata, M., Raiteri, C. M., Ostorero, L., Tosti, G., & Teräsranta, H. 2003. "The connection between X-ray and radio-optical emission in blazars", in ASP Conf. Ser. Vol. 299, High Energy Blazar Astronomy, ed. L. O. Takalo, & E. Valtaoja (San Francisco: ASP), in press.

Marscher, A. P., & Travis, J. P. 1996. "Synchrotron self-Compton Interpretation of Multi-waveband Observations of Gamma-Ray Bright Blazars", A&ASS, 120, 537–540.

Marshall, H. L., Canizares, C. R., & Schulz, N. S. 2002. "The High-Resolution X-Ray Spectrum of SS 433 Using the *CHANDRA* HETGS", ApJ, 564, 941–952.

Martocchia, A., Karas, V., & Matt, G. 2000. "Effects of Kerr space-time on spectral features from X-ray illuminated accretion discs", MNRAS, 312, 817–826.

Mattox, J. R., Hallum, J. C., Marscher, A. P., Jorstad, S. G., Waltman, E. B., Teräsranta, H., Aller, H. D., & Aller, M. F. 2001. "A Gamma-Ray Flare of Quasar CTA 26", ApJ, 549, 906–914.

Matveenko, L. I. 1977. "Radio astronomy", Itogi Nauki i Tekhniki (Adv. Sci. Tech.) Ser. Astron., 1–156.

———. 1981. "The fine structure of radio sources", Sov. Sc. Rev. Sec. E, Astroph. and Space Phys. rev., ed. R. A. Syunyaev, V.1, 83–122.

———. 1981. "The bright H_2O maser source in the Orion A", Soviet Ast. Lett., 7, 2, 100–105.

———. 1993. "The low frequency variability of the 3C 345 quasar", Soviet Ast. Lett., 19, 4, 291–294.

Matveenko, L. I., Diamond, P. J., Graham, D. A., & Junor, W. 1992. "VLBI observations of the H_2O supermaser outburst in the Orion Nebula", in Proceed. of Conference Held in Arlington, Virginia, 9–11 March, 1992, 176–179.

Matveenko, L. I., Graham, D. A., & Diamond, P. J. 1988. "The H_2O maser flare in the Orion–KL Nebula", Soviet Ast. Lett., 6, 468–477.

Matveenko, L. I., Kardashev, N. S., & Sholomitsky, G. B. 1965. "On a long baseline radio inteferometer", Radiofizika, 8, 4, 651–654.

Matveenko, L. I., Kellermann, K. I., Pauliny-Toth, I. I. K., Kostenko, V. I., Moiseev, I. G., Kogan, L. R., Witzel, A., Rönnäng, B. O., & Shaffer, D. B. 1980. "Structure of the nucleus of the NGC 1275 Seyfert galaxy", Soviet Ast. Lett., 6, 2, 77–86.

Matveenko, L. I., Kogan, L. R., Kostenko, V. I., Moiseev, I. G., Efanov, V. A., Clark, B. G., Kellermann, K. I., Grove, J. J., Cohen, M. H., Broderick, J. J., & Jauncey, D. L. 1973. "Superfine structure of some compact radiosources", AZh, 50, 6, 1157–1165.

Matveenko, L. I., & Kostenko, V. I. 1979. "On the structure of the Crab Nebula", Aust. J. Phys., 32, 105–111.

Matveyenko, L. I. 1984. "Limitation of the VLBI angular resolution", Proc. of Workshop on *Quasat*, Gr Ensersdorf, Austria, 18–22 June, 1984, (ESA SP-213), 119–121.

Matveyenko, L. I., Sagdeev, R. Z., Balebanov, V. M., Shevchenko, V. I., Kostenko, V. I., Grishmanovskiy, V. A., Velikhov, V. E., Ignatov, S. P., Kasnevskij, B. Z., Kogan, L. R., Kozlov, A. N., Kopelyanskij, G. D., Molodyanu, A. P., Molotov, E. P., Papatsenko, A. H., Romanov, A. M., Strukov, I. A., Timofeev, V. V., Shevchenko, A. V., Severnyi, A. B., Moiseyev, I. G., Sorochenko, R. L., Tsivillev, A. P., Martirosyan, R. M., Aslanyan, A. M., Gulyan, A. G., Yatskiv, Ya. S., & Golovnya, M. V. 1986. "VLBI Network at 18 cm wavelength", Soviet Ast. Lett., 12, 1, 59–65.

McCarthy, P. J., van Breugel, W. J. M., Spinrad, H., & Djorgovski, S. 1987. "A correlation between the radio and optical morphologies of distant 3CR radio galaxies", ApJ, 321, L29–L33.

McCollough, M. L., Robinson, C. R., Zhang, S. N., Harmon, B. A., Hjellming, R. M., Waltman, E. B., Foster, R. S., Ghigo, F. D., Briggs, M. S., Pendleton, G. N., & Johnston, K. J. 1999. "Discovery of Correlated Behavior between the Hard X-Ray and the Radio Bands in Cygnus X-3", ApJ, 517, 951–955.

McCormick, D. G., Spencer, R. E., & Fender, R. P. 2003. "Circular Polarisation Observations of SS 433*", in: Circular Polarization of Relativistic Jet Sources, ed. R. P. Fender & J.-P. Macquart, Ap&SS, in press.

McCready, L. L., Pawsey, J. L., & Payne-Scott, R. 1947. "Solar Radiation at Radio Frequencies and its Relation to Sunspots", Proc. Roy. Soc. of London, Series A, 190, 357–375.

McHardy, I. M., Lawson, A., Newsam, A., Marscher, A., Robson, E. I., & Stevens, J. 1999. "Simultaneous X-ray and infrared variability in the quasar 3C 273", MNRAS, 310, 571–576.

McLure, R. J., & Jarvis, M. J. 2002. "Measuring the black hole masses of high-redshift quasars", MNRAS, 337, 109–116.

Meier, D. L., Edgington, S., Godon, P., Payne, D. G., & Lind, K. R. 1997. "A magnetic switch that determines the speed of astrophysical jets", Nature, 388, 350–352.

Melia, F., & Falcke, H. 2001. "The Supermassive Black Hole at the Galactic Center", ARA&A, 39, 309–352.

Mellema, G., Kurk, J. D., & Röttgering, H. J. A. 2002. "Evolution of clouds in radio galaxy cocoons", A&A, 395, 13–16.

Migliari, S., Fender, R., & Méndez, M. 2002. "Iron Emission Lines from Extended X-ray Jets in SS 433: Reheating of Atomic Nuclei", Science, 297, 1673–1676.

Mirabel, I. F., Dhawan, V., Chaty, S., Rodriguez, L. F., Martí, J., Robinson, C. R., Swank, J., & Geballe, T. 1998. "Accretion instabilities and jet formation in GRS 1915+105", A&A, 330, L9–L12.

Mirabel, I. F., Dhawan, V., Mignani, R. P., Rodrigues, I., & Guglielmetti, F. 2001. "A high-velocity black hole on a Galactic-halo orbit in the solar neighbourhood", Nature, 413, 139–141.

Mirabel, I. F., Mignani, R., Rodríguez, I., Combi, J. A., Rodríguez, L. F., & Guglielmetti, F. 2002. "The runaway black hole GRO J1655−40", A&A, 395, 595–599.

Mirabel, I. F., & Rodríguez, L. F. 1998. "Microquasars in our Galaxy", Nature, 392, 673–676.

———. 1999. "Sources of Relativistic Jets in the Galaxy", ARA&A, 37, 409–443.

Miyaji, T., Hasinger, G., & Schmidt, M. 2000. "Soft X-ray AGN luminosity function from it *ROSAT* surveys. I. Cosmological evolution and contribution to the soft X-ray background", A&A, 353, 25–40.

Miyazaki, S., Sekiguchi, M., Imi, K., Okada, N., Nakata, F., & Komiyama, Y. 1998. "Characterization and mosaicking of CCDs and the applications to the Subaru wide-field camera (Suprime-Cam)", in Proc. SPIE 3355: Optical Astronomical Instrumentation, ed. S. D'Odorico, (Bellingham, WA, USA:Society of Photo-optical Instrumentation Engineers), 363–374.

Miyoshi, M., Moran, J., Herrnstein, J., Greenhill, L., Nakai, N., Diamond, P., & Inoue, M. 1995. "Evidence for a black-hole from high rotation velocities in a sub-parsec region of NGC 4258", Nature, 373, 127–129.

Moellenbrock, G. A., Fujisawa, K., Preston, R. A., Gurvits, L. I., Dewey, R. J., Hirabayashi, H., Inoue, M., Kameno, S., Kawaguchi, M., Iwata, T., Jauncey, D. L., Migenes, V., Roberts, D. H., Schilizzi, R. T., & Tingay, S. J. 1996. "A 22 GHz VLBI Survey of 140 Compact Extragalactic Radio Sources", AJ, 111, 2174–2186.

Moran, J. M., Barrett, A. H., Rogers, A. E. E., Burke, B. F., Zuckerman, B., Penfield, H., & Meeks, M. L. 1967. "Observations of OH emission in the H II region W3 with a 74400λ interferometer", ApJ, 148, L69–L72.

Moran, J. M., Burke, B. F., Barrett, A. H., Rogers, A. E. E., Ball, J. A., Carter, J. C., & Cudaback, D. D. 1968. "The structure of the OH source in W3", ApJ, 152, L97–L100.

Morrison, P., Billingham, J., & Wolfe, J. 1977. "The Search for Extraterrestrial Intelligence", (Washington, DC: NASA).

Mundell, C. G., Pedlar, A., Baum, S. A., O'Dea, C. P., Gallimore, J. F., & Brinks, E. 1994. "MERLIN observations of neutral hydrogen absorption in the Seyfert nucleus of NGC 4151", MNRAS, 272, 355–362.

Mundell, C. G., Wilson, A. S., Ulvestad, J. S., & Roy, A. L. 2000. "Parsec-scale images of flat-spectrum radio sources in Seyfert galaxies", ApJ, 529, 816–831.

Mundell, C. G., Wrobel, J. M., Pedlar, A., & Gallimore, J. F. 2003. "The nuclear regions of the Seyfert galaxy NGC 4151 — parsec-scale H I absorption and a remarkable radio jet", ApJ, 583, 192–204.

Muno, M. P., Morgan, E. H., & Remillard, R. A. 1999. "Quasi-periodic Oscillations and Spectral States in GRS 1915+105", ApJ, 527, 321–340.

Murgia, M. 2003. "Spectral ages of CSOs and CSSs", PASA, 20, 19–24.

Murgia, M., Fanti, C., Fanti, R., Gregorini, L., Klein, U., Mack, K.-H., & Vigotti, M. 1999. "Synchrotron spectra and ages of compact steep spectrum radio sources", A&A, 345, 769–777.

Myers, S. T., Fassnacht, C. D., Djorgovski, S. G., Blandford, R. D., Matthews, K., Neugebauer, G., Pearson, T. J., Readhead, A. C. S., Smith, J. D., Thompson, D. J., Womble, D. S., Browne, I. W. A., Wilkinson, P. N., Nair, S., Jackson, N., Snellen, I. A. G., Miley, G. K., de Bruyn, A. G., & Schilizzi, R. T. 1995. "1608+656: A Quadruple-Lens System Found in the CLASS Gravitational Lens Survey", ApJ, 447, L5–L8.

Nagar, N. M., Falcke, H., Wilson, A. S., & Ulvestad, J. S. 2002. "Radio sources in low-luminosity active galactic nuclei. III. 'AGNs' in a distance-limited sample of 'LLAGNs'", A&A, 392, 53–82.

Nakamura, M., Uchida, Y., & Hirose, S. 2001. "Production of Wiggled Structure of AGN Radio Jets in the Sweeping Magnetic Twist Mechanism", New Astron., 6, 61–78.

Narayan, R. 1992. "Physics of Pulsar Scintillation", Phil. Trans. Roy. Soc., 341, 151.

Narayan, R., Mahadevan, R., & Quataert, E. 1998. "Advection-dominated accretion around black holes", in Theory of Black Hole Accretion Disks, ed. M. A. Abramowicz, G. Björnsson, & J. E. Pringle (Cambridge: CUP), 148–182.

Nayakshin, S., Rappaport, S., & Melia, F. 2000. "Time-dependent Disk Models for the Micro-quasar GRS 1915+105", ApJ, 535, 798–814.

Nilsson, K., Valtonen, M. I., Kotilainen, I., & Jaakkola, T. 1993. "On the redshift-apparent size diagram of double radio sources", ApJ, 413, 453–476.

O'Dea, C. P. 1998. "The Compact Steep-Spectrum and Gigahertz Peaked-Spectrum Radio Sources", PASP, 110, 493–532.

O'Dea, C. P., & Baum, S. A. 1997. "Constraints on Radio Source Evolution from the Compact Steep Spectrum and GHz Peaked Spectrum Radio Sources", AJ, 113, 148–161.

O'Dea, C. P., Baum, S. A., & Stanghellini, C. 1991. "What are the gigahertz peaked-spectrum radio sources?", ApJ, 380, 66–77.

Oh, S. P. 1999. "Observational Signatures of the First Luminous Objects", ApJ, 527, 16–30.

———. 2002. "Probing the dark ages with metal absorption lines", MNRAS, 336, 1021–1029.

Oh, S. P., Haiman, Z., & Rees, M. J. 2001. "He II Recombination Lines from the First Luminous Object", ApJ, 553, 73–77.

Oren, A. L., & Wolfe, A. M. 1995. "A Faraday rotation search for magnetic fields in quasar damped Lyα absorption systems", ApJ, 445, 624–641.

Owen, F. N., Eilek, J. A., & Kassim, N. E. 2000. "M 87 at 90 centimeters: a different picture", ApJ, 543, 611–619.

Owsianik, I., & Conway, J. E. 1998. "First detection of hot spot advance in Compact Symmetric Objects. Evidence for a class of very young extragalactic radio sources", A&A, 337, 69–79.

Owsianik, I., Conway, J. E., & Polatidis, A. G. 1999. "The youngest lobe-dominated radio sources", New Astron. Rev., 43, 669–673.

Pacholczyk, A. G. 1970. "Radio Astrophysics" (San Francisco: W. H. Freeman).

Pacini, F., & Rees, M. J. 1970. "Nature of Pulsar Radiation", Nature, 226, 622–624.

Paczynski, B. 1998. "Are Gamma-Ray Bursts in Star-Forming Regions?", ApJ, 494, L45–L48.

Padovani, P., Costamante, L., Ghisellini, G., Giommi, P., & Perlman, E. 2002. "*Beppo*SAX Observations of Synchrotron X-Ray Emission from Radio Quasars", ApJ, 581, 895–911.

Palmer, H. P., Rowson, B., Anderson, B., Donaldson, W., Miley, G. K., Gent, H., Adgie, R. L., Slee, O. B., & Crowther, J. H. 1967. "Radio diameter measurements with interferometer baselines of one million and two million wavelengths", Nature, 213, 789–790.

Paragi, Z., Vermeulen, R. C., Fejes, I., Schilizzi, R. T., Spencer, R. E., & Stirling, A. M. 1999. "The inner radio jet region and the complex environment of SS433", A&A, 348, 910–916.

Patnaik, A. R., Browne, I. W. A., Wilkinson, P. N., & Wrobel, J. M. 1992. "Interferometer phase calibration sources. I - The region 35–75 deg", MNRAS, 254, 655–676.

Pauliny-Toth, I. I. K., & Kellermann, K. I. 1966. "Variations in the Radio-Frequency Spectra of 3C 84, 3C 273, 3C 279, and Other Radio Sources", ApJ, 146, 634–645.

———. 1968. "Repeated outbursts in the radio galaxy 3C 120", ApJ, 152, L169–L175.

———. 1972. "The NRAO 5-GHz radio source survey. III. The 140-ft "strong" source survey", AJ, 77, 797–809.

Pauliny-Toth, I. I. K., Preuß, E., Witzel, A., Genzel, R., Kellermann, K. I., Shaffer, D. B., Matveenko, L. I., Moiseev, I. G., Kogan, L. R., Kostenko, V. I., Rönnäng, B. O. 1978. "High resolution observations of compact radio sources at 1.35 cm wavelength", Soviet Ast. Lett., 4, 2, 64–69.

Peacock, J. A. 1999. "Radio galaxies and structure formation", in The Most Distant Radio Galaxies, ed. H. J. A. Röttgering, P. N. Best, & M. D. Lehnert, (Amsterdam: Royal Netherlands Academy of Arts and Sciences), 377–398.

Pearson, T. J., & Readhead, A. C. S. 1988. "The milliarcsecond structure of a complete sample of radio sources. II. First-epoch maps at 5 GHz", ApJ, 328, 114–142.

Peacock, J. A., & Wall, J. V. 1982. "Bright extragalactic radio sources at 2.7 GHz. II - Observations with the Cambridge 5–km telescope", MNRAS, 198, 843–860.

Pearson, T. J., Unwin, S. C., Cohen, M. H., Linfield, R. P., Readhead, A. C. S., Seielstad, G. A., Simon, R. S., & Walker, R. C. 1981. "Superluminal expansion of quasar 3C 273", Nature, 290, 365–368.

Peck, A. B., Henkel, C., Ulvestad, J. S., Brunthaler, A., Falcke, H., Elitzur, M., Menten, K. M., & Gallimore, J. F. 2003. "The flaring H_2O megamaser and compact radio source in Mrk 348", ApJ, 590, in press (astro-ph/0303423).

Peck, A. B., & Taylor, G. B. 2000. "The COINS Sample: VLBA Identifications of Compact Symmetric Objects", ApJ, 534, 90–103.

Pedlar, A., Ghataure, H. S., Davies, R. D., Harrison, B. A., Perley, R., Crane, P. C., & Unger, S. W. 1990. "The Radio Structure of NGC 1275", MNRAS, 246, 477–489.

Peng, B., Kraus, A., Krichbaum, T. P., Müller, S. A. H., Qian, S. J., Quirrenbach, A., Wagner, S. J., Witzel, A., Zensus, J. A., Jin, C., & Bock, H. 2000. "Infrared, radio and optical variability of the BL Lacertae object 2007+777", A&A, 353, 937–943.

Penrose, R. 1969. Nuovo Cimento, 1, 252.

Pentericci, L., Röttgering, H. J. A., Miley, G. K., McCarthy, P., Spinrad, H., van Breugel, W. J. M., & Macchetto, F. 1999. "*HST* images and properties of the most distant radio galaxies", A&A, 341, 329–347.

Perlman, E. S., Padovani, P., Giommi, P., Sambruna, R., Jones, Laurence, R., Tzioumis, A., & Reynolds, J. 1998. "The Deep X-Ray Radio Blazar Survey. I. Methods and First Results", AJ, 115, 1253–1294.

Pettengill, G. H., & Dyce, B. R. 1965. "A radar determination of the rotation of the planet Mercury". Nature206, 1240.

Pettini, M., Shapley, A. E., Steidel, C. C., Cuby, J.-G., Dickinson, M., Moorwood, A. F. M., Adelberger, K. L., & Giavalisco, M. 2001. "The rest-frame optical spectra of Lyman break galaxies: star formation, extinction, abundances, and kinematics", ApJ, 554, 981–1000.

Phillips, R. B., & Mutel, R. L. 1980. "High-resolution observations of the compact radio sources CTD 93 and 3C 395 at 1671 megahertz", ApJ, 236, 89–98.

———. 1982. "On symmetric structure in compact radio sources", A&A, 106, 21–24.

Phillips, R. B., & Shaffer, D. B. 1983. "VLBI maps of 3C 147, 3C 286, 3C 380, NRAO 150, CTD 93, and 3C 395 at 2.3 GHz", ApJ, 271, 32–43.

Phinney, E. S. 1987. "How fast can a blob go?", in Superluminal Radio Sources, ed. J. A. Zensus & T. J. Pearson (Cambridge: CUP), 301–305.

Podariu, S., Daly, R. A., Mory, M. P. & Ratra, B. 2003. "Radio galaxy redshift-angular size data constraints on dark energy", ApJ, 584, 577–579.

Polatidis, A. G., & Conway, J. E. 2002. "Proper motions in Compact Symmetric objects", PASA, 20, 69–74.

Polatidis, A. G., Wilkinson, P. N., Xu, W., Readhead, A. C. S., Pearson, T. J., Taylor, G. B., & Vermeulen, R. C. 1995. "The First Caltech–Jodrell Bank VLBI Survey. I. $\lambda =$ 18 Centimeter Observations of 87 Sources", ApJS, 98, 1–32.

Preston, R. A., Morabito, D. D., Williams, J. G., Faulkner, J., Jauncey, D. L., & Nicolson, G. D. 1985. "A VLBI Survey at 2.29 GHz", AJ, 90, 1599–1641.

Pringle, J. E. 1997. "Self-induced warping of accretion discs: non-linear evolution and application to AGN", MNRAS, 292, 136–147.

Purcell, G. H. 1973. "The Structure of Compact Radio Sources at 606 MHz", Ph.D. thesis, California Inst. of Tech. (Pasadena, CA, USA: CIT).

Pushkarev, A. B., & Gabuzda, D. C. 2002. "Evidence for Interaction with a Surrounding Medium in Several BL Lacertae Objects", in ASP Conf. Ser. Vol. 250, Particles and Fields in Radio Galaxies, ed. R. Laing & K. Blundell (San Francisco: ASP), 200–203.

Pushkarev, A. B., Gabuzda, D. C., & Cawthorne, T. V. 2001. "Frequency Dependence of the Parsec-Scale Polarization Structures of BL Lacertae Objects", in IAU Symp. 205, Galaxies and Their Constituents at the Highest Angular Resolution, ed. R. T. Schilizzi Schilizzi, S. N. Vogel, F. Paresce, & M. S. Elvis, (San Francisco: ASP), 140–141.

Quataert, E., & Gruzinov, A. 2000. "Constraining the Accretion Rate onto Sagittarius A* Using Linear Polarization", ApJ, 545, 842–846.

Quirrenbach, A., Witzel, A., Kirchbaum, T. P., Hummel, C. A., Wegner, R., Schalinski, C. J., Ott, M., Alberdi, A., & Rioja, M. 1992. "Statistics of intraday variability in extragalactic radio sources", A&A, 258, 279–284.

Quirrenbach, A., Witzel, A., Krichbaum, T., Hummel, C. A., & Alberdi, A. 1989. "Rapid variability of extragalactic radio sources", Nature, 337, 442–444.

Rayner, D. P., Norris, R. P., & Sault, R. J. 2000. "Radio circular polarization of active galaxies", MNRAS, 319, 484–496.

Razin, V. A. 1960. Radiofizika, 3, 584–595.

———. 1960. Izvestia Vys. Ucheb. Zaved. Radiofiz., 3, 921.

Readhead, A. C. S. 1994. "Equipartition brightness temperature and the inverse Compton catastrophe", ApJ, 426, 51–59.

Readhead, A. C. S., Taylor, G. B., Xu, W., Pearson. T. J., Wilkinson, P. N., & Polatidis, A. G. 1996. "The Statistics and Ages of Compact Symmetric Objects", ApJ, 460, 612–633.

Readhead, A. C. S., Xu, W., Pearson, T. J., Wilkinson, P. N., & Polatidis, A. G. 1994. "Compact Symmetric Objects", in Compact Extragalactic Radio Sources, ed. J. A. Zensus & K. I. Kellermann (Socorro, NM: NRAO), 17–22.

Rees, M. J. 1966. "Appearance of Relativistically Expanding Radio Sources", Nature, 211, 468–470.

———. 1967. "Studies in radio source structure", MNRAS, 135, 345–360.

———. 1984. "Black Hole Models for Active Galactic Nuclei", ARA&A, 22, 471–506.

———. 1989. "The radio/optical alignmnet of high-z radio galaxies: triggering of star formation in radio lobes", MNRAS, 239, 1P–4P.

———. 1994. "Origin of the seed magnetic field for a galactic dynamo", in Cosmical Magnetism, ed. D. Lynden-Bell (Dordrecht: Kluwer), 155–160.

———. 1998. "Probing the 'Dark Age' with *NGST*", in The *Next Generation Space Telescope*: Science Drivers and Technological Challenges, 34th Liege Astr. Coll., ESA SP-429, 5

Reid, M. J., Biretta, J. A., Junor, W., Muxlow, T. W. B., & Spencer, R. E. 1989. "Subluminal motion and limb brightning in the nuclear jet of M 87", ApJ, 336, 112–120.

Reid, M. J., Readhead, A. C. S., Vermeulen, R. C., & Treuhaft, R. N. 1999. "The Proper Motion of Sagittarius A*. I. First VLBA Results", ApJ, 524, 816–823.

Reimers, D., Kohler, S., Wisotzki, L., Groote, D., Rodríguez-Pascual, P., & Wamsteker, W. 1997. "Patchy intergalactic He II absorption in HE 2347−4342. II. The possible discovery of the epoch of He-reionization", A&A 327, 890–900.

Richards, E. A., Fomalont, E. B., Kellermann, K. I., Windhorst, R. A., Partridge, R. B., Cowie, L. L., & Barger, A. J. 1999. "Optically Faint Microjansky Radio Sources", ApJ, 526, L73–L76.

Richards, E. A., Kellermann, K. I., Fomalont, E. B., Windhorst, R. A., & Partridge, R. B. 1998. "Radio Emission from Galaxies in the Hubble Deep Field", AJ, 116, 1039–1054.

Rickett, B. J. 1969. "Frequency Structure of Pulsar Intensity Variations", Nature, 221, 158–159.

———. 1986. "Refractive interstellar scintillation of radio sources", ApJ, 307, 564–574.

———. 1990. "Radio propagation through the turbulent interstellar plasma", ARA&A, 28, 561–605.

———., Kedziora-Chudczer, L., & Jauncey, D. L. 2002. "Interstellar Scintillation of the Polarized Flux Density in Quasar PKS 0405−385", ApJ, 581, 103–126.

Rickett, B. J., Witzel, A., Kraus, A., Krichbaum, T. P., & Qian, S. J. 2001. "Annual Modulation in the Intraday Variability of quasar 0917+624 due to Interstellar Scintillation", ApJ, 550, L11–L14.

Robinson, I., Schild, A., & Schucking, E. (ed.) 1964. Proc. First Texas Conference on Relativistic Astrophysics (Chicago: Univ. of Chicago Press).

Rogers, A. E. E., Doeleman, S., Wright, M. C. H., Bower, G. C., Backer, D. C., Padin, S., Philips, J. A., Emerson, D. T., Greenhill, L., Moran, J. M., & Kellermann, K. I. 1994. "Small-scale structure and position of Sagittarius A(*) from VLBI at 3 millimeter wavelength", ApJ, 434, L59–L62.

Rogstad, D. H., & Shostack, G. S. 1971. "Aperture Synthesis Study of Neutral Hydrogen in the Galaxy M 101: I. Observations", A&A, 13, 99–107.

Romney, J., Padrielli, L., Bartel, N., Weiler, K. W., Ficarra, A., Mantovani, F., Bååth, L. B., Matveyenko, L., Moiseev, I. G., & Nicholson, G. 1984. "The milliarcsecond scale structure of low frequency variable sources", ApJ, 99, 289–299.

Ros, E. 2003. "The VLBA 2cm Survey: Kinematics of pc-Scale Structures in Active Galactic Nuclei", in Highlights in Spanish Astrophysics (III), ed. J. Gallego, J. Zamorano, & N. Cardiel, (Dordrecht: Kluwer), in press (astro-ph/0211199).

Ros, E., Kellermann, K. I., Lister, M. L., Zensus, J. A., Cohen, M. H., Vermeulen, R. C., Kadler, M., & Homan, D. C. 2002. "Kinematics of parsec-scale structures in AGN: the 2cm VLBA Survey", in Proc. of the 6th European VLBI Network Symp., ed. E. Ros, R. W. Porcas, A. P. Lobanov, & J. A. Zensus, (Bonn: MPIfR), 105–106.

Ros, E., Zensus, J. A., & Lobanov, A. P. 2000. "Total Intensity and Polarized Emission of the Parsec-Scale Jet in 3C 345", A&A, 354, 55–66.

Roy, A. L., Wilson, A. S., Ulvestad, J. S., & Colbert, E. J. M. 2000. "Slow jets in Seyfert galaxies: NGC 1068", in EVN Symp. 2000, Proceedings of the 5th European VLBI Network Symp., ed. J. E. Conway, A. G. Polatidis, R. S. Booth, & Y. M. Pihlström (Onsala, Sweden: Onsala Space Obs.), 7–10.

Roy, A. L., Wrobel, J. M., Wilson, A. S., Ulvestad, J. S., Norris, R. P., Mundell, C. G., Krichbaum, T. P., Falcke, H., & Colbert, E. J. M. 2001. "Parsec-scale jets and tori in Seyfert galaxies", in IAU Symp. No. 205, Galaxies and Their Constituents at the Highest Angular Resolutions, ed. R. T. Schilizzi, S. N. Vogel, F. Paresce, & M. S. Elvis, (San Francisco: ASP), 70–71.

Rudnick, L., & Jones, T. W. 1983. "Rotation measures for compact variable radio sources", AJ, 88, 518–526.

Ruszkowski, M., & Begelman, M. C. 2002. "Circular Polarization from Stochastic Synchrotron Sources", ApJ, 573, 485–495.

Ryle, M. 1962. "The New Cambridge Radio Telescope". Nature194, 517–518.

Ryu, D., Kang, H., & Biermann, P. L. 1998. "Cosmic magnetic fields in large scale filaments and sheets" A&A, 335, 19–25.

Sagdeev, R. Z. 1984. "Some prospects of space VLBI", in Proc. of Workshop on *Quasat*, Gr Ensersdorf, Austria, 18–22 June, 1984, (ESA SP-213), 19–21.

Sagdeev, R. Z., Kerzhanovitch, V. V., Kogan, L. R., Kostenko, V. I., Linkin, V. M., Matveyenko, L. I., Nazirov, R. R., Pogrebenko, S. V., Strukov, I. A., Preston, R. A., Purcel, J., Hildebrand, C. E., Grishmanovskiy, V. A., Kozlov, A. N., Molotov, E. P., Blamont, J. E., Boloh, I., Laurance, G., Kaufmann, P., Galt, J., Biraud, F., Boischot, A., Ortega-Molina, A., Rosolen, C., Petit, G., Mezger, P. G., Schwartz, R., Rönnäng, B. O., Spencer, R. E., Nicolson, G., Rogers, A. E. E., Cohen, M. H., Martirosyan, R. M., Moiseyev, I. G., & Jatskiv, J. S. 1992. "Differential VLBI measurements of the Venus atmosphere dynamics by balloons: VEGA project", A&A, 254, 387–392.

Salpeter, E. E. 1969. "Pulsar Amplitude Variations", Nature, 221, 31–33.

Sargent, W. L. W., Schechter, P. L., Boksenberg, A., & Shortridge, K. 1977. "Velocity dispersions for 13 galaxies", ApJ, 212, 326–334.

Savolainen, T., Wiik, K., Valtaoja, E., Jorstad, S. G., & Marscher, A. P. 2002. "Connections between millimetre continuum variations and VLBI structure in 27 AGN", A&A, 394, 851–861.

Schaye, J., Rauche, M., Sargent, W. L. W., & Kim, T.-S. 2000. "The detection of oxygen in the low-density intergalactic medium", ApJ, 541, L1–L4.

Scheuer, P. A. G. 1968. "Amplitude Variations in Pulsed Radio Sources", Nature, 218, 920–922.

Schoenmakers, A. P., de Bruyn, A. G., Röttgering, H. J. A., & van der Laan, H. 1999. "The Mpc-scale radio source associated with the GPS galaxy B1144+352", A&A, 341, 44–57.

Schödel, R., Ott, T., Genzel, R., Hofmann, R., Lehnert, M., Eckart, A., Mouawad, N., Alexander, T., Reid, M. J., Lenzen, R., Hartung, M., Lacombe, F., Rouan, D., Gendron, E., Rousset, G., Lagrange, A.-M., Brandner, W., Ageorges, N., Lidman, C., Moorwood, A. F. M., Spyromilio, J., Hubin, N., & Menten, K. M. 2002. "A star in a 15.2-year orbit around the supermassive black hole at the centre of the Milky Way", Nature, 419, 694–696.

Seyfert, C. K. 1943. "Nuclear emission in spiral nebulae", ApJ, 97, 28–40.

Shaffer, D. B., Cohen, M. H., Jauncey, D. L., & Kellermann, K. I. 1972. " Rapid Change in the Visibility Function of the Radio Galaxy 3C 120", ApJ, 173, L147–L150.

Shaffer, D. B., Kellermann, K. I., & Cornwell, T. J. 1999. "CTD 93 and the Nature of Gigahertz Peaked Spectrum Radio Sources", ApJ, 515, 558–566.

Shapiro, S. L., & Teukolsky, S. A. 1983. "Black Holes White Dwarfs and Neutron Stars", (New York: John Wiley).

Shapirovskaya, N. Y. 1982. "An interpretation of the decimeter flux variability of extragalactic radio sources", AZh, 59, 246–252 (translation in Soviet Ast., 26, 151–154).

Shaver, P. A. 1978. "Extragalactic Radio Recombination Lines", A&A, 68, 97–105.

Shaver, P. A., & Pierre, M. 1989. "Large-scale anisotropy in the sky distribution of extragalactic radio sources", A&A, 220, 35–41.

Shaver, P. A., Wall, J. V., & Kellermann, K. I. 1996. "PKS 1251−407: a radio-loud quasar at z=4.46", MNRAS, 278, L11–L15.

Shaver, P. A., Wall, J. V., Kellermann, K. I., Jackson, C. A., & Hawkins, M. R. S. 1996. "Decrease in the space density of quasars at high redshift", Nature, 384, 439–441.

Shaver, P. A., Windhorst, R. A., Madau, P., & de Bruyn, A. G. 1999. "Can the reionization epoch be detected as a global signature in the cosmic background?", A&A, 345, 380–390.

Shchekinov, Yu. A. 2002. "Metal enrichment of the high-z intergalactic medium", Astron. Astrophys. Transact., in press (astro-ph/0205320).

Shibai, H., Takeuchi, T., Rengarajan, T. N., & Hirashita, H. 2001. "Absorption Measurement of Hydrogen Molecules in the Early Universe", PASJ, 53, 589–593.

Shimmins, A. J., & Wall, J. V. 1973. "Accurate flux densities at 8.87 GHz of 195 radio sources", Austr. J. Phys, 26, 93–109.

Shklovsky, J. 1965. "Possible Secular Variation of the Flux and Spectra of Radio Emissions of Source 1934−63", Nature, 206, 176–177.

Sholomitskii, G. B. 1965. "Variability of the radio source CTA-102", Inform. Bull. on Var. Stars, 83, 1–2.

Sholomitsky, G. B. 1965. "Fluctuations in the 32.5-cm Flux of CTA 102", AZh, 42, 673–674.

Sholomitsky, G. B., Kurilchik, V. N., Matveenko, L. I., & Khromov, G. S. 1964. "Observations of some weak radio emission sources at 32 cm", Soviet Ast., 41, 823–827.

Sholomitsky, G. B., Sleptsova, N. F., Matveenko, L. I. 1966. "Spectra of the components of 3C 273", AZh, 9, 1135–1137.

Siemiginowska, A., Bechtold, J., Aldcroft, T. L., Elvis, M., Harris, D. E., & Dobrzycki, A. 2002. "*CHANDRA* Discovery of a 300 kpc X-ray Jet in the GPS Quasar PKS 1127−145", ApJ, 570, 543–556.

Silver, C. S., Taylor, G. B., & Vermeulen, R. C. 1998. "A Radio Milli-Halo in the Nucleus of NGC 1275", ApJ, 502, 229–234.

Sincell, M. W., & Krolik, J. H. 1994. "Relativistic induced Compton scattering in synchrotron self-absorbed sources", ApJ, 430, 550–567.

Singal, A. K. 1986. "Magnetization effects in an incoherent synchrotron source", A&A, 155, 242–246.

Slysh, V. I. 1963. "Angular Size of Radio Stars", Nature, 199, 682–690.

———. 1973. "Resonance Stark Effect in OH and H_2O masers", Astroph. Lett., 14, 213–216.

———. 1992. "The synchro-Compton limit of the brightness temperature of nonstationary radio sources", ApJ, 391, 453–455.

Slysh, V. I., Voronkov, M. A., Migenes, V., Shibata, K. M., Umemoto, T., Altunin, V. I., Val'tts, I. E., Kanevsky, B. Z., Popov, M. V., Kovalenko, A. V., Fomalont, E. B., Poperechenko, B. A., Gorshenkov, Yu. N., Carlson, B. R., Dougherty, S. M., Reynolds, J. E., Jiang, D. R., Smirnov, A. I., & Grachev, V. G. 2001. "Space-VLBI observations of the OH maser OH 34.26+0.15: low interstellar scattering", MNRAS, 320, 217–223.

Smail, I., Owen, F. N., Morrison, G. E., Keel, W. C., Ivison, R. J., & Ledlow, M. J. 2002. "The Diversity of Extremely Red Objects", ApJ, 581, 844–864.

Smoot, G. F., Gorenstein, M. V., & Muller, R. A. 1977. "Detection of anisotropy in the cosmic blackbody radiation", Phys. Rev. Lett, 39, 898–901.

Snellen, I. A. G., & Best, P. N. 2001. "Distant FR I radio galaxies in the Hubble deep field: implications for the cosmological evolution of radio-loud AGN", MNRAS, 328, 897–902.

Snellen, I. A. G., Schilizzi, R. T., Miley, G. K., Bremer, M. N., Röttgering, H. J. A., & van Langevelde, H. J. 1999. "Faint Gigahertz Peaked Spectrum sources and the evolution of young radio sources", New Astron. Rev., 43, 675–679.

Snellen, I. A. G., Schilizzi, R. T., Miley, G. K., de Bruyn, A. G., Bremer, M. N., & Rottgering, H. J. A. 2000. "On the evolution of young radio-loud AGN", MNRAS, 319, 445–456.

Sokolov, A., & Marscher, A. P. 2003. "A theoretical study of multifrequency variability in blazar jets", in ASP Conf. Ser. Vol. 299, High Energy Blazar Astronomy, ed. L. O. Takalo, & E. Valtaoja (San Francisco: ASP), in press.

Spaans, M., & Norman, C. A. 1997. "Hydrogen Recombination Line Masers at the Epochs of Recombination and Reionization", ApJ, 488, 27–34.

Stanghellini, C. 2003. "Radio properties of GPS and CSS sources. Galaxies versus Quasars", PASA, 20, 118–122.

Stanghellini, C., Baum, S. A., O'Dea, C. P., & Morris, G. B. 1990. "Extended radio emission associated with GHz-peaked-spectrum radio sources", A&A, 233, 379–384.

Stanghellini, C., Dallacasa, D., O'Dea, C. P., Baum, S. A., Fanti, R., & Fanti, C. 2001. "VLBA observations of GHz-Peaked-Spectrum radio sources at 15 GHz", A&A, 377, 377–388.

Stanghellini, C., O'Dea, C. P., Baum, S. A., & Laurikainen, E. 1993. "Optical CCD imaging of GHz-peaked-spectrum radio sources", ApJ, 88, 1–21.

Stanghellini, C., O'Dea, C. P., Dallacasa, D., Baum, S. A., Fanti, R., & Fanti, C. 1998. "A complete sample of GHz-peaked-spectrum radio sources and its radio properties", A&AS, 131, 303–315.

Steidel, C. C., Adelberger, K. L., Giavalisco, M., Dickinson, M., & Pattini, M. 1999. "Lyman-break galaxies at $z \gtrsim 4$ and the evolution of the ultraviolet luminosity density at high redshift", ApJ, 519, 1–17.

Stelmah, J. 1994. "VLBI data for compact radio sources versus exotic forms of matter in the Universe", ApJ, 428, 61–64.

Stepanas, P. G., & Saha, P. 1995. "Estimating q_0 from angular size statistics", MNRAS, 272, L13–L15.

Stickel, M., Meisenheimer, K., & Kühr, H. 1994. "The optical identification status of the 1 Jy radio source catalogue", A&AS, 105, 211–234.

Tavecchio, F., Maraschi, L., Sambruna, R. M., & Urry, C. M. 2000. "The X-ray Jet of PKS 0637−752: Inverse Compton Radiation from the Cosmic Microwave Background?", ApJ, 544, L23–L26.

Taylor, A. R., & Braun, R. 1998. "Science with the Square Kilometre Array", `http://www.skatelescope.org/ska_science.shtml`.

Taylor, G. B., Readhead, A. C. S., & Pearson, T. J. 1996."Pinpointing the Center of Activity in Compact Symmetric Objects. I. Sources from the Pearson-Readhead Survey", ApJ, 463, 95–104.

Taylor, G. B., Vermeulen, R. C., Pearson, T. J., Readhead, A. C. S., Henstock, D. R., Browne, I. W. A., & Wilkinson, P. N. 1994. "The second Caltech-Jodrell Bank VLBI survey. 1: Observations of 91 of 193 sources", ApJS, 95, 345–369.

Taylor, G. B., Vermeulen, R. C., Readhead, A. C. S., Pearson, T. J., Henstock, D. R., & Wilkinson, P. N. 1996. "A Complete Flux-Density–limited VLBI Survey of 293 Flat-Spectrum Radio Sources", ApJS, 107, 37–68.

Taylor, J. H., & Cordes, J. M. 1993. "Pulsar distances and the galactic distribution of free electrons", ApJ., 411, 674–684.

Terashima, Y., & Wilson, A. S. 2003. "*CHANDRA* snapshot observations of low-luminosity AGNs with a compact radio source", ApJ, 583, 145-158.

Teräsranta, H., & Valtaoja, E. 1994. "Brightness temperatures and viewing angles for extragalactic radio sources: A test of unification schemes for active galactic nuclei", A&A, 283, 51–58.

Teräsranta, H., Urpo, S., Wiren, S., & Valtonen, M. 2001. "Search for new flat-spectrum radio sources", A&A, 368, 431–439.

Teräsranta, H., Tornikoski, M., Mujunen, A., Karlamaa, K., Valtonen, T., Henelius, N., Urpo, S., Lainela, M., Pursimo, T., Nilsson, K., Wiren, S., Lähteenmäki, A., Korpi, M., Rekola, R., Heinämäki, P., Hanski, M., Nurmi, P., Kokkonen, K., Keinänen, P., Joutsamo, O., Oksanen, J., Pietilä, H., Valtaoja, E., Valtonen, M., & Könönen, P. 1998. "Fifteen years monitoring of extragalactic radio sources at 22, 37, and 87 GHz", A&AS, 132, 305–335.

Thakkar, D. D., Xu, W., Readhead, A. C. S., Pearson, T. J., Taylor, G. B., Vermeulen, R. C., Polatidis, A. G., & Wilkinson, P. N. 1995. "The First Caltech–Jodrell Bank VLBI Survey. II. $\lambda = 18$ Centimeter Observations of 25 Sources", ApJS, 98, 33–40.

Theuns, T., Schaye, J., Zaroubi, S., Kim, T.-S., Tzanavaris, P., & Carswell, B. 2002. "Constraints on Reionization from the Thermal History of the Intergalactic Medium", ApJ, 567, L103–L106.

Theuns, T., Viel, M., Kay, S., Schaye, J., Carswell, R. F., & Tzanavaris, P. 2002. "Galactic winds in the intergalactic medium", ApJ, 578, L5–L8.

Thorne, K. S., Price, R. H., & MacDonald, D. A. 1986. "Black-Holes - The Membrane Paradigm", (New Haven: Yale Univ. Press).

Tingay, S. J., Jauncey, D. L., King, E. A., Tzioumis, A. K., Lovell, J. E. J., & Edwards, P. G. 2003. "ATCA Monitoring Observations of 202 Compact Radio Sources in Support of the VSOP AGN Survey", PASJ, submitted.

Tingay, S. J., Preston, R. A., Lister, M. L., Piner, B. G., Murphy, D. W., Jones, D. L., Meier, D. L., Pearson, T. J., Readhead, A. C. S., Hirabayashi, H., Murata, Y., Kobayashi, H., & Inoue, M. 2001. "Measuring the Brightness Temperature Distribution of Extragalactic Radio Sources with Space VLBI", ApJ, 549, L55–L58.

Tolman, R. C. 1930. "On the estimation of distances in a curved universe with a non-static line element", Proc. Nat. Acad. Sci., 16, 511–520.

Tornikoski, M., Jussila, I., Johansson, P., Lainela, M., & Valtaoja, E. 2001. "Radio Spectra and Variability of Gigahertz-Peaked Spectrum Radio Sources and Candidates", AJ, 121, 1306–1318.

Tozzi, P., Madau, P., Meiksin, A., & Rees, M. J. 2000. "Radio Signatures of H I at High Redshift: Mapping the End of the "Dark Ages"", ApJ, 528, 597–606.

Tsuboi, M., Miyazaki, A., & Tsutsumi, T. 1999. "Flare of Sgr A* at Short Millimeter Wavelengths", in ASP Conf. Ser. Vol. 186, The Central Parsecs of the Galaxy, ed. H. Falcke, A. Cotera, W. T. Duschl, F. Melia, & Rieke, M. (San Francisco: ASP), 105–112.

Ulvestad, J. S., Roy, A. L., Colbert, E. J. M., & Wilson, A. S. 1998. "A subparsec radio jet or disk in NGC 4151", ApJ, 496, 196–202.

Ulvestad, J. S., Wrobel, J. M., & Carilli, C. L. 1999. "Radio continuum evidence for outflow and absorption in the Seyfert 1 galaxy Markarian 231", ApJ, 516, 127–140.

Ulvestad, J. S., Wrobel, J. M., Roy, A. L., Wilson, A. S., Falcke, H., & Krichbaum, T. P. 1999. "Subrelativistic radio jets and parsec-scale absorption in two Seyfert galaxies", ApJ, 517, L81–L84.

Ulvestad, J. S., & Ho, L. C. 2001. "Statistical properties of radio emission from the Palomar Seyfert galaxies", ApJ, 558, 561–577.

———. 2001. "The origin of radio emission in low-luminosity active galactic nuclei: jets, accretion flows, or both?", ApJ, 562, L133–L136.

Ulvestad, J. S., & Wilson, A. S. 1989. "Radio structures of Seyfert galaxies. VII. Extension of a distance-limited sample", ApJ, 343, 659–671.

Unwin, S. C., Cohen, M. H., Hodges, M. W., Zensus, J. A., & Biretta, J. A. 1989. "Superluminal motion in the quasar 3C 279", ApJ, 340, 117–128.

Unwin, S. C., Wehrle, A. E., Urry, C. M., Gilmore, D. M., Barton, E. J., Kjerulf, B. C., Zensus, J. A., & Rabaca, C. R. 1994. "Inverse Compton X-ray emission from the superluminal quasar 3C 345", ApJ, 432, 103–113.

Urry, C. M., & Padovani, P. 1995. "Unified schemes for radio-loud active galactic nuclei", PASP, 107, 803–845.

Ustyugova, G. V., Lovelace, R. V. E., Romanova, M. M., Li, H., & Colgate, S. A. 2000. "Poynting Jets from Accretion Disks: Magnetohydrodynamic Simulations", ApJ, 541, L21–L24.

Valageas, P., & Silk, J. 1999. "The reheating and reionization history of the universe", A&A, 347, 1–20.

Valtaoja, E., Lähteenmäki, A., & Teräsranta, H. 1992. "A northern 2 Jy sample of compact extragalactic radio sources suitable for millimeter and space VLBI observations", A&AS, 95, 73–85.

Valtaoja, E., Lähteenmäki, A., Teräsranta, H., & Lainela, M. 1999. "Total Flux Density Variations in Extragalactic Radio Sources. I. Decomposition of Variations into Exponential Flares", ApJS, 120, 95–99.

Valtaoja, E., & Teräsranta, H. 1996. "Unified Schemes for Radio-Loud Active Galactic Nuclei", A&AS, 120, 491–494.

van der Klis, M. 2000. "Millisecond Oscillations in X-ray Binaries", ARA&A, 38, 717–760.

van Ojik, R., Röttgering, H. J. A., Miley, G. K., & Hunstead, R. W. 1997. "The gaseous environments of radio galaxies in the early Universe: kinematics of the Lyman α emission and spatially resolved H I absorption", A&A, 317, 358–384.

Venemans, B. P., Kurk, J. D., Miley, G. K., Röttgering, H. J. A., van Breugel, W., Carilli, C. L., De Breuck, C., Ford, H., Heckman, T., McCarthy, P., & Pentericci, L. 2002. "The Most Distant Structure of Galaxies Known: A Protocluster at $z = 4.1$", ApJ, 569, L11–L14.

Vermeulen, R. C. 1995. "Superluminal sources", Proc. Natl. Acad. Sci. USA, 92, 11385–11389 (correction in Proc. Natl. Acad. Sci. USA, 93, 6846).

Vermeulen, R. C., & Cohen, M. H. 1994. "Superluminal motion statistics and cosmology", ApJ, 430, 467–494.

Vermeulen, R. C., Readhead, A. C. S., & Backer, D. C. 1994. "Discovery of a Nuclear Counterjet in NGC 1275: A New Way to Probe the Parsec-scale Environment", ApJ, 430, L41–L44.

Vermeulen, R. C., Ros, E., Kellermann, K. I., Cohen, M. H., Zensus, J. A., & van Langevelde, H. J. 2003. "The Shroud Around the Twin Radio Jets in NGC 1052", A&A, 401, 113–127.

———. 2003. "The shroud around the "compact, symmetric" radio jets in NGC 1052", PASA, 20, 65–68.

Vermeulen, R. C., Schilizzi, R. T., Spencer, R. E., Romney, J. D., & Fejes, I. 1993. "A series of VLBI images of SS433 during the outbursts in May/June 1987", A&A, 270, 177–188.

Villata, M., Raiteri, C. M., Kurtanidze, O. M., Nikolashvili, M. G., Ibrahimov, M. A., Papadakis, I. E., Tsinganos, K., Sadakane, K., Okada, N., Takalo, L. O., Sillanpää, A., Tosti, G., Ciprini, S., Frasca, A., Marilli, E., Robb, R. M., Noble, J. C., Jorstad, S. G., Hagen-Thorn, V. A., Larionov, V. M., Nesci, R., Maesano, M., Schwartz, R. D., Basler, J., Gorham, P. W., Iwamatsu, H., Kato, T., Pullen, C., Benítez, E., J., de Diego, A., Moilanen, M., Oksanen, A., Rodriguez, D., Sadun, A. C., Kelly, M., Carini, M. T., Miller, H. R., Catalano, S., Dultzin-Hacyan, D., Fan, J. H., Ishioka, R., Karttunen, H., Keinänen, P., Kudryavtseva, N. A., Lainela, M., Lanteri, L., Larionova, E. G., Matsumoto, K., Mattox, J. R., Montagni, F., Nucciarelli, G., Ostorero, L., Papamastorakis, J., Pasanen, M., Sobrito, G., & Uemura, M. 2002. "The WEBT BL Lac Campaign 2000", A&A, 390, 407–421.

Vishwakarma, R. G. 2001. "Consequences on variable Lambda-models from distant Type Ia supernovae and compact radio sources", Class. Quant. Grav., 18, 1159–1172.

von Montigny, C., Bertsch, D. L., Chiang, J., Dingus, B. L., Esposito, J. A., Fichtel, C. E. Fierro, J. M., Hartman, R. C., Hunter, S. D., Kanbach, G., Kniffen, D. A., Lin, Y. C., Mayer-Hasselwander, H. A., Michelson, P. F., Nolan, P. L., Radecke, H. D., Schneid, E., Sreekumar, P., Thompson, D. J., & Willis, T. 1995. "Studies of some superluminal blazars and strong flat-spectrum radio quasars, that are not seen in high energy gamma-rays by EGRET", A&A, 299, 680–688.

Véron-Cetty, M.-P., & Véron, P. 2001. "A catalogue of quasars and active nuclei: 10th edition", A&A, 374, 92–94.

Wagner, S. J., & Witzel, A. 1995. "Intraday Variability In Quasars and BL Lac Objects", ARA&A, 33, 163–197.

Walker, M. A. 1998. "Interstellar scintillation of compact extragalactic radio sources", MNRAS, 294, 307–311.

Walker, R. C., & Anantharamaiah, K. R. 2003. "A VLBA Search for a Stimulated Recombination Line from The Accretion Region in NGC 1275", AJ, 125, 1756–1761.

Walker, R. C., Dhawan, V., Romney, J. D., Kellermann, K. I., & Vermeulen, R. C. 2000. "VLBA absorption imaging of ionized gas associated with the accretion disk in NGC 1275", ApJ, 530, 233–244.

Walker, R. C., Romney, J. D., & Benson, J. M. 1994. "Detection of a VLBI Counterjet in NGC 1275: A Possible Probe of the Parsec Scale Accretion Region", ApJ, 430, L45–L48.

Wall, J. V., & Jackson, C. A. 1997. "Dual-population radio source unification", MNRAS, 290, L17–L22.

Wall, J. V., & Peacock, J. A. 1985. "Bright extragalactic radio sources at 2.7 GHz. III - The all-sky catalogue", MNRAS, 216, 173–192.

Wall, J. V., Shimmins, A. J., & Merkelijn, J. K. 1971. "The Parkes 2700 MHz Survey", Aust. Jou. Phys., Astroph. Supp., 19, 1–68.

Wandel, A., Peterson, B. M., & Malkan, M. A. 1999. "Central Masses and Broad-Line Region Sizes of Active Galactic Nuclei. I. Comparing the Photoionization and Reverberation Techniques", ApJ, 526, 579–591.

Wang, Y. P., Yamada, T., & Taniguchi, Y. 2003. "Cosmic star formation history associated with QSO activity: an approach using the black hole to bulge mass correlation", ApJ, in press (astro-ph/0212534).

Wardle, J. F. C., & Homan, D. C. 2002. "The nature of jets: evidence from circular polarization observations", in ASP Conf. Ser. Vol. 250, Particles and Fields in Radio Galaxies, ed. R. A. Laing & K. M. Blundell (San Francisco: ASP), 152–163.

Wardle, J. F. C., Homan, D. C., Ojha, R., & Roberts, D. H. 1998. "Electron-positron jets associated with the quasar 3C 279", Nature, 395, 457–461.

Wehrle, A. E., Piner, B. G., Unwin, S. C., Zook, A. C., Xu, W., Marscher, A. P., Teräsranta, H., & Valtaoja, E. 2001. "Kinematics of the Parsec-Scale Relativistic Jet in Quasar 3C 279: 1991–1997", ApJS, 133, 297–320.

Weiler, K. W., & de Pater, I. 1983. "A catalog of high accuracy circular polarization measurements", ApJS, 52, 293–297.

White, R. L, Becker, R. H., Helfand, D. J., & Gregg, M. D. 1997. "A Catalog of 1.4 GHz Radio Sources from the FIRST Survey", ApJ, 475, 479–493.

Whitney, A. R., Shapiro, I. I., Rogers, A. E. E., Robertson, D. S., Knight, C. A., Clark, T. A., Goldstein, R. M., Marandino, G. E., & Vandenberg, N. R. 1971. "Quasars Revisited: Rapid Time Variations Observed via Very Long Baseline Interferometry", Science, 173, 225–230.

Wiik, K. & Valtaoja, E. 2001. "The geometry of the universe from high resolution LBI data of AGN shocks", A&A, 366, 1061–1064.

Wilkinson, P. N., Browne, I. W. A., Alcock, D., Myles, S., Henstock, D., Readhead A. C. S., Pearson T. J., Xu, W., Taylor, G. B., Polatidis, A. G., & Vermeulen, R. C. 1998. "The angular size–redshift test for compact radio sources in the Caltech–Jodrell Bank surveys", in Observational Cosmology with the New Radio Surveys, ed. M. N. Bremer, N. Jackson, & I. Pérez-Fournon, (Dordrecht: Kluwer), 221–226.

Wilkinson, P. N., Browne, I. W. A., Patnaik, A. K., Wrobel, J. M., & Sorathia, B. 1998. "Interferometer phase calibration sources - III. The regions $+20 \leq \delta_{\rm B1950} \leq +35$ deg and $+75 \leq \delta_{\rm B1950} \leq +90$ deg", MNRAS, 300, 790–816.

Wilkinson, P. N., Polatidis, A. G., Readhead, A. C. S., Xu, W., & Pearson, T. J. 1994. "Two-sided ejection in powerful radio sources: The compact symmetric objects", ApJ, 432, L87–L90.

Willott, C. J., Rawlings, S., Blundell, K. M., Lacy, M., & Eales, S. A. 2001. MNRAS, 322, 536–554.

Wilson, A. S., Roy, A. L., Ulvestad, J. S., Colbert, E. J. M., Weaver, K. A., Braatz, J. A., Henkel, C., Matsuoka, M., Xue, S., Iyomoto, N., & Okada, K. 1998. "The ionization fraction in the obscuring 'torus' of an active galactic nucleus", ApJ, 505, 587–593.

Wilson, A. S., & Yang, Y. 2002. "*CHANDRA* X-ray Imaging and Spectroscopy of the M 87 Jet and Nucleus", ApJ, 568, 133–140.

Wilson, A. S., Young, A. J., & Shopbell, P. L. 2001. "*CHANDRA* X-ray Observations of Pictor A: High Energy Cosmic Rays in a Radio Galaxy?", ApJ, 547, 740–753.

Windhorst, R. A., Mathis, D., & Neuschaefer, L. 1990. "The evolution of weak radio galaxies at radio and optical wavelengths", in ASP Conf. Ser. Vol. 10, Evolution of the Universe of Galaxies, ed. R. G. Kron, (San Francisco: ASP), 389–403.

Witzel, A., Heeschen, D. S., Schalinski, C., & Krichbaum, T. 1986. "Kurzzeit-Variabilität extragalaktischer Radioquellen", Mitteilungen der Astronomischen Gesellschaft, 65, 239–241.

Wootten, A., Claussen, M., Marvel, K. B., & Wilking, B. 1999. "Temporal evolution of maser structures within 1 au of IRAS 16293", BAAS, 31, 898.

Wright, A. E., Wark, R. M., Troup, E., Otrupcek, R., Hunt, A., & Cooke, D. J. 1990. "Flux Densities at 8400-MHz for a Large Sample of Parkes Radio Sources", PASA, 8, 261–262.

Wright, A. E., & Otrupcek, R. E. 1990. "Parkes Catalogue", (Sydney: ATNF, CSIRO).

Wright, M. C. H., & Backer, D. C. 1993. "Flux Density of Sagittarius A at $\lambda = 3\,\mathrm{mm}$", ApJ, 417, 560–564.

Wrobel, J. M. 1984. "Radio continuum activity in the elliptical galaxy NGC 1052", ApJ, 284, 531–543.

———. 1993. "Faraday-rotation measures and intrinsic polarization position angles of very long baseline interferometry core-jet sources", AJ, 106, 444–454.

Xu, W., Readhead, A. C. S., Pearson, T. J., Polatidis, A. G., & Wilkinson, P. N. 1995. "The First Caltech–Jodrell Bank VLBI Survey. III. VLBI and MERLIN Observations at 5 GHz and VLA Observations at 1.4 GHz", ApJS, 99, 297–348.

Yuan, F., Markoff, S., & Falcke, H. 2002. "A Jet-ADAF model for Sgr A*", A&A, 383, 854–863.

Zensus, J. A. 1997. "Parsec-Scale Jets in Extragalactic Radio Sources", ARA&A, 305, 607–696.

Zensus, J. A., & Pearson, T. J. (ed.) 1987. "Superluminal Radio Sources", (Cambridge: CUP).

Zensus, J. A., Ros, E., Kellermann, K. I., Cohen, M. H., Vermeulen, R. C., & Kadler, M. 2002. "Sub-Milliarcsecond Imaging of Quasars and Active Galactic Nuclei. II. Additional Sources", AJ, 124, 662–674.

Zhao, J.-H., Bower, G. C., & Goss, W. M. 2001. "Radio Variability of Sagittarius A*-a 106 Day Cycle", ApJ, 547, L29–L32.

Zhao, J.-H., Ekers, R. D., Goss, W. M., Lo, K. Y., & Narayan, R. 1989., "Long-Term Variations of the Compact Radio Source Sgr A at the Galactic Center", in IAU Symp. 136, The Center of the Galaxy, ed. M. Morris (Dordrecht: Kluwer), 535–542.

Zhao, J.-H., Goss, W. M., Lo, K. Y., & Ekers, R. D. 1992. "Evidence for Intrinsic Activity of Sgr A at the Galactic Center", in ASP Conf. Ser. Vol. 31, Relationship between Active Galactic Nuclei and Starburst Galaxies, ed. A. V. Filippenko, (San Francisco: ASP), 295–299.

Zhao, J.-H., Young, K. H., Herrnstein, R. M., Ho, P. T. P., Tsutsumi, T., Lo., K. Y., Goss, W. M., & Bower, G. C. 2003. "Variability of Sagittarius A*–Flares at 1 Millimeter", ApJ, 586, L29–L32.

Zhu, Z.-H., & Fujimoto, M.-K. 2002. "Cardesian expansion: constraints from compact radio source angular size versus redshift data", ApJ, 581, 1–4.

Radio Astronomy at the Fringe
ASP Conference Series, Vol. 300, 2003
J.A. Zensus, M.H. Cohen, & E. Ros, eds.

Index of Authors

Object Index

IAU designation (B1950.0)

Miscellaneous designation

Radio Astronomy
at the Fringe

Designed by Eduardo Ros and J. Anton Zensus
from series style.

Composed by the editors
with LaTeX 2.09 by Leslie Lamport
(based on TeX by Donald E. Knuth),
using the packages
Epsf by Tomas Rockicki,
Graphicx by David Carlisle,
and Makeidx by Johannes Braams.

Camera copy
with Dvips by Tomas Rockicki
at 600 dots per inch.

A LISTING OF THE VOLUMES

Published
by

THE ASTRONOMICAL SOCIETY OF THE PACIFIC
(ASP)

An international, nonprofit, scientific and educational organization founded in 1889

All book orders or inquiries concerning

ASTRONOMICAL SOCIETY OF THE PACIFIC
CONFERENCE SERIES
(ASP - CS)

and

INTERNATIONAL ASTRONOMICAL UNION VOLUMES
(IAU)

should be directed to the:

Astronomical Society of the Pacific Conference Series
390 Ashton Avenue
San Francisco CA 94112-1722 USA

Phone: 800-335-2624 (Within USA)
Phone: 415-337-2126
Fax: 415-337-5205

E-mail: service@astrosociety.org
Web Site: http://www.astrosociety.org

Complete lists of proceedings of past IAU Meetings are maintained at the IAU Web site at the URL: http://www.iau.org/publicat.html

Volumes 32 - 189 in the IAU Symposia Series may be ordered from:

Kluwer Academic Publishers
P. O. Box 117
NL 3300 AA Dordrecht
The Netherlands

Kluwer@wKap.com

ASP CONFERENCE SERIES VOLUMES

Published by the Astronomical Society of the Pacific

PUBLISHED: 1988 (* asterisk means OUT OF STOCK)

Vol. CS -1 PROGRESS AND OPPORTUNITIES IN SOUTHERN HEMISPHERE OPTICAL ASTRONOMY: CTIO 25TH Anniversary Symposium
eds. V. M. Blanco and M. M. Phillips
ISBN 0-937707-18-X

Vol. CS-2 PROCEEDINGS OF A WORKSHOP ON OPTICAL SURVEYS FOR QUASARS
eds. Patrick S. Osmer, Alain C. Porter, Richard F. Green, and Craig B. Foltz
ISBN 0-937707-19-8

Vol. CS-3 FIBER OPTICS IN ASTRONOMY
ed. Samuel C. Barden
ISBN 0-937707-20-1

Vol. CS-4 THE EXTRAGALACTIC DISTANCE SCALE:
Proceedings of the ASP 100th Anniversary Symposium
eds. Sidney van den Bergh and Christopher J. Pritchet
ISBN 0-937707-21-X

Vol. CS-5 THE MINNESOTA LECTURES ON CLUSTERS OF GALAXIES AND LARGE-SCALE STRUCTURE
ed. John M. Dickey
ISBN 0-937707-22-8

PUBLISHED: 1989

Vol. CS-6 * SYNTHESIS IMAGING IN RADIO ASTRONOMY: A Collection of Lectures from the Third NRAO Synthesis Imaging Summer School
eds. Richard A. Perley, Frederic R. Schwab, and Alan H. Bridle
ISBN 0-937707-23-6

PUBLISHED: 1990

Vol. CS-7 PROPERTIES OF HOT LUMINOUS STARS: Boulder-Munich Workshop
ed. Catharine D. Garmany
ISBN 0-937707-24-4

Vol. CS-8 * CCDs IN ASTRONOMY
ed. George H. Jacoby
ISBN 0-937707-25-2

Vol. CS-9 COOL STARS, STELLAR SYSTEMS, AND THE SUN: Sixth Cambridge Workshop
ed. George Wallerstein
ISBN 0-937707-27-9

Vol. CS-10 * EVOLUTION OF THE UNIVERSE OF GALAXIES:
Edwin Hubble Centennial Symposium
ed. Richard G. Kron
ISBN 0-937707-28-7

Vol. CS-11 CONFRONTATION BETWEEN STELLAR PULSATION AND EVOLUTION
eds. Carla Cacciari and Gisella Clementini
ISBN 0-937707-30-9

Vol. CS-12 THE EVOLUTION OF THE INTERSTELLAR MEDIUM
ed. Leo Blitz
ISBN 0-937707-31-7

PUBLISHED: 1991

Vol. CS-13 THE FORMATION AND EVOLUTION OF STAR CLUSTERS
ed. Kenneth Janes
ISBN 0-937707-32-5

ASP CONFERENCE SERIES VOLUMES

Published by the Astronomical Society of the Pacific

PUBLISHED: 1991 (* asterisk means OUT OF STOCK)

Vol. CS-14 ASTROPHYSICS WITH INFRARED ARRAYS
ed. Richard Elston
ISBN 0-937707-33-3

Vol. CS-15 LARGE-SCALE STRUCTURES AND PECULIAR MOTIONS IN THE UNIVERSE
eds. David W. Latham and L. A. Nicolaci da Costa
ISBN 0-937707-34-1

Vol. CS-16 Proceedings of the 3rd Haystack Observatory Conference on ATOMS, IONS, AND MOLECULES: NEW RESULTS IN SPECTRAL LINE ASTROPHYSICS
eds. Aubrey D. Haschick and Paul T. P. Ho
ISBN 0-937707-35-X

Vol. CS-17 LIGHT POLLUTION, RADIO INTERFERENCE, AND SPACE DEBRIS
ed. David L. Crawford
ISBN 0-937707-36-8

Vol. CS-18 THE INTERPRETATION OF MODERN SYNTHESIS OBSERVATIONS OF SPIRAL GALAXIES
eds. Nebojsa Duric and Patrick C. Crane
ISBN 0-937707-37-6

Vol. CS-19 RADIO INTERFEROMETRY: THEORY, TECHNIQUES, AND APPLICATIONS, IAU Colloquium 131
eds. T. J. Cornwell and R. A. Perley
ISBN 0-937707-38-4

Vol. CS-20 FRONTIERS OF STELLAR EVOLUTION: 50th Anniversary McDonald Observatory (1939-1989)
ed. David L. Lambert
ISBN 0-937707-39-2

Vol. CS-21 THE SPACE DISTRIBUTION OF QUASARS
ed . David Crampton
ISBN 0-937707-40-6

PUBLISHED: 1992

Vol. CS-22 NONISOTROPIC AND VARIABLE OUTFLOWS FROM STARS
eds. Laurent Drissen, Claus Leitherer, and Antonella Nota
ISBN 0-937707-41-4

Vol CS-23 * ASTRONOMICAL CCD OBSERVING AND REDUCTION TECHNIQUES
ed. Steve B. Howell
ISBN 0-937707-42-4

Vol. CS-24 COSMOLOGY AND LARGE-SCALE STRUCTURE IN THE UNIVERSE
ed. Reinaldo R. de Carvalho
ISBN 0-937707-43-0

Vol. CS-25 ASTRONOMICAL DATA ANALYSIS, SOFTWARE AND SYSTEMS I - (ADASS I)
eds. Diana M. Worrall, Chris Biemesderfer, and Jeannette Barnes
ISBN 0-937707-44-9

Vol. CS-26 COOL STARS, STELLAR SYSTEMS, AND THE SUN: Seventh Cambridge Workshop
eds. Mark S. Giampapa and Jay A. Bookbinder
ISBN 0-937707-45-7

Vol. CS-27 THE SOLAR CYCLE: Proceedings of the National Solar Observatory/Sacramento Peak 12th Summer Workshop
ed. Karen L. Harvey
ISBN 0-937707-46-5

ASP CONFERENCE SERIES VOLUMES

Published by the Astronomical Society of the Pacific

PUBLISHED: 1992 (asterisk means OUT OF STOCK)

Vol. CS-28 AUTOMATED TELESCOPES FOR PHOTOMETRY AND IMAGING
eds. Saul J. Adelman, Robert J. Dukes, Jr., and Carol J. Adelman
ISBN 0-937707-47-3

Vol. CS-29 Viña del Mar Workshop on CATACLYSMIC VARIABLE STARS
ed. Nikolaus Vogt
ISBN 0-937707-48-1

Vol. CS-30 VARIABLE STARS AND GALAXIES
ed. Brian Warner
ISBN 0-937707-49-X

Vol. CS-31 RELATIONSHIPS BETWEEN ACTIVE GALACTIC NUCLEI AND STARBURST GALAXIES
ed. Alexei V. Filippenko
ISBN 0-937707-50-3

Vol. CS-32 COMPLEMENTARY APPROACHES TO DOUBLE AND MULTIPLE STAR RESEARCH, IAU Colloquium 135
eds. Harold A. McAlister and William I. Hartkopf
ISBN 0-937707-51-1

Vol. CS-33 * RESEARCH AMATEUR ASTRONOMY
ed. Stephen J. Edberg
ISBN 0-937707-52-X

Vol. CS-34 ROBOTIC TELESCOPES IN THE 1990's
ed. Alexei V. Filippenko
ISBN 0-937707-53-8

PUBLISHED: 1993

Vol. CS-35 * MASSIVE STARS: THEIR LIVES IN THE INTERSTELLAR MEDIUM
eds. Joseph P. Cassinelli and Edward B. Churchwell
ISBN 0-937707-54-6

Vol. CS-36 PLANETS AROUND PULSARS
ed. J. A. Phillips, S. E. Thorsett, and S. R. Kulkarni
ISBN 0-937707-55-4

Vol. CS-37 FIBER OPTICS IN ASTRONOMY II
ed. Peter M. Gray
ISBN 0-937707-56-2

Vol. CS-38 NEW FRONTIERS IN BINARY STAR RESEARCH: Pacific Rim Colloquium
eds. K. C. Leung and I.-S. Nha
ISBN 0-937707-57-0

Vol. CS-39 THE MINNESOTA LECTURES ON THE STRUCTURE AND DYNAMICS OF THE MILKY WAY
ed. Roberta M. Humphreys
ISBN 0-937707-58-9

Vol. CS-40 INSIDE THE STARS, IAU Colloquium 137
eds. Werner W. Weiss and Annie Baglin
ISBN 0-937707-59-7

Vol. CS-41 ASTRONOMICAL INFRARED SPECTROSCOPY: FUTURE OBSERVATIONAL DIRECTIONS
ed. Sun Kwok
ISBN 0-937707-60-0

ASP CONFERENCE SERIES VOLUMES

Published by the Astronomical Society of the Pacific

PUBLISHED: 1993 (* asterisk means OUT OF STOCK)

Vol. CS-42 GONG 1992: SEISMIC INVESTIGATION OF THE SUN AND STARS
ed. Timothy M. Brown
ISBN 0-937707-61-9

Vol. CS-43 SKY SURVEYS: PROTOSTARS TO PROTOGALAXIES
ed. B. T. Soifer
ISBN 0-937707-62-7

Vol. CS-44 PECULIAR VERSUS NORMAL PHENOMENA IN A-TYPE AND RELATED STARS,
IAU Colloquium 138
eds. M. M. Dworetsky, F. Castelli, and R. Faraggiana
ISBN 0-937707-63-5

Vol. CS-45 LUMINOUS HIGH-LATITUDE STARS
ed. Dimitar D. Sasselov
ISBN 0-937707-64-3

Vol. CS-46 THE MAGNETIC AND VELOCITY FIELDS OF SOLAR ACTIVE REGIONS,
IAU Colloquium 141
eds. Harold Zirin, Guoxiang Ai, and Haimin Wang
ISBN 0-937707-65-1

Vol. CS-47 THIRD DECENNIAL US-USSR CONFERENCE ON SETI --
Santa Cruz, California, USA
ed. G. Seth Shostak
ISBN 0-937707-66-X

Vol. CS-48 THE GLOBULAR CLUSTER-GALAXY CONNECTION
eds. Graeme H. Smith and Jean P. Brodie
ISBN 0-937707-67-8

Vol. CS-49 GALAXY EVOLUTION: THE MILKY WAY PERSPECTIVE
ed. Steven R. Majewski
ISBN 0-937707-68-6

Vol. CS-50 STRUCTURE AND DYNAMICS OF GLOBULAR CLUSTERS
eds. S. G. Djorgovski and G. Meylan
ISBN 0-937707-69-4

Vol. CS-51 OBSERVATIONAL COSMOLOGY
eds. Guido Chincarini, Angela Iovino, Tommaso Maccacaro, and Dario Maccagni
ISBN 0-937707-70-8

Vol. CS-52 ASTRONOMICAL DATA ANALYSIS SOFTWARE AND SYSTEMS II - (ADASS II)
eds. R. J. Hanisch, R. J. V. Brissenden, and Jeannette Barnes
ISBN 0-937707-71-6

Vol. CS-53 BLUE STRAGGLERS
ed. Rex A. Saffer
ISBN 0-937707-72-4

PUBLISHED: 1994

Vol. CS-54 * THE FIRST STROMLO SYMPOSIUM: THE PHYSICS OF ACTIVE GALAXIES
eds. Geoffrey V. Bicknell, Michael A. Dopita, and Peter J. Quinn
ISBN 0-937707-73-2

Vol. CS-55 OPTICAL ASTRONOMY FROM THE EARTH AND MOON
eds. Diane M. Pyper and Ronald J. Angione
ISBN 0-937707-74-0

Vol. CS-56 INTERACTING BINARY STARS
ed. Allen W. Shafter
ISBN 0-937707-75-9

ASP CONFERENCE SERIES VOLUMES

Published by the Astronomical Society of the Pacific

PUBLISHED: 1994 (* asterisk means OUT OF STOCK)

Vol. CS-57	STELLAR AND CIRCUMSTELLAR ASTROPHYSICS eds. George Wallerstein and Alberto Noriega-Crespo ISBN 0-937707-76-7
Vol. CS-58 *	THE FIRST SYMPOSIUM ON THE INFRARED CIRRUS AND DIFFUSE INTERSTELLAR CLOUDS eds. Roc M. Cutri and William B. Latter ISBN 0-937707-77-5
Vol. CS-59	ASTRONOMY WITH MILLIMETER AND SUBMILLIMETER WAVE INTERFEROMETRY, IAU Colloquium 140 eds. M. Ishiguro and Wm. J. Welch ISBN 0-937707-78-3
Vol. CS-60	THE MK PROCESS AT 50 YEARS: A POWERFUL TOOL FOR ASTROPHYSICAL INSIGHT, A Workshop of the Vatican Observatory --Tucson, Arizona, USA eds. C. J. Corbally, R. O. Gray, and R. F. Garrison ISBN 0-937707-79-1
Vol. CS-61	ASTRONOMICAL DATA ANALYSIS SOFTWARE AND SYSTEMS III - (ADASS III) eds. Dennis R. Crabtree, R. J. Hanisch, and Jeannette Barnes ISBN 0-937707-80-5
Vol. CS-62	THE NATURE AND EVOLUTIONARY STATUS OF HERBIG Ae/Be STARS eds. Pik Sin Thé, Mario R. Pérez, and Ed P. J. van den Heuvel ISBN 0-9837707-81-3
Vol. CS-63	SEVENTY-FIVE YEARS OF HIRAYAMA ASTEROID FAMILIES: THE ROLE OF COLLISIONS IN THE SOLAR SYSTEM HISTORY eds. Yoshihide Kozai, Richard P. Binzel, and Tomohiro Hirayama ISBN 0-937707-82-1
Vol. CS 64 *	COOL STARS, STELLAR SYSTEMS, AND THE SUN: Eighth Cambridge Workshop ed. Jean-Pierre Caillault ISBN 0-937707-83-X
Vol. CS-65 *	CLOUDS, CORES, AND LOW MASS STARS: The Fourth Haystack Observatory Conference eds. Dan P. Clemens and Richard Barvainis ISBN 0-937707-84-8
Vol. CS-66 *	PHYSICS OF THE GASEOUS AND STELLAR DISKS OF THE GALAXY ed. Ivan R. King ISBN 0-937707-85-6
Vol. CS-67	UNVEILING LARGE-SCALE STRUCTURES BEHIND THE MILKY WAY eds. C. Balkowski and R. C. Kraan-Korteweg ISBN 0-937707-86-4
Vol. CS-68 *	SOLAR ACTIVE REGION EVOLUTION: COMPARING MODELS WITH OBSERVATIONS eds. K. S. Balasubramaniam and George W. Simon ISBN 0-937707-87-2
Vol. CS-69	REVERBERATION MAPPING OF THE BROAD-LINE REGION IN ACTIVE GALACTIC NUCLEI eds. P. M. Gondhalekar, K. Horne, and B. M. Peterson ISBN 0-937707-88-0
Vol. CS-70 *	GROUPS OF GALAXIES eds. Otto-G. Richter and Kirk Borne ISBN 0-937707-89-9

ASP CONFERENCE SERIES VOLUMES

Published by the Astronomical Society of the Pacific

PUBLISHED: 1995 (* asterisk means OUT OF STOCK)

Vol. CS-71 TRIDIMENSIONAL OPTICAL SPECTROSCOPIC METHODS IN ASTROPHYSICS, IAU Colloquium 149
eds. Georges Comte and Michel Marcelin
ISBN 0-937707-90-2

Vol. CS-72 MILLISECOND PULSARS: A DECADE OF SURPRISE
eds. A. S Fruchter, M. Tavani, and D. C. Backer
ISBN 0-937707-91-0

Vol. CS-73 AIRBORNE ASTRONOMY SYMPOSIUM ON THE GALACTIC ECOSYSTEM: FROM GAS TO STARS TO DUST
eds. Michael R. Haas, Jacqueline A. Davidson, and Edwin F. Erickson
ISBN 0-937707-92-9

Vol. CS-74 PROGRESS IN THE SEARCH FOR EXTRATERRESTRIAL LIFE: 1993 Bioastronomy Symposium
ed. G. Seth Shostak
ISBN 0-937707-93-7

Vol. CS-75 MULTI-FEED SYSTEMS FOR RADIO TELESCOPES
eds. Darrel T. Emerson and John M. Payne
ISBN 0-937707-94-5

Vol. CS-76 GONG '94: HELIO- AND ASTERO-SEISMOLOGY FROM THE EARTH AND SPACE
eds. Roger K. Ulrich, Edward J. Rhodes, Jr., and Werner Däppen
ISBN 0-937707-95-3

Vol. CS-77 ASTRONOMICAL DATA ANALYSIS SOFTWARE AND SYSTEMS IV - (ADASS IV)
eds. R. A. Shaw, H. E. Payne, and J. J. E. Hayes
ISBN 0-937707-96-1

Vol. CS-78 ASTROPHYSICAL APPLICATIONS OF POWERFUL NEW DATABASES: Joint Discussion No. 16 of the 22nd General Assembly of the IAU
eds. S. J. Adelman and W. L. Wiese
ISBN 0-937707-97-X

Vol. CS-79 * ROBOTIC TELESCOPES: CURRENT CAPABILITIES, PRESENT DEVELOPMENTS, AND FUTURE PROSPECTS FOR AUTOMATED ASTRONOMY
eds. Gregory W. Henry and Joel A. Eaton
ISBN 0-937707-98-8

Vol. CS-80 * THE PHYSICS OF THE INTERSTELLAR MEDIUM AND INTERGALACTIC MEDIUM
eds. A. Ferrara, C. F. McKee, C. Heiles, and P. R. Shapiro
ISBN 0-937707-99-6

Vol. CS-81 LABORATORY AND ASTRONOMICAL HIGH RESOLUTION SPECTRA
eds. A. J. Sauval, R. Blomme, and N. Grevesse
ISBN 1-886733-01-5

Vol. CS-82 * VERY LONG BASELINE INTERFEROMETRY AND THE VLBA
eds. J. A. Zensus, P. J. Diamond, and P. J. Napier
ISBN 1-886733-02-3

Vol. CS-83 * ASTROPHYSICAL APPLICATIONS OF STELLAR PULSATION, IAU Colloquium 155
eds. R. S. Stobie and P. A. Whitelock
ISBN 1-886733-03-1

ATLAS INFRARED ATLAS OF THE ARCTURUS SPECTRUM, 0.9 - 5.3 μm
eds. Kenneth Hinkle, Lloyd Wallace, and William Livingston
ISBN: 1-886733-04-X

ASP CONFERENCE SERIES VOLUMES

Published by the Astronomical Society of the Pacific

PUBLISHED: 1995 (* asterisk means OUT OF STOCK)

Vol. CS-84 THE FUTURE UTILIZATION OF SCHMIDT TELESCOPES, IAU Colloquium 148
eds. Jessica Chapman, Russell Cannon, Sandra Harrison, and Bambang Hidayat
ISBN 1-886733-05-8

Vol. CS-85 * CAPE WORKSHOP ON MAGNETIC CATACLYSMIC VARIABLES
eds. D. A. H. Buckley and B. Warner
ISBN 1-886733-06-6

Vol. CS-86 FRESH VIEWS OF ELLIPTICAL GALAXIES
eds. Alberto Buzzoni, Alvio Renzini, and Alfonso Serrano
ISBN 1-886733-07-4

PUBLISHED: 1996

Vol. CS-87 NEW OBSERVING MODES FOR THE NEXT CENTURY
eds. Todd Boroson, John Davies, and Ian Robson
ISBN 1-886733-08-2

Vol. CS-88 * CLUSTERS, LENSING, AND THE FUTURE OF THE UNIVERSE
eds. Virginia Trimble and Andreas Reisenegger
ISBN 1-886733-09-0

Vol. CS-89 ASTRONOMY EDUCATION: CURRENT DEVELOPMENTS, FUTURE COORDINATION
ed. John R. Percy
ISBN 1-886733-10-4

Vol. CS-90 THE ORIGINS, EVOLUTION, AND DESTINIES OF BINARY STARS IN CLUSTERS
eds. E. F. Milone and J. -C. Mermilliod
ISBN 1-886733-11-2

Vol. CS-91 BARRED GALAXIES, IAU Colloquium 157
eds. R. Buta, D. A. Crocker, and B. G. Elmegreen
ISBN 1-886733-12-0

Vol. CS-92 * FORMATION OF THE GALACTIC HALO .... INSIDE AND OUT
eds. Heather L. Morrison and Ata Sarajedini
ISBN 1-886733-13-9

Vol. CS-93 RADIO EMISSION FROM THE STARS AND THE SUN
eds. A. R. Taylor and J. M. Paredes
ISBN 1-886733-14-7

Vol. CS-94 MAPPING, MEASURING, AND MODELING THE UNIVERSE
eds. Peter Coles, Vicent J. Martinez, and Maria-Jesus Pons-Borderia
ISBN 1-886733-15-5

Vol. CS-95 SOLAR DRIVERS OF INTERPLANETARY AND TERRESTRIAL DISTURBANCES: Proceedings of 16th International Workshop National Solar Observatory/Sacramento Peak
eds. K. S. Balasubramaniam, Stephen L. Keil, and Raymond N. Smartt
ISBN 1-886733-16-3

Vol. CS-96 HYDROGEN-DEFICIENT STARS
eds. C. S. Jeffery and U. Heber
ISBN 1-886733-17-1

Vol. CS-97 POLARIMETRY OF THE INTERSTELLAR MEDIUM
eds. W. G. Roberge and D. C. B. Whittet
ISBN 1-886733-18-X

ASP CONFERENCE SERIES VOLUMES

Published by the Astronomical Society of the Pacific

PUBLISHED: 1996 (* asterisk means OUT OF STOCK)

Vol. CS-98 FROM STARS TO GALAXIES: THE IMPACT OF STELLAR PHYSICS ON GALAXY EVOLUTION
eds. Claus Leitherer, Uta Fritze-von Alvensleben, and John Huchra
ISBN 1-886733-19-8

Vol. CS-99 COSMIC ABUNDANCES:
Proceedings of the 6th Annual October Astrophysics Conference
eds. Stephen S. Holt and George Sonneborn
ISBN 1-886733-20-1

Vol. CS-100 ENERGY TRANSPORT IN RADIO GALAXIES AND QUASARS
eds. P. E. Hardee, A. H. Bridle, and J. A. Zensus
ISBN 1-886733-21-X

Vol. CS-101 ASTRONOMICAL DATA ANALYSIS SOFTWARE AND SYSTEMS V – (ADASS V)
eds. George H. Jacoby and Jeannette Barnes
ISBN 1080-7926

Vol. CS-102 THE GALACTIC CENTER, 4th ESO/CTIO Workshop
ed. Roland Gredel
ISBN 1-886733-22-8

Vol. CS-103 THE PHYSICS OF LINERS IN VIEW OF RECENT OBSERVATIONS
eds. M. Eracleous, A. Koratkar, C. Leitherer, and L. Ho
ISBN 1-886733-23-6

Vol. CS-104 PHYSICS, CHEMISTRY, AND DYNAMICS OF INTERPLANETARY DUST, IAU Colloquium 150
eds. Bo Å. S. Gustafson and Martha S. Hanner
ISBN 1-886733-24-4

Vol. CS-105 PULSARS: PROBLEMS AND PROGRESS, IAU Colloquium 160
ed. S. Johnston, M. A. Walker, and M. Bailes
ISBN 1-886733-25-2

Vol. CS-106 THE MINNESOTA LECTURES ON EXTRAGALACTIC NEUTRAL HYDROGEN
ed. Evan D. Skillman
ISBN 1-886733-26-0

Vol. CS-107 COMPLETING THE INVENTORY OF THE SOLAR SYSTEM:
A Symposium held in conjunction with the 106th Annual Meeting of the ASP
eds. Terrence W. Rettig and Joseph M. Hahn
ISBN 1-886733-27-9

Vol. CS-108 M.A.S.S. -- MODEL ATMOSPHERES AND SPECTRUM SYNTHESIS:
5th Vienna - Workshop
eds. Saul J. Adelman, Friedrich Kupka, and Werner W. Weiss
ISBN 1-886733-28-7

Vol. CS-109 COOL STARS, STELLAR SYSTEMS, AND THE SUN: Ninth Cambridge Workshop
eds. Roberto Pallavicini and Andrea K. Dupree
ISBN 1-886733-29-5

Vol. CS-110 BLAZAR CONTINUUM VARIABILITY
eds. H. R. Miller, J. R. Webb, and J. C. Noble
ISBN 1-886733-30-9

Vol. CS-111 MAGNETIC RECONNECTION IN THE SOLAR ATMOSPHERE:
Proceedings of a Yohkoh Conference
eds. R. D. Bentley and J. T. Mariska
ISBN 1-886733-31-7

ASP CONFERENCE SERIES VOLUMES

Published by the Astronomical Society of the Pacific

PUBLISHED: 1996 (* asterisk means OUT OF STOCK)

Vol. CS-112 THE HISTORY OF THE MILKY WAY AND ITS SATELLITE SYSTEM
eds. Andreas Burkert, Dieter H. Hartmann, and Steven R. Majewski
ISBN 1-886733-32-5

PUBLISHED: 1997

Vol. CS-113 EMISSION LINES IN ACTIVE GALAXIES: NEW METHODS AND TECHNIQUES, IAU Colloquium 159
eds. B. M. Peterson, F.-Z. Cheng, and A. S. Wilson
ISBN 1-886733-33-3

Vol. CS-114 YOUNG GALAXIES AND QSO ABSORPTION-LINE SYSTEMS
eds. Sueli M. Viegas, Ruth Gruenwald, and Reinaldo R. de Carvalho
ISBN 1-886733-34-1

Vol. CS-115 GALACTIC CLUSTER COOLING FLOWS
ed. Noam Soker
ISBN 1-886733-35-X

Vol. CS-116 THE SECOND STROMLO SYMPOSIUM: THE NATURE OF ELLIPTICAL GALAXIES
eds. M. Arnaboldi, G. S. Da Costa, and P. Saha
ISBN 1-886733-36-8

Vol. CS-117 DARK AND VISIBLE MATTER IN GALAXIES
eds. Massimo Persic and Paolo Salucci
ISBN-1-886733-37-6

Vol. CS-118 FIRST ADVANCES IN SOLAR PHYSICS EUROCONFERENCE: ADVANCES IN THE PHYSICS OF SUNSPOTS
eds. B. Schmieder. J. C. del Toro Iniesta, and M. Vázquez
ISBN 1-886733-38-4

Vol. CS-119 PLANETS BEYOND THE SOLAR SYSTEM AND THE NEXT GENERATION OF SPACE MISSIONS
ed. David R. Soderblom
ISBN 1-886733-39-2

Vol. CS-120 LUMINOUS BLUE VARIABLES: MASSIVE STARS IN TRANSITION
eds. Antonella Nota and Henny J. G. L. M. Lamers
ISBN 1-886733-40-6

Vol. CS-121 ACCRETION PHENOMENA AND RELATED OUTFLOWS, IAU Colloquium 163
eds. D. T. Wickramasinghe, G. V. Bicknell, and L. Ferrario
ISBN 1-886733-41-4

Vol. CS-122 FROM STARDUST TO PLANETESIMALS: Symposium held as part of the 108th Annual Meeting of the ASP
eds. Yvonne J. Pendleton and A. G. G. M. Tielens
ISBN 1-886733-42-2

Vol. CS-123 THE 12th 'KINGSTON MEETING': COMPUTATIONAL ASTROPHYSICS
eds. David A. Clarke and Michael J. West
ISBN 1-886733-43-0

Vol. CS-124 DIFFUSE INFRARED RADIATION AND THE IRTS
eds. Haruyuki Okuda, Toshio Matsumoto, and Thomas Roellig
ISBN 1-886733-44-9

Vol. CS-125 ASTRONOMICAL DATA ANALYSIS SOFTWARE AND SYSTEMS VI
eds. Gareth Hunt and H. E. Payne
ISBN 1-886733-45-7

ASP CONFERENCE SERIES VOLUMES

Published by the Astronomical Society of the Pacific

PUBLISHED: 1997 (* asterisk means OUT OF STOCK)

Vol. CS-126 FROM QUANTUM FLUCTUATIONS TO COSMOLOGICAL STRUCTURES
eds. David Valls-Gabaud, Martin A. Hendry, Paolo Molaro, and Khalil Chamcham
ISBN 1-886733-46-5

Vol. CS-127 PROPER MOTIONS AND GALACTIC ASTRONOMY
ed. Roberta M. Humphreys
ISBN 1-886733-47-3

Vol. CS-128 MASS EJECTION FROM AGN (Active Galactic Nuclei)
eds. N. Arav, I. Shlosman, and R. J. Weymann
ISBN 1-886733-48-1

Vol. CS-129 THE GEORGE GAMOW SYMPOSIUM
eds. E. Harper, W. C. Parke, and G. D. Anderson
ISBN 1-886733-49-X

Vol. CS-130 THE THIRD PACIFIC RIM CONFERENCE ON
RECENT DEVELOPMENT ON BINARY STAR RESEARCH
eds. Kam-Ching Leung
ISBN 1-886733-50-3

PUBLISHED: 1998

Vol. CS-131 BOULDER-MUNICH II: PROPERTIES OF HOT, LUMINOUS STARS
ed. Ian D. Howarth
ISBN 1-886733-51-1

Vol. CS-132 STAR FORMATION WITH THE INFRARED SPACE OBSERVATORY (ISO)
eds. João L. Yun and René Liseau
ISBN 1-886733-52-X

Vol. CS-133 SCIENCE WITH THE NGST (Next Generation Space Telescope)
eds. Eric P. Smith and Anuradha Koratkar
ISBN 1-886733-53-8

Vol. CS-134 BROWN DWARFS AND EXTRASOLAR PLANETS
eds. Rafael Rebolo, Eduardo L. Martin, and Maria Rosa Zapatero Osorio
ISBN 1-886733-54-6

Vol. CS-135 A HALF CENTURY OF STELLAR PULSATION INTERPRETATIONS:
A TRIBUTE TO ARTHUR N. COX
eds. P. A. Bradley and J. A. Guzik
ISBN 1-886733-55-4

Vol. CS-136 GALACTIC HALOS: A UC SANTA CRUZ WORKSHOP
ed. Dennis Zaritsky
ISBN 1-886733-56-2

Vol. CS-137 WILD STARS IN THE OLD WEST: PROCEEDINGS OF THE 13th NORTH
AMERICAN WORKSHOP ON CATACLYSMIC VARIABLES
AND RELATED OBJECTS
eds. S. Howell, E. Kuulkers, and C. Woodward
ISBN 1-886733-57-0

Vol. CS-138 1997 PACIFIC RIM CONFERENCE ON STELLAR ASTROPHYSICS
eds. Kwing Lam Chan, K. S. Cheng, and H. P. Singh
ISBN 1-886733-58-9

Vol. CS-139 PRESERVING THE ASTRONOMICAL WINDOWS:
Proceedings of Joint Discussion No. 5 of the 23rd General Assembly of the IAU
eds. Syuzo Isobe and Tomohiro Hirayama
ISBN 1-886733-59-7

ASP CONFERENCE SERIES VOLUMES

Published by the Astronomical Society of the Pacific

PUBLISHED: 1998 (* asterisk means OUT OF STOCK)

Vol. CS-140 SYNOPTIC SOLAR PHYSICS --18th NSO/Sacramento Peak Summer Workshop
eds. K. S. Balasubramaniam, J. W. Harvey, and D. M. Rabin
ISBN 1-886733-60-0

Vol. CS-141 ASTROPHYSICS FROM ANTARCTICA:
A Symposium held as a part of the 109th Annual Meeting of the ASP
eds. Giles Novak and Randall H. Landsberg
ISBN 1-886733-61-9

Vol. CS-142 THE STELLAR INITIAL MASS FUNCTION: 38th Herstmonceux Conference
eds. Gerry Gilmore and Debbie Howell
ISBN 1-886733-62-7

Vol. CS-143 * THE SCIENTIFIC IMPACT OF THE GODDARD HIGH RESOLUTION SPECTROGRAPH (GHRS)
eds. John C. Brandt, Thomas B. Ake III, and Carolyn Collins Petersen
ISBN 1-886733-63-5

Vol. CS-144 RADIO EMISSION FROM GALACTIC AND EXTRAGALACTIC COMPACT SOURCES, IAU Colloquium 164
eds. J. Anton Zensus, G. B. Taylor, and J. M. Wrobel
ISBN 1-886733-64-3

Vol. CS-145 ASTRONOMICAL DATA ANALYSIS SOFTWARE AND SYSTEMS VII – (ADASS VII)
eds. Rudolf Albrecht, Richard N. Hook, and Howard A. Bushouse
ISBN 1-886733-65-1

Vol. CS-146 THE YOUNG UNIVERSE GALAXY FORMATION AND EVOLUTION AT INTERMEDIATE AND HIGH REDSHIFT
eds. S. D'Odorico, A. Fontana, and E. Giallongo
ISBN 1-886733-66-X

Vol. CS-147 ABUNDANCE PROFILES: DIAGNOSTIC TOOLS FOR GALAXY HISTORY
eds. Daniel Friedli, Mike Edmunds, Carmelle Robert, and Laurent Drissen
ISBN 1-886733-67-8

Vol. CS-148 ORIGINS
eds. Charles E. Woodward, J. Michael Shull, and Harley A. Thronson, Jr.
ISBN 1-886733-68-6

Vol. CS-149 SOLAR SYSTEM FORMATION AND EVOLUTION
eds. D. Lazzaro, R. Vieira Martins, S. Ferraz-Mello, J. Fernández, and C. Beaugé
ISBN 1-886733-69-4

Vol. CS-150 NEW PERSPECTIVES ON SOLAR PROMINENCES, IAU Colloquium 167
eds. David Webb, David Rust, and Brigitte Schmieder
ISBN 1-886733-70-8

Vol. CS-151 COSMIC MICROWAVE BACKGROUND AND LARGE SCALE STRUCTURES OF THE UNIVERSE
eds. Yong-Ik Byun and Kin-Wang Ng
ISBN 1-886733-71-6

Vol. CS-152 FIBER OPTICS IN ASTRONOMY III
eds. S. Arribas, E. Mediavilla, and F. Watson
ISBN 1-886733-72-4

Vol. CS-153 LIBRARY AND INFORMATION SERVICES IN ASTRONOMY III -- (LISA III)
eds. Uta Grothkopf, Heinz Andernach, Sarah Stevens-Rayburn, and Monique Gomez
ISBN 1-886733-73-2

ASP CONFERENCE SERIES VOLUMES

Published by the Astronomical Society of the Pacific

PUBLISHED: 1998 (* asterisk means OUT OF STOCK)

Vol. CS-154 COOL STARS, STELLAR SYSTEMS AND THE SUN: Tenth Cambridge Workshop
eds. Robert A. Donahue and Jay A. Bookbinder
ISBN 1-886733-74-0

Vol. CS-155 SECOND ADVANCES IN SOLAR PHYSICS EUROCONFERENCE: THREE-DIMENSIONAL STRUCTURE OF SOLAR ACTIVE REGIONS
eds. Costas E. Alissandrakis and Brigitte Schmieder
ISBN 1-886733-75-9

PUBLISHED: 1999

Vol. CS-156 HIGHLY REDSHIFTED RADIO LINES
eds. C. L. Carilli, S. J. E. Radford, K. M. Menten, and G. I. Langston
ISBN 1-886733-76-7

Vol. CS-157 ANNAPOLIS WORKSHOP ON MAGNETIC CATACLYSMIC VARIABLES
eds. Coel Hellier and Koji Mukai
ISBN 1-886733-77-5

Vol. CS-158 SOLAR AND STELLAR ACTIVITY: SIMILARITIES AND DIFFERENCES
eds. C. J. Butler and J. G. Doyle
ISBN 1-886733-78-3

Vol. CS-159 BL LAC PHENOMENON
eds. Leo O. Takalo and Aimo Sillanpää
ISBN 1-886733-79-1

Vol. CS-160 ASTROPHYSICAL DISCS: An EC Summer School
eds. J. A. Sellwood and Jeremy Goodman
ISBN 1-886733-80-5

Vol. CS-161 HIGH ENERGY PROCESSES IN ACCRETING BLACK HOLES
eds. Juri Poutanen and Roland Svensson
ISBN 1-886733-81-3

Vol. CS-162 QUASARS AND COSMOLOGY
eds. Gary Ferland and Jack Baldwin
ISBN 1-886733-83-X

Vol. CS-163 STAR FORMATION IN EARLY-TYPE GALAXIES
eds. Jordi Cepa and Patricia Carral
ISBN 1-886733-84-8

Vol. CS-164 ULTRAVIOLET–OPTICAL SPACE ASTRONOMY BEYOND HST
eds. Jon A. Morse, J. Michael Shull, and Anne L. Kinney
ISBN 1-886733-85-6

Vol. CS-165 THE THIRD STROMLO SYMPOSIUM: THE GALACTIC HALO
eds. Brad K. Gibson, Tim S. Axelrod, and Mary E. Putman
ISBN 1-886733-86-4

Vol. CS-166 STROMLO WORKSHOP ON HIGH-VELOCITY CLOUDS
eds. Brad K. Gibson and Mary E. Putman
ISBN 1-886733-87-2

Vol. CS-167 HARMONIZING COSMIC DISTANCE SCALES IN A POST-HIPPARCOS ERA
eds. Daniel Egret and André Heck
ISBN 1-886733-88-0

Vol. CS-168 NEW PERSPECTIVES ON THE INTERSTELLAR MEDIUM
eds. A. R. Taylor, T. L. Landecker, and G. Joncas
ISBN 1-886733-89-9

ASP CONFERENCE SERIES VOLUMES

Published by the Astronomical Society of the Pacific

PUBLISHED: 1999 (* asterisk means OUT OF STOCK)

Vol. CS-169 11th EUROPEAN WORKSHOP ON WHITE DWARFS
eds. J.-E. Solheim and E. G. Meištas
ISBN 1-886733-91-0

Vol. CS-170 THE LOW SURFACE BRIGHTNESS UNIVERSE, IAU Colloquium 171
eds. J. I. Davies, C. Impey, and S. Phillipps
ISBN 1-886733-92-9

Vol. CS-171 LiBeB, COSMIC RAYS, AND RELATED X- AND GAMMA-RAYS
eds. Reuven Ramaty, Elisabeth Vangioni-Flam, Michel Cassé, and Keith Olive
ISBN 1-886733-93-7

Vol. CS-172 ASTRONOMICAL DATA ANALYSIS SOFTWARE AND SYSTEMS VIII
eds. David M. Mehringer, Raymond L. Plante, and Douglas A. Roberts
ISBN 1-886733-94-5

Vol. CS-173 THEORY AND TESTS OF CONVECTION IN STELLAR STRUCTURE: First Granada Workshop
ed. Álvaro Giménez, Edward F. Guinan, and Benjamín Montesinos
ISBN 1-886733-95-3

Vol. CS-174 CATCHING THE PERFECT WAVE: ADAPTIVE OPTICS AND INTERFEROMETRY IN THE 21st CENTURY,
A Symposium held as a part of the 110th Annual Meeting of the ASP
eds. Sergio R. Restaino, William Junor, and Nebojsa Duric
ISBN 1-886733-96-1

Vol. CS-175 STRUCTURE AND KINEMATICS OF QUASAR BROAD LINE REGIONS
eds. C. M. Gaskell, W. N. Brandt, M. Dietrich, D. Dultzin-Hacyan, and M. Eracleous
ISBN 1-886733-97-X

Vol. CS-176 OBSERVATIONAL COSMOLOGY: THE DEVELOPMENT OF GALAXY SYSTEMS
eds. Giuliano Giuricin, Marino Mezzetti, and Paolo Salucci
ISBN 1-58381-000-5

Vol. CS-177 ASTROPHYSICS WITH INFRARED SURVEYS: A Prelude to SIRTF
eds. Michael D. Bicay, Chas A. Beichman, Roc M. Cutri, and Barry F. Madore
ISBN 1-58381-001-3

Vol. CS-178 STELLAR DYNAMOS: NONLINEARITY AND CHAOTIC FLOWS
eds. Manuel Núñez and Antonio Ferriz-Mas
ISBN 1-58381-002-1

Vol. CS-179 ETA CARINAE AT THE MILLENNIUM
eds. Jon A. Morse, Roberta M. Humphreys, and Augusto Damineli
ISBN 1-58381-003-X

Vol. CS-180 SYNTHESIS IMAGING IN RADIO ASTRONOMY II
eds. G. B. Taylor, C. L. Carilli, and R. A. Perley
ISBN 1-58381-005-6

Vol. CS-181 MICROWAVE FOREGROUNDS
eds. Angelica de Oliveira-Costa and Max Tegmark
ISBN 1-58381-006-4

Vol. CS-182 GALAXY DYNAMICS: A Rutgers Symposium
eds. David Merritt, J. A. Sellwood, and Monica Valluri
ISBN 1-58381-007-2

Vol. CS-183 HIGH RESOLUTION SOLAR PHYSICS: THEORY, OBSERVATIONS, AND TECHNIQUES
eds. T. R. Rimmele, K. S. Balasubramaniam, and R. R. Radick
ISBN 1-58381-009-9

ASP CONFERENCE SERIES VOLUMES

Published by the Astronomical Society of the Pacific

PUBLISHED: 1999 (* asterisk means OUT OF STOCK)

Vol. CS-184 THIRD ADVANCES IN SOLAR PHYSICS EUROCONFERENCE:
MAGNETIC FIELDS AND OSCILLATIONS
eds. B. Schmieder, A. Hofmann, and J. Staude
ISBN 1-58381-010-2

Vol. CS-185 PRECISE STELLAR RADIAL VELOCITIES, IAU Colloquium 170
eds. J. B. Hearnshaw and C. D. Scarfe
ISBN 1-58381-011-0

Vol. CS-186 THE CENTRAL PARSECS OF THE GALAXY
eds. Heino Falcke, Angela Cotera, Wolfgang J. Duschl, Fulvio Melia, and Marcia J. Rieke
ISBN 1-58381-012-9

Vol. CS-187 THE EVOLUTION OF GALAXIES ON COSMOLOGICAL TIMESCALES
eds. J. E. Beckman and T. J. Mahoney
ISBN 1-58381-013-7

Vol. CS-188 OPTICAL AND INFRARED SPECTROSCOPY OF CIRCUMSTELLAR MATTER
eds. Eike W. Guenther, Bringfried Stecklum, and Sylvio Klose
ISBN 1-58381-014-5

Vol. CS-189 CCD PRECISION PHOTOMETRY WORKSHOP
eds. Eric R. Craine, Roy A. Tucker, and Jeannette Barnes
ISBN 1-58381-015-3

Vol. CS-190 GAMMA-RAY BURSTS: THE FIRST THREE MINUTES
eds. Juri Poutanen and Roland Svensson
ISBN 1-58381-016-1

Vol. CS-191 PHOTOMETRIC REDSHIFTS AND HIGH REDSHIFT GALAXIES
eds. Ray J. Weymann, Lisa J. Storrie-Lombardi, Marcin Sawicki, and Robert J. Brunner
ISBN 1-58381-017-X

Vol. CS-192 SPECTROPHOTOMETRIC DATING OF STARS AND GALAXIES
ed. I. Hubeny, S. R. Heap, and R. H. Cornett
ISBN 1-58381-018-8

Vol. CS-193 THE HY-REDSHIFT UNIVERSE:
GALAXY FORMATION AND EVOLUTION AT HIGH REDSHIFT
eds. Andrew J. Bunker and Wil J. M. van Breugel
ISBN 1-58381-019-6

Vol. CS-194 WORKING ON THE FRINGE:
OPTICAL AND IR INTERFEROMETRY FROM GROUND AND SPACE
eds. Stephen Unwin and Robert Stachnik
ISBN 1-58381-020-X

PUBLISHED: 2000

Vol. CS-195 IMAGING THE UNIVERSE IN THREE DIMENSIONS:
Astrophysics with Advanced Multi-Wavelength Imaging Devices
eds. W. van Breugel and J. Bland-Hawthorn
ISBN 1-58381-022-6

Vol. CS-196 THERMAL EMISSION SPECTROSCOPY AND ANALYSIS OF DUST, DISKS, AND REGOLITHS
eds. Michael L. Sitko, Ann L. Sprague, and David K. Lynch
ISBN: 1-58381-023-4

Vol. CS-197 XVth IAP MEETING DYNAMICS OF GALAXIES:
FROM THE EARLY UNIVERSE TO THE PRESENT
eds. F. Combes, G. A. Mamon, and V. Charmandaris
ISBN: 1-58381-24-2

ASP CONFERENCE SERIES VOLUMES

Published by the Astronomical Society of the Pacific

PUBLISHED: 2000 (* asterisk means OUT OF STOCK)

Vol. CS-198	EUROCONFERENCE ON "STELLAR CLUSTERS AND ASSOCIATIONS: CONVECTION, ROTATION, AND DYNAMOS" eds. R. Pallavicini, G. Micela, and S. Sciortino ISBN: 1-58381-25-0
Vol. CS-199	ASYMMETRICAL PLANETARY NEBULAE II: FROM ORIGINS TO MICROSTRUCTURES eds. J. H. Kastner, N. Soker, and S. Rappaport ISBN: 1-58381-026-9
Vol. CS-200	CLUSTERING AT HIGH REDSHIFT eds. A. Mazure, O. Le Fèvre, and V. Le Brun ISBN: 1-58381-027-7
Vol. CS-201	COSMIC FLOWS 1999: TOWARDS AN UNDERSTANDING OF LARGE-SCALE STRUCTURES eds. Stéphane Courteau, Michael A. Strauss, and Jeffrey A. Willick ISBN: 1-58381-028-5
Vol. CS-202 *	PULSAR ASTRONOMY – 2000 AND BEYOND, IAU Colloquium 177 eds. M. Kramer, N. Wex, and R. Wielebinski ISBN: 1-58381-029-3
Vol. CS-203	THE IMPACT OF LARGE-SCALE SURVEYS ON PULSATING STAR RESEARCH, IAU Colloquium 176 eds. L. Szabados and D. W. Kurtz ISBN: 1-58381-030-7
Vol. CS-204	THERMAL AND IONIZATION ASPECTS OF FLOWS FROM HOT STARS: OBSERVATIONS AND THEORY eds. Henny J. G. L. M. Lamers and Arved Sapar ISBN: 1-58381-031-5
Vol. CS-205	THE LAST TOTAL SOLAR ECLIPSE OF THE MILLENNIUM IN TURKEY eds. W. C. Livingston and A. Özgüç ISBN: 1-58381-032-3
Vol. CS-206	HIGH ENERGY SOLAR PHYSICS – *ANTICIPATING HESSI* eds. Reuven Ramaty and Natalie Mandzhavidze ISBN: 1-58381-033-1
Vol. CS-207	NGST SCIENCE AND TECHNOLOGY EXPOSITION eds. Eric P. Smith and Knox S. Long ISBN: 1-58381-036-6
ATLAS	VISIBLE AND NEAR INFRARED ATLAS OF THE ARCTURUS SPECTRUM 3727-9300 Å eds. Kenneth Hinkle, Lloyd Wallace, Jeff Valenti, and Dianne Harmer ISBN: 1-58381-037-4
Vol. CS-208	POLAR MOTION: HISTORICAL AND SCIENTIFIC PROBLEMS, IAU Colloquium 178 eds. Steven Dick, Dennis McCarthy, and Brian Luzum ISBN: 1-58381-039-0
Vol. CS-209	SMALL GALAXY GROUPS, IAU Colloquium 174 eds. Mauri J. Valtonen and Chris Flynn ISBN: 1-58381-040-4
Vol. CS-210	DELTA SCUTI AND RELATED STARS: Reference Handbook and Proceedings of the 6th Vienna Workshop in Astrophysics eds. Michel Breger and Michael Houston Montgomery ISBN: 1-58381-043-9

ASP CONFERENCE SERIES VOLUMES

Published by the Astronomical Society of the Pacific

PUBLISHED: 2000 (* asterisk means OUT OF STOCK)

Vol. CS-211 MASSIVE STELLAR CLUSTERS
eds. Ariane Lançon and Christian M. Boily
ISBN: 1-58381-042-0

Vol. CS-212 FROM GIANT PLANETS TO COOL STARS
eds. Caitlin A. Griffith and Mark S. Marley
ISBN: 1-58381-041-2

Vol. CS-213 BIOASTRONOMY `99: A NEW ERA IN BIOASTRONOMY
eds. Guillermo A. Lemarchand and Karen J. Meech
ISBN: 1-58381-044-7

Vol. CS-214 THE Be PHENOMENON IN EARLY-TYPE STARS, IAU Colloquium 175
eds. Myron A. Smith, Huib F. Henrichs and Juan Fabregat
ISBN: 1-58381-045-5

Vol. CS-215 COSMIC EVOLUTION AND GALAXY FORMATION:
STRUCTURE, INTERACTIONS AND FEEDBACK
The 3rd Guillermo Haro Astrophysics Conference
eds. José Franco, Elena Terlevich, Omar López-Cruz, and Itziar Aretxaga
ISBN: 1-58381-046-3

Vol. CS-216 ASTRONOMICAL DATA ANALYSIS SOFTWARE AND SYSTEMS IX
eds. Nadine Manset, Christian Veillet, and Dennis Crabtree
ISBN: 1-58381-047-1 ISSN: 1080-7926

Vol. CS-217 IMAGING AT RADIO THROUGH SUBMILLIMETER WAVELENGTHS
eds. Jeffrey G. Mangum and Simon J. E. Radford
ISBN: 1-58381-049-8

Vol. CS-218 MAPPING THE HIDDEN UNIVERSE: THE UNIVERSE BEHIND THE MILKY WAY
THE UNIVERSE IN HI
eds. Renée C. Kraan-Korteweg, Patricia A. Henning, and Heinz Andernach
ISBN: 1-58381-050-1

Vol. CS-219 DISKS, PLANETESIMALS, AND PLANETS
eds. F. Garzón, C. Eiroa, D. de Winter, and T. J. Mahoney
ISBN: 1-58381-051-X

Vol. CS-220 AMATEUR - PROFESSIONAL PARTNERSHIPS IN ASTRONOMY:
The 111th Annual Meeting of the ASP
eds. John R. Percy and Joseph B. Wilson
ISBN: 1-58381-052-8

Vol. CS-221 STARS, GAS AND DUST IN GALAXIES: EXPLORING THE LINKS
eds. Danielle Alloin, Knut Olsen, and Gaspar Galaz
ISBN: 1-58381-053-6

PUBLISHED: 2001

Vol. CS-222 THE PHYSICS OF GALAXY FORMATION
eds. M. Umemura and H. Susa
ISBN: 1-58381-054-4

Vol. CS-223 COOL STARS, STELLAR SYSTEMS AND THE SUN:
Eleventh Cambridge Workshop
eds. Ramón J. García López, Rafael Rebolo, and María Zapatero Osorio
ISBN: 1-58381-056-0

Vol. CS-224 PROBING THE PHYSICS OF ACTIVE GALACTIC NUCLEI
BY MULTIWAVELENGTH MONITORING
eds. Bradley M. Peterson, Ronald S. Polidan, and Richard W. Pogge
ISBN: 1-58381-055-2

ASP CONFERENCE SERIES VOLUMES

Published by the Astronomical Society of the Pacific

PUBLISHED: 2001 (* asterisk means OUT OF STOCK)

Vol. CS-225 VIRTUAL OBSERVATORIES OF THE FUTURE
eds. Robert J. Brunner, S. George Djorgovski, and Alex S. Szalay
ISBN: 1-58381-057-9

Vol. CS-226 12th EUROPEAN CONFERENCE ON WHITE DWARFS
eds. J. L. Provencal, H. L. Shipman, J. MacDonald, and S. Goodchild
ISBN: 1-58381-058-7

Vol. CS-227 BLAZAR DEMOGRAPHICS AND PHYSICS
eds. Paolo Padovani and C. Megan Urry
ISBN: 1-58381-059-5

Vol. CS-228 DYNAMICS OF STAR CLUSTERS AND THE MILKY WAY
eds. S. Deiters, B. Fuchs, A. Just, R. Spurzem, and R. Wielen
ISBN: 1-58381-060-9

Vol. CS-229 EVOLUTION OF BINARY AND MULTIPLE STAR SYSTEMS
A Meeting in Celebration of Peter Eggleton's 60th Birthday
eds. Ph. Podsiadlowski, S. Rappaport, A. R. King, F. D'Antona, and L. Burderi
IBSN: 1-58381-061-7

Vol. CS-230 GALAXY DISKS AND DISK GALAXIES
eds. Jose G. Funes, S. J. and Enrico Maria Corsini
ISBN: 1-58381-063-3

Vol. CS-231 TETONS 4: GALACTIC STRUCTURE, STARS, AND
THE INTERSTELLAR MEDIUM
eds. Charles E. Woodward, Michael D. Bicay, and J. Michael Shull
ISBN: 1-58381-064-1

Vol. CS-232 THE NEW ERA OF WIDE FIELD ASTRONOMY
eds. Roger Clowes, Andrew Adamson, and Gordon Bromage
ISBN: 1-58381-065-X

Vol. CS-233 P CYGNI 2000: 400 YEARS OF PROGRESS
eds. Mart de Groot and Christiaan Sterken
ISBN: 1-58381-070-6

Vol. CS-234 X-RAY ASTRONOMY 2000
eds. R. Giacconi, S. Serio, and L. Stella
ISBN: 1-58381-071-4

Vol. CS-235 SCIENCE WITH THE ATACAMA LARGE MILLIMETER ARRAY (ALMA)
ed. Alwyn Wootten
ISBN: 1-58381-072-2

Vol. CS-236 ADVANCED SOLAR POLARIMETRY: THEORY, OBSERVATION, AND
INSTRUMENTATION, The 20th Sacramento Peak Summer Workshop
ed. M. Sigwarth
ISBN: 1-58381-073-0

Vol. CS-237 GRAVITATIONAL LENSING: RECENT PROGRESS AND FUTURE GOALS
eds. Tereasa G. Brainerd and Christopher S. Kochanek
ISBN: 1-58381-074-9

Vol. CS-238 ASTRONOMICAL DATA ANALYSIS SOFTWARE AND SYSTEMS X
eds. F. R. Harnden, Jr., Francis A. Primini, and Harry E. Payne
ISBN: 1-58381-075-7

Vol. CS-239 MICROLENSING 2000: A NEW ERA OF MICROLENSING ASTROPHYSICS
ed. John Menzies and Penny D. Sackett
ISBN: 1-58381-076-5

ASP CONFERENCE SERIES VOLUMES

Published by the Astronomical Society of the Pacific

PUBLISHED: 2001 (* asterisk means OUT OF STOCK)

Vol. CS-240 GAS AND GALAXY EVOLUTION,
A Conference in Honor of the 20th Anniversary of the VLA
eds. J. E. Hibbard, M. P. Rupen, and J. H. van Gorkom
ISBN: 1-58381-077-3

Vol. CS-241 CS-241 THE 7TH TAIPEI ASTROPHYSICS WORKSHOP ON
COSMIC RAYS IN THE UNIVERSE
ed. Chung-Ming Ko
ISBN: 1-58381-079-X

Vol. CS-242 ETA CARINAE AND OTHER MYSTERIOUS STARS:
THE HIDDEN OPPORTUNITIES OF EMISSION SPECTROSCOPY
eds. Theodore R. Gull, Sveneric Johannson, and Kris Davidson
ISBN: 1-58381-080-3

Vol. CS-243 FROM DARKNESS TO LIGHT:
ORIGIN AND EVOLUTION OF YOUNG STELLAR CLUSTERS
eds. Thierry Montmerle and Philippe André
ISBN: 1-58381-081-1

Vol. CS-244 YOUNG STARS NEAR EARTH: PROGRESS AND PROSPECTS
eds. Ray Jayawardhana and Thomas P. Greene
ISBN: 1-58381-082-X

Vol. CS-245 ASTROPHYSICAL AGES AND TIME SCALES
eds. Ted von Hippel, Chris Simpson, and Nadine Manset
ISBN: 1-58381-083-8

Vol. CS-246 SMALL TELESCOPE ASTRONOMY ON GLOBAL SCALES, IAU Colloquium 183
eds. Wen-Ping Chen, Claudia Lemme, and Bohdan Paczyński
ISBN: 1-58381-084-6

Vol. CS-247 SPECTROSCOPIC CHALLENGES OF PHOTOIONIZED PLASMAS
eds. Gary Ferland and Daniel Wolf Savin
ISBN: 1-58381-085-4

Vol. CS-248 MAGNETIC FIELDS ACROSS THE HERTZSPRUNG-RUSSELL DIAGRAM
eds. G. Mathys, S. K. Solanki, and D. T. Wickramasinghe
ISBN: 1-58381-088-9

Vol. CS-249 THE CENTRAL KILOPARSEC OF STARBURSTS AND AGN:
THE LA PALMA CONNECTION
eds. J. H. Knapen, J. E. Beckman, I. Shlosman, and T. J. Mahoney
ISBN: 1-58381-089-7

Vol. CS-250 PARTICLES AND FIELDS IN RADIO GALAXIES CONFERENCE
eds. Robert A. Laing and Katherine M. Blundell
ISBN: 1-58381-090-0

Vol. CS-251 NEW CENTURY OF X-RAY ASTRONOMY
eds. H. Inoue and H. Kunieda
ISBN: 1-58381-091-9

Vol. CS-252 HISTORICAL DEVELOPMENT OF MODERN COSMOLOGY
eds. Vicent J. Martínez, Virginia Trimble, and María Jesús Pons-Bordería
ISBN: 1-58381-092-7

PUBLISHED: 2002

Vol. CS-253 CHEMICAL ENRICHMENT OF INTRACLUSTER AND INTERGALACTIC MEDIUM
eds. Roberto Fusco-Femiano and Francesca Matteucci
ISBN: 1-58381-093-5

ASP CONFERENCE SERIES VOLUMES

Published by the Astronomical Society of the Pacific

PUBLISHED: 2002 (* asterisk means OUT OF STOCK)

Vol. CS-254 EXTRAGALACTIC GAS AT LOW REDSHIFT
eds. John S. Mulchaey and John T. Stocke
ISBN: 1-58381-094-3

Vol. CS-255 MASS OUTFLOW IN ACTIVE GALACTIC NUCLEI: NEW PERSPECTIVES
eds. D. M. Crenshaw, S. B. Kraemer, and I. M. George
ISBN: 1-58381-095-1

Vol. CS-256 OBSERVATIONAL ASPECTS OF PULSATING B AND A STARS
eds. Christiaan Sterken and Donald W. Kurtz
ISBN: 1-58381-096-X

Vol. CS-257 AMiBA 2001: HIGH-*Z* CLUSTERS, MISSING BARYONS, AND CMB POLARIZATION
eds. Lin-Wen Chen, Chung-Pei Ma, Kin-Wang Ng, and Ue-Li Pen
ISBN: 1-58381-097-8

Vol. CS-258 ISSUES IN UNIFICATION OF ACTIVE GALACTIC NUCLEI
eds. Roberto Maiolino, Alessandro Marconi, and Neil Nagar
ISBN: 1-58381-098-6

Vol. CS-259 RADIAL AND NONRADIAL PULSATIONS AS PROBES OF STELLAR PHYSICS, IAU Colloquium 185
eds. Conny Aerts, Timothy R. Bedding, and Jørgen Christensen-Dalsgaard
ISBN: 1-58381-099-4

Vol. CS-260 INTERACTING WINDS FROM MASSIVE STARS
eds. Anthony F. J. Moffat and Nicole St-Louis
ISBN: 1-58381-100-1

Vol. CS-261 THE PHYSICS OF CATACLYSMIC VARIABLES AND RELATED OBJECTS
eds. B. T. Gänsicke, K. Beuermann, and K. Reinsch
ISBN: 1-58381-101-X

Vol. CS-262 THE HIGH ENERGY UNIVERSE AT SHARP FOCUS: CHANDRA SCIENCE, held in conjunction with the 113th Annual Meeting of the ASP
eds. Eric M. Schlegel and Saeqa Dil Vrtilek
ISBN: 1-58381-102-8

Vol. CS-263 STELLAR COLLISIONS, MERGERS AND THEIR CONSEQUENCES
ed. Michael M. Shara
ISBN: 1-58381-103-6

Vol. CS-264 CONTINUING THE CHALLENGE OF EUV ASTRONOMY: CURRENT ANALYSIS AND PROSPECTS FOR THE FUTURE
eds. Steve B. Howell, Jean Dupuis, Daniel Golombek, Frederick M. Walter, and Jennifer Cullison
ISBN: 1-58381-104-4

Vol. CS-265 ω CENTAURI, A UNIQUE WINDOW INTO ASTROPHYSICS
eds. Floor van Leeuwen, Joanne D. Hughes, and Giampaolo Piotto
ISBN: 1-58381-105-2

Vol. CS-266 ASTRONOMICAL SITE EVALUATION IN THE VISIBLE AND RADIO RANGE, IAU Technical Workshop
eds. J. Vernin, Z. Benkhaldoun, and C. Muñoz-Tuñón
ISBN: 1-58381-106-0

Vol. CS-267 HOT STAR WORKSHOP III: THE EARLIEST STAGES OF MASSIVE STAR BIRTH
ed. Paul A. Crowther
ISBN: 1-58381-107-9

Vol. CS-268 TRACING COSMIC EVOLUTION WITH GALAXY CLUSTERS
eds. Stefano Borgani, Marino Mezzetti, and Riccardo Valdarnini
ISBN: 1-58381-108-7

ASP CONFERENCE SERIES VOLUMES

Published by the Astronomical Society of the Pacific

PUBLISHED: 2002 (* asterisk means OUT OF STOCK)

Vol. CS-269	THE EVOLVING SUN AND ITS INFLUENCE ON PLANETARY ENVIRONMENTS eds. Benjamín Montesinos, Álvaro Giménez, and Edward F. Guinan ISBN: 1-58381-109-5
Vol. CS-270	ASTRONOMICAL INSTRUMENTATION AND THE BIRTH AND GROWTH OF ASTROPHYSICS: A Symposium held in honor of Robert G. Tull eds. Frank N. Bash and Christopher Sneden ISBN: 1-58381-110-9
Vol. CS-271	NEUTRON STARS IN SUPERNOVA REMNANTS eds. Patrick O. Slane and Bryan M. Gaensler ISBN: 1-58381-111-7
Vol. CS-272	THE FUTURE OF SOLAR SYSTEM EXPLORATION, 2003-2013 Community Contributions to the NRC Solar System Exploration Decadal Survey ed. Mark V. Sykes ISBN: 1-58381-113-3
Vol. CS-273	THE DYNAMICS, STRUCTURE AND HISTORY OF GALAXIES eds. G. S. Da Costa and H. Jerjen ISBN: 1-58381-114-1
Vol. CS-274	OBSERVED HR DIAGRAMS AND STELLAR EVOLUTION eds. Thibault Lejeune and João Fernandes ISBN: 1-58381-116-8
Vol. CS-275	DISKS OF GALAXIES: KINEMATICS, DYNAMICS AND PERTURBATIONS eds. E. Athanassoula, A. Bosma, and R. Mujica ISBN: 1-58381-117-6
Vol. CS-276	SEEING THROUGH THE DUST: THE DETECTION OF HI AND THE EXPLORATION OF THE ISM IN GALAXIES eds. A. R. Taylor, T. L. Landecker, and A. G. Willis ISBN: 1-58381-118-4
Vol. CS 277	STELLAR CORONAE IN THE CHANDRA AND XMM-NEWTON ERA eds. Fabio Favata and Jeremy J. Drake ISBN: 1-58381-119-2
Vol. CS 278	NAIC–NRAO SCHOOL ON SINGLE-DISH ASTRONOMY: TECHNIQUES AND APPLICATIONS eds. Snezana Stanimirovic, Daniel Altschuler, Paul Goldsmith, and Chris Salter ISBN: 1-58381-120-6
Vol. CS 279	EXOTIC STARS AS CHALLENGES TO EVOLUTION, IAU Colloquium 187 eds. Christopher A. Tout and Walter Van Hamme ISBN: 1-58381-122-2
Vol. CS 280	NEXT GENERATION WIDE-FIELD MULTI-OBJECT SPECTROSCOPY eds. Michael J. I. Brown and Arjun Dey ISBN: 1-58381-123-0
Vol. CS 281	ASTRONOMICAL DATA ANALYSIS SOFTWARE AND SYSTEM XI eds. David A. Bohlender, Daniel Durand, and Thomas H. Handley ISBN: 1-58381-124-9 ISSN: 1080-7926
Vol. CS 282	GALAXIES: THE THIRD DIMENSION eds. Margarita Rosado, Luc Binette, and Lorena Arias ISBN: 1-58381-125-7
Vol. CS 283	A NEW ERA IN COSMOLOGY eds. Nigel Metcalfe and Tom Shanks ISBN: 1-58381-126-5

ASP CONFERENCE SERIES VOLUMES

Published by the Astronomical Society of the Pacific

PUBLISHED: 2002 (* asterisk means OUT OF STOCK)

Vol. CS 284 AGN SURVEYS
eds. R. F. Green, E. Ye. Khachikian, and D. B. Sanders
ISBN: 1-58381-127-3

Vol. CS 285 MODES OF STAR FORMATION AND THE ORIGIN OF FIELD POPULATIONS
eds. Eva K. Grebel and Walfgang Brandner
ISBN: 1-58381-128-1

PUBLISHED: 2003

Vol. CS 286 CURRENT THEORETICAL MODESL AND HIGH RESOLUTION SOLAR OBSERVATIONS: PREPARING FOR ATST
eds. Alexei A. Pevtsov and Han Uitenbroek
ISBN: 1-58381-129-X

Vol. CS 287 GALACTIC STAR FORMATION ACROSS THE STELLAR MASS SPECTRUM
eds. J.M. De Buizer and N.S. van der Bliek
ISBN:1-58381-130-3

Vol. CS 288 STELLAR ATMOSPHERE MODELING
eds. I. Hubeny, D. Mihalas and K. Werner
ISBN: 1-58381-131-1

Vol. CS 289 THE PROCEEDINGS OF THE IAU 8TH ASIAN-PACIFIC REGIONAL MEETING, VOLUME 1
eds. Satoru Ikeuchi, John Hearnshaw and Tomoyuki Hanawa
ISBN: 1-58381-134-6

Vol. CS 290 ACTIVE GALACTIC NUCLEI: FROM CENTRAL ENGINE TO HOST GALAXY
eds. S. Collin, F. Combes and I. Shlosman
ISBN: 1-58381-135-4

Vol. CS-291 HUBBLE'S SCIENCE LEGACY:
FUTURE OPTICAL/ULTRAVIOLET ASTRONOMY FROM SPACE
eds. Kenneth R. Sembach, J. Chris Blades, Garth D. Illingworth and Robert C. Kennicutt, Jr.
ISBN: 1-58381-136-2

Vol. CS-292 INTERPLAY OF PERIODIC, CYCLIC AND STOCHASTIC VARIABILITY IN SELECTED AREAS OF THE H-R DIAGRAM
ed. Christiaan Sterken
ISBN: 1-58381-138-9

Vol. CS-293 3D STELLAR EVOLUTION
eds. S. Turcotte, S. C. Keller and R. M. Cavallo
ISBN: 1-58381-140-0

Vol. CS-294 SCIENTIFIC FRONTIERS IN RESEARCH ON EXTRASOLAR PLANETS
eds. Drake Deming and Sara Seager
ISBN: 1-58381-141-9

Vol. CS-295 ASTRONOMICAL DATA ANALYSIS SOFTWARE AND SYSTEMS XII
eds. Harry E. Payne, Robert I. Jedrzejewski and Richard N. Hook
ISBN: 1-58381-142-7

Vol. CS-296 NEW HORIZONS IN GLOBULAR CLUSTER ASTRONOMY
eds. Giampaolo Piotto, Georges Meylan, S. George Djorgovski and Marco Riello
ISBN: 1-58381-143-5

Vol. CS-297 STAR FORMATION THROUGH TIME
eds. Enrique Perez, Rosa M. Gonzalez Delgado and Guillermo Tenorio-Tagle
ISBN: 1-58381-144-3

ASP CONFERENCE SERIES VOLUMES

Published by the Astronomical Society of the Pacific

PUBLISHED: 2003 (* asterisk means OUT OF STOCK)

Vol. CS-298 GAIA SPECTROSCOPY: SCIENCE AND TECHNOLOGY
ed. Ulisse Munari
ISBN: 1-58381-145-1

Vol. CS-299 HIGH ENERGY BLAZAR ASTRONOMY
eds. Leo O. Takalo and Esko Valtaoja
ISBN: 1-58381-146-X

Vol. CS-300 RADIO ASTRONOMY AT THE FRINGE
eds. J. Anton Zensus, Marshall H. Cohen and Eduardo Ros
ISBN: 1-58381-147-8

A LISTING OF IAU VOLUMES MAY BE FOUND ON THE NEXT PAGE

INTERNATIONAL ASTRONOMICAL UNION (IAU) VOLUMES

Published by the Astronomical Society of the Pacific

PUBLISHED: 1999 (* asterisk means OUT OF STOCK)

Vol. No. 190	NEW VIEWS OF THE MAGELLANIC CLOUDS eds. You-Hua Chu, Nicholas B. Suntzeff, James E. Hesser, and David A. Bohlender ISBN: 1-58381-021-8
Vol. No. 191	ASYMPTOTIC GIANT BRANCH STARS eds. T. Le Bertre, A. Lèbre, and C. Waelkens ISBN: 1-886733-90-2
Vol. No. 192	THE STELLAR CONTENT OF LOCAL GROUP GALAXIES eds. Patricia Whitelock and Russell Cannon ISBN: 1-886733-82-1
Vol. No. 193	WOLF-RAYET PHENOMENA IN MASSIVE STARS AND STARBURST GALAXIES eds. Karel A. van der Hucht, Gloria Koenigsberger, and Philippe R. J. Eenens ISBN: 1-58381-004-8
Vol. No. 194	ACTIVE GALACTIC NUCLEI AND RELATED PHENOMENA eds. Yervant Terzian, Daniel Weedman, and Edward Khachikian ISBN: 1-58381-008-0

PUBLISHED: 2000

Vol. XXIVA	TRANSACTIONS OF THE INTERNATIONAL ASTRONOMICAL UNION REPORTS ON ASTRONOMY 1996-1999 ed. Johannes Andersen ISBN: 1-58381-035-8
Vol. No. 195	HIGHLY ENERGETIC PHYSICAL PROCESSES AND MECHANISMS FOR EMISSION FROM ASTROPHYSICAL PLASMAS eds. P. C. H. Martens, S. Tsuruta, and M. A. Weber ISBN: 1-58381-038-2
Vol. No. 197	ASTROCHEMISTRY: FROM MOLECULAR CLOUDS TO PLANETARY SYSTEMS eds. Y. C. Minh and E. F. van Dishoeck ISBN: 1-58381-034-X
Vol. No. 198	THE LIGHT ELEMENTS AND THEIR EVOLUTION eds. L. da Silva, M. Spite, and J. R. de Medeiros ISBN: 1-58381-048-X

PUBLISHED: 2001

IAU SPS	ASTRONOMY FOR DEVELOPING COUNTRIES Special Session of the XXIV General Assembly of the IAU ed. Alan H. Batten ISBN: 1-58381-067-6
Vol. No. 196	PRESERVING THE ASTRONOMICAL SKY eds. R. J. Cohen and W. T. Sullivan, III ISBN: 1-58381-078-1
Vol. No. 200	THE FORMATION OF BINARY STARS eds. Hans Zinnecker and Robert D. Mathieu ISBN: 1-58381-068-4
Vol. No. 203	RECENT INSIGHTS INTO THE PHYSICS OF THE SUN AND HELIOSPHERE: HIGHLIGHTS FROM SOHO AND OTHER SPACE MISSIONS eds. Pål Brekke, Bernhard Fleck, and Joseph B. Gurman ISBN: 1-58381-069-2
Vol. No. 204	THE EXTRAGALACTIC INFRARED BACKGROUND AND ITS COSMOLOGICAL IMPLICATIONS eds. Martin Harwit and Michael G. Hauser ISBN: 1-58381-062-5

INTERNATIONAL ASTRONOMICAL UNION (IAU) VOLUMES

Published by the Astronomical Society of the Pacific

PUBLISHED: 2001 (* asterisk means OUT OF STOCK)

Vol. No. 205 GALAXIES AND THEIR CONSTITUENTS
AT THE HIGHEST ANGULAR RESOLUTIONS
eds. Richard T. Schilizzi, Stuart N. Vogel, Francesco Paresce, and Martin S. Elvis
ISBN: 1-58381-066-8

Vol. XXIVB TRANSACTIONS OF THE INTERNATIONAL ASTRONOMICAL UNION
REPORTS ON ASTRONOMY
ed. Hans Rickman
ISBN: 1-58381-087-0

PUBLISHED: 2002

Vol. No. 12 HIGHLIGHTS OF ASTRONOMY
ed. Hans Rickman
ISBN: 1-58381-086-2

Vol. No. 199 THE UNIVERSE AT LOW RADIO FREQUENCIES
eds. A. Pramesh Rao, G. Swarup, and Gopal-Krishna
ISBN: 58381-121-4

Vol. No. 206 COSMIC MASERS: FROM PROTOSTARS TO BLACKHOLES
eds. Victor Migenes and Mark J. Reid
ISBN: 1-58381-112-5

Vol. No. 207 EXTRAGALACTIC STAR CLUSTERS
eds. Doug Geisler, Eva K. Grebel, and Dante Minniti
ISBN: 1-58381-115-X

PUBLISHED: 2003

Vol. XXVA TRANSACTIONS OF THE INTERNATIONAL ASTRONOMICAL UNION
REPORTS ON ASTRONOMY 1999-2002
ed. Hans Rickman
ISBN: 1-58381-137-0

Vol. No. 208 ASTROPHYSICAL SUPERCOMPUTING USING PARTICLE SIMULATIONS
eds. Junichiro Makino and Piet Hut
ISBN: 1-58381-139-7

Vol. No. 209 PLANETARY NEBULAE: THEIR EVOLUTION AND ROLE IN THE UNIVERSE
eds. Sun Kwok, Michael Dopita and Ralph Sutherland
ISBN: 1-58381-148-6

Vol. No. 211 BROWN DWARFS
ed. Eduardo Martín
ISBN: 1-58381-132-X

Vol. No. 212 A MASSIVE STAR ODYSSEY: FROM MAIN SEQUENCE TO SUPERNOVA
eds. Karel A. van der Hucht, Artemio Herrero and César Esteban
ISBN: 1-58381-133-8

Ordering information is available at the beginning of the listing

DURHAM UNIVERSITY LIBRARY
3 0104 01205378 5